Techniques in Confocal Microscopy

Reliable Lab Solutions

Techniques in Confocal Microscopy

Reliable Lab Solutions

Edited by

P. Michael Conn

Director, Office of Research Advocacy (OHSU)
Senior Scientist, Divisions of Reproductive Sciences
and Neuroscience (ONPRC)

Professor, Departments of Pharmacology and Physiology,
Cell and Developmental Biology, and Obstetrics and Gynecology (OHSU)
Beaverton, Oregon

AMSTERDAM • BOSTON • HEIDELBERG • LONDON
NEW YORK • OXFORD • PARIS • SAN DIEGO
SAN FRANCISCO • SINGAPORE • SYDNEY • TOKYO
Academic Press is an imprint of Elsevier

Academic Press is an imprint of Elsevier
Linacre House, Jordan Hill, Oxford OX2 8DP, UK
30 Corporate Drive, Suite 400, Burlington, MA 01803, USA
525 B Street, Suite 1900, San Diego, CA 92101-4495, USA
32 Jamestown Road, London NW1 7BY, UK

Material in the work originally appeared in Volumes 302, 307 and 356
of *Methods in Enzymology* (1999, 2002, Elsevier Inc.)

Notice
No responsibility is assumed by the publisher for any injury and/or damage to persons
or property as a matter of products liability, negligence or otherwise, or from any use
or operation of any methods, products, instructions or ideas contained in the material
herein. Because of rapid advances in the medical sciences, in particular, independent
verification of diagnoses and drug dosages should be made

ISBN: 978-0-12-384658-7

For information on all Academic Press publications
visit our website at elsevierdirect.com

Printed and bound by CPI Group (UK) Ltd, Croydon, CR0 4YY
10 11 12 10 9 8 7 6 5 4 3 2 1

CONTENTS

PART I   Practical Considerations and Equipment

PART II   Functional Approaches

11. Receptor–Ligand Internalization

Guido Orlandini, Rita Gatti, Gian Carlo Gazzola, Alberico Borghetti, and Nicoletta Ronda

12. Quantitative Imaging of Metabolism by Two-Photon Excitation Microscopy

David W. Piston and Susan M. Knobel

13. Trafficking of the Androgen Receptor

Virginie Georget, Béatrice Terouanne, Jean-Claude Nicolas, and Charles Sultan

17. *In Vivo* Imaging of Mammalian Central Nervous System Neurons with the
In Vivo Confocal Neuroimaging (ICON) Method

Sylvia Prilloff, Petra Henrich-Noack, Ralf Engelmann, and Bernhard A. Sabel

18. Identification of Viral Infection by Confocal Microscopy

David N. Howell and Sara E. Miller

19. Membrane Trafficking

Sabine Kupzig, San San Lee, and George Banting

PART III   Green Fluorescent Protein

20. Monitoring of Protein Secretion with Green Fluorescent Protein

Christoph Kaether and Hans-Hermann Gerdes

25. Fluorescent Proteins in Single- and Multicolor Flow Cytometry

Lonnie Lybarger and Robert Chervenak

26. Jellyfish Green Fluorescent Protein: A Tool for Studying Ion Channels and Second-Messenger Signaling in Neurons

L. A. C. Blair, K. K. Bence, and J. Marshall

27. Expression of Green Fluorescent Protein and Inositol 1,4,5-Triphosphate Receptor in *Xenopus laevis* Oocytes

*Atsushi Miyawaki, Julie M. Matheson, Lee G. Sayers, Akira Muto,
Takayuki Michikawa, Teiichi Furuichi, and Katsuhiko Mikoshiba*

31. Fluorescence *In Situ* Hybridization of LCM–Isolated Nuclei from Paraffin Sections

Douglas J. Demetrick, Lisa M. DiFrancesco, and Sabita K. Murthy

CONTRIBUTORS

Numbers in parentheses indicate the pages on which the authors' contributions begin.

George F. Babcock (163), Department of Surgery and Cell Biology, University of Cincinnati College of Medicine; Shriners Hospitals for Children, Cincinnati Burns Institute, Cincinnati, Ohio, USA

George Banting (335), Department of Biochemistry, School of Medical Sciences, University of Bristol, University Walk Bristol, United Kingdom

Larry S. Barak (391), Howard Hughes Medical Institute, Duke University Medical Center, Durham, North Carolina, USA

K. K. Bence (421), Department of Molecular Pharmacology, Physiology, and Biotechnology, Brown University Providence, Rhode Island, USA

Miguel Berrios (47), Department of Pharmacological Sciences, School of Medicine, Stony Brook University Medical Center; University Microscopy Imaging Center, State University of New York, Stony Brook, New York, USA

Kanti D. Bhoola (221), Lung Institute of Western Australia, University of Western Australia, Nedlands, Perth Western Australia, Australia

Ghassan Bkaily (105), Department of Anatomy and Cell Biology, Faculty of Medicine, University of Sherbrooke, Sherbrooke, Quebec, Canada

L. A. C. Blair (421), Department of Molecular Pharmacology, Physiology, and Biotechnology, Brown University Providence, Rhode Island, USA

Matthias Böhnke (283), Arlington, Virginia, USA

Alberico Borghetti (175), Department of Experimental Medicine, Via Volturno 39, Parma, Italy

Dieter Brocksch (479), Humangenetik für Biologen, Universität Frankfurt, Frankfurt/Main, Germany

Christof Buehler (261), Arlington, Virginia, USA

Marc G. Caron (391), Howard Hughes Medical Institute, Duke University Medical Center, Durham, North Carolina, USA

Wayne E. Cascio (443), Department of Cell Biology and Anatomy, University of North Carolina at Chapel Hill, Chapel Hill, North Carolina, USA

Kam Tai Chan (21), Department of Electronic Engineering, Center for Advanced Research in Photonics, The Chinese University of Hong Kong, Shatin, NT, Hong Kong

Robert Chervenak (409), Department of Microbiology and Immunology, LSU Health Sciences Center-Shreveport, Shreveport, Louisiana, USA

David E. Colflesh (47), University Microscopy Imaging Center, State University of New York, Stony Brook, New York, USA

Kimberly A. Conlon (47), Department of Pharmacological Sciences, School of Medicine, Stony Brook University Medical Center, State University of New York, Stony Brook, New York, USA

Guy Cox (27), Electron Microscope Unit, University of Sydney, Sydney, NSW, Australia

Stephen R. Cronin (363), Department of Biology, UCSD, La Jolla, California, USA

Douglas J. Demetrick (489), The Departments of Pathology and Laboratory Medicine; Oncology; Biochemistry and Molecular Biology; Medical Genetics; The University of Calgary, and Calgary Laboratory Services

Lisa M. DiFrancesco (489), The Departments of Pathology and Laboratory Medicine; The University of Calgary, and Calgary Laboratory Services

Tommy Duong (355), Department of Cell Biology, CLONTECH Laboratories, Inc., Palo Alto, California, USA

Ralf Engelmann (307), Medical Faculty, Institute of Medical Psychology, Otto-von-Guericke University of Magdeburg, Magdeburg, Germany, email: Bernhard.Sabel@med.ovgu.de; Carl Zeiss MicroImaging GmbH (Jena), Product Management BioSciences, Laser Scanning Microscopy

Stephen S. G. Ferguson (391), Howard Hughes Medical Institute, Duke University Medical Center, Durham, North Carolina, USA

Carlos D. Figueroa (221), Institute of Anatomy, Histology and Pathology, Faculty of Medicine, Universidad Austral de Chile, Valdivia, Chile

Teiichi Furuichi (433), Department of Molecular Neurobiology, Institute of Medical Science, University of Tokyo, Minato-ku, Tokyo, Japan

Rita Gatti (175), Department of Experimental Medicine, Via Volturno 39, Parma, Italy

Gian Carlo Gazzola (175), Department of Experimental Medicine, Via Volturno 39, Parma, Italy

Virginie Georget (205), INSERM U439, Pathologie Moleculaire des Recepteurs Nucleaires, Rue de Navacelles Montpellier, France

Hans-Hermann Gerdes (345), Institute for Neurobiology, University of Heidelberg, Heidelberg, Germany

Enrico Gratton (261), Arlington, Virginia, USA

Gordon L. Hager (379), Head, Hormone Action and Oncogenesis Section, Laboratory Chief, Bethesda, Maryland, USA

Randolph Y. Hampton (363), Department of Biology, UCSD, La Jolla, California, USA

Hisashi Hashimoto (79), Department of Anatomy, The Jikei University School of Medicine, Nishishinbashi, Minato-ku, Tokyo, Japan

Hao He (21), Department of Electronic Engineering, Center for Advanced Research in Photonics, The Chinese University of Hong Kong, Shatin, NT, Hong Kong

Petra Henrich-Noack (307), Medical Faculty, Institute of Medical Psychology, Otto-von-Guericke University of Magdeburg, Magdeburg, Germany, email: Bernhard.Sabel@med.ovgu.de

Ho Pui Ho (21), Department of Electronic Engineering, Center for Advanced Research in Photonics, The Chinese University of Hong Kong, Shatin, NT, Hong Kong

David N. Howell (317), Department of Pathology, Duke University Medical Center, Durham, North Carolina, USA

Chiao-Chian Huang (355), Department of Cell Biology, CLONTECH Laboratories, Inc., Palo Alto, California, USA

Hiroshi Ishikawa (79), Laboratory of Regenerative Medical Science, The Nippon Dental University School of Life Dentistry at Tokyo, Fujimi, Chiyoda-ku, Tokyo, Japan

Danielle Jacques (105), Department of Anatomy and Cell Biology, Faculty of Medicine, University of Sherbrooke, Sherbrooke, Quebec, Canada

Manabu Kagayama (73), Department of Anatomy, Tohoku University School of Dentistry, Sendai, Japan

Steven R. Kain (355), Department of Cell Biology, CLONTECH Laboratories, Inc., Palo Alto, California, USA

Ki Hean Kim (261), Arlington, Virginia, USA

Susan M. Knobel (189), Department of Molecular Physiology and Biophysics, Vanderbilt University Medical School, Nashville, Tennessee, USA

Siu Kai Kong (21), Department of Biochemistry, The Chinese University of Hong Kong, Shatin, NT, Hong Kong

Christoph Kaether (345), Institute for Neurobiology, University of Heidelberg, Heidelberg, Germany

Sabine Kupzig (335), Department of Biochemistry, School of Medical Sciences, University of Bristol, University Walk Bristol, United Kingdom

Moriaki Kusakabe (79), Research Center for Food Safety, Graduate School of Agricultural and Life Sciences, The University of Tokyo, Yayoi, Bunkyo-ku, Tokyo, Japan

Georgia Lahr (479), Humangenetik für Biologen, Universität Frankfurt, Frankfurt/Main, Germany

Stephane A. Laporte (391), Howard Hughes Medical Institute, Duke University Medical Center, Durham, North Carolina, USA

San San Lee (335), Department of Biochemistry, School of Medical Sciences, University of Bristol, University Walk Bristol, United Kingdom

John J. Lemasters (443), Department of Cell Biology and Anatomy, University of North Carolina at Chapel Hill, Chapel Hill, North Carolina, USA

Xianqiang Li (355), Department of Cell Biology, CLONTECH Laboratories, Inc., Palo Alto, California, USA

Lonnie Lybarger (409), Department of Microbiology and Immunology, LSU Health Sciences Center-Shreveport, Shreveport, Louisiana, USA

Carlos B. Mantilla (143), Departments of Anesthesiology, and Physiology & Biomedical Engineering, Mayo Clinic, Rochester, Minnesota, USA

J. Marshall (421), Department of Molecular Pharmacology, Physiology, and Biotechnology, Brown University Providence, Rhode Island, USA

Barry R. Masters (261, 283), Arlington, Virginia, USA

Julie M. Matheson (433), Department of Molecular Neurobiology, Institute of Medical Science, University of Tokyo, Minato-ku, Tokyo, Japan

Anette Mayer (479), Humangenetik für Biologen, Universität Frankfurt, Frankfurt/Main, Germany

Takayuki Michikawa (433), Department of Molecular Neurobiology, Institute of Medical Science, University of Tokyo, Minato-ku, Tokyo, Japan

Katsuhiko Mikoshiba (433), Department of Molecular Neurobiology, Institute of Medical Science, University of Tokyo, Minato-ku, Tokyo, Japan

Sara E. Miller (317), Department of Pathology, Duke University Medical Center, Durham, North Carolina, USA

Atsushi Miyawaki (433), Department of Molecular Neurobiology, Institute of Medical Science, University of Tokyo, Minato-ku, Tokyo, Japan

Sabita K. Murthy (489), Consultant Geneticist and Head of Molecular Cytogenetics Unit, Genetics Center, Al Wasl Hospital, Dubai, UAE

Akira Muto (433), Department of Molecular Neurobiology, Institute of Medical Science, University of Tokyo, Minato-ku, Tokyo, Japan

Jean-Claude Nicolas (205), INSERM U439, Pathologie Moleculaire des Recepteurs Nucleaires, Rue de Navacelles Montpellier, France

Hisayuki Ohata (443), Department of Cell Biology and Anatomy, University of North Carolina at Chapel Hill, Chapel Hill, North Carolina, USA

Guido Orlandini (175), Department of Experimental Medicine, Via Volturno 39, Parma, Italy

David W. Piston (189), Department of Molecular Physiology and Biophysics, Vanderbilt University Medical School, Nashville, Tennessee, USA

Y. S. Prakash (143), Departments of Anesthesiology, and Physiology & Biomedical Engineering, Mayo Clinic, Rochester, Minnesota, USA

Sylvia Prilloff (307), Medical Faculty, Institute of Medical Psychology, Otto-von-Guericke University of Magdeburg, Magdeburg, Germany, email: Bernhard.Sabel@med.ovgu.de

Ting Qian (443), Department of Cell Biology and Anatomy, University of North Carolina at Chapel Hill, Chapel Hill, North Carolina, USA

Chad T. Robinson (163), Department of Surgery and Cell Biology, University of Cincinnati College of Medicine; Shriners Hospitals for Children, Cincinnati Burns Institute, Cincinnati, Ohio, USA

Nicoletta Ronda (175), Department of Clinical Medicine, Nephrology and Health Sciences Via Gramsci 14, University of Parma, Parma, Italy

Bernhard A. Sabel (307), Medical Faculty, Institute of Medical Psychology, Otto-von-Guericke University of Magdeburg, Magdeburg, Germany, email: Bernhard.Sabel@med.ovgu.de

Yasuyuki Sasano (73), Department of Anatomy, Tohoku University School of Dentistry, Sendai, Japan

Lee G. Sayers (433), Department of Molecular Neurobiology, Institute of Medical Science, University of Tokyo, Minato-ku, Tokyo, Japan

Karin Schütze (479), Humangenetik für Biologen, Universität Frankfurt, Frankfurt/Main, Germany

Gary C. Sieck (143), Departments of Anesthesiology, and Physiology & Biomedical Engineering, Mayo Clinic, Rochester, Minnesota, USA

Celia J. Snyman (221), Department of Biochemistry, Genetics and Microbiology, University of KwaZulu-Natal, Pietermariztburg, South Africa

Peter T. C. So (261), Arlington, Virginia, USA

Monika Stich (479), Humangenetik für Biologen, Universität Frankfurt, Frankfurt/Main, Germany

Charles Sultan (205), INSERM U439, Pathologie Moleculaire des Recepteurs Nucleaires, Rue de Navacelles Montpellier, France

Xuejun Sun (123), Department of Oncology, Cross Cancer Institute, University of Alberta, Edmonton, Alberta, Canada

Béatrice Terouanne (205), INSERM U439, Pathologie Moleculaire des Recepteurs Nucleaires, Rue de Navacelles Montpellier, France

Donna R. Trollinger (443), Department of Cell Biology and Anatomy, University of North Carolina at Chapel Hill, Chapel Hill, North Carolina, USA

Robert H. Webb (3), Wellman Laboratories of Photomedicine, Bartlett 703, Boston, Massachusetts, USA

James L. Wittliff (463), Hormone Receptor Laboratory, Department of Biochemistry and Molecular Biology, Brown Cancer Center, University of Louisville, Louisville, Kentucky, USA

Rose Chik Ying Ong (21), Department of Biochemistry, The Chinese University of Hong Kong, Shatin, NT, Hong Kong

Jie Zhang (391), Howard Hughes Medical Institute, Duke University Medical Center, Durham, North Carolina, USA

Xiaoning Zhao (355), Department of Cell Biology, CLONTECH Laboratories, Inc., Palo Alto, California, USA

PREFACE

This volume deals with the rapidly evolving topic of confocal microscopy and other important microscopic techniques. The (now outdated) OVID database (which included MEDLINE, Current Contents, and other sources) listed 76 references using keywords, "confocal microscopy" for the 5-year period 1985–1989. In contrast, for the 4-year period 1995–1998, nearly 3600 references are listed. Presently, PubMed lists over 41,000 references! This is certainly a testament to the growing value and interest in this topic.

This volume documents many diverse uses for confocal microscopy in disciplines that broadly span biology. The methods presented include shortcuts and conveniences not included in the initial publications. The techniques are described in a context that allows comparisons to other related methodologies. The authors were encouraged to do this in the belief that such comparisons are valuable to readers who must adapt extant procedures to new systems. Also, so far as possible, methodologies are presented in a manner that stresses their general applicability and reports their potential limitations. Although, for various reasons, some topics are not covered, the volume provides a substantial and current overview of the extant methodology in the field and a view of its rapid development.

Particular thanks to the authors for revising their chapters promptly and to the staff at Academic Press for maintaining high standards of production quality and for timely publication of this work.

P. Michael Conn

PART I

Practical Considerations and Equipment

CHAPTER 1

Theoretical Basis of Confocal Microscopy

Robert H. Webb

Wellman Laboratories of Photomedicine
Bartlett 703, Boston
Massachusetts, USA

I. A Simple View

A confocal microscope is most valuable in seeing clear images inside thick samples. To demonstrate this, I want to start with a conventional wide-field epifluorescence microscope shown in Fig. 1. The left diagram (Fig. 1) demonstrates the illumination light, and the right shows light collected from the sample. In the right diagram we see that a broad field of illumination is imaged into the thick sample. Although the illumination is focused at one plane of the sample, it lights up

DOI: 10.1016/B978-0-12-384658-7.00001-1

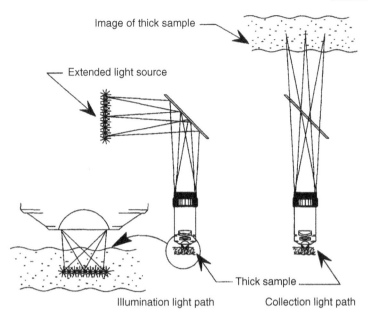

Image of thick sample

Extended light source

Thick sample

Illumination light path   Collection light path

Fig. 1 A conventional (wide-field) microscope for fluorescence in epitaxial configuration.

all of the samples. In the right diagram we see that the microscope objective has formed the image of the whole thick sample at the image plane of the microscope. If we put a film, charge-coupled device (CCD), or retina at the image plane, it will record the in-focus image of one plane within the thick sample, but it will also record all of those out-of-focus images of the other planes.

In Fig. 2 I show an alternative arrangement. Instead of a broad light source, I use a single point source of light and image it inside the thick sample. That focused light illuminates a single point inside the sample very brightly, but of course it also illuminates the rest of the sample at least weakly. On the right (Fig. 2), the image of the thick sample is very bright where the sample was brightly illuminated and dimmer where it was weakly illuminated. Since my intention is to look only at one point inside the thick sample, I will now put a pinhole in the image plane. The pinhole lets through only the light that is forming the bright part of the image. Behind the pinhole I put a detector, as shown in Fig. 3. That detector registers the brightness of the part of the thick sample that is illuminated by the focused light and ignores the rest of the sample. What we have here is a point source of light, a point focus of light inside the object or sample, and a pinhole detector, all three confocal with each other. That is a confocal microscope (Pawley, 1996; Wilson, 1990; Webb, 1996).

This confocal microscope has all the features we need for looking at a point inside a thick sample. However, it is not very interesting to look at a single point. So we have to find a way to map out the whole sample point by point.

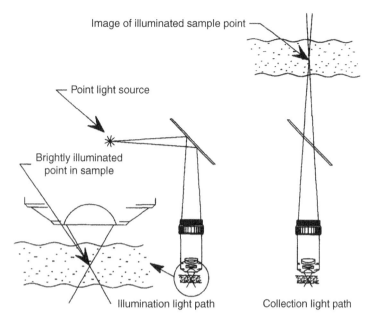

Fig. 2 The microscope of Fig. 1 with point illumination.

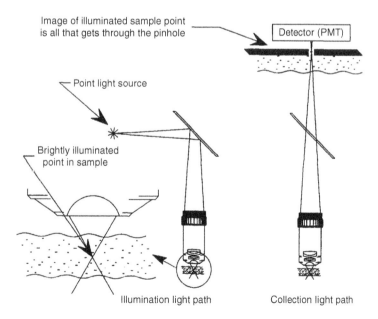

Fig. 3 The microscope of Fig. 2 becomes confocal when a pinhole blocks light from all parts of the sample outside the focus.

Most laser-scanning confocal microscopes look at one point of the sample at a time. Other varieties look at many well-separated points at once, but locally they are imaging one point at a time.

The easiest way to look around in the sample is to move the sample, a technique called stage scanning. More complex scanning means allow the sample to be stationary while we move the illuminated spot(s) over the sample. But those are engineering details. Instead of concerning ourselves with them at the moment, let us assume that they are solved and investigate what properties this confocal microscope has.

II. Optical Sectioning

Our microscope discriminates against points near, but not in, the focal spot. When the unwanted points are beside the focal spot, the contrast has improved. However, this device also discriminates against points above and below the focus, a feature we call optical sectioning. Instead of using a microtome to slice a thin section out of a thick sample, we can now image that thin section inside the sample. Parts of the sample that are above the imaged point or below it will be illuminated weakly, and light from those parts will be mostly rejected by the pinhole. With scanning, this microscope can image a whole plane inside a thick sample and then be focused deeper into the sample to image a different layer, and those two images do not interfere with each other. With proper controls, the microscope can image a whole stack of optical sections, which can later be assembled into a three-dimensional display (Brakenhoff *et al.*, 1988; Morgan *et al.*, 1992).

Figure 4 shows an even more abstract sketch of a confocal microscope that emphasizes optical sectioning. An object in the sample that lies above the focal point is imaged above the pinhole. Light going toward that image is mostly blocked by the pinhole mask.

The confocal microscope also rejects light from points adjacent to the one illuminated. That increases the contrast, even for thin samples. Contrast enhancement is always desirable, particularly when we need to look at something dim next to something bright. This fact explains why confocal microscopes are used so often for conventional (thin sections) fluorescence microscopy applications.

One thing our confocal microscope cannot do is look through walls. By that I mean that if a layer absorbs light, then deeper layers will be harder to see. That drop-off of visibility limits the sample thickness to about 50 μm in many cases, although there are many instances of looking 0.5 mm into tissue.

III. Point–Spread Function

Now I want to discuss the physics of the effects just observed. In Fig. 2 we saw that light from a point source is imaged inside the sample. In the sample, that light forms a double cone, as shown in Fig. 5A. Figure 5B uses gray scale to show where

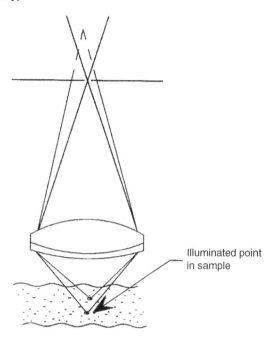

Fig. 4 Another schematic view of the confocal microscope. The point of interest is imaged in the pinhole, while light from the more proximate point is largely blocked by the pinhole. This organization is called "optical sectioning."

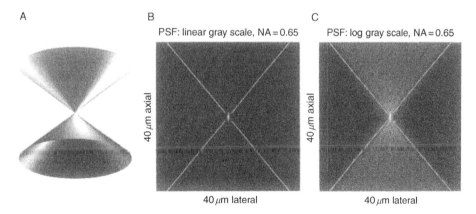

Fig. 5 (A) Light from an objective lens fills a (double) cone. (B) The actual light intensity is plotted as a linear gray scale. The same presentation, with a logarithmic gray scale, is shown (C), with the lightest value being 10^5 times the darkest.

the light is most intense. The scale is logarithmic so that the peak is 10^5 times brighter than the darkest areas. On a linear scale, shown in Fig. 5C, only the peak has any intensity. The cross section of the cone, shown as lines in the gray scale images, represents a numerical aperture (NA) of 0.65 for the sample in air or 0.86 if the index of the sample is 1.33 (water).

Looking close to the focal point, as shown in Fig. 6, there is structure to the intensity distribution. The lines show the geometric edges of the light seen in Fig. 5, but the pattern is much more complex. Physicists call this pattern the point-spread function. One way to think of the point-spread function is in terms of probability: the probability that a photon from the point source will reach some point (**q**) is prob (**q**). A photon is 10^5 times more likely to reach the focal point than some point far from it, still within the light cone. Keep this probability picture in mind because we will use it again to understand the confocal arrangement.

The complexity in the pattern in Fig. 6 is due to the effects of diffraction, a consequence of the wave nature of light (Born and Wolf, 1991). It is the objective lens that causes the diffraction pattern and that pattern is displayed perfectly by using a point source of light. In the transverse plane (perpendicular to the symmetry axis of the lens), the central part of this pattern is called the Airy disk (Hecht, 1990). Figure 7 shows an Airy disk in linear and logarithmic gray scale. The radius of the inner dark ring in this pattern is a measure of the resolving power of the microscope. Such a resolution element (or resel) is nearly the same as the diameter of the disk at 50% intensity. Both are somewhat arbitrary measures of resolution.

It is important to understand that these point-spread functions (the patterns we have been looking at) are due to the lens. The smaller the lens aperture, the wider the pattern (and the lower the resolution). A lens with a small aperture has a large point-spread function and a low resolution. A lens with a large aperture can resolve much smaller things because its point-spread function is smaller. The term "numerical aperture" refers to a measure of the lens aperture in angle and takes into account the smaller wavelengths in a refractive medium (NA $= n \sin\theta$ and $\lambda = \lambda_0/n$).

The image, our final result, is due to light returning from points in the sample. I will use the term "remittance" to include reflection, scattering, refraction, diffraction, and fluorescence—all the things that the sample does to redirect light. To analyze an image, we need to know how bright the illumination was at

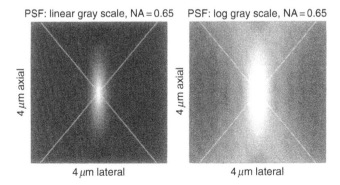

Fig. 6   The point-spread function close to the focus. These are the patterns of Fig. 5, magnified 10×.

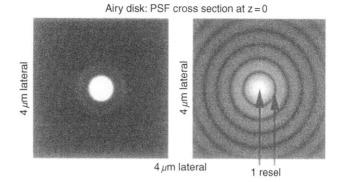

Fig. 7 The point-spread function in the focal plane: the pattern of Fig. 6 rotated about the *z*-axis at *z* = 0. *Left*: linear gray scale. *Right*: logarithmic gray scale.

every point in the sample. For instance, a strong remitter illuminated by dim light will return the same amount of light as a weak remitter illuminated by bright light. Multiply the remittance by the illumination intensity at every point, which is what the point-spread function describes, and that describes the image brightness distribution in the image.

The image that we want is the image of the illuminated object, viewed through the objective lens. We know that the objective lens causes point sources of light to image as point-spread functions, so we should expect that to happen again. Here I am going to use a trick. I could take every point in the illuminated object, find its point-spread function at the image plane, and then add those all up—a process called convolution. A simpler way is to use the fact that the laws of optics work in both directions: it does not matter which way the light goes. So I will start with a (point) pinhole and know that its image in the sample is a point-spread function—in fact, the same point-spread function that we saw before. Because I only care about light that gets through the pinhole, I can use the distribution (the point-spread function) of the intensity of light that comes from the pinhole and use that to evaluate how much light goes to (and through) the pinhole. To distinguish the two, I will call the point-spread function for the illumination source PSF_S and the point-spread function for the collection pinhole PSF_P (S for source and P for pinhole).

Now imagine that the sample is featureless: its remittance is everywhere the same, as it would be in a fluorescent liquid. The detector is going to register light coming through the pinhole, from the part of the sample illuminated by PSF_S and sampled by PSF_P. Every point feels the influence of both point-spread functions, which means that the two should be multiplied. The product point-spread function is a point-spread function for the whole microscope, a confocal point-spread function.

$$PSF_{CF} = PSF_S \times PSF_P \qquad (1)$$

Every point on those gray scale plots is an intensity, and I need only to multiply the intensities at each point to find the mutual intensity—the intensity of light that came from the point source was remitted by the sample and passed through the pinhole.

One interpretation of the point-spread function is that of a probability. PSF_S is the pattern of prob (**q**) for every point q in the sample, and PSF_P is another (independent) pattern of prob (**q**) for every point q. The probability of detecting a photon is the probability that a photon goes from the point source to the sample and goes from the sample to and through the pinhole. These are independent probabilities, so the mutual probability is their product, as stated in Eq. (1).

Figure 8 shows how much sharper the confocal point-spread function is than either the source or the pinhole point-spread function. Subsidiary peaks that were 0.01 times the main peak become only 0.0001. That reduction is the source of the increased contrast and the optical sectioning.

There is, however, another hidden fact that helps with sectioning. Go back for a moment to Fig. 5, where we started with a double cone of light. Although the illuminated point in the sample gets concentrated light, the same total amount of light passes through each plane perpendicular to the axis. So we might worry that the out-of-focus remission could add up to a lot of light. That is just what happens in a wide-field microscope, where that extra returned light obscures the view of interior planes of the sample where the microscope is focused. However, with the point source and pinhole, those out-of-focus planes contribute so little light (see Fig. 4) that the interior planes are sharp and clear. Mathematically, we could predict that this is so by integrating over the confocal point-spread function, but Fig. 8 shows what to expect, with its drastic reduction of the intensity away from the central peak.

The formalism for this is

$$\int_{\text{Plane}} \text{PSF} = \int_{\text{Plane}} \text{light} = \text{constant} \tag{2}$$

for the wide-field (single) point-spread function.

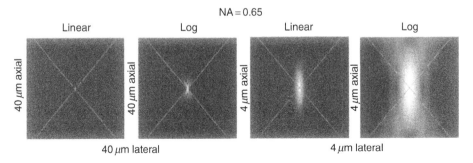

Fig. 8 The confocal point-spread function. Figures 5 and 6, for the wide-field microscope, show much larger point-spread functions.

For the confocal point-spread function, however, the integral is not of "light," it is of "light that reached a sample point *and* got back through the pinhole." Then the integral over the focal plane is much larger than the integral over any other plane.

$$\int_{\text{Focal plane}} \text{PSF}_{\text{CF}} \gg \int_{\text{Any other plane}} \text{PSF}_{\text{CF}} \tag{3}$$

Better yet, the sum of all those integrals is still much less than the amount of intensity at the focus.

$$\int_{\text{Focal plane}} \text{PSF}_{\text{CF}} \gg \int\int_{\text{Any other planes}} \text{PSF}_{\text{CF}} \tag{4}$$

That is just a complicated way of saying that the pinhole excludes almost all the light from anywhere but the focus. This is really a remarkable thing: the amount of light that gets through the pinhole from everywhere away from the focus is much less than what comes from the focus. The mathematics shows this and real confocal microscopes confirm it.

What makes a confocal microscope is a point-spread function that is the product of two individual point-spread functions.

IV. Pinhole

Now I need to go back to the "point" pinhole and be a little more realistic.

What is a point source? A point is a mathematical fiction, so tiny that no light would come from it. A star, however, makes a pretty good real point source. That is because the image of the star is smaller than the point-spread function of the lens we use to observe it. The same is true of a point pinhole. If the extent of the pinhole is less than the point-spread function of the lens, then that is a point pinhole. It turns out that we can use a pinhole that is about three resels across and still get almost all of the confocal effects (Pawley, 1996; Wilson, 1990). An even bigger pinhole will blur the point-spread function enough to degrade the optical sectioning and contrast enhancement. So my ideal confocal microscope will use a three resel pinhole, that is, a pinhole that is three times the size of the Airy disk.

V. Magnification

There is a trick here that may be confusing. Every lens has two point-spread functions: one on each side. If the lens magnifies by a factor of 60, then the point-spread function on one side is small and that on the other side is roughly 60 times as big (roughly, because the diffraction peak details are not exactly images of each other and magnification is a concept from geometric optics). A 60× objective lens with NA = 0.85 forms a resel at the sample that is 0.4 μm across, but at the image plane of the microscope, the NA is 0.014 and the resel is 24 μm. So there is no need to make submicron pinholes; use one that is comfortably in the 50- to 100-μm range.

New objective lenses are becoming available that have high NA and low magnification, so pinholes need to be adjusted to compensate. We generally say that magnification is of no interest in confocal microscopy, as no image is ever really formed (Piston, 1998). However, in this case, it is important to adjust pinhole size to suit the objective lens.

As an example, the microscope I use has a 100/1.2 and a 40/1.2 objective lens. The two lenses have identical resolutions, but the 40× has a larger field of view, so it is the one I use. The three resel pinholes for the 100× would be 28 μm at the usual 150-mm image point, while for the 40× would be 9 μm there. Of course the actual pinhole location will probably not be a 150-mm point, as modern objective lenses work with an infinite conjugate ("infinity corrected"), and some extra magnification before the pinhole is used to make the physical device of manageable size. Also, I have control of the pinhole size, but I am not told how many resels it is, which would be useful information. These really are details, but it might be well to pay attention to them when running a confocal microscope.

VI. Complete Microscope

Figure 9 shows a complete confocal microscope in which the point-spread function is the product of two individual point-spread functions. Notice that the engineering details are still hidden in a box called "scanning engine." That box may have moving mirrors that sweep the laser beam over the sample or it might have a disk full of holes whose rotation sweeps many illumination spots over the sample. There are many varieties of scanning engine too, and the engineering details are in fact truly important. However, the theory of the confocal microscope

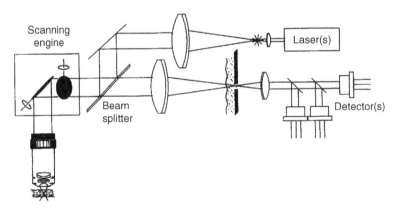

Fig. 9 A generic confocal microscope.

does not require us to understand scanning engines. Rather, we should look at the two point-spread functions that go into making up the confocal point-spread function.

The microscope in Fig. 9 is usually used to detect fluorescence. The beam splitter is a dichroic mirror that reflects the fluorescent light and passes the excitation light. It can, however, be used equally well to detect light remitted without a change of wavelength by making the beam splitter a partially silvered mirror. If there is no wavelength change, the two point-spread functions are identical, except for the convolution on the pinhole. Even with the wavelength change due to fluorescence, there is not much change of point-spread function. Equation (5) gives the lateral resolution in terms of NA and wavelength for the confocal microscope. The quantity Δr is the full-width at half-maximum intensity of the confocal point-spread function. I give this measure because there is some confusion in the literature as to how to use the Rayleigh criterion for resolution in the confocal situation. Equation (6) gives the axial resolution (the optical section) (Pawley, 1996):

$$\Delta r = \frac{0.32\lambda}{NA} \tag{5}$$

$$\Delta z = \frac{1.26n\lambda}{(NA)^2} \tag{6}$$

where NA is the numerical aperture, n is the index of refraction, and λ is the wavelength in vacuum (no medium). For fluorescence, λ should be replaced by the (geometric) mean of the two wavelengths.

As an example, suppose we use an objective of NA = 0.9 with fluorescein isothiocyanate (FITC). The excitation wavelength might be 488 nm, and the fluorescence will be centered around 530 nm. So the lateral resolution is 0.18 μm and the optical section is 1 μm.

Fluorescence may be very weak, so we probably want to use as big a pinhole as possible. Three (Rayleigh) resels would be 1 μm, at the sample, or somewhere around 80 μm in the first image plane of the objective lens. A bigger pinhole will let through even more light, but at the price of reducing the resolution and contrast.

There is another consequence of the wavelength shift of fluorescence. The confocal microscope requires that the pinhole be optically conjugate to the illuminated spot in the thick sample. That means that both the point source and the pinhole have to be imaged at the same place. Such a confocal arrangement is possible if the microscope objective is highly achromatic or if the pinhole position is adjusted to compensate for chromaticity. Some confocal microscopes use a single pinhole for all colors, so they need good apochromats or similar objectives. Other confocal microscopes separate their colors before the pinholes, so each pinhole can be adjusted separately and less expensive objectives can be used. This multipinhole design risks misalignment by a factor of the number of pinholes (usually three). There are trade-offs in both price and convenience.

Most confocal microscopists want the option of exciting fluorophores with two lasers at once. That demands good achromaticity so that both colors focus in the same plane. So do not try to save money on objective lenses!

Figure 9 sketches the generic confocal microscope for use primarily in fluorescence. The point source in this case is a laser that has been brought to a point focus before expansion to fill the objective lens with light. That point focus has to be less than the (magnified) point-spread function of the objective. The scanning engine in this design may be a pair of mirrors mounted on galvanometer motors, which are optically conjugate to the pupils of the objective lens. Notice that the scanning mirrors tilt the laser beam back and forth and untilt the remitted light back to a stationary beam. That stationary beam is then focused on a pinhole (here I have chosen the single pinhole design). After the pinhole, a series of dichroic beam splitters separate the various colors and send them to detectors appropriate to each color. Generally, the detectors are photomultiplier tubes, although avalanche photodiodes are also used. For further discrimination against the excitation light, filters are placed in front of the detectors. Much of the cost of confocal microscopes has to do with changing those filters, the pinhole size, the choice of detector, the size of the scan (the field of view), and other parameters necessary to a useful picture.

VII. Varieties of Confocal Microscope

One of the engineering details I have been ignoring is the scanning engine. Confocal microscopes come in two versions, O and P. The O version puts the scanning in an object plane, the P in a pupil plane.

In the P-confocal microscope (CM-P or variously CSLM, CLSM, and other permutations), the scanning occurs in a plane optically conjugate to the pupil of the objective lens. A deflection device, usually moving mirrors, changes the angle of a light beam, usually a laser beam, causing one or a few illumination spots to scan over the object. The same mirrors (usually) then descan the remitted beam to keep it stationary on a detection pinhole (Webb, 1984). There are many variants to all this, but the theory of the confocal microscope applies to all.

The most common O-confocal microscope is the disk scanner or tandem scanning microscope, in which a disk full of holes spins in a plane optically conjugate to the object, thus causing the images of those many holes to scan over the object. Then either the same set of holes or a different set on the same disk serve as detection pinholes (Kino and Corle, 1989). The CM-O can use nonlaser light and provides a live image to the eye or a camera.

VIII. Multiphoton Microscope

There is another way to have two point-spread functions multiply so that the microscope becomes confocal (Denke et al., 1990). A very intense light source, focused to a very small spot, can deliver two or more photons at once to an absorber. For instance, a single photon at 488 nm can excite fluorescence in fluorescein dye at around 530 nm. The same energy could also come from two photons at twice the wavelength (976 nm) if they arrive at very nearly the same time. They will arrive at very nearly the same time if the light is intense enough and the focus is tight enough. The position of each photon is controlled independently by the point-spread function of the objective lens, so we have two identical point-spread functions. These are, again, independent probabilities, so they multiply to give a confocal point-spread function just like that of the confocal microscope.

There is an extra benefit to the multiphoton configuration. The illumination light is of much longer wavelength than the single photon excitation light, so it is less likely to damage the sample. Furthermore, the excitation light can only cause multiphoton processes in the very intense focus, so the light passing through out-of-focus planes does not bleach the fluorophore.

Finally, no pinhole is needed to achieve this confocal arrangement. Any light at the fluorescent wavelength has to originate in the focal volume, so any remitted light at that wavelength coming out of the objective lens will contribute to a good image. Figure 10 shows the generic multiphoton microscope that uses no pinhole. There are no alignment problems at the detector!

There is also a cost to the multiphoton configuration. First, it only works in fluorescence. Second, it is not simple to get all that light concentrated in space and time. Typically, the source for this microscope is a pulsed laser pumped by some other big light source. That is both expensive and difficult to maintain.

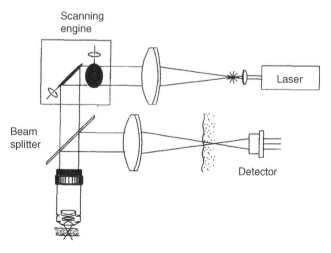

Fig. 10 A multiphoton microscope. The laser delivers all its energy in a very short pulse, so two photons can reach the sample nearly simultaneously. No pinhole is needed.

IX. Light Sources

A. Laser

Real confocal microscopes have a lot of engineering details. In general, these are not appropriate for this discussion, but one of the details is—the light source. Confocal microscopes of the CM-P flavor use lasers as their source. Disk-scanning confocal microscopes (CM-O) have the advantage of being able to use almost any bright source. I will discuss briefly the properties of the lasers used in point-scanning microscopes and what the implications of the other light sources are.

Lasers are bright monochromatic sources whose light emerges in a tightly collimated coherent beam. What that means for a confocal microscope is that the point source is very nearly perfect. Lasers are also monochromatic and coherent, but the coherence is rather incidental for the use we make of the laser. Coherent light is light whose waves are all exactly in phase. That can be useful in confocal microscopy, but it is not necessary.

B. Speckle

In fact, coherent light makes very bad illumination for a broad field because it produces the phenomenon known as speckle. Speckle arises from the interference of light scattered from nearby points within the illumination field. That interference produces dark and bright spots next to each other in the image. Speckle does not arise in a confocal microscope because we look at only one illuminated point (one point-spread function) at a time. If we look more carefully, we find that speckle does affect the confocal image as out-of-focus planes scatter coherently into the pinhole. However, that is a second-order effect that we need not worry about here and that does not occur in a fluorescent image.

C. Monochromaticity

A monochromatic light source is perfect for fluorescence imaging. A simple long-pass filter rejects the excitation light and allows the fluorescence through.

X. Incoherent Sources

Nonlaser sources are generally grouped under the heading incoherent. Both monochromaticity and good collimation are necessary for coherence. Incoherent sources include arc lamps and other bright sources that are too big to act as points. Often we limit the color with narrow band filters so that only useful colors fall on the sample. The result of such limiting is to reduce the intensity of the light at the sample. Sometimes that is a good idea if we are having trouble with bleaching, but generally nonlaser sources are light starved.

For a disk-scanning confocal microscope, we need to use a light source that covers the object field (the sample) with uniform light. Köhler illumination accomplishes just that, but requires some care (Inoue and Spring, 1997). Thus, although a nonlaser source may seem simpler, it often requires as much attention as a laser.

One of the useful aspects of disk-scanning confocal microscopes is that one may obtain a true-color image. If we do not know what is in a sample it is often useful to illuminate it with white light. That also may give us a more familiar looking image, closer to what the eye sees in a simple bench microscope.

XI. Bleaching

One of the major uses of confocal microscopy is for fluorescence imaging. The fluorophores used tend to bleach out when exposed to too much light. Bleaching is generally thought to be proportional to light dose, although there are some examples of nonlinearities, both favorable and unfavorable to the high intensities of the laser-scanning confocal microscopes (Pawley, 1996). Fluorophore saturation certainly occurs, and that limits the useful intensity, but bleaching is permanent, as is photodamage to biological structures. There really is not much one can do about a process proportional to light dose, as the fluorescence has the same dependence, but one can avoid exposure to light that is not giving a fluorescent signal. That means not exposing the sample to light during "fly-back"—the time when the scanning engine is merely returning to zero. It also means avoiding exposure to one laser while another is being used. These are just points that a careful microscope builder pays attention to.

A major benefit of the multiphoton microscope is that only light in the focal volume participates in the bleaching. All single photon microscopes expose the layers above and below the focal plane, even though they image only the focal plane. The advantage of the multiphoton microscope is clear and is one of the major reasons for using it.

XII. Colocality

It is often desirable to test whether two fluorophores are attached to the exact same point on a cell, that is, the same molecule. To do this the microscope displays an image in each of the two colors, and the observer checks whether the redder image exactly overlies the bluer. This condition will be met if the two pinholes (if there are two) are imaged at exactly the same point of the sample and also if two laser sources (if there are two) are imaged at exactly the same point of the sample. To check, use a small particle such as a polystyrene bead (used in flow cytometry) that fluoresces in two colors (Molecular Probes, Inc., http://www.probes.com/). The two-color images should be coincident in position and size.

XIII. Numerical Aperture

In a conventional microscope, NA is well defined. It depends on the diameter of the objective lens' pupil, the focal length, and the wavelength in the sample medium. In the cone of Fig. 5A, these quantities are unambiguous, and even when diffraction is included, the term NA is clear. We use it to describe the resolution obtainable with a microscope, as used in Eqs. (5) and (6).

$$NA = n \sin\theta \qquad (7)$$

Snell's law tells us that $n \sin\theta$ does not change across interfaces, so NA is a handy quantity. But resolution really depends on the angle θ, the most extreme ray in the light cone of Fig. 5A, and the index of refraction only enter because the reduced wavelength λ/n is the proper distance metric. The term aperture enters because $\sin\theta$ is governed by the aperture of the lens: the pupil. A big aperture is needed for a big NA, but you do not get the benefit of a big aperture if you are sending a small beam of light through it.

Whether we use NA or $\sin\theta$ as a synonym for resolution, we assume that the objective lens pupil is uniformly filled with light. In a confocal microscope, that might well not be true! The profile of a laser beam is generally Gaussian, so there is less light at the pupil edge. The profile of the beams in a CM-O will be a diffraction pattern, so the same is true. The microscope manufacturer will make a choice as to how much light can be wasted by overfilling the pupil, and that will affect the resolution of the microscope. I do not see much that any of us can do about this situation, but it might be a good idea to test the microscope resolution.

XIV. Summary

A confocal microscope forms its image by recording light primarily from a small focal volume, largely ignoring points to the side or above or below. That volume, described as a point-spread function, is the product of two similar functions that are generated by the objective lens. Because of that multiplication, the recorded light is greater than even the integrated total of the light from all other points in a thick sample. Some of the implications of implementing this theory are reflected in the choices available to users of confocal microscopes.

Acknowledgment

This work was supported in part by DE-FG02-91ER61229 from the Office of Health and Environmental Research of the Department of Energy.

References

Born, M., and Wolf, E. (1991). "Principles of Optics," Pergamon Press, New York.
Brakenhoff, G. J., et al. (1988). Scann. Microsc. 2, 1831.
Denke, W. J., et al. (1990). Science 248, 73.

Hecht, E. (1990). "Optics," 2nd edn. Addison-Wesley, Redding, MA.

Inoue, S., and Spring, K. R. (1997). *In* "Video Microscopy, the Fundamentals," p. 22. Plenum, New York.

Kino, G. S., and Corle, T. R. (1989). *Phys. Today* **42,** 55.

Morgan, F. E., *et al.* (1992). *Scann. Microsc.* **6,** 345.

Pawley, J. (ed.) (1996). "Handbook of Biological Confocal Microscopy," 3rd edn. Plenum, New York.

Piston, D. W. (1998). *Biol. Bull.* **195,** 1.

Wilson, T. (ed.) (1990). "Confocal Microscopy," Academic Press, London.

Webb, R. H. (1984). *Appl. Opt.* **23,** 3680.

Webb, R. H. (1996). *Rep. Prog. Phys.* **59,** 427.

CHAPTER 2

Practical Considerations in Acquiring Biological Signals from Confocal Microscope

Hao He,[*] Rose Chik Ying Ong,[†] Kam Tai Chan,[*] Ho Pui Ho,[*] and Siu Kai Kong[†]

[*]Department of Electronic Engineering
Center for Advanced Research in Photonics
The Chinese University of Hong Kong
Shatin, NT, Hong Kong

[†]Department of Biochemistry
The Chinese University of Hong Kong
Shatin, NT, Hong Kong

I. Introduction

With the development of fluorescent indicators and recombinant proteins such as Fluo-3 and green fluorescent protein-tagged chimeras, fluorescence microscopic imaging (FMI) offers unparalleled opportunities to study biochemical events in living cells with a minimum of perturbation. In the conventional wide view, FMI not only is a sharp image generated from an in-focus area, blurred images from all out-of-focus regions are also acquired and thereby distorting and degrading the contrast of the image. However, in a confocal microscope, the excitation light generated by laser is focused to a discrete point of the specimen to reduce the wide view illumination and a pinhole is set in front of detector to eliminate the out-of-focus signals from the nonfocal points. By this improvement, clear signals from the focal point can be obtained and a sharp image from the confocal plane can be built up with scanning mirrors and be stored in computer for further analysis. With this technical advancement, a number of cellular activities have been observed and their significance elucidated. For example, depolarization of the mitochondrial membrane potential and release of mitochondrial proteins into the cytoplasm during apoptosis have been discovered (Goldstein *et al.*, 2005). However, potential problems and flaws in the use of confocal microscopy have seldom been documented. We present here several potential problems that we have encountered which either reduced the efficiency of the experiment process or gave rise erroneous interpretations.

II. Laser–Induced Rise of Fluorescence Signals

The development of lipophilic membrane-permeant acetoxymethyl (AM) ester form of fluorescent indicators provides a nondestructive method of introducing probes into the cytoplasm for various biochemical studies without perforating the cells with micropipettes and injectors. After crossing the cell membrane, the AM form of the indicator is hydrolyzed by esterases and converted back to its hydrophilic form. Practically, the esterified indicator is first dissolved in organic solvents like dimethyl sulfoxide (DMSO) before cell loading. A technical difficulty encountered is that in some cell types, indicators have been shown to be compartmentalized in endosomes as a consequence of endocytosis or in other organelles such as mitochondria and endoplasmic reticulum. As the probe compartmentalization is a temperature-dependent process, the temperature for dye loading should be low enough to slow down endocytosis but high enough to hydrolyze all intracellular indicators. The compartmentalization problem can be further circumvented by dispersing indicators with detergents such as Pluronic F-127 and fetal calf serum to improve the dye solubility and avoid the accumulation of dye on the cell surface.

In the course of measuring Fluo-3 fluorescence, we sometimes observed a spontaneous and sustained increase in fluorescence signal intensity. For example, when HeLa cells loaded with Fluo-3/AM (1.5 μM) for 1 h at room temperature were exposed to laser scanning without the addition of stimuli, Fluo-3 signals increased progressively with time. It was initially suspected that the spontaneous cellular activity had led to the fluorescence burst. However, when it was discovered that only illuminated cells were affected in this way, while all other cells outside of the area exposed to the laser light had only background fluorescence, we concluded that laser irradiation might have converted residual nonfluorescent Fluo-3/AM molecules into its fluorescent form by heat generated by laser irradiation.

In our experimental setup and those of many other laboratories, the lipophilic membrane-permeant form of Fluo-3 was admitted into the cell by simple diffusion. As mentioned above, esterases remove the AM group and convert the indicator into the hydrophilic and fluorescent form which is trapped intracellularly. We surmised that the hydrolytic reaction might need time for completion and, should there be sufficient unconverted Fluo-3/AM in the cytoplasm when the experiment began, the heat generated by laser irradiation or fluorescent energy loss would promote the enzymatic reaction and lead to the apparently spontaneous fluorescence emission. The confirmation of this hypothesis would be possible with a ratiometric indicator by measuring the fluorescence emission at the isosbestic point. A different cell system, L929 cells, has been used in this experiment which further shows that the erroneous signal arises from the hardware setup rather than that of cells.

To prevent the problem of unwanted fluorescence, we might increase the temperature during dye loading to speed up conversion but this may enhance indicator compartmentation. Alternatively, we recommend that cells under study should first be illuminated until the fluorescence has reached a stable level indicating that the unhydrolyzed dye has been exhausted before beginning the experiment. Care should of course also be taken to prevent photobleaching because of overstrong excitation. Too frequent excitation provokes photobleaching.

Therefore, it is best to test a few dye-loading concentrations and loading times before starting experiments. Longer loading time will create the problem mentioned above. Normally, the dye concentrations of 5–10 μM are used and loading times can be in the range of 30 min to 1 h depending on the cell type. An ideal method in studying cellular activity is to employ dual emission probes such as Indo-1/AM for Ca^{2+} with a UV laser for excitation. Emissions at two different wavelengths can then be used ratiometrically. The advantage of using this semi-quantitatively ratiometric determination is that information obtained in this way is independent of intracellular dye concentration. The limitation of this suggestion is that it requires a UV excitation and the UV light source is not as widely available as visible laser lines in the commercial setups.

III. Water Content of DMSO and Efficiency of Dye Loading

In our experience, the dryness of DMSO also affects the dye-loading efficiency. Using calcium green-1 as an example, when cells were loaded with calcium green-1 which had been dissolved in dry DMSO at room temperature, a relatively homogeneous cytosolic fluorescence was observed. By contrast, when cells were incubated with calcium green-1 dissolved in nondry DMSO, localized bright spots were seen in the optical sections. Our experience reveals that the higher the water content in DMSO, the more pronounced that anomaly. We surmised that low dye solubility in water-contained DMSO gives rise to insoluble indicators which produced bright spots after endocytosis. Therefore, the dryness of DMSO used is important for a successful experiment. With wet DMSO, the hot spots can saturate the detector and the wide range of fluorescence intensity creates difficulties in data acquisition and image analysis, for example, background subtraction. But most importantly, the artifact hot spots might be misinterpreted and mask genuine local responses.

To obtain dry DMSO, molecular sieves (can be obtained from Sigma) were first prepared by leaving the particles in an oven at 150 °C overnight. DMSO from the supplier was then incubated with the dry molecular sieve in an air-tight bottle for at least 2 h. These procedures can remove the water in the wet DMSO.

A. "Temperature Effect"—On Focus

Temperature is one of the most important physical parameters that controls physiological processes. To mimic the *in vivo* system, temperature for experiments is usually set to the body temperature of the host animal, which is normally higher than the room temperature. In single cell studies, cells are seeded on coverslips which are mounted in chamber or a holder with physiological buffer. One approach to control temperature is to heat up the holder to the desired temperature by commercially available temperature control unit. Unfortunately, many temperature control units usually rely on a negative feedback system to keep the temperature at a narrow range near the desired temperature rather than keeping the temperature precisely and constantly at a fixed point. In the case of confocal microscopic studies, these temperature fluctuations affect the refractive index of the immersion oil between the objectives and the coverglass. As a result of the fluctuations of the refractive index of immersion oil, different sections would have come into focus even the microscope is trained on a fixed x–y plane. For example, with the use of a low fluorescence immersion oil ($\eta^{23\ °C} = 1.515$ and $\Delta\eta = 0.385 \times 10^{-3}\ °C^{-1}$), a temperature change of 1 °C could cause a focal plane shift of 0.25 μm (Lee *et al.*, 1996). In relation to cellular dimension, this introduces a very significant error in confocal studies.

In view of the serious error that can result from even very small temperature fluctuations reported here, it is imperative that in confocal microscopy work, the temperature be rigidly controlled. Recently, delicate instrument, which provides an accurate temperature control not only in the bath holder but also in the coverslip and microscopic objective, is commercially available to meet this purpose. However, one should check with the supplier whether the objectives you use could be heated. Alternatively, one can perform experiments in a thermostatically controlled laboratory at a fixed temperature to avoid temperature fluctuations, temperature gradient, and temperature leak between the cell holder and the microscope.

B. Technical Tips

1. Many fluorochromes have overlapping emission spectra. To obtain valid information from multicolor experiments, use fluorescent dyes with discrete emission spectra or those with the least spectral cross-talk. There are several tools available on the World Wide Web to help investigators analyze the fluorophore spectra. For example, "Fluorescence Spectrum Viewer" is a valuable interactive tool that allows investigators to check the spectral compatibility of fluorophores, as well as the selection of appropriate filters and excitation wavelengths.

Fluorescence Spectrum Viewer from different companies:

http://www.bdbiosciences.com/external_files/media/spectrumviewer/index.jsp

http://www.invitrogen.com/site/us/en/home/support/Research-Tools/Fluores-cence-SpectraViewer.html

http://www.omegafilters.com/curvo2/index.php?

2. DMSO is hygroscopic. Dissolving fluorescent dyes in wet DMSO reduces dye-loading efficiency and introduces undesirable readouts. Always dissolve the AM ester indicators in anhydrous DMSO. Pluronic F-127 (20%, w/v), a nonionic surfactant, has been shown to facilitate the solubilization of water-insoluble AM ester dyes such as Fluo-3 or Fluo-4 in physiological buffer. For loading dyes into large cells, or cells with fine processes like neurons, it is imperative to test the optimal loading time and dye concentration for the cells to be investigated.

3. While fluorescent indicators with AM ester are generally useful in introducing dyes to intracellular space of living cells, these dyes may not work in a similar fashion in nonmammalian cells and tissues. For example, AM indicators for calcium measurement do not work well in moth neurons. Also, fluorescent calcium indicators may perturb physiological processes by acting as a buffer. Therefore, low-affinity reporter dyes in minimal concentrations are recommended.

4. Phenol red is a pH indicator in many culture media such as RPMI 1640 or DMEM (Dulbecco's modified Eagle medium). It can be excited by argon laser (488 nm) and emits fluorescence over a wide spectrum. When present in the experimental system, phenol red will generate a lot of noisy signals. In this connection, it is advised to use phenol red-free medium or buffer for fluorescence studies. Also, many natural products or compounds extracted from plants or medicinal herbs carry intrinsic fluorescence that substantially interferes the fluorescence readouts when they are used in the cell-based assays. It is important to scan the emission spectrum of these compounds to avoid spectral overlap with the probes to be used in the study.

5. For acquiring signals from surface nonadherent live cells through confocal microscopy, we can use polylysine to provide surface positive charges to help cells to stick to glass coverslips for imaging studies. To prepare the polylysine-coated coverslips, acid-washed coverglass can be coated with poly-L-lysine (1 mg/ml in double distilled (dd) water) for 30 min at room temperature. After soaking, wash the coverslips extensively with dd water to get rid of the free poly-L-lysine as they are cytotoxic. Dry the coverslips in a sterile condition. Acid-washed coveslips can be prepared by soaking coverglass in HCl (4 M) for 3 days. After washing with dd water, soak the coverslips in ethanol (70%, v/v) overnight. After rinsing with dd water, sterilize the coverslips by autoclave. Make sure the coverslips are not sticking to each other.

6. Whenever possible, start imaging with low laser power. Increase laser power only when necessary as higher laser power leads to greater photobleaching rate. For the imaging of thicker tissues, or *in vivo* imaging of tissues below surface,

two-photon (2P) imaging may be the method of choice because of better tissue penetration. The concept of 2P excitation is based on the fact that two photons of low energy can excite a fluorochrome in a quantum event, resulting in the emission of a fluorescence photon. Practically, most of the fluorophores for biological imaging have excitation spectra in the UV–visible range, whereas the spectral content of a femotosecond (fs) laser used to excite the fluorophores contains a wide range of wavelengths in the near infrared (NIR) region of 700–1100 nm. When fs laser is used, it should be noted that the refractive indices for the visible and NIR light are different, and as a result of this, the focal point of the NIR laser beam is usually not on the same plane of the imaging plane. To get good confocal images, the focal point of the fs laser should be adjusted to coincide with the focus of the imaging plane as obtained from using visible excitation. This can be achieved by using an optical design similar to those for collimation, usually a pair of lens with tunable distance between them to change the focal point. Apart from imaging, fs laser working in the NIR region has many other applications in biological field such as DNA transfection or cell–cell fusion (He *et al.*, 2008a,b).

References

Goldstein, J. C., Muñoz-Pinedo, C., Ricci, J. E., Adams, S. R., Kelekar, A., Schuler, M., Tsien, R. Y., and Green, D. R. (2005). *Cell Death Differ.* **12,** 453–462.

He, H., Chan, K. T., Kong, S. K., and Lee, R. K. Y. (2008a). *Appl. Phys. Lett.* **93,** 163901–163903.

He, H., Kong, S. K., Lee, R. K., Suen, Y. K., and Chan, K. T. (2008b). *Opt. Lett.* **33,** 2961–2963.

Lee, M. F., Kong, S. K., Fung, K. P., Lui, C. P., and Lee, C. Y. (1996). *Biol. Signals* **5,** 291–300.

CHAPTER 3

Equipment for Mass Storage and Processing of Data

Guy Cox

Electron Microscope Unit
University of Sydney
Sydney, NSW, Australia

Confocal microscopes do not produce images that are patterns of light—their "images" are numerical data in the memory of a computer. Such an image has many advantages—it can be transmitted and copied simply, with 100% fidelity, but its longevity depends entirely on the medium on which it is stored. Initially it is probably stored in the Random Access Memory (RAM) of the computer—a very insecure location from which it will be lost if the computer is turned off, or if that area of memory is overwritten. To store that image with all its information intact, we must write it in digital form on some less transient medium—a reproduction on paper or film of the image encoded by this data, however good, cannot contain all the information of the original.

An image represented by a series of numbers in a computer is not directly accessible to human senses. When the image is planar—a series of values or "*pixels*" representing intensities in a 2D array—it normal to display it as corresponding intensities on a monitor. Even this may be less than trivial. Most confocal microscopes, or CCD cameras, can store 12 or 16-bit images, capable of storing 4096 or 65,536 intensity levels, respectively. Monitors, or more accurately the display boards driving them, typically can display only 256 levels in each color channel. The human eye, however, can distinguish no more than 64 different intensity levels. There can be a lot of information lost in the transition from numerical data to human perception.

A confocal microscope does not produce a planar image, it produces a 3D array of sample values or "*voxels*," which is even harder to present to the eye in a meaningful form. Much of the image processing associated with confocal microscopy is devoted to overcoming the problem—it is visualization software, dedicated to making the electronic data comprehensible to the human eye and brain. A further step is to take advantage of the numerical nature of the information to extract measurements about our sample, and here we are entering the realm of image analysis or morphometry rather than image processing.

This chapter reviews some of the available solutions to these two problems. This is a rapidly moving area, and new alternatives will doubtlessly become available almost as soon as this is printed. Computers become more powerful daily, and tasks which require supercomputers one year are simple on personal computers the next. As well as assessing currently available technologies, therefore, I will try to provide enough background information to enable users to assess what the future has to offer in an informed and rational way.

I. Mass Storage

Confocal images demand substantial storage capacity. The smallest image one is likely to acquire is a single-plane, single-channel image measuring 512×512 pixels. If it is saved as an 8-bit image (1 byte per pixel), each point in the image can have one of 256 possible gray levels. This image will require one quarter of a megabyte (Mbyte) to store, with an additional overhead to store basic information about the picture, either in a "header" at the start of the file, or at the end, or even in a separate file. Confocal microscopes will capture much larger images than this, and the image may contain more than one channel. A 3-channel, 1024×1024 pixel image would be what most users would acquire, at 16 bits per pixel (65,536 gray levels) to allow for future manipulation and analysis. We are therefore storing 6 Mbytes per plane. A 3D image dataset through 50 μm at 200 nm per slice (Nyquist sampling) would contain 250 planes and thus require 1.5 gigabytes (Gbyte) of storage space. Time-course datasets (XYZT) from high-speed confocals could be much larger. At the time of writing, personal computers likely to be used on microscopes typically have 500 Gbyte hard disks, and 2 terabyte (Tbyte) is the largest available, and still (relatively) expensive. A hard-working confocal microscope could fill a 500 Gbyte hard drive in a week, so permanent image storage will require either some form of removable storage media or a major file server.

A. Removable Storage Media

The key requirement for image storage is generally permanence. Silver-based black and white film negatives provide an archival medium proved to last longer than a century, dye-based color photographic media are less durable but are still demonstrably capable of surviving for decades away from daylight. One would hope for at least equal permanence from digital storage, but only one of the storage media available to a present-day computer user has a history longer than 20 years. That medium is magnetic tape, and the experience from its relatively long history in that its longevity cannot match photographic film. Estimates of durability of other types of media are based on accelerated aging tests—at best these can only be an indication.

The permanence of different media types is only part of the story, however, since a digital record is of no use without devices to read it. Hence it is important to assess the likelihood of the survival of appropriate hardware in a functional form. Popularity can be relevant—popular hardware is more likely to survive than something which sold in small quantities, regardless of technical merit. Popularity brings the additional benefit that images saved on a popular medium are readable at home, and in other laboratories. For years Unix systems used 1/4″ tape in DC-6000 format cassettes for backup and software distribution (Cox, 2006), but it would now be hard to find a drive to read these tapes (even if the tapes were still readable). The need for high data capacities has often driven the confocal microscope user to "high-end" devices which have a much higher price and smaller user base, such as the WORM and MO disks which were popular in the 1990s, but are now virtually impossible to read. Even the once ubiquitous Iomega ZIP disks—a popular consumer product—are very much a legacy item. This factor, as well as the cost per megabyte, needs to be taken into account when choosing a storage system.

1. Sequential Devices

Magnetic tapes are referred to as sequential devices since they can be read from, and written to, only by moving sequentially from one end. Although they are erasable and rewritable, this means that one cannot generally replace a single file or group of files. New data can only be added to the end of the tape; if something is to be erased the entire tape, or at least one writing session, must be erased and rewritten. Cost-wise, tape is one of the cheapest bulk storage media available, but as an image storage system it suffers from the time taken to locate and recover any one file. Also, although it is rewritable it will not stand an infinite number of uses. Linear tape open (LTO) tapes quote around 200 read–write cycles. For image storage this may not be an issue, and read-only cycles are less wearing. Long-term storage life is quoted as 15–30 years, with the caveat that tape cassettes are very sensitive to being dropped.

Tape comes into its own as backup medium, where a whole file system is recorded at one time, and can therefore be recovered in the event of a system crash. Modern backup software provides powerful tools for managing backups, including user-friendly interfaces for finding individual files to restore (invaluable in a confocal facility where files always seem to get deleted accidentally) and incremental backups so that only files which have changed are added to the tape.

LTO tapes are now the system of choice. "Open" means that the standard is open, not proprietary, so tapes and drives from many different manufacturers are available, which should help future availability. LTO-3 (400 Mbyte) is the commonest in current use—LTO-4 (800 Mbyte) is now available and an even higher standard, LTO-5, is proposed. However, the standard only requires drives to read tapes from the previous two versions, so your ability to read a 30-year-old tape, whatever its condition, is not guaranteed.

2. Random Access Devices

Random access devices are either discs (optical or magnetic), or solid-state "flash" memory. These use a directory structure to read any individual file without reference to what else is stored on the disk. In magnetic or flash drives this is also true when recording, but optical media must be recorded sequentially and are only random access on playback. There are both erasable and nonerasable random access devices—in the past this meant also a distinction between magnetic and optical media, but that line is now distinctly blurred.

Since confocal images are always large, the speed of the device can be an important factor, particularly its sustained transfer rate. Speeds of individual systems are considered later, but with external drives one also needs to bear in mind the speed of the connection to the computer. Universal serial bus (USB) 2 has a nominal speed of 480 Mbyte/s, while the alternative FireWire has a nominal 400 Mbyte/s speed. However, this is misleading since FireWire has a more sophisticated control system, and will usually deliver faster overall speeds, particularly with large files. Some computers now have 786 Mbyte/s FireWire 800, which will easily outperform USB 2. PC-card (formally called PCMCIA) interfaces are another option, most often found on notebook computers. The speed is hard to quantify since devices can be either 16 or 32 bit, and there are various bus-mastering protocols, but in general USB 2 or FireWire will be faster.

a. Optical Disks

CD-ROM disks have been around for 25 years, and digital video disk (DVD) for 12 or so. Any recent computer will have a CD/DVD reader and writer, so no extra hardware is needed. They are relatively slow to write—the exact speed depending on the burner and the blank disc. "Single speed" for a CD is a rather pedestrian 150 Kbyte/s, so the actual speed can be calculated from this—10 speed will be 1.5 Mbyte/s, and so on. 24× is assumed for Table I. DVD single speed is 1.32 Mbyte/s—equivalent to 9× CD speed (http://www.osta.org/technology/dvdqa/dvdqa4.htm). Attempting to burn a CD or DVD that is not rated for the speed of the writer is a recipe for absolutely certain disaster!

Table I
Costs and Characteristics of Various Bulk Storage Media

Media	Capacity	Sustained speed	Cost per Gbyte
Tape			
LTO-3	400 Gbyte	60 Mbyte/s[a]	12c
LTO-4	800 Gbyte	120 Mbyte/s[b]	
LTO-5 (pending)	1.2 Tbyte	180 Mbyte/s[b]	
Disks			
CD-R	650 Mbyte	3.6 Mbyte/s ($24\times$)[c]	40c
DVD $\pm$ R	4.7 Gbyte	3.6 Mbyte/s ($6\times$)[c]	10c
DVD + R DL	8.7 Gbyte		40c
Blu-Ray BD-R	25 Gbyte	72 Mbyte/s ($2\times$)	20c
Blu-Ray BD-R 2L	50 Gbyte		40c
USB hard drives	$\leq$2 Tbyte	480 Mbyte/s[d]	15c
Flash media			
USB sticks	$\leq$64 Gbyte	<120 Mbyte/s	$3.00
CF	$\leq$64 Gbyte	<120 Mbyte/s	$7.00
SD	$\leq$32 Gbyte	<30 Mbyte/s	$4.00

Speeds quoted are approximate only and may not match real-world results. Prices are only very approximate; they are quoted in US cents but are based on figures from USA and Australia. Typically newer higher capacity media such as 64 Gbyte flash drives and 50 Gbyte Blu-Ray disks will be more expensive per Gbyte at the time of their introduction but this differential will rapidly disappear.

[a]User reports.
[b]Manufacturer's spec.
[c]Writing speed–reading speed is much faster.
[d]Speed defined by USB 2.0 standard—actual performance will be lower.

Recordable CDs have a dye layer (which is bleached by the writing laser) in front of a reflective layer of evaporated metal. The metal in turn is protected only by a thin layer of lacquer, and it is not advisable to write on this with spirit-based pens or to affix adhesive labels, either of which could damage the protection offered by the lacquer. Archival stability for up to a century is claimed by the makers, and the wide-spread availability of CD readers should ensure that hardware will still be extant well into the next century. The nature of the metal used has also been a topic of some discussion—gold, used in the more expensive blank disks, is claimed (plausibly enough) to be less likely to suffer from corrosion than the aluminum of cheaper disks (Cox, 2006). Silver is another alternative. Exposure of the dye to light will inevitably reduce life. The capacity of a CD is only 650 Mbyte, not much in terms of modern confocal images.

DVDs use essentially the same technology, so similar considerations should apply. There are two different standards—DVD−R and DVD + R—both holding 4.7 Gbyte. Most drives will read both, though some only write one or the other. Since DVD drives have confocal optics, it is possible to use two recording layers. The DVD + R DL offers a capacity of 8.7 Gbyte—they are less common so single layer DVDs are cheaper per Gbyte, but the larger capacity of the DL might be worth the extra cost for the sake of convenience. In principle double-sided DVDs are also permitted by the standard, but they do not seem to have reached the marketplace. Since blank DVDs, if bought by the 100, cost little more than CDs, they are the cheapest storage medium at the present time.

Both CDs and DVDs are also available in rewritable form, but this is not really of interest from the point of view of long-term image storage. The erasing process is slow, and an erasable disk must have a shorter archival life.

b. Blu-Ray Disks (BD-R)

Blu-Ray disks use the shorter wavelength of a blue laser to write smaller spots and thus increase capacity. At the time of writing this is new technology so prices are relatively high, but they are certain to fall. Being able to store 25 Gbyte (single layer) or 50 Gbyte (double layer) on a CD-sized disk is a very attractive proposition. $1\times$ for BD-R is 36 Mbyte/s (http://www.blu-ray.com/faq/#bluray_speed), and $2\times$ is already available, with $6\times$ scheduled to appear about the time this is being written.

Most BD-R disks use an inorganic material as the writing medium, which should make for excellent archival performance. A version called BD-R LTH, however, uses an azo dye (also used in CD and DVD media)—this means that the disks can be made on similar machines to DVDs, but is probably detrimental from the storage point of view.

c. Hard Disks

Hard drives are now extraordinarily cheap—the cost per Gbyte is comparable with optical or tape storage (Table I), yet this cost includes the drive mechanism! External hard drives come in two forms, pocket size ones which take their power from the computer, and larger (but still compact) ones which have a separate power supply. The latter contain the same drives as are used internally, and hence are fast and high capacity (up to 2 Tbyte at the time of writing). Pocket drives are typically smaller in capacity, and may be slower. However, the speed of access is usually limited by the connection, which may be USB or FireWire. It is very tempting to write images to a terabyte external drive, and just buy another when it is full. The problem here is that one is dependent on the mechanism remaining functional, so for really long-term storage it would be a risky approach.

A safer approach would be to set up a large multidisk server with full backup facilities. For a lab which has the resources to manage such a system it could well be the option of choice. By the time the storage is even approaching full one can be confident that a system with 10 times the capacity will be available at less than the original cost. Typically servers use redundant array of individual disks (RAID) technology, in which surplus information is written so that failure of one disk will not cause data loss. Coupled with effective backup (either by tape or by "mirroring" the data on a second disk system) one can have a high level of confidence in the integrity of data.

d. Flash Drives

"Flash" memory is solid-state memory, like the system RAM of a computer, but unlike RAM it retains its data when switched off. This works by trapping electrons behind an insulating oxide layer, where they will remain until a voltage high enough to move them is applied (Heber, 2009). It comes in various formats, but all use the same technology. The ubiquitous USB memory sticks are the most common, and are (at the time of writing) available in capacities up to 64 Gbyte. Card formats such as CF (compact flash) and SD (secure digital) are used mostly in cameras, but do have the advantage that they can easily be filed in a compact and organized way. SD cards in particular are extremely small. This can be useful if one has a large collection.

The lifetime of stored data in flash memory seems to be unreported, though charge leakage is known to be a design issue. More often discussed is the finite number or reads and writes each location can support (Heber, 2009), but one is most unlikely to reach this limit storing image data. The speeds of flash memory devices typically range from 15 Mbyte/s to 120 Mbyte/s. The cost per gigabyte is huge, compared to other options, so they are unlikely to be a long-term storage solution, but their simplicity and extreme robustness makes them ideal for image transfer and transport.

B. Data Storage Conclusions

Tape drives have a place primarily as backup devices. In this role, they provide fast and cost-effective data storage. LTO, which is an open format and thus supported by a range of vendors, currently seems to be the only game in town.

For long-term archival storage of confocal images optical disks offer very low cost, good anticipated archival integrity, and universal readability. However, optical technology remains inefficient in its writing arrangements. For effective management of a confocal lab a fast, rewritable storage medium is also needed and both portable hard drives and flash USB sticks can provide this.

II. Image Processing

A. Image Formats

In the early days of confocal and other digital microscopy, manufacturers all used proprietary formats—and often did not even provide export facilities. Part of the reason for this was the need to store metadata—information about the image acquisition parameters, which standard formats such as tiff did not have space for. A proposal for a "universal" format for biological microscopy was put forward in 1990 (Dean et al., 1990), the Image Cytometry Standard. This has two files for each image—a simple text file (.ics) containing the metadata in a standard format and a separate file (.ids) for the image data. Public domain, open source software (Libics) is available for the format.

Nikon used the ics format in its confocal microscopes for many years. This was part of a general trend among manufacturers to use nonproprietary formats. Leica, for example, used tiff files, with a separate database containing the metadata. The database was also exported as a text file so that the imaging conditions were accessible to the microscopist. Some other manufacturers took the approach of using vendor-specific tags in the tiff format to contain metadata. In the past 5 years, the pendulum has swung the other way, and now almost everyone uses proprietary formats, providing free readers so that the files can be handled on other computers.

This creates a problem for a confocal lab where several different microscopes are used. Users may well use different microscopes in one project and want to compare results in the same software. To address this problem a collaborative project between Scottish and American researchers set up the Open Microscopy Environment, to provide transparent image and metadata manipulation from all formats (http://openmicroscopy.org/). The main outcomes of this initiative are OMERO, a client–server application which will open all vendors' formats and provide simple visualization tools, and Bio-Formats, a Java library which can be used as a plug-in for the popular freeware Image-J program.

B. Data Compression

Having looked at the considerable data storage problem presented by confocal images, a natural question to ask is "Can we reduce the size of the problem? Is it possible to compress the image data, to make it smaller?" There are many general-purpose data compression systems around, but most of these work on some variation or combination of three well-known algorithms. Run-length encoding (RLE) looks for sequences of identical values and replaces them with one copy of the value and a multiplier. Lempel–Ziv–Welch (LZW) and Huffman encoding look for repeated sequences, assign each a token, then replace each occurrence by its token. None of these works well with real images so that if you compress confocal images using one of the popular algorithms such as Zip you will find that gains are very modest—at best the files will shrink by 10–20%, which does little to alleviate the problem.

The most effective compression techniques for real-world images are "lossy"—part of the data is discarded. The philosophy behind this approach is that not all the data in an image is relevant to the impression it makes on the human eye. In terms of a confocal image, for example, there will always be noise in the background caused by statistical fluctuations in the numbers of photons arriving at the detector and random release of electrons in the photomultiplier and amplifier. The eye will typically not even see these, and certainly will not be perturbed if they are missing. Using this approach, very large file compressions can be achieved with losses which are undetectable to the eye.

The almost universal compression system in use today is the Joint Photographic Experts' Group (JPEG) compression protocol (Avinash, 1993; Pennebaker and Mitchell, 1993), which is supported by many painting and image manipulation programs. This breaks the image into blocks of 8×8 pixels, each of which is then processed through a Discrete Cosine Transform. This is similar to a Fourier transform, but much faster to implement, and gives an 8×8 array in frequency space. The frequency components with the lowest information content are then eliminated, discarding high-frequency information (fine detail) preferentially, and the remaining components are stored (using Huffman encoding) in the compressed image. The amount to be discarded in frequency space can be specified, which gives the user control over the trade-off between image quality and degree of compression. Typically monochrome images can be compressed down to one-fifth or less of their original size with no visible loss of quality (Avinash, 1993) (Fig. 1). Color images can be compressed further, since luminance (brightness) and chrominance (color) are treated separately, and the eye can tolerate a greater loss of information in the chrominance signal. Compression and decompression are similar

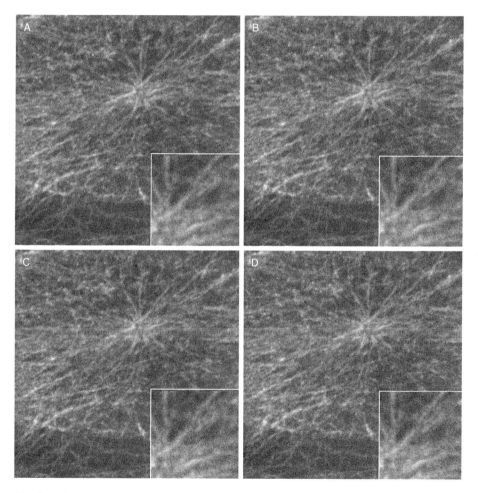

Fig. 1 Effects of image compression on a confocal fluorescence image of a rat aortic smooth muscle cell labeled with FITC antitubulin. (A) The original, uncompressed image. (B) JPEG compressed to 50% of the original size. (C) JPEG compressed to 20% of the original size. (D) Compressed to 10% of the original size. The insets show a small part of the image at a $2\times$ magnification.

operations, and require similar amounts of computer time. A consequence of the separation of chrominance and luminance information is that JPEG can only operate on gray-scale or true color (RGB) images. If (as is common in confocal microscopy) a false-color palette is added to a gray-scale image it would have to be converted to a full color image, tripling its size, before compression could start.

A more recent alternative to DCT is wavelet compression, of which the best known implementation is the JPEG 2000 format. With microscope images it does not seem to offer any improvement over DCT (Cox, 2006). Fractal compression, a proprietary technique developed by Iterative Systems Inc. (Anson, 1993; Barnsley and Hurd, 1993) take a totally different approach: it creates mathematical expressions which, when iterated, recreate the original image. It can give spectacular levels of compression. Unlike JPEG compression, creating the compressed image is a very time-consuming process but decompression is very quick. This has made it most useful for such items as CD-Rom encyclopedias, where the images are compressed once, for mass distribution and frequent decompression. It does not seem to have gained much acceptance in more general markets.

The effectiveness of—and the sacrifices in—JPEG compression of a confocal image is explored in Fig. 1. The original image, of microtubules in a smooth muscle cell, was 512 × 512 pixels, so the image data occupied 262,144 bytes. As a TIFF file, the file size was 263,866 bytes, the increase being the size of the TIFF header. A compressed TIFF file using lossless LZW encoding reduced the file size to 225,844 bytes, 85% of the original. With RLE encoding the file size actually increased to 275,848 bytes. Compressing to half the original size with JPEG the loss in quality is barely noticeable (Fig. 1B). At 20% of the original size (Fig. 1C) the difference is becoming apparent but is not obtrusive, and even reducing the file to only 10% of the original size (Fig. 1D) gives an image which is at least usable.

JPEG compression is a viable approach to reducing the size of confocal files, giving high compression ratios before the loss of quality becomes visible. However, storing archival copies as jpeg files is not a good idea, since information is lost. Many journals now run integrity tests on submitted images to identify any undisclosed manipulations, and these will reveal if lossy compression has been applied. The scope for further analysis is also much reduced—for example, edge-finding algorithms for detecting features will instead usually find the edges of the 8 × 8 pixel blocks.

The only effective lossless image compression format is Portable Net Graphics (PNG), a format originally designed with particular features to aid decompression over the Internet. The format and algorithms are open, and while it gives much poorer compression performance than JPEG it does offer useful reductions in size without any loss of quality.

C. 3D Rendering

Confocal microscopes collect images in three dimensions, which should make them appear more realistic than the 2D images generated by conventional fluorescence microscopes. Achieving that goal is far from easy. We believe that we see the world in three dimensions, but this is not really so—we *interpret* the world in three dimensions. The difference in parallax between the views seen by our two eyes is one factor in this—and it can be the major factor—but in everyday life it is but one among many. Experiments have shown that people can tell one object is in front of another even when the parallax is below the resolution of the eye. Other cues we use are motion parallax when we move our head, convergence and focus of our eyes, perspective, concealment of one object by another and our knowledge of the size and shape of everyday things. All together they are extremely effective—you need only watch a game of tennis to realize just how good our spatial perception is.

The last item on this list—our intuitive knowledge of the size and shape of everyday objects—is useless in the microscopic world. In the everyday world, we are also not used to looking at totally transparent objects. In the microscope, looking at wide field or confocal fluorescence images, or even at a histological section, everything is transparent. Another of our essential visual cues—objects obscuring their background—is therefore lost.

It is quite straightforward to generate binocular parallax—or at least an approximation to it—from our dataset. Stereoscopy involves producing one view for each eye from our series of slices. In a stereo pair, depth is converted into displacement between two images. Distant objects appear at the same place in both pictures, near objects appear displaced toward the centre line between the pictures. There is a limit to how much displacement the eye can tolerate—in a real specimen we converge and focus our eyes differently when looking at near and far objects. In a stereo pair picture everything is actually at the same plane, so we cannot do this. If a dataset extends for some considerable depth into a sample we may therefore have to artificially restrict the displacement—that is, foreshorten the depth—for the sake of comfortable viewing. The resulting stereoscopic image will still suffer from the disadvantage that all objects are transparent, and the lack of motion parallax. If we move our head when looking at a stereo pair the entire image appears to swing. Nevertheless it can be a dramatically effective way of presenting confocal images and confocal microscopes all offer stereoscopic views of confocal datasets without requiring additional software.

Stereo pairs are generally produced in confocal microscopy by projecting the series of images on to a plane, displacing each plane by one pixel from its predecessor. This is computationally simple but has the disadvantage that the perceived depth is essentially arbitrary, a function of pixel size and section thickness. An alternative approach is to rotate the dataset to generate the two views, and if we were generating stereo pairs as part of a full 3D reconstruction this tactic would usually be employed. Neither method is geometrically ideal but in practice both are extremely effective.

The resulting stereo pair can be presented on screen as an "anaglyph" with one image in red and the other in green or cyan, and viewed through appropriately colored glasses. This means that the original images must be monochrome—single channel without false-color palettes. This is rather limiting, and special viewers are available for viewing stereo pairs placed side-by-side on the computer screen, thereby avoiding these limitations. A more sophisticated approach requires the viewer to wear special glasses containing liquid-crystal shutters and linked to the computer by an infrared receptor. The left and right images are rapidly alternated on the monitor while the glasses switch between the two eyes in synchronism.

To bring in more of the visual cues into the perception of 3D datasets we need to produce a series of computer-generated views of our specimen, mimicking the views we would get if we looked all around it as a real object. This is computer 3D reconstruction, or volume rendering: the application of formulas or algorithms that can be implemented on a computer to create 3D projections from datasets. If several projections are created from different viewpoints, these can be played in quick succession to form rotations or animations. This brings in the element of motion parallax, which often can give us a clearer understanding of the 3D structure of the sample.

1. Rendering Techniques

There are several approaches to rendering volume datasets on a computer. In general, there is always a trade-off between speed and preservation of accuracy and detail. The amount of data which must be manipulated is very large, so that both processor speed and memory are important.

a. Surface Extraction

The original confocal datasets will consist of a set or grid of xyz coordinates of points in space with an intensity value for each point. We can reduce this to a simpler case by deciding which intensity values actually correspond to the "object" or objects we wish to render, and the extracting a reduced dataset containing just the x, y, and z coordinates of our objects. By geometric rendering—fitting tiny polygons to the points defining the surface of the "object"—we can produce a view of the surfaces of the objects we have extracted. This approach is fast, but we can only see the surface of our object—internal detail is lost. It can give an excellent view of a specimen if it is something that can be represented just by a surface, and presents it to the eye in a way which is easy to comprehend since transparency is no longer an issue. Often directional "lighting" effects are added to enhance the "realism."

b. Simple Projections

In many cases, especially in cell biology, we are concerned with more complex relationships than can be represented with surface extraction. In this case our reconstructions must make use of all the information in the original dataset. The simplest approach is to project the 3D array, from a given angle, on to a plane. If we take the average of all the intensity values (voxels) along each line the end result will be an average brightness projection (Fig. 2A). Alternatively, if we find and record the maximum brightness along each line rather than taking the average we will form a maximum brightness projection (Fig. 2B). Which will be more

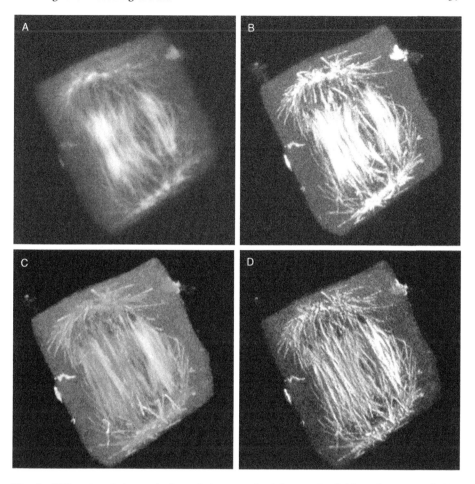

Fig. 2 Different rendering methods applied to a confocal dataset of a dividing wheat root cell. The specimen is stained with FITC antitubulin and shows the spindle microtubules at late anaphase. The dataset is $256 \times 256 \times 88$ (extracted from a larger field) and was acquired on a Bio-Rad Micro-Radiance system. In each case the projection is a top view of the data—looking straight down the stack of sections with no rotation. All projections were scaled to a uniform contrast range. (A) shows an average brightness projection and (B) shows a maximum brightness projection, both created with Confocal Assistant from Bio-Rad. The average brightness projection gives a very good impression of the overall distribution of microtubules but with very little impression of depth—in fact a view from the opposite direction would be identical. The maximum brightness projection gives much more prominence to individual microtubules and small groups. However, noise is much more apparent since there has been no averaging. (C) is a weighted reconstruction created in Voxel View—the tonal range gives a much clearer indication of what is near the top and what is further down the stack. (D), also created in Voxel View, has a lighting model applied. The individual microtubules (or bundles of microtubules) are much easier to distinguish thanks to the "shadows" cast.

effective will depend on the sample but in many samples of interest to the biologist the end results will be surprisingly similar.

With these projections the computation is simplified since we do not need to calculate the z-coordinates as we rotate the dataset—it is enough to know where each point is in x and y. Typically we would keep either x or y fixed and rotate about the other axis, so that we are tracking translations in one direction only. This approach has the advantage that all internal detail contributes to the final image, but suffers from the major problem that there is no distinction between front and back—both projections are identical. This can be alleviated if we create a stereo pair rather than a single image at each viewpoint. Nevertheless, if the image is complex, the problem of visualizing totally transparent objects may make it hard to comprehend.

c. Weighted Projection (Alpha Blending)

In this technique, a projection is again created by summing all the voxels along each line of sight, but with the added refinement that the voxels in front are made to contribute more than those behind. To do this the z position of each voxel must also be calculated at each rotation angle, which makes it the most computationally demanding technique. In this way, at each viewpoint the front is distinct from the back, yet internal detail can still be seen (Fig. 2C). The level of transparency allocated to a voxel will be based on its brightness, but the exact relationship is under the user's control—and in practice needs considerable experimentation to give the best reconstruction (Fig. 3). The end result is inevitably a compromise between seeing internal detail and giving a clear front/back differentiation, but it can be a very effective compromise.

Most software which offers weighted projection will also be able to add a lighting model to the reconstruction. In this case each voxel is treated not as an intensity value but as a reflective point on to which light "shines" from a source which the user can position. This will generate shadows which can make surface detail much clearer as well as helping with front/back relationships (Fig. 2D). It will also strongly emphasize noise, which will appear as granular detail on the surface; it is important not to mistake this for true surface structure. If noise is obtrusive it may be beneficial to preprocess the data with a smoothing filter (see below). Applying a lighting model adds hugely to the processing requirements and will slow down even a fast computer.

When rotating a dataset there will be some positions in which the one sampling line will "hit" a much higher number of voxels than its neighbor. The reconstruction will then show alternating dark and bright stripes in a parallel or criss-cross pattern (Fig. 4A). This problem is known as "aliasing" and can be avoided by treating each voxel as a cube in space rather than as a point (Fig. 4B). This is obviously a better approximation to the truth, but equally presents another large increase in the amount of processing required and so will slow down the process of generating the reconstruction. In general, the user will first experiment to obtain all

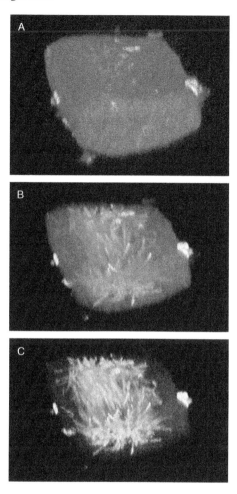

Fig. 3 The control of transparency. The dataset is the same as in Fig. 2 but is now reconstructed in an oblique view. If we keep even midlevel voxels fairly opaque (A) we see mostly the cytoplasm of the cell, looking like a suitcase with little inkling of the contents. Bringing the opacity curve closer to linear (B) we begin to see the spindle through the cytoplasm. Setting the opacity so that mid to low intensity voxels have little effect shows the spindle almost exclusively (C). Projections generated in Voxel View.

the correct parameters for the reconstruction and only then turn on antialiasing to generate the final view or series of views.

If we create a movie from a series of views generated by a weighted projection algorithm we have recovered the two key visual cues of motion parallax and objects in front obscuring those behind. Making the movie as a series of stereoscopic views also brings in binocular parallax. High-end systems also offer perspective—introducing a scaling difference between objects at the front and the back.

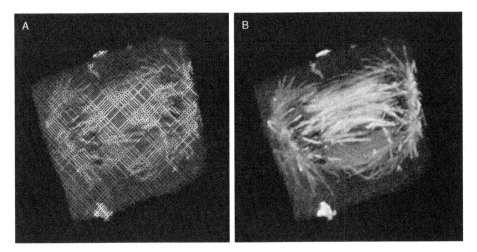

Fig. 4 Aliasing. (A) At certain orientations aliasing can be extremely apparent, not to say distracting, and information about the sample can be almost totally obscured. (B) Antialiasing does a very good job of rectifying the situation at the expense of a substantial increase in rendering time. Projections generated in Voxel View.

2. Software

Software for simple maximum brightness and average brightness projections, and for producing pixel-displacement stereo pairs, is bundled with all confocal microscopes from major manufacturers. Some include considerably more than this minimum, though often as a more or less expensive extra. When purchasing third-party software, therefore, one is generally looking for something more sophisticated — something capable of producing weighted (alpha-blended) projections, with antialiasing, lighting, perspective and maybe surface extraction. Cutting planes, scripting of complex movements and 3D measurements are other typical features.

The major players in the confocal market are VoxBlast, Imaris, Huygens, and Volocity. These are all products targeted at the fluorescence microscopy market—there are other rendering packages which cater for the X-ray tomography field, where the structures are dark rather than bright. Most of these products offer additional modules for deconvolution, image acquisition, and so on, and the final purchase decision may therefore be complex. However, most offer some form of free trial evaluation, and it is well worth taking advantage of this—before making an expensive purchase decision it is important to know that your choice will give you the results you want, and with an interface you can handle.

It is also worth considering what is available in the freeware sector. For straightforward, very high quality, volume rendering without any measurement facilities, VOXX, developed at Indiana University is the author's favorite. Measurements are promised in a future release. This is freeware, but not open source. Then there is the ubiquitous ImageJ—an image analysis package developed at the National Institute of Health, which is both free and open source. Plug-ins for every conceivable function have been written in other labs, so that it can be configured to meet practically any need. For 3D visualization and measurement it is best to get the FIJI package, which includes a rich set of relevant plug-ins.

D. Image Filtering and Preprocessing

Confocal datasets tend to have much worse signal/noise ratios than wide-field images. This is a matter of simple statistics. If one acquires a wide field, 512×512 pixel, image with a CCD camera, and a 1-s accumulated exposure, then each point in the image is being sampled for 1 s. If one acquires a confocal image of the same size over *10*s, each point in the image is sampled for just *0.04 ms*. This means that a noise-reducing filter can often be advantageously applied to our data, particularly if it is to be used for further processing or visualization.

A *convolution filter* moves through a dataset pixel by pixel, or voxel by voxel, modifying each value by reference to its immediate neighbors. Simple smoothing filters carry out a weighted average, averaging (for example) nine pixels but giving the centre one more weight, then replacing the original value of the central pixel with the new average. Thus a pixel which is much brighter than its neighbors will become darker, one that is darker will become lighter. The new value is placed in the output image—the original data are unchanged as the filter moves on to the next pixel and its neighbors. Smoothing filters of this type are not computationally demanding and give effective noise reduction with the penalty of some loss of detail.

A *median filter* is similar but typically gives a better compromise between noise reduction and preservation of detail. It is more computer intensive and therefore slower. It looks at all the pixels within, typically, a 3×3 region. The central pixel is then replaced (in the output image) by the median value of these nine pixels. The diagram shows this in action:

Ranking our original nine pixels we have: 56, 75, 90, 97, **100**, 101, 105, 127, 136. The new value for the pixel being processed is 100, the median of the list with four values above it and four below.

Confocal microscopes normally offer smoothing and median filters, but sometimes only in two dimensions. Most 3D reconstruction and deconvolution software offers filters which operate in three dimensions and this is definitely preferable.

Confocal microscopes do not have equal resolution in all three dimensions. The lateral resolution has a linear relationship with the numerical aperture (NA) of the lens, whereas the depth resolution varies as the *square* of the NA. With the highest NA lenses (NA 1.3–1.4) the lateral resolution will be around 200 nm whereas the depth resolution will be about 500 nm. With lower NA lenses the difference becomes progressively larger. This can spoil the appearance of 3D reconstructions,

127	136	56		127	136	56
100	75	90	becomes	100	**100**	90
101	105	97		101	105	97

and taking too small a number of optical sections in the interests of conserving disk space will matters worse. To fully exploit the depth resolution of a particular objective we should take at least two slices within the minimum resolved distance (the Nyquist criterion). As 3D reconstructions become more powerful, the problem of poor z resolution in confocal sections becomes more serious—images become fuzzier as they rotate toward a side-on view. We may be able to improve the situation by preprocessing of our image data.

The most thorough (and time-consuming) approach is to deconvolve the images with the point-spread function of the confocal microscope. In principle this could offer good results, since deconvolution of confocal images is a less intractable problem than deconvolution of wide-field ones. Most deconvolution packages offer the ability to handle confocal datasets as well as wide field. Plug-ins are also available for ImageJ. However, the relatively poor signal/noise ratio of confocal images limits the application of deconvolution algorithms. Whether for this reason, or because of the cost, or the time required, this approach has not achieved any great popularity.

A much simpler alternative was proposed by Cox and Sheppard (1993), who showed that a simple inverse filter based on an edge-enhancing algorithm, applied in the z direction only, was enough to dramatically improve the appearance of 3D confocal datasets. A similar feature was later added to Autoquant.

E. 3D Morphometry

Morphometry—segmentation and measurement of 2D features in images—is a routine part of many imaging labs, and many 2D measurement features are included among the standard software of most confocal microscopes. 3D measurement facilities are less common, but are in the author's experience often in demand. Surface area and volume of 3D structures are fundamental questions for anatomists and histologists to ask. To some extent stereological techniques—which lie beyond the scope of this chapter—can provide the answer, but tools for direct measurement of morphometric parameters still have an important part to play.

Volume is one of the simplest measurements to implement since a simple count of the voxels within the object is all that is needed. With some labor it could even be done, slice by slice, in a 2D image analysis program. It is provided by the built-in software in most recent confocal microscopes, and by 3D reconstruction software such as Imaris, VoxBlast, and Volocity. The typical approach is that the user will "seed" a volume by selecting a point within the volume of interest and then select the range of gray values which segment out the desired volume. More advanced systems will automatically find the edges of a structure by looking for abrupt changes of contrast. The voxels within the volume are then counted and converted to absolute units.

Surface area is less simple since a simple count of surface voxels does *not* give the correct result. However, it is routine in high-end software such as VoxBlast and Imaris. Another measurement feature implemented by most high-end 3D packages is 3D line length—the length of a line in 3D space. While this is a less commonly required measurement than area and volume, it would be difficult to obtain any other way. Once again, there are also plug-ins in ImageJ for this purpose.

F. Image Processing Conclusions

The prime need in the confocal lab is for rendering systems for 3D datasets. High-end 3D rendering systems offer genuinely worthwhile advantages over the simple projections offered with all confocal systems. Most of these will also offer extensive 3D measurement parameters. The range of facilities bundled with a confocal microscope will vary greatly, depending on the microscope and the chosen options. It is worthwhile taking time to learn and test the capabilities of the bundled software first, before purchasing additional rendering and analysis software. One is then in a better position to choose the packages which will best fill the additional needs of the laboratory.

Appendix

Manufacturers/sources of software mentioned in the text

Autoquant. Media Cybernetics Inc, 4340 East-West Hwy, Suite 400, Bethesda, MD 20814–4411 USA. http://www.mediacy.com/index.aspx?page=AutoDeblur Visualize

Bioformats. http://openmicroscopy.org/

Fiji. http://pacific.mpi-cbg.de/wiki/index.php/Main_Page

Huygens. Scientific Volume Imaging, Laapersveld 63, 1213 VB Hilversum, The Netherlands. http://www.svi.nl/

ImageJ. http://rsbweb.nih.gov/ij/

Imaris. Bitplane AG. Badenerstrasse 682, CH-8048, Zürich, Switzerland http://www.bitplane.com

Libics. http://libics.sourceforge.net/

OMERO. http://openmicroscopy.org/

Volocity. Perkin-Elmer Inc, 940 Winter Street, Waltham, Massachusetts 02451, USA http://www.cellularimaging.com/

VoxBlast. Vaytek Inc, 505 North 3rd St. Suite 200, Fairfield, Iowa 52556, USA http://www.vaytek.com vaytek@vaytek.com

VOXX. http://www.nephrology.iupui.edu/imaging/voxx/

References

Anson, L. F. (1993). Fractal image compression. *Byte* **19**(11), 195–202.

Avinash, G. B. (1993). Image compression and data integrity in confocal microscopy. *In* "Proc. 51st Annual Meeting," pp. 206–207. Microscopical Society of America.

Barnsley, M. F., and Hurd, L. P. (1993). Fractal Image Compression A.K. Peters, Wellelsey.

Cox, G. C. (2006). Mass storage, display and hard copy. *In* "Handbook of Biological Confocal Microscopy," 3rd edn. (J. Pawley, ed.), pp. 580–594. Springer, New York.

Cox, G. C., and Sheppard, C. (1993). Effects of image deconvolution on optical sectioning in conventional and confocal microscopes. *Bio-Imaging* **1,** 82–95.

Dean, P., Mascio, L., Ow, D., Sudar, D., and Mullikin, J. (1990). Proposed standard for image cytometry data files. *Cytometry* **11,** 561–569.

Heber, J. (2009). Supermemories. *New Sci.* **2737,** 40–43.

Pennebaker, W. B., and Mitchell, J. (1993). "JPEG Still Image Compression Standard." Van Nostrand Rheinhold, New York.

CHAPTER 4

Antifading Agents for Confocal Fluorescence Microscopy

Miguel Berrios,[*,†] Kimberly A. Conlon,[*] and David E. Colflesh[†]

[*]Department of Pharmacological Sciences
School of Medicine
Stony Brook University Medical Center
State University of New York, Stony Brook
New York, USA

[†]University Microscopy Imaging Center
State University of New York, Stony Brook
New York, USA

I. Introduction to the Revised Chapter

During the past decade and a half, major advances in light optics and photo-activatable (photoswitchable) fluorescent probes have been combined in a number of strategies to overcome the resolution limit of lens-based optical microscopes from about 200 nm to less than 20 nm (Betzig *et al.*, 2006; Juette *et al.*, 2008). The various strategies directed at improving the resolution limit of the wide field and confocal fluorescent microscopes are grouped under an umbrella known as super-resolution systems (Huang *et al.*, 2009).

The breaking of the resolution barrier imposed upon lens-based optical microscopes by the diffraction of light has come as the result of great talent, systematic experimentation, and immense effort and economic investment (Betzig *et al.*, 1991; Hell, 2003, 2007; Hell and Wichmann, 1994). Unfortunately, adequate specimen preparative methods to take full advantage of these new confocal and super-resolution fluorescence microscopes have not received equal attention and consequently are lagging well behind.

It is now evident that refinement of specimen preparation is urgently needed to take advantage of confocal and super-resolution fluorescence microscopy. Efforts toward this goal will require investments mimicking those made to take advantage of the spatial resolution offered by transmission electron microscopes in the 1950s. Similarly, once appropriate specimen preparations are available, super-resolution fluorescence microscopy is likely to generate vast new information on cell structure matching or surpassing that obtained by electron optics during the 1950s, 1960s, and 1970s. The new expanded resolving power of fluorescence microscopy combined with the ever increasing number of highly selective fluorescently tagged probes is likely to bring about a new revolution of our understanding of cell fine structure and function.

Improving current specimen preparation methods for highly sensitive confocal and super-resolution fluorescence microscopy would require the systematic, and at time tedious, reevaluation of rather routine procedures ranging from adequate coverslip mounting to the effectiveness of strategies dealing with autofluorescence and the fading of fluorochromes.

Although material in the following chapter was originally published about a decade ago, the strategies it presents to counteract autofluorescence and the fading of fluorochromes under laser light have proven effective and may be as valid now for newer confocal and super-resolution fluorescence microscopy as when first published. The following text has been revised to include up-to-date information and additional details for mounting specimens.

II. Introduction

Advances in fluorescent probe chemistry, more economical and powerful lasers, improvements in confocal and image acquisition systems together with faster and inexpensive computers have made confocal microscopes available to a wider range of biological applications and users. Currently, single and dual labeling immuno-fluorescence (e.g., Furuse *et al.*, 1998; Meller *et al.*, 1995; Nathke *et al.*, 1996), along with imaging of green fluorescent protein (e.g., Chalfie, 1995; Cubitt *et al.*, 1995; Darsow *et al.*, 1998; Gerdes and Kaether, 1996; Larrick *et al.*, 1995; Prasher, 1995; Roberts and Goldfarb, 1998) and fluorescent *in situ* hybridization (FISH) (e.g., Dernburg and Sedat, 1998; Lemieux *et al.*, 1992; Loidl *et al.*, 1998; Moens and Pearlman, 1991), are the most widely used applications of the confocal microscope. Single or dual labeling immunofluorescence may involve either a direct or an indirect method (Kawamura, 1977; Springall and Polak, 1995). The greatest advantage of confocal and nanoscale microscopy systems equipped with digital image acquisition devices is their ability to discriminate between signals originating from in-focus and out-of-focus optical planes to produce digital images that can be merged pixel-by-pixel to generate perfectly registered two-dimensional and three-dimensional renditions.

To obtain the maximum information from these renditions, however, each optical section must provide a good quality image. Optical aberrations may be reduced when all objects in the microscope's light path, including the objective lens, coverslip, and immersion oil, are used following manufacturer's specifications. In confocal fluorescence, the quality of signal from fluorescent probes is dependent not only on the properties of the confocal system, the microscope's objective lens, and the image acquisition device itself but on the condition of the specimen as well. Specimens should be prepared with care to avoid introducing artifacts that, aside from some other deleterious effects, may increase background fluorescence levels. Important considerations in this regard are the fixation procedure, antibody probes, and incubation conditions, and ultimately the method used for mounting the specimen between slide and coverslip. In fact, immunofluorescence and coverslip mounting protocols that may be suitable for the wide-field microscope may prove inadequate for confocal and super-resolution fluorescence microscopy.

In addition, a pervasive phenomenon-restricting image acquisition is the deterioration of fluorescence emitted by fluorochrome-conjugated antibodies during laser scanning of optical planes. Fluorescence fading may be very pronounced for fluorescein isothiocyanate (FITC)-conjugated antibodies (Bock *et al.*, 1985). This fading phenomenon is particularly striking during imaging of dual-labeled specimens because of the differential deterioration of fluorescence emitted by several fluorochromes (e.g., FITC vs. tetramethylrhodamine isothiocyanate (TRITC)). Increasing the concentration of fluorescent probes including fluorochrome-conjugated antibodies is not a viable solution since this usually results in higher background fluorescence. Alternatively, the selection of fluorescent probes which are less susceptible to fading is often the best option, if available, for multilabeled specimens. In most situations, mounting biological specimens with antifading agents can solve fluorescence fading problems.

This chapter discusses strategies to reduce specimen-associated background fluorescence and the properties and performance of several antifading agents during acquisition of multiple optical sections. Although discussions are focused on indirect immunofluorescence microscopy, most sections are applicable to other *in situ* localization methods. In our case, several antifading agents, including the commercial preparations Vectashield® (Vector Laboratories, Burlingame, CA), SlowFade® and ProLong® (Invitrogen, Carlsbad, CA), have fluorochrome antifading properties that alleviate the above problems, making acquisition by confocal and super-resolution fluorescent microscopy of optical Z-series from a multilabeled specimen reproducible.

III. Specimen Preparation

In our experience, several aspects of specimen preparation should receive special consideration when attempting to obtain fluorescent images with low backgrounds and reduced fluorochrome fading. In general, we place as much effort into reducing fluorochrome fluorescence fading as we place into avoiding artifacts that may result in background fluorescence. Attention to fixation and antibody probing procedures, and when necessary, introduction of procedures directed at reducing naturally occurring background fluorescence during specimen preparation has helped us obtain good quality fluorescent images (Fig. 1). These three aspects of specimen preparation directly affect capturing images with low backgrounds and reduced fluorochrome fading and are discussed in the following sections.

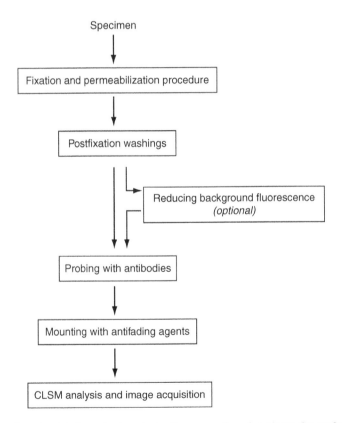

Fig. 1 Flowchart showing the main steps during the preparation of specimens for confocal immunofluorescence. Reduction of specimen-associated background fluorescence is optional.

A. Fixation

There are at least as many fixation procedures available in the literature as there are immunofluorescence protocols. All of them, however, are at best a compromise between preservation of structure and preservation of antigenic determinants. Examples of fixation procedures for immunofluorescence can be found in comprehensive reviews by Kawamura (1977) and Springall and Polak (1995) and articles by Perez *et al.* (1995), Poot *et al.* (1996), Dernburg and Sedat (1998), and Solari (1998). Although fixatives containing aldehydes have a tendency to increase background fluorescence, we have found that these compounds provide superior morphological preservation over, for example, fixation with either ethanol or methanol. We found that fixatives based on buffered salt solutions containing 3–4% (w/v) paraformaldehyde or combinations of paraformaldehyde and 0.1% (v/v) glutaraldehyde worked well for most confocal applications (e.g., Berrios and Colflesh, 1995; Colflesh *et al.*, 1999; Meller *et al.*, 1995). In fact, buffered paraformaldehyde alone may be an adequate fixative for most confocal immunofluorescence applications. To prepare these fixatives at an appropriate pH, we prefer to dissolve paraformaldehyde powder in a buffer rather than the more traditional resuspension of this reagent in hot water and pH titration with 6.25 N NaOH. A procedure that uses 1,4-piperazinediethanesulfonic acid (Pipes) buffer instead of sodium hydroxide to produce an 8% (w/v) paraformaldehyde (pH 7.5) stock solution is shown in Table I (Berrios and Colflesh, 1995). This stock solution should be kept at room temperature and is stable for many weeks. Another key step to reduce background fluorescence staining is the inclusion of one or two postfixation washes, preferably with buffered solutions containing primary amines (e.g., tris-[hydroxymethyl]-aminomethane (tris)-glycine, tris-Cl, glycine, 2,2′,2″-nitrilotriethanol (triethanolamine) buffer solutions, etc.). These postfixation washes reduce the risk of cross-linking antibodies by quenching available aldehyde groups.

B. Reducing Specimen-Associated Background Fluorescence

The detection of antigens by confocal immunofluorescence has been limited by two phenomena: (1) fluorochrome fading and (2) specimen-associated background fluorescence (also known as autofluorescence). Both phenomena act together to

Table I

Preparation of an 8% (w/v) Paraformaldehyde 154 mM Pipes (pH 7.5) Stock[a]

1	Weigh 2 g of paraformaldehyde powder
2	Dissolve the above in 15 ml of 256 mM Pipes (pH 7.5) while stirring in a water bath at 70 °C (~20 min)
3	Once in solution, complete volume to 25 ml with cold water
4	Clarify the resulting solution by centrifugation at 5000 × g for 10 min
5	Store supernatant in a capped glass bottle at room temperature[b]

[a]Berrios and Colflesh (1995).
[b]Discard stock after 6 weeks.

reduce the effectiveness of this technique, particularly when antigens are found in low abundance. As discussed in Sections IV and V, fluorochrome fading may be reduced effectively using a combination of antifading agents and rapid scanning lasers (short dwelling time) (Berrios and Colflesh, 1995). Much less has been done, however, to reduce background fluorescence associated with the specimen. This phenomenon still remains as an unsurmountable threshold to the limit of detection of immunofluorescent assays. We have devised a procedure in which reducing agents are used to lower specimen-associated background fluorescence before detection of subcellular antigens by confocal immunofluorescence microscopy (Colflesh *et al.*, 1999).

Reduction of specimen-associated background fluorescence is performed immediately after paraformaldehyde fixation of the sample (see Table II). We found that it is best to prepare the solution of reducing agent fresh just before use (Colflesh *et al.*, 1999). The reducing agent (sodium borohydride or sodium cyanoborohydride) is dissolved in a defined buffered saline solution, such as 140 mM NaCl; 10 mM phosphate buffer, pH 7.4 (PBS). These solutions must be discarded after use because their ability to reduce background fluorescence diminishes with time. The final or working concentration of the reducing agent is defined by the amount of specimen-associated background fluorescence of the sample, and it is determined by incubating the specimen in increasing concentrations of reducing agent. We have successfully used concentrations ranging from 10 μM to 100 mM of sodium borohydride with no detectable deterioration of specimen morphology or its antigenic determinants (Colflesh *et al.*, 1999). At higher concentrations (e.g., 10–100 mM), the pH of a solution of sodium borohydride in PBS increases from 7.2 to about 9.0–9.5. If this increase in pH is a concern, sodium cyanoborohydride can be used instead. We found that the pH of solutions of sodium cyanoborohydride increased to about 8.5. To maximize reproducibility from experiment to experiment, the specimen is fixed, then, if necessary, specimen-associated background fluorescence reduced and antibody incubations and washes performed in 5.4 mm deep $\times$ 20 mm diameter wells of two-well immunofluorescence chambers (Berrios, 1989; Fisher and Berrios, 1998). These chambers (#70330, Electron Microscopy Sciences, Hatfield, PA) are assembled around the

Table II

Buffered Solutions and Reagents for Reducing Specimen–Associated Background Fluorescence

Buffered solutions and reagents
1 MSM-Pipes
2 MSM-Pipes containing nonionic detergents
3 4% (w/v) paraformaldehyde in MSM-Pipes
4 PBS
5 Sodium borohydride or sodium cyanoborohydride[a]

[a]Fresh bottle (or store powder dry in a dessicator).

microscope slide without any bonding substance and are taken apart before microscopical examination. Previous experience using hydrophobic substances such as nail polish, bee's wax, dental wax, rubber cement, or silicone glue to confine solutions over specimens during incubations proved difficult to reproduce and, in general, these substances had the tendency to increase fluorescence backgrounds. Other devices, such as chamber slides (Lab-Tek, Nalgene Nunc International (a division of Thermo Fisher Scientific), Rochester, NY), have a tendency to leave adhesive residues and their use precludes examination of specimens such as tissue smears and sections. Observations described in this chapter were obtained using these immunofluorescence chambers to fix specimens, to reduce specimen-associated background fluorescence, if necessary, and to probe specimens with antibodies.

Typically, a general procedure to reduce specimen-associated background fluorescence is as follows: after fixation, the sample is washed twice with a buffered defined salt solution (e.g., 18 mM $MgSO_4$, 5 mM $CaCl_2$, 40 mM KCl, 24 mM NaCl, 0.5% (v/v) Triton X-100®, 0.5% (v/v) Nonidet P-40®, and 5 mM Pipes, pH 6.8 (MSM-Pipes); Smith *et al.*, 1987) and then incubated with 400 μl of a solution containing the desired reducing agent at 37 °C for 10 min in a humidified chamber (a humidified chamber can be assembled using a plastic box 110 mm × 160 mm × 40 mm deep (e.g., #64322, Electron Microscopy Sciences) containing an 8–10 mm thick water-soaked plastic sponge insert at the bottom). We have found these to be the optimum conditions to reduce specimen-associated background fluorescence. When the incubation is performed at 23 °C, no reduction of specimen-associated background fluorescence was observed. Incubation is followed by two washes with a buffered defined salt solution (e.g., MSM-Pipes) to completely remove all traces of the reducing agent. A general procedure is outlined below.

1. Reduction of Background Autofluorescence

1. Place slide containing tissue culture cells or frozen section into a two-well immunofluorescence chamber (Berrios, 1989; Fisher and Berrios, 1998).
2. Fill each well with freshly prepared 4% (w/v) paraformaldehyde in MSM-Pipes (Berrios and Colflesh, 1995; Colflesh *et al.*, 1999).
3. Incubate at room temperature with gentle shaking for 3 min.
4. Remove fixative.
5. Fix again by adding 400 μl/well of 4% (w/v) paraformaldehyde in MSM-Pipes (Berrios and Colflesh, 1995; Smith *et al.*, 1987).
6. Incubate at room temperature with gentle shaking for 3 min.
7. During second fixation, weigh out reducing agent (sodium borohydride or sodium cyanoborohydride) and dilute to desired concentration with PBS.
8. Remove fixative and wash twice (1 min each time, with gentle shaking) with 800 μl MSM-Pipes containing nonionic detergents.

9. Remove second wash and add 400 μl/well of reducing agent in PBS solution.
10. Incubate for 10 min at 37 °C in a humidified chamber.
11. Remove reducing agent and wash twice (1 min each time, with gentle shaking) with 800 μl MSM-Pipes containing nonionic detergents.

We have tested several strategies for their ability to reduce naturally occurring background fluorescence associated with insect (*Drosophila* female abdomen) and mammalian (mouse liver) 7-μm-thick cryosections (Colflesh *et al.*, 1999). Freshly prepared cryosections were incubated with either sodium borohydride or sodium cyanoborohydride (Table II). Figure 2 contains results of one of these experiments. Sodium cyanoborohydride was slightly more effective in reduction of autofluorescence than sodium borohydride, possibly because the former is less reactive and consequently more stable in aqueous solutions (compare Fig. 2A with B). When mouse liver cryosections were incubated in 10 mM sodium borohydride for 10 min at 37 °C, no reduction of tissue autofluorescence was observed (Fig. 3A). Mouse liver tissue autofluorescence was reduced by about 48% when cryosections were incubated in 100 mM sodium borohydride for 10 min at 37 °C (Fig. 3B). The inability of 10 mM sodium borohydride to quench autofluorescence in this tissue may be a result of the tissue's high fluorescence intensity. A comparison of specimen-associated background fluorescence between *Drosophila* adult female abdomen and that of mouse liver sections showed that the fluorescence intensity of the former is lower than that of the latter (Figs. 2 and 3).

C. Probing with Antibodies

Once the specificity of an antibody has been determined using enzyme-linked immunosorbant assay (ELISA), competitive ELISA, immunoblots, etc., focus should be placed on identifying an appropriate fluorochrome-conjugated secondary antibody. In our experience, using fluorochrome-conjugated secondary antibodies for visualization of first specific antibody–antigen complexes has several practical advantages over the direct labeling of specific (primary) antibodies. Secondary antibodies are widely available commercially. They are relatively inexpensive (currently ranging from about US$ 0.08/μg to US$ 0.8/μg). They are available as IgG fractions from different hosts (e.g., chicken, donkey, goat, mouse, rabbit, etc.) conjugated to a wide range of fluorochromes. In addition, the use of secondary antibodies allows for signal amplification and eliminates possible detrimental effects of the chemical conjugation procedure upon specific antibodies.

Fluorochrome-conjugated secondary antibodies are most often available as specific anti-IgG (antiheavy and/or antilight chains) affinity-purified antibody fractions. Comprehensive lists of unconjugated and fluorochrome-conjugated antibody fractions is beyond the scope of this chapter and may be found in USBiological's Antibody Reference (7th Edition, 2009, USBiological, Swampscott, MA) or at http://www.usbio.net/subcategories.php?category=antibodies or in well-organized charts included in catalogs from several commercial vendors (see Table III).

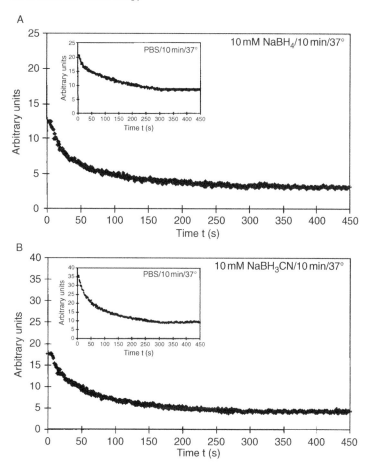

Fig. 2 Autofluorescence of *Drosophila* cryosections after incubation in either sodium borohydride or sodium cyanoborohydride. Graphs represent actual recordings of autofluorescence intensity measured at 1 s intervals during 450 s of continuous laser illumination. Each time point presents measurements from three random fields. (A, B) samples incubated for 10 min at 37 °C in either PBS (insets in A, B), freshly prepared 10 mM sodium borohydride (A), or 10 mM sodium cyanoborohydride (B). X-axis is laser exposure time (s) and Y-axis is autofluorescence intensity expressed in arbitrary units. Adapted from Colflesh *et al.* (1999). Reproduced with permission.

We have had success performing confocal immunofluorescence using commercially available fluorochrome-conjugated secondary antibodies (affinity-purified fractions) containing approximately 3.6 mol of fluorochrome per mole of IgG. To reduce nonspecific fluorescence, we have found that it is best to obtain affinity-purified antibody fractions and to use them at relatively low working concentrations, usually at concentrations that are lower than those recommended by the commercial supplier. Most often we obtain optimal staining using working dilutions ranging anywhere from 1:1000 to 1:10,000 (1.5–0.2 μg (IgG)/ml) of the

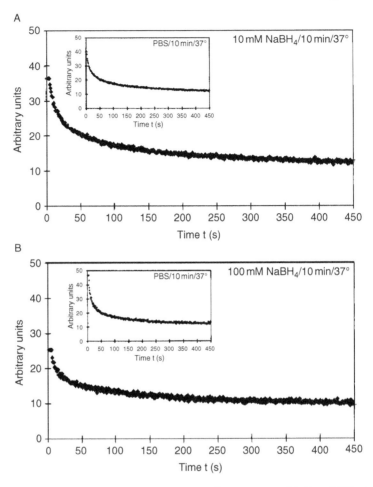

Fig. 3 Autofluorescence of mouse liver cryosections after treatment with 10 and 100 mM sodium borohydride. Graphs represent actual recordings of autofluorescence intensity measured at 1 s intervals during 450 s of continuous laser illumination. Each time point represents measurements from three random fields. (A, B) Samples incubated for 10 min at 37 °C in either PBS (insets in A, B), freshly prepared 10 mM sodium borohydride (A), or 100 mM sodium borohydride (B). *X*-axis is laser exposure time (s) and *Y*-axis is fluorescence intensity in arbitrary units. Adapted from Colflesh *et al.* (1999). Reproduced with permission.

original stock. We also prefer to keep fluorochrome-conjugated secondary antibodies as 20 μl stocks in 50% (v/v) glycerol at -20 °C. Under these conditions these reagents do not show signs of deterioration for up to 4 years. These antibodies are diluted into buffered salt solutions (e.g., PBS) supplemented with nonionic detergents (e.g., 0.5% (v/v) of either Triton X-100®, Nonidet P-40®, or Tween 20®) just before use, discarding the remainder. As stated earlier, to maximize staining

Table III
Suppliers of Fluorochrome–Conjugated Antibodies in the United States[a]

Name	Address
Aves Labs, Inc.	16200 South West Pacific Highway
	Tigard, OR 97224, USA
	Phone: 1-503-245-1858
	Fax: 1-503-245-8784
	http://www.aveslab.com
E. Y. Laboratories	107 North Amphlett Boulevard
	San Mateo, CA 94401, USA
	Phone: 1-650-342-3296
	Fax: 1-650-342-2648
	http://www.eylabs.com
Jackson ImmunoResearch Laboratories	P.O. Box 9, West Grove, PA 19390, USA
	Phone: 1-800-367-5296 or 1-610-869-4024
	Fax: 1-610-869-0171
	E-mail: curserjaxn@aol.com
Millipore Corporation	290 Concord Road
	Billerica, MA 01821, USA
	Phone: 1-800-645-5476 or 1-978-715-4321
	http://www.millipore.com
MP Biomedicals, Life Sciences Division	29525 Fountain Parkway
	Solon, OH 44139, USA
	Phone: 1-800-854-0530
	Fax: 1-800-334-6999
	http://www.mpbio.com
Life Technologies Corporation	5791 Van Allen Way
Invitrogen (Molecular Probes)	Carlsbad, CA 92008, USA
	Phone: 1-800-955-6288 (option 3, Ext. 46029)
	Fax: 1-800-331-2286
	http://www.invitrogen.com
Santa Cruz Biotechnology	2145 Delaware Avenue
	Santa Cruz, CA 95060, USA
	Phone: 1-800-457-3800
	Fax: 1-831-457-3801
	E-mail: scbt@scbt.com
Thermo Scientific Pierce	Phone: 1-800-874-3723 or 1-815-968-0747
	E-mail: pierce.cs@thermofisher.com
	www.thermo.com/pierce
USBiological	P.O. Box 261, Swampscott, MA 01907, USA
	Phone: 1-800-520-3011 or 1-781-639-5092
	Fax: 1-781-639-1768
	E-mail: service@usbio.net
	http://www.usbio.net

[a]Partial list.

reproducibility from experiment to experiment, we perform antibody incubations and washes in wells of immunofluorescence chambers (Berrios, 1989; Fisher and Berrios, 1998).

IV. Antifading Agents

A. Mounting Specimens with Antifading Agents

This is a simple procedure but in our experience some care must be taken to avoid shearing the specimen, introducing air bubbles and/or fluorescent contaminants and/or leaving too much mounting media between the slide and the coverslip. An excess of mounting media between the slide and the coverslip does not improve the antifading properties of the agent and may hamper the gathering of confocal slices when using objective lenses with short working distances. To maintain better control over this operation, we prefer to use individual glass coverslips over each specimen (e.g., 18 mm × 18 mm × (0.13–0.17 mm) thick) instead of one large coverslip over the whole slide or most of it. When using the immunofluorescence chambers (Berrios, 1989; Fisher and Berrios, 1998), we generate two samples per slide; this leaves sufficient space for placing an adequate amount of seal (e.g., colorless nail polish) around each coverslip. We recommend adding, over the wet specimen, approximately 30 nl of mounting media per square millimeter (mm^2) of coverslip (i.e., 10 μl for an 18 mm^2 coverslip). Alternatively, to help standardize this procedure, excess mounting media may be squeezed out by gently pressing the coverslip against the slide for 3 min using the attraction force of two small permanent (Alnico) magnets (e.g., #3041947; Edmund Scientifics, Tonawanda, NY) (Berrios and Colflesh, 1995). Excess mounting media may be removed using either a DNase/RNase-free 10 μl ultra-micropipette tip (Biohit Oyj, Finland), a 200 μl natural/yellow pipette tip (Eppendorf, Germany), or a similar quality disposable plastic pipette tip attached to a low vacuum line (e.g., -10 psi (0.68 atm)).

- To avoid light refraction artifacts during the acquisition of fluorescent optical sections (Z-series or stacks), great care should be placed when mounting specimens between or under coverslips. To avoid artifacts of this nature, the surface of coverslips should be checked for flatness and tilted other than 90 ° or perpendicular to the optical axis of the objective lens. In practice, small area coverslips, round or square, have the flattest surfaces.

- To reduce void space as well as avoid fluorescent signal from the mounting media itself, coverslips should be mounted with the smallest amount of mounting media. For instance, for tissue culture cells growing as a monolayer on slides or coverslips, the value given above (30 nl/mm^2 of coverslip surface) may be reduced to approximately 20 nl/mm^2 of coverslip surface (at this ratio, the distance separating two coverslips or a slide and a coverslip would be on average approximately 20 μm).

Care must be taken to avoid the drying of specimens before mounting the coverslip. In our experience, the antifading properties of mounting agents are not affected by wet specimens. Each coverslip may be sealed with clear nail polish and kept in the refrigerator (at approximately 4 °C) and in the dark until use to reduce evaporation.

- In practice, regardless of specimens, fluorescence signals tend to deteriorate with time, consequently, it is best to avoid storing probed specimens (including under refrigeration conditions). Specimen examination should occur as soon as the staining procedure has been completed.

- Moreover, when capturing images from multilabeled specimens, fluorescent signals from faster fading fluorescent probes should be acquired first if possible.

B. Agents and Their Antifading Properties

There are several antifading preparations reported in the literature, and a number of these are available commercially under various Trade names (Table IV). Antifading preparations are most often used as coverslip mounting media to reduce the fading of fluorescence after UV light or laser excitation of fluorochromes. Although it is not our aim to evaluate the antifading properties of all mounting media available, we have surveyed, using a confocal laser scanning microscope (CLSM), the properties of several common antifading agents both reported in the literature and commercially available.

To evaluate the antifading properties of several mounting media, we selected nuclei from *Drosophila melanogaster* adult male accessory glands (Schmidt *et al.*, 1985)

Table IV
Commercial Antifading Agents[a]

Trade name	Supplier
1 ProLong®	Invitrogen
2 SlowFade®	Carlsbad, CA 92008, USA
3 SlowFade Light®	Phone: 1-800-955-6288
	Fax: 1-800-331-2286
	E-mail: catalog@invitrogen.com
4 Vectasheild® (H-1000)	Vector Laboratories
	Burlingame, CA 94010, USA
	Phone: 1-800-227-6666
	Fax: 1-650-697-0339
	E-mail: vector@vectorlabs.com
	http://www.vectorlabs.com
5 Citifluor®	Citifluor Limited
AF1, AF2, and AF3	London, UK
	Phone: 44-0-20-7739-6561
	Fax: 44-0-20-7729-2936
	E-mail: enquiries@citifluor.co.uk
	http://www.citifluor.co.uk

[a]Partial list.

and histone proteins as targets. Cells in this *Drosophila* tissue are arrested at interphase, and nuclei are homogeneous in size and morphology. The histones are an abundant and widely distributed group of nuclear proteins. Briefly, *Drosophila* accessory glands were dissected out, transferred to a sodium dodecylsulfate-cleansed microscope slide and nuclei extruded from cells exactly as described (Berrios, 1994). Indirect immunofluorescence was as described previously (Berrios and Colflesh, 1995). Cell nuclei were probed with a mouse monoclonal antibody (mAb052, clone F152.C25.WJJ, Millipore, Billerica, MA) directed against histones (H1 and core proteins H2a, H2b, H3, and H4) followed by either FITC-conjugated donkey antimouse IgG antibodies, TRITC-conjugated donkey antimouse IgG antibodies, or Texas red sulfonyl chloride (TRSC)-conjugated donkey antimouse IgG antibodies. Stained nuclei were mounted with either MSM-Pipes without detergents (Berrios and Fisher, 1986; Smith *et al.*, 1987) as a control or this defined buffered salt solution supplemented with one of the following antifading mounting media: 0.1% (w/v) *p*-phenylenediamine (Johnson and Araujo, 1981), 0.3% (w/v) gallic acid *n*-propyl ester (*n*-propyl gallate) (Giloh and Sedat, 1982) and the commercial preparations Vectashield®, SlowFade®, or ProLong® (Table IV). The pH of mounting solutions was kept between 6.8 and 7.2, and the initial nuclear fluorescence staining was similar for all mounting media.

• Stocks of commercially available or noncommercial antifading preparations should be stored in several aliquots at 4 °C and in the dark for a period of no more than 6 months. Aliquots significantly reduce the chances of cross-contamination with fluorochromes and/or bacteria.

Fluorescence signal intensity data from each nuclei was collected simultaneously at 60 s intervals during 15 min of continuous laser epi-illumination using an Odyssey® confocal 200 mW argon-ion multiline laser scanning system (Noran Instruments, Middleton, WI) attached to a Nikon® Diaphot inverted microscope (Nikon, Melville, NY) equipped with a Nikon® 60X/1.4 CFN planapochromat oil immersion objective lens (Berrios and Colflesh, 1995). The objective lens was focused near the diameter of 12 randomly selected nuclei. On going through this objective lens, the laser scanned an area of about 630 μm^2. Laser scanning rate was 30 frames/s and dwell time was 0.1 μs/pixel. To detect fluorescence emission from FITC, specimens were exposed to a laser excitation wavelength of 488 nm and images captured through a 500 nm barrier filter. To detect fluorescence emission from either TRITC or TRSC, specimens were exposed to a laser excitation wavelength of 529 nm and images captured through a 550 nm barrier filter. The confocal detector aperture was adjusted to 15 μm, laser power output was set at either 20% or 50% (see Table V). Photomultiplier gain was set at 88% (1.1 kV), a value at which no signal was detected from specimens probed in the absence of first specific antibodies (Berrios and Colflesh, 1995; Colflesh *et al.*, 1999).

FITC fluorescence faded rapidly when specimens were mounted without antifading agent and exposed to CLSM (Fig. 4). Fluorescence signal from FITC was reduced to about 5% after 1 min of exposure to 50% laser power (Fig. 4A) and to

Table V
Laser Scanning[a] Power Measured[b] at the Specimen

Nominal laser power[c]	Excitation wavelength	
	488 nm (FITC)	529 nm (TRITC/TRSC)
20%	90 nW/μm^2	20 nW/μm^2
50%	500 nW/μm^2	120 nW/μm^2

[a]Laser scanning rate was 30 frames/s.

[b]Measurements were made with an 840 C Optical Power Meter (Newport Corporation, Irvine, CA).

[c]Power settings were regulated through an acousto-optic deflector driven by the Odyssey® software package (Noran Instruments). Adapted from Berrios and Colflesh (1995). Reproduced with permission.

about 50% after 1 min of exposure to 20% laser power (Fig. 4B). In contrast, TRITC fluorescence faded to about 55% and TRSC fluorescence faded to about 80% after 15 min of exposure to 50% laser power (Fig. 4A). At 20% laser power, TRITC and TRSC fluorescence remained nearly unchanged (Fig. 4B).

When specimens were mounted in either p-phenylenediamine or n-propyl gallate, two widely used antifading agents (Battaglia et al., 1994; Bock et al., 1985; Jensen et al., 1995; Lemieux et al., 1992; Scheynius and Lundahl, 1990; Soliman, 1988), FITC fluorescence faded to about 10% after 15 min of exposure to 50% (Figs. 5A and 6A) or 20% (Figs. 5B and 6B) laser power. TRITC fluorescence in the presence of either p-phenylenediamine or n-propyl gallate was reduced to about 70% after 15 min of exposure to 50% laser power (Figs. 5A and 6A). Although TRITC fluorescence in the presence of p-phenylenediamine remained nearly unchanged after 15 min of exposure to 20% laser power (Fig. 5B), in the presence of n-propyl gallate TRITC fluorescence was reduced to 68% (Fig. 6B). TRSC fluorescence in the presence of p-phenylenediamine faded to 70% and 80% after 15 min of exposure to 50% and 20% laser power, respectively (Fig. 5). In contrast, TRSC fluorescence in the presence of n-propyl gallate remained nearly unchanged after 15 min of exposure to either 50% or 20% laser power (Fig. 6).

The commercial antifading preparations, Vectashield®, SlowFade®, and ProLong® were also tested for their ability to quench fading of fluorescence signals under the above conditions. FITC fluorescence was reduced between 20% and 40% when specimens were mounted in Vectashield®, SlowFade®, or ProLong®, after 15 min of exposure to 50% laser power (Figs. 7A, 8A, and 9A). However, when specimens were mounted in either Vectashield® or SlowFade® and exposed to 20% laser power, about 50–70% of FITC fluorescence signal remained after 15 min of exposure (Figs. 7B and 8B). FITC fluorescence was reduced to 20% under 20% laser power when using ProLong® (Fig. 9B). In the presence of ProLong®, TRITC and TRSC fluorescence were reduced to about 88% and 70%, respectively after 15 min of exposure to 50% laser power (Fig. 9A) and to 95% and 90%, respectively after 15 min of exposure to 20% laser power (Fig. 9B).

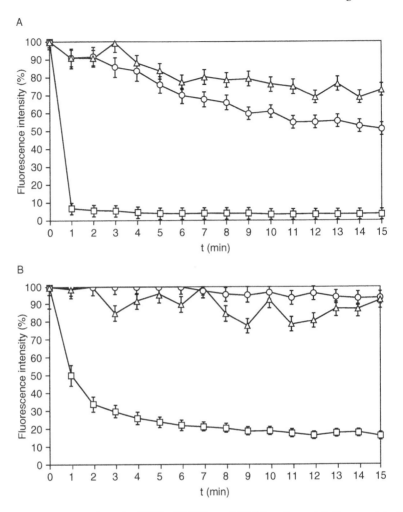

Fig. 4 Fading characteristics for FITC-, TRITC-, and TRSC-conjugated IgG exposed to CLSM without antifading agents. X-axis is laser exposure time (min) and Y-axis is fluorescence intensity expressed as a percentage relative to the initial signal (i.e., $t = 0$ min fluorescence intensity is 100%). Each point is an average of 20 measurements at 1 min intervals. (A) Fading during exposure to 50% laser power. (B) Fading during exposure to 20% laser power. FITC-conjugated IgG (open squares), TRITC-conjugated IgG (open circles), and TRSC-conjugated IgG (open triangles). Unpublished results of Colflesh and Berrios (1998) and adapted from Berrios and Colflesh (1995). Reproduced with permission.

As shown above, 0.1% (w/v) *p*-phenylenediamine and 0.3% (w/v) *n*-propyl gallate provide a level of protection to fluorochrome fluorescence fading over mounting specimens in PBS or MSM-Pipes (Berrios and Fisher, 1986; Smith *et al.*, 1987). Commercially available preparations such as the ones tested here,

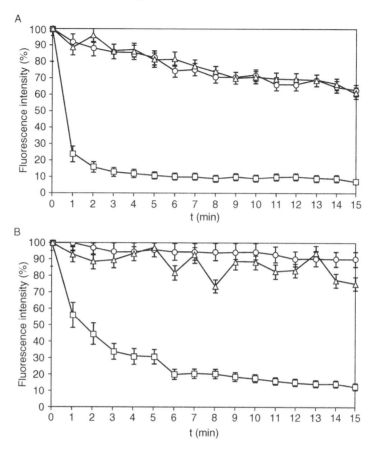

Fig. 5 Fading characteristics for FITC-, TRITC-, and TRSC-conjugated IgG exposed to CLSM in the presence of 0.1% (w/v) *p*-phenylenediamine. *X*-axis is laser exposure time (min) and *Y*-axis is fluorescence intensity expressed as a percentage relative to the initial signal (i.e., $t = 0$ min fluorescence intensity is 100%). Each point is an average of 20 measurements at 1 min intervals. (A) Fading during exposure to 50% laser power. (B) Fading during exposure to 20% laser power. FITC-conjugated IgG (open squares), TRITC-conjugated IgG (open circles), and TRSC-conjugated IgG (open triangles). Unpublished results of Colflesh and Berrios (1998) and adapted from Berrios and Colflesh (1995). Reproduced with permission.

however, are superior in reducing fluorochrome fluorescence fading and provide protection for longer periods of time. In our case, antifading solutions containing 0.1% (w/v) *p*-phenylenediamine (Johnson and Araujo, 1981) or 0.3% (w/v) *n*-propyl gallate (Giloh and Sedat, 1982) had shorter shelf-lives than commercial preparations and some had the tendency to generate nonspecific fluorescence as the preparation aged. Nonspecific fluorescence staining of cell nuclei was a common observation when using *p*-phenylenediamine (e.g., see also Kittelberger *et al.*, 1989).

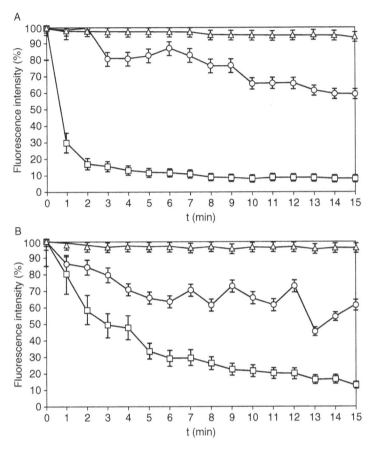

Fig. 6 Fading characteristics for FITC-, TRITC-, and TRSC-conjugated IgG exposed to CLSM in the presence of 0.3% (w/v) *n*-propyl gallate. *X*-axis is laser exposure time (min) and *Y*-axis is fluorescence intensity expressed as a percentage relative to the initial signal (i.e., $t = 0$ min fluorescence intensity is 100%). Each point is an average of 20 measurements at 1 min intervals. (A) Fading during exposure to 50% laser power. (B) Fading during exposure to 20% laser power. FITC-conjugated IgG (open squares), TRITC-conjugated IgG (open circles), and TRSC-conjugated IgG (open triangles). Unpublished results of Colflesh and Berrios (1998) and adapted from Berrios and Colflesh (1995). Reproduced with permission.

V. Acquisition of Fading Fluorescent Images

Attempts to circumvent problems associated with differential deterioration of fluorescence emitted by separate fluorochromes by reducing laser dwelling time and laser power output (Kaufman *et al.*, 1971; Rigaut and Vassy, 1991) severely limit the ability to capture comparable image pairs from, for example, FITC and TRITC or TRSC fluorescence. Although in our case reducing laser power and dwelling time has not proven practical, these strategies applied in conjunction with

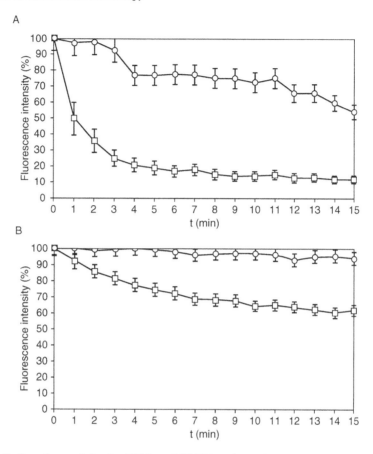

Fig. 7 Fading characteristics for FITC- and TRITC-conjugated IgG exposed to CLSM in the presence of Vectashield®. X-axis is laser exposure time (min) and Y-axis is fluorescence intensity expressed as a percentage relative to the initial signal (i.e., $t = 0$ min fluorescence intensity is 100%). Each point is an average of 20 measurements at 1 min intervals. (A) Fading during exposure to 50% laser power. (B) Fading during exposure to 20% laser power. FITC-conjugated IgG (open squares) and TRITC-conjugated IgG (open circles). Adapted from Berrios and Colflesh (1995). Reproduced with permission

reducing autofluorescence and using antifading agents are effective to limit deterioration of fluorescence signals. Furthermore, as a consequence of the selectiveness of the confocal microscope, the signal intensity at each plane of focus tends to be lower than that observed with the wide field fluorescence microscope. This reduction in brightness must also be taken into consideration during specimen preparation, antibody probing, and image capturing. In addition, prior to image capturing, the acquisition system should be adjusted to employ the full dynamic range of the gray scale (i.e., 256 gray levels). This precaution will not only improve the signal-to-noise ratio but also prevent pixel saturation. More detailed

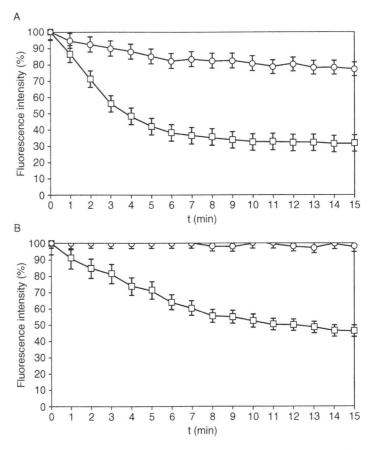

Fig. 8 Fading characteristics for FITC- and TRITC-conjugated IgG exposed to CLSM in the presence of SlowFade®. *X*-axis is laser exposure time (min) and *Y*-axis is fluorescence intensity expressed as a percentage relative to the initial signal (i.e., $t = 0$ min fluorescence intensity is 100%). Each point is an average of 20 measurements at 1 min intervals. (A) Fading during exposure to 50% laser power. (B) Fading during exposure to 20% laser power. FITC-conjugated IgG (open squares) and TRITC-conjugated IgG (open circles). Adapted from Berrios and Colflesh (1995). Reproduced with permission.

discussions on this and related subjects may be found in works edited by Shotton (1993) and Wootton *et al.* (1995) and in comprehensive reviews by Majlof and Forsgren (1993), Brelje *et al.* (1993), and Paddy (1998).

Recently, we found these strategies essential to acquire comparable image pairs from *Drosophila* accessory gland nuclei to show the distribution of *Drosophila* chromatin remodeling protein 1 (CRP1) (Crevel *et al.*, 1997; Kawasaki *et al.*, 1994) and chromatin. We took advantage of the reduction of tissue autofluorescence after treatment with sodium borohydride and the antifading properties exhibited by ProLong® to perform dual optical Z-series on *Drosophila* accessory gland nuclei (Crevel *et al.*, 1997) immunostained with mixtures containing anti-CRP1

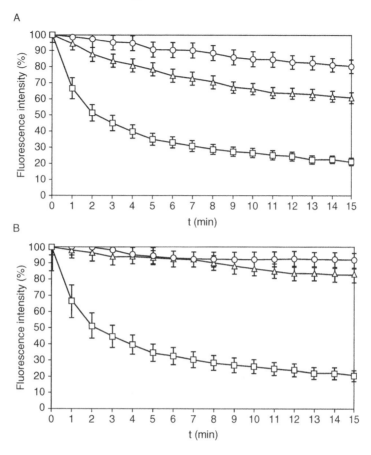

Fig. 9 Fading characteristics for FITC-, TRITC-, and TRSC-conjugated IgG exposed to CLSM in the presence of ProLong®. X-axis is laser exposure time (min) and Y-axis is fluorescence intensity expressed as a percentage relative to the initial signal (i.e., $t = 0$ min fluorescence intensity is 100%). Each point is an average of 20 measurements at 1 min intervals. (A) Fading during exposure to 50% laser power. (B) Fading during exposure to 20% laser power. FITC-conjugated IgG (open squares), TRITC-conjugated IgG (open circles), and TRSC-conjugated IgG (open triangles). Adapted from Colflesh *et al.* (1999) Reproduced with permission.

antibodies and antihistone antibodies (Colflesh *et al.*, 1999). *Drosophila* adult male accessory glands were prepared, fixed, incubated with sodium borohydride and probed with antibodies for indirect immunofluorescence as previously (Colflesh *et al.*, 1999) and described in this chapter. To colocalize CRP1 with histones, nuclei were probed with a mixture containing antihistone mouse monoclonal antibody (mAb052; Millipore) and affinity-purified rabbit anti-CRP1 antibodies (Crevel *et al.*, 1997). Specimens were mounted with ProLong® and examined by CLSM. To minimize fading, images derived from FITC and TRSC wavelengths were obtained using 20% and 50% laser power, respectively. Staining with antihistone monoclonal antibody mAb052 followed by FITC-conjugated secondary antibodies was assigned the color green. Staining with anti-CRP1 antibodies followed by TRSC-conjugated secondary antibodies was assigned the color red. Figure 10 shows two sets each containing three immunofluorescent optical sections from two accessory gland nuclei. Figure 10A shows the TRSC image, Fig. 10B shows the corresponding FITC image, and Fig. 10C shows a pixel-by-pixel superimposition of images shown in Fig. 10A and B. Yellow indicates areas of colocalization (Fig. 10C). Anti-CRP1 antibodies and antihistone antibodies stained the nuclear interior of accessory gland nuclei with different patterns (compare Fig. 10A with B). Antihistone antibodies showed areas of dense chromatin around the nuclear

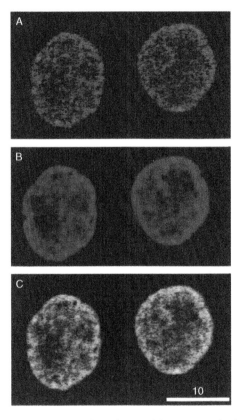

Fig. 10 Confocal immunofluorescent staining of *Drosophila* accessory gland nuclei probed with anti-CRP1 and antihistone antibodies. (A–C) Optical sections through the diameter of two accessory gland nuclei stained with a mixture containing antihistone mouse monoclonal antibody and affinity-purified rabbit anti-CRP1 antibodies followed by a mixture containing FITC-conjugated donkey antimouse IgG and TRSC-conjugated donkey antirabbit IgG. Specimen-associated autofluorescence was reduced with sodium borohydride. Specimens were mounted with ProLong®. (A) TRSC fluorescence signal (CRP1). (B) FITC fluorescence signal (histones). (C) Pixel-by-pixel superimposition of (A) and (B). Bar = 10 μm. Adapted from Colflesh *et al.* (1999). Reproduced with permission. (See Plate no.1 in the Color Plate Section.)

periphery (Fig. 10A). In contrast, CRP1 was distributed as a diffused network that spanned the nuclear interior including the nucleolus (Fig. 10B). In these nuclei, anti-CRP1 antibodies showed areas of overlap with antihistone antibody staining, particularly at or near the peripheral chromatin (Fig. 10C).

Acknowledgments

We thank Drs. Colin Dingwall and Farshid Guilak for valuable suggestions and critical reading of the original work. We also thank Wen Hui Feng and Laura Rosenberger and Noran Instruments' engineers Dennis Maier and Adam Myerov for their expert technical assistance and Joan Martin for help in assembling the manuscript. This work was supported by research grants PO1CA47995 from the National Cancer Institute, DCB-8615969 from The National Science Foundation and an American Cancer Society institutional grant.

Appendix: Vendors

Materials and reagents (except for antifading agents, see Table IV) included in this chapter for which the supplier was not specified may be purchased from the following sources:

Sigma-Aldrich Chemical Company
P.O. Box 14508
St. Louis, MO 63178, USA
Phone: 1-800-325-3010
Fax: 1-800-325-5052
http://www.sigmaaldrich.com

Edmund Scientifics
60 Pearce Avenue
Tonawanda, NY 14150, USA
Phone: 1-800-728-6999
Fax: 1-800-828-3299
E-mail: scientifics@edsci.com
http://scientificsonline.com

Electron Microscopy Sciences
1560 Industry Road
Hatfield, PA 19440, USA
Phone: 1-215-412-8400
Fax: 1-215-412-8450 or 1-215-412-8451
http://www.emsdiasum.com

Fisher Scientific (a division of Thermo Fisher Scientific)
2000 Park Lane Drive
Pittsburgh, PA 15275, USA
Phone: 1-800-766-7000
Fax: 1-800-926-1166
http://www.fishersci.com

Lancaster Synthesis, Inc. (a division of Johnson Matthey Company)
26 Parkridge Road
Ward Hill, MA 01835, USA
Phone: 1-978-521-6300
Fax: 1-978-521-6350
www.lancastersynthesis.com

Millipore Corporation
290 Concord Road
Billerica, MA 01821, USA
Phone: 1-800-645-5476 or 1-978-715-4321
http://www.millipore.com

Mallinckrodt Baker, Inc.
222 Red School Lane
Phillipsburg, NJ 08865, USA
Phone: 1-908-859-2151
Fax: 1-908-859-9318
http://www.solvitcenter.com

Nalgene Nunc International (a division of Thermo Fisher Scientific)
75 Panorama Creek Drive
Rochester, NY 14625, USA
Phone: 1-800-625-4327
Fax: 1-585-586-8987
E-mail: nnitech@nalgenunc.com
http://www.nalgenunc.com

Thomas Scientific
1654 High Hill Road
Swedesboro, NJ 08085, USA
Phone: 1-856-467-2000
Fax: 1-856-467-3087
http://www.thomassci.com

VWR Scientific
Bridgeport, NJ 08014, USA
Phone: 1-800-932-5000 or 1-888-897-5463
Fax: 1-609-467-3336
E-mail: http://www.vwrsp.com/customer/forms/index.cgi?tmpl=form_contactus
http://www.vwrsp.com

References

Battaglia, M., Pozzi, D., Grimaldi, S., and Parasassi, T. (1994). *Biotech. Histochem.* **69,** 152.
Berrios, M. (1989). *Med. Lab. Sci.* **46,** 276.
Berrios, M. (1994). *Biotech. Histochem.* **69,** 78.
Berrios, M., and Colflesh, D. E. (1995). *Biotech. Histochem.* **70,** 40.
Berrios, M., and Fisher, P. A. (1986). *J. Cell Biol.* **103,** 711.
Betzig, E., Trautman, J. K., Harris, T. D., Weiner, J. S., and Kostelak, R. L. (1991). *Science* **251,** 1468.
Betzig, E., Patterson, G. H., Sougrat, R., Lindwasser, O. W., Olenych, S. T., Bonifacino, J. S., Davidson, M. W., Lippincott-Schwartz, J., and Hess, H. F. (2006). *Science* **313,** 1642.
Bock, G., Hilchenbach, M., Schauenstein, K., and Wick, G. (1985). *J. Histochem. Cytochem.* **33,** 669.

Brelje, T. C., Wessendorf, M. W., and Sorenson, R. L. (1993). *In* "Methods in Cell Biology," (B. Matsumoto, ed.), p. 97. Academic Press, San Diego.

Chalfie, M. (1995). *Photochem. Photobiol.* **62,** 651.

Colflesh, D. E., and Berrios, M. (1998). (unpublished).

Colflesh, D. E., Conlon, K. A., and Berrios, M. (1999). *J. Histotech.* **22,** 23.

Crevel, G., Huikeshoven, H., Cotterill, S., Simon, M., Wall, J., Philpott, A., Laskey, R. A., McConnell, M., Fisher, P. A., and Berrios, M. (1997). *J. Struct. Biol.* **118,** 9.

Cubitt, A. B., Heim, R., Adams, S. R., Boyd, A. E., Gross, L. A., and Tsien, R. Y. (1995). *Trends Biochem. Sci.* **20,** 448.

Darsow, T., Burd, C. G., and Emr, S. D. (1998). *J. Cell Biol.* **142,** 913.

Dernburg, A. F., and Sedat, J. W. (1998). *In* "Methods in Cell Biology," (M. Berrios, ed.), p. 187. Academic Press, San Diego, CA.

Fisher, P. A., and Berrios, M. (1998). *In* "Methods in Cell Biology," (M. Berrios, ed.), p. 397. Academic Press, San Diego, CA.

Furuse, M., Fujita, K., Hiiragi, T., Fujimoto, K., and Tsukita, S. (1998). *J. Cell Biol.* **141,** 1539.

Gerdes, H. H., and Kaether, C. (1996). *FEBS Lett.* **389,** 44.

Giloh, H., and Sedat, J. W. (1982). *Science* **217,** 1252.

Hell, S. W. (2003). *Nat. Biotech.* **21,** 1347.

Hell, S. W. (2007). *Science* **316,** 1153.

Hell, S. W., and Wichmann, J. (1994). *Opt. Lett.* **19,** 280.

Huang, B., Bates, M., and Zhuang, X. (2009). *Annu. Rev. Biochem.* **78,** 993.

Jensen, H., Broholm, N., and Norrilb, B. (1995). *J. Histochem. Cytochem.* **43,** 507.

Johnson, G. D., and Araujo, G. M. (1981). *Immunol. Methods* **43,** 349.

Juette, M. F., Gould, T. J., Lessard, M. D., Mlodzianoski, M. J., Nagpure, B. S., Bennet, B. T., Hess, S. T., and Bewersdorf, J. (2008). *Nat. Methods* **6,** 527.

Kaufman, G. I., Nester, J. F., and Wasserman, D. E. (1971). *J. Histochem. Cytochem.* **19,** 469.

Kawamura, A. (1977). Fluorescent Antibody Techniques and Their Applications. University of Tokyo Press, Tokyo.

Kawasaki, K., Philpott, A., Avilion, A. A., Berrios, M., and Fisher, P. A. (1994). *J. Biol. Chem.* **269,** 10169.

Kittelberger, R., Davis, P. F., and Stehbens, W. E. (1989). *Acta Histochem.* **86,** 137.

Larrick, J. W., Balint, R. F., and Youvan, D. C. (1995). *Immunotechnology* **1,** 83.

Lemieux, N., Dutrillaux, B., and Viegas-Pequignot, E. (1992). *Cytogenet. Cell Genet.* **59,** 311.

Loidl, J., Klein, F., and Engebrecht, J. (1998). *In* "Methods in Cell Biology," (M. Berrios, ed.), p. 257. Academic Press, New York.

Majlof, L., and Forsgren, P. (1993). *In* "Methods in Cell Biology," (B. Matsumoto, ed.), p. 79. Academic Press, San Diego, CA.

Meller, V. H., Fisher, P. A., and Berrios, M. (1995). *Chromosome Res.* **3,** 255.

Moens, P. B., and Pearlman, R. E. (1991). *In* "Methods in Cell Biology," (B. A. Hamkalo, and S. G. Elgin, eds.), p. 101. Academic Press, San Diego, CA.

Nathke, I. S., Adams, C. L., Polakis, P., Sellin, J. H., and Nelson, W. J. (1996). *J. Cell Biol.* **134,** 165.

Paddy, M. R. (1998). *In* "Methods in Cell Biology," (M. Berrios, ed.), p. 49. Academic Press, San Diego, CA.

Perez, J. L., De Ona, M., Niubo, J., Villar, H., Melon, S., Garcia, A., and Martin, R. (1995). *J. Clin. Microbiol.* **33,** 1646.

Poot, M., Zhang, Y. Z., Kramer, J. A., Wells, K. S., Jones, L. J., Hanzel, D. K., Lugade, A. G., Singer, V. L., and Haugland, R. P. (1996). *J. Histochem. Cytochem.* **44,** 1363.

Prasher, D. C. (1995). *Trends Genet.* **11,** 320.

Rigaut, J. P., and Vassy, J. (1991). *Anal. Quant. Cytol. Histol.* **13,** 223.

Roberts, P. M., and Goldfarb, D. S. (1998). *In* "Methods in Cell Biology," (M. Berrios, ed.), p. 545. Academic Press, San Diego, CA.

Scheynius, A., and Lundahl, P. (1990). *Arch. Dermatol. Res.* **281,** 521.

Schmidt, T., Chen, P. S., and Pellagrini, M. (1985). *J. Biol. Chem.* **260,** 7645.

Shotton, D. (1993). Electronic Light Microscopy: Techniques in Modern Biomedical Microscopy. Wiley-Liss, Inc, New York.

Smith, D. E., Gruenbaum, Y., Berrios, M., and Fisher, P. A. (1987). *J. Cell Biol.* **105,** 771.

Solari, A. F. (1998). *In* "Methods in Cell Biology," (M. Berrios, ed.), p. 236. Academic Press, San Diego, CA.

Soliman, A. M. (1988). *Arch. Otorhinolaryngol.* **245,** 28.

Springall, D. R., and Polak, J. M. (1995). *In* "Image Analysis in Histology: Conventional Confocal Microscopy," (R. Wooton, D. R. Springall, and J. M. Polak, eds.), p. 123. Cambridge University Press, New York.

Wootton, R., Springall, D. R., and Polak, J. M. (1995). Image Analysis in Histology: Conventional Confocal Microscopy. Cambridge University Press, New York.

CHAPTER 5

Mounting Techniques for Confocal Microscopy

Manabu Kagayama and Yasuyuki Sasano

Department of Anatomy
Tohoku University School of Dentistry
Sendai, Japan

I. Introduction

The confocal microscope is a valuable research tool for imaging biological specimens labeled with fluorochromes. Images produced by confocal microscopy are different from those by usual light microscopy in that the former is cut at the confocal plane of the objective lens. This offers a higher resolution of the image. Further, it is possible to cut optically serial sections and reconstruct three-dimensional structure in thick biological specimens using a confocal microscope (Paddock, 1997; Salisbury, 1994).

The quality of images taken by the confocal microscope is dependent on the specimen preparation procedure, such as fixation, embedding, cutting, staining, and mounting. Fluorescein isothiocyanate (FITC) and tetraethylrhodamine isothiocyanate (TRITC) are the most commonly used fluorochromes in immunohistochemistry and *in situ* hybridization. Fluorescence decay is rapid for these

DOI: 10.1016/B978-0-12-384658-7.00005-9

fluorochromes when glycerin or buffer solution is used as the mounting medium. *p*-Phenylenediamine and 1,4-diazobicyclo[2.2.2]octane are well-known additives that retard fading considerably (Krenik *et al.*, 1989; Platt and Michael, 1983). Commercial mounting media of unknown chemical composition are available to preserve the intensity of fluorescence. These mounting media work well in thin sections. When applying these mounting media to thick sections, air bubbles occur frequently between coverslips and glass slides. This results in an increase in background noises by confocal microscopy.

Using a confocal microscope, a series of optical sections can be produced from a thick specimen labeled with fluorochromes. The number of optical sections collected is dependent on the characteristics of the specimens and is a function of objective lenses. A confocal microscope is usually equipped with oil immersion-type objective lenses for maximum resolution. Because the working distance of these lenses is very short (0.14–0.16 mm), the thickness of the coverslip impedes the observation of deeper parts of thick sections.

This chapter describes mounting methods for confocal microscopy.

II. Materials and Methods

A. Fixation

Tissues from rats are used for observations. Under pentobarbital anesthesia, the tissues are fixed by perfusion through arteries with 4% (w/v) paraformaldehyde solution adjusted at pH 7.4 with 0.1 M phosphate buffer. For the vital staining of calcified tissues, some of the animals are injected with calcein (2 mg/kg body weight) and alizarin red (12 mg/kg) before fixation. After perfusion fixation, tissues are dissected and fixed in the same fixative for 10–20 h and stored in phosphate-buffered saline (PBS) at 4 °C. Because most of the tissues used are calcified, part of the tissues are decalcified in 10% (w/v) EDTA at 4 °C for 1–3 months.

B. Methods for Thin Sections

After decalcification, tissues are dehydrated, immersed in xylene, and embedded in paraffin. Sections (3–5 μm) cut with a sliding microtome are placed on slide glasses coated with adhesive (APS slide glass, Matunami Glass Co., Osaka, Japan) and are allowed to dry on a hot plate. The sections are stored at 4 °C. For frozen sections, the tissues are immersed in 10–20% (w/v) sucrose solution and are frozen by a mixture of dry ice and acetone. Sections are cut with a cryostat microtome at a thickness of 10–30 μm, attached on the slide glasses, and allowed to dry.

The sections are stained with FITC- or TRITC-conjugated phalloidin (Sigma Chemical Co., St. Louis, MO) diluted to 50 μg/ml by PBS containing 0.05% (v/v) Triton X-100 at 4 °C for 1–3 h. After staining, the sections are rinsed with PBS and mounted. To mount the sections, one to two drops of mounting medium are dispensed, and coverslips are placed onto the sections. The mounted specimens are stored at 4 °C in the dark. A mounting medium containing polyvinyl alcohol (PVA), Permafluor (Lipshaw, Pittsburgh, PA), and Vectashield (Vector Laboratories, Inc., Burlingame, CA) is used.

The PVA solution is prepared by adding 20 g of PVA (average molecular weight 30,000–70,000, Sigma Chemical Co.) to 80 ml of Tris–HCl buffer (65 mM) with continuous stirring to form a slurry. Then 15 ml of fluorescence-free glycerol (Merck, Darmstadt, Germany), 100 mg of chlorobutanol, and 100 mg of p-phenylenediamine are added to this solution. The solution is adjusted to pH 8.2 with 1 M HCl. The slurry is stirred and heated on a hot plate to the boiling point for about 15 min until the particles are dissolved. The mixture is degassed with a rotary vacuum pump and stored at 4 °C in the dark.

C. Methods for Thick Sections Cut with Microslicer

After fixation and decalcification, the tissue blocks are affixed with cyanoacrylate resin to a metal block holder, and sections are cut at the thickness of 100–200 μm within PBS using a microslicer (Dosaka, Kyoto, Japan) with continuous visual monitoring. The sections are transferred onto the slide glasses with a small brush and allowed to dry. The sections are stained with FITC- or TRITC-conjugated phalloidin (Sigma Chemical Co.) diluted to 50 μg/ml by PBS containing 0.05% Triton X-100 at 4 °C for 1–3 days, depending on the section thickness. After staining, the sections are rinsed thoroughly with PBS for 6–10 h and mounted as described earlier.

D. Methods for Whole Mount Specimens and Thick Sections Cut with Saw Microtome

After fixation, the tissues stained vitally with calcein and alizarin red are dehydrated, immersed in propylene oxide, and embedded in Spurr's epoxy resin (Taab Laboratories Equipment, Aldermaston, Berkshire, UK). The tissues are sliced with a saw microtome (SP1600, Leica, Heidelberg, Germany) at a thickness of 100–200 μm. As whole mount specimens, calvaria of 8-day-old rats are stained directly with FITC- or TRITC-conjugated phalloidin, rinsed, dehydrated, and embedded in epoxy resin as described earlier.

Thick sections and whole mount specimens are ground with sandpaper to expose the desired surface. After washing with distilled water, these specimens are dried and attached on the slides with cyanoacrylate. Specimens embedded with resins can be mounted temporally by immersion oil and covered with glasses for the observation with dry-type objective lenses. If the structure of interest is present at a deeper part than the working distance of oil immersion objectives, the surface of specimens can be ground again.

E. Confocal Microscopy

Confocal microscopy is performed using a microscope (CLSM, Leica) equipped with objective lenses 10×, 25×, 40×, 63×, and 100× (all were oil immersion type, except 10×, which was dry type) with aperture settings of 0.25, 0.75, 1.30, 1.40, and 1.32, respectively. The working distance of oil immersion objectives is 0.14–0.16 mm. The epifluorescence mode is used for the observation. To obtain optimum images of three-dimensional structures, images are cut serially and reconstructed with the computer of CLSM. The wavelength of excitation light used is 488 nm for FITC and calcein and 512 nm for TRITC and alizarin red.

III. Results and Discussions

Mounting methods using water-soluble media for thin sections worked well for confocal microscopy. Fluorescence of the specimens mounted with Permafluor and PVA was preserved for more than 1 year (Figs. 1 and 2).

As in the thin sections, images obtained from thick sections mounted with water-soluble media were very clear, even in objects deep within the thick sections. However, in most of the specimens mounted with Permafluor, air bubbles occurred near the perimeter of coverslips after prolonged storage. No bubbles occurred in the sections mounted with PVA and stored more than 1 year. Air bubbles may be formed possibly by the shrinkage of the hardened medium, because shrinkage of Permafluor was larger than that of PVA when mounting media were placed onto the glasses and allowed to dry. We have not used Vectashield until recently and performance of the thick sections mounted with it is not known. The manufacturer of Vectashield recommends sealing the coverslips with clear nail polish or a plastic sealant for prolonged storage.

In sections thicker than 150 μm, the deeper part of the specimens could not be focused because of the short working distance of oil immersion objective lenses. One of the possible methods to overcome this difficulty is the use of a thinner coverslip (0.04–0.06) instead of a standard one (0.17 mm).

Tissues embedded in epoxy resin were transparent to the light used, and labeling lines of calcein and alizarin red were clearly visible at depths greater than 200 μm below the surface of the preparation. Figure 3 is a confocal micrograph of osteonal lacunae and canaliculi stained vitally with calcein in the thick section embedded by epoxy resin. Although the staining of osteonal lacunae and canaliculi was less intense than that of the surface of bone at the time of fluorochrome injection (calcification front), confocal microscopy can reveal those structures very clearly. As reported previously (Kagayama *et al.*, 1996), the rate of fading appeared to be less in specimens mounted with the resin than with PVA. An advantage of this method is that whole mount specimens can be examined by confocal microscopy. Without the use of a coverslip, there was no interface in which air bubbles could occur, and deeper parts of thick specimens could be analyzed than in the specimen covered with glasses.

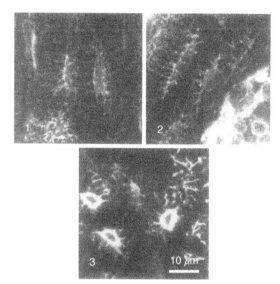

Figs. 1–3 Confocal micrographs were reconstructed from 30 serial images taken by a 0.3-μm step using 100× objective lens. Scale of the micrographs as indicated in Fig. 3.

Figs. 1 and 2 Confocal micrographs of osteocytes taken using frozen sections stained with FITC-conjugated phalloidin, mounted with Permafluor, and stored for 1.5 years (Fig. 1) and 2 weeks (Fig. 2). Note that fine cytoplasmic processes of osteocytes are similarly visible in Figs. 1 and 2.

Fig. 3 Confocal micrographs of osteonal lacunae and canaliculi stained vitally with calcein in thick sections embedded in epoxy resin.

References

Kagayama, M., Sasano, Y., Hirata, M., Mizoguchi, I., and Takahashi, I. (1996). *Biotech. Histochem.* **71,** 231.

Krenik, K. D., Kephart, G. M., Offord, K. P., Dunette, S. L., and Gleich, G. J. (1989). *J. Immunol. Methods* **117,** 91.

Paddock, S. W. (1997). *Curr. Biol.* **7,** R182.

Platt, J. L., and Michael, A. F. (1983). *J. Histochem. Cytochem.* **31,** 840.

Salisbury, J. R. (1994). *Histol. Histopathol.* **9,** 773.

CHAPTER 6

Preparation of Whole Mounts and Thick Sections for Confocal Microscopy

Hisashi Hashimoto,* Hiroshi Ishikawa,[†] and Moriaki Kusakabe[‡]

*Department of Anatomy
The Jikei University School of Medicine
Nishishinbashi, Minato-ku, Tokyo, Japan

[†]Laboratory of Regenerative Medical Science
The Nippon Dental University School of Life Dentistry at Tokyo
Fujimi, Chiyoda-ku, Tokyo, Japan

[‡]Research Center for Food Safety
Graduate School of Agricultural and Life Sciences
The University of Tokyo
Yayoi, Bunkyo-ku, Tokyo, Japan

I. Update

Ten years have passed since we reported our procedures for preparation of whole mounts and thick sections for confocal microscopy. However, little has changed from the original procedures except the pH of the gelatin solution for

79
DOI: 10.1016/B978-0-12-384658-7.00006-0

being labeled with fluorochrome. The pH of the solution was lowered from 11 to 10 because gelatin might break down in a high pH solution. The procedures were applied to investigate the development of vascular networks during the morphogenesis of intestinal villi (Hashimoto *et al.*, 1999), the development of the mucosal vascular system in the distal colon (Ito *et al.*, 2007), and the development of vascular networks during the odontogenesis.

In recent years, micro CT and magnetic resonance imaging microscopy (μMRI) were developed and three-dimensional digital mouse atlas was made with a μMRI (Dhenain *et al.*, 2001). We have developed a novel method to obtain serial vascular images from the whole mouse brain and reconstructed a high-resolution three-dimensional image of the vascular networks in the whole brain (Hashimoto *et al.*, 2009). These methods can provide a volume-rendered image, but do not suit for three-dimensional analysis of tissue architecture. Nowadays, various attempts are made to regenerate an organ from pluripotent stem cells, such as embryonic stem (ES) cells and induced pluripotent stem cells. It is important to evaluate three-dimensionally the tissue structure and vascular networks in the organ. Therefore, our method will contribute in the regenerative medicine as well as in the developmental biology.

II. Introduction

Numerous researchers have attempted to reveal a three-dimensional relationship in tissue and cellular architectures because it is fundamental to a proper understanding of the biological system. It is very difficult to examine the distribution of nerve fibers and capillary nets in a tissue, the morphological changes of the epithelial tissue in a developing organ, and the shape and appearance of a cell from tissue sections. Distribution of a molecule in a tissue and interrelationships between tissue components may not be revealed until a three-dimensional reconstruction is made. One three-dimensional image may give us far more information than hundreds of tissue sections.

Three-dimensional analysis has mainly been performed by reconstructing from serial tissue sections, but there are technical difficulties and problems during the process due to damage or artifacts caused by the sectioning process. The confocal laser scanning microscope (CLSM) has been used for three-dimensional analysis. The CLSM possesses some attractive features: it enables us to get a sharp and vivid optical tomogram that is free from glaring fluorescence in out-of-focus planes and to digitalize and store the image in a permanent form. Consequently, serial optical images from thick specimens can be obtained easily by relocating focus planes at an arbitrary interval and can be reconstructed into a three-dimensional image with a high-powered personal computer.

Specimens subjected to three-dimensional investigation with CLSM must satisfy certain criteria: they should have little autofluorescence and little absorbance of light for excitation and emission and they should not have obstacles for penetration of a fluorochrome tracer. This chapter presents a method (Procedure I, see Appendix) that was developed primarily to observe the three-dimensional distribution of the extracellular matrixes: laminin and tenascin (Hashimoto and Kusakabe, 1997). Laminin is a noncollagenous glycoprotein and is one of the main constituents of basement membranes in both fetal and adult organs (Figs. 1–4, 16, and 18). Three-dimensional visualization revealed branching in the fetal lung (Figs. 1 and 2) and the development of the pituitary gland (Hashimoto *et al.*, 1998a). The expression of tenascin is spatiotemporally restricted and its heterogenous distribution is well indicated by three-dimensional images (Figs. 5, 6, and 16). Our method is composed simply of fixation, sectioning if necessary, pretreatment, and immunostaining as usual, but each step was reexamined. This method is now applicable to other tissue components, such as neurotransmitters, including vasoactive intestinal peptide (VIP; Fig. 17) (Lee *et al.*, 1997), somatostatin (Figs. 7 and 8), and substance P, peptide hormones, neurofilaments, and interleukins. In addition, this chapter also introduces a novel method for the three-dimensional investigation of vascular nets (Figs. 12–14 and 17–20) (Hashimoto *et al.*, 1998a,b,c).

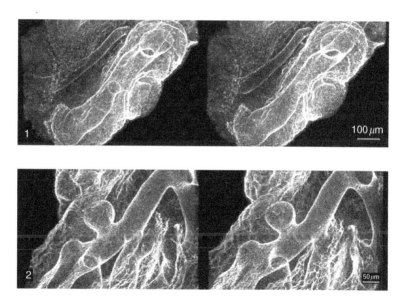

Figs. 1 and 2 Stereo pair images of laminin in a fetal mouse lung on day 11 (Fig. 1) and day 12 (Fig. 2) of gestation. Fetal mouse lungs were removed, fixed, pretreated, and immunostained for laminin as described in Procedure I. The antibodies used were rabbit antimouse laminin IgG (AT-2404, E.Y. Labs, Inc., San Mateo, CA) diluted 1:500 and FITC-conjugated goat antirabbit IgG antibody (No. 454, MBL Co., Nagoya, Japan) diluted 1:100. Three-dimensional stereo pair images were reconstructed with optional software for Carl Zeiss LSM-510 from 25 images taken at a 5-μm interval (Fig. 1) and from 43 images taken at a 3-μm interval (Fig. 2) with a Carl Zeiss LSM-10 and a Plan NeoFLUOR 20× objective lens (NA 0.5).

82

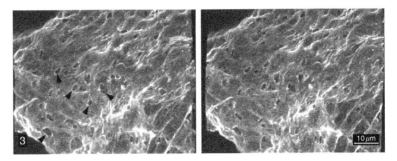

Fig. 3 Stereo pair images of a laminin sheet in the basement membrane zone of an adult mouse small intestinal villus. A proximal small intestine was removed from an adult mouse, fixed, sectioned at 150 μm, pretreated and immunostained for laminin as described in Procedure I. The specimen was dehydrated and mounted in Rigolac as mentioned in Procedure IV. The antibodies used were rabbit antimouse laminin IgG (AT-2404, E.Y. Labs, Inc.) diluted 1:3000 and BodipyFL-conjugated goat antirabbit IgG antibody (B-2766, Molecular Probes, Inc., Eugene, OR) diluted 1:200. Three-dimensional stereo pair images were reconstructed from 74 images taken at a 0.45-μm interval with an LSM-510 CLSM and a Plan NeoFLUOR 100× objective lens (NA 1.3). Arrowheads indicate pores in the laminin sheet.

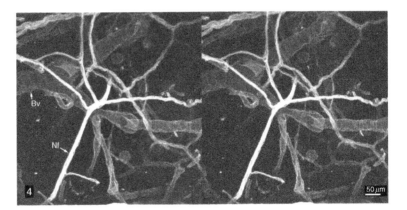

Fig. 4 Stereo pair images of laminin in subcutaneous tissue of an adult mouse ear. An adult mouse ear was fixed as usual and treated with 0.1 M disodium ethylenediaminetetraacetic acid (EDTA) in 0.1 M sodium phosphate buffer at 4 °C for 1 week. The skin and subcutaneous tissue of the ear were mechanically peeled off and immunostained for laminin after pretreatment with sodium deoxycholate. The antibodies used were rabbit antimouse laminin IgG (AT-2404, E.Y. Labs) diluted 1:500 and FITC-conjugated goat antirabbit IgG antibody (No. 454, MBL Co.) diluted 1:100. Three-dimensional stereo pair images were reconstructed with an LSM-510 from 12 images taken at a 3-μm interval with an LSM-310 and a 20× objective lens. In the subcutaneous tissue of the mouse ear, immunofluorescence of laminin localized at the basement membrane indicates the three-dimensional distribution of numerous blood vessels (Bv) and nerve fibers (Nf).

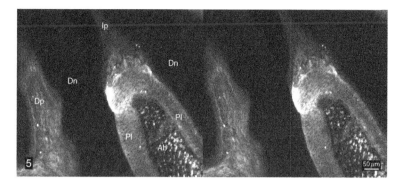

Fig. 5 Stereo pair images of tenascin in the periodontal ligament and tooth pulp of a mouse. A molar portion in the mandible of a mouse was fixed as usual and decalcified with 0.1 M disodium ethylenediaminetetetraacetic acid (EDTA) in 0.1 M sodium phosphate buffer at 4 °C for 1 week. The specimen was sliced at 150 μm, pretreated with sodium deoxycholate and immunostained for tenascin. The antibodies used were a rat antihuman tenascin monoclonal antibody (developed in our laboratory, RIKEN) at a concentration of 10 μg/ml and FITC-conjugated goat antirat IgG antibody (No. 6270, TAGO, Inc., Burlingame, CA) diluted 1:100. Three-dimensional images were reconstructed with an LSM-510 from 25 images taken at a 3-μm interval with an LSM-310 and a 20× objective lens. Tenascin is observed in the periodontal ligament, the odontoblast layer in the dental pulp (Dp) and around the osteocyte. Prominent heterogeneity in tenascin distribution is found in the periodontal ligament (Pl). Dense tenascin is localized at the periodontal ligament adjoining the dental cervix. Dn, dentine; Ab, alveolar bone; Ip, interdental papilla.

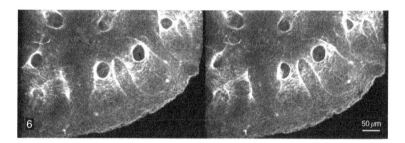

Fig. 6 Stereo pair images of tenascin in a fetal mouse lung at day 14 of gestation. Fetal mouse lungs were removed, fixed, pretreated and immunostained for tenascin as described in Procedure I. The antibodies used were a rat antihuman tenascin monoclonal antibody (developed in our laboratory, RIKEN) at a concentration of 10 μg/ml and FITC-conjugated goat antirat IgG antibody (No. 6270, TAGO, Inc.) diluted 1:100. Three-dimensional images were reconstructed with an LSM-510 from 19 images taken at a 3-μm interval with an LSM-10 and a 20× objective lens. Tenascin is found encompassing the last branching portion of bronchia.

III. Fixation

The strongly fixed tissue is not suited for three-dimensional observation with CLSM. Various fixatives, such as formaldehyde, picric acid, mercuric chloride, ethanol, methanol, and acetone, have been used for preparing microscopic sections.

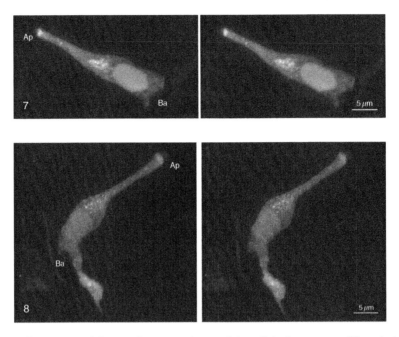

Figs. 7 and 8   Stereo pair images of somatostatin-containing cells in the mouse small intestinal villus. A proximal portion of an adult mouse small intestine was fixed, sectioned at 150 μm, pretreated and immunostained for somatostatin as described in Procedure I. The specimen was dehydrated and mounted in Rigolac as described in Procedure IV. The antibodies used were rabbit antisomatostatin serum (JIMRO, Gunma, Japan) and BodipyFL-conjugated goat antirabbit IgG antibody (B-2766) diluted 1:200. Three-dimensional stereo pair images were reconstructed from 35 images taken at a 0.25-μm interval (Fig. 7) and from 58 images taken at a 0.27-μm interval (Fig. 8) with an LSM-510 and a 100× objective lens. These three-dimensional images clearly indicate shapes and forms of the somatostatin cell situated among epithelial cells in the villus. The apex (Ap) of the somatostatin cell is exposed to the lumen of the small intestine and its base (Ba) lies on the basement membrane. Some somatostatin cells have a basal process traveling along the base of the epithelium as reported previously by Larsson *et al.* (1979) and Sjölund *et al.* (1983).

However, each fixative more or less induces autofluorescence in biological samples by the denaturation or chemical modification of macromolecules. Strong fixatives, such as formaldehyde solution, surpasses in preserving tissue and cellular architectures and immobiling antigenic molecules, but coincidentally induce intense autofluorescence. In a preliminary study, the autofluorescence of rat liver and kidney fixed by calcium-formalin solution, 4% paraformaldehyde in sodium phosphate buffer, Zamboni's fixative, Carnoy's fixative, or Methacarn fixative was examined. Although each tissue showed intense autofluorescence in the cytoplasm, fluorescence from the liver fixed by Methacarn fixative was the brightest. The autofluorescence in these fixed tissues was not reduced by pretreatment with Triton X-100 or

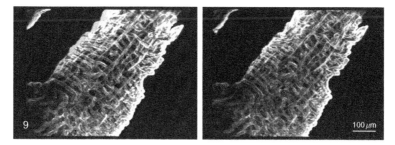

Fig. 9 An example of pepsin pretreatment (immunofluorescence staining for laminin in the small intestinal villus). The adult mouse small intestine was fixed with 4% paraformaldehyde in 0.1 M sodium phosphate buffer at 4 °C overnight. After several rinses with phosphate buffer, the small intestine was sliced between villi with razor blades under a dissection microscope to obtain intact villi. The specimen was pretreated with 3% sodium deoxycholate solution at room temperature for 4 h and with 0.1% pepsin in 0.01 N HCl at room temperature for 4 h. The specimen was then immunostained for laminin with rabbit antimouse laminin antibody (AT-2404) diluted 1:500 and FITC-conjugated goat antirabbit IgG serum (No. 454) diluted 1:100. Three-dimensional images were reconstructed with an LSM-10 from eight images taken at a 3-μm interval. The pepsin pretreatment digested laminin in the epithelial basement membrane, whereas intense immunofluorescence of laminin clearly indicate the capillary basement membrane in the lamina propria.

deoxycholic acid. Pretreatment with pepsin reduced the autofluorescence of paraformaldehyde-fixed specimens and improved the penetration of antibodies, but this pretreatment was not appropriate for a whole mount specimen, as stated in the following section.

We scanned previous reports for immunostaining against laminin and found a report by Belford *et al.* (1987). Belford *et al.* (1987) immunostained laminin in a whole mount retina of rats and mice fixed by 0.5% paraformaldehyde in 15% saturated picric acid solution for 1–2 h. They mentioned that fixation with formaldehyde concentrations greater than 0.5% reduced specific immunofluorescence and increased background fluorescence in whole mount preparations. Picric acid is one of the chief constituents in Bouin's fixative and Zamboni's fixative (Stefani *et al.*, 1967) and is thought to produce intermolecular salt links by inducing polarity in the amino acid side chains of proteins (Pearse, 1980). Since the introduction of a buffered picric acid–formaldehyde mixture by Stefani *et al.* (1967), some researchers applied this mixture to the fixation of peptide-containing endocrine cells and nerves. Costa *et al.* (1980) stated that the picric acid–formaldehyde mixture preserved the structure of the tissue and the antigenicity of peptide-containing structures better than the other fixatives tested. Somogyi and Takagi (1982) presumed that the use of picric acid not only allows the use of low-concentration glutaraldehyde but also improves fine structural detail and, through the preservation of immunogenicity, provides better immunostaining than aldehydes alone. Therefore, we tried a 0.1 M sodium phosphate-buffered (SPB, pH 7.0) mixture of 0.5% paraformaldehyde and 15% (v/v) saturated picric acid solution for a whole mount

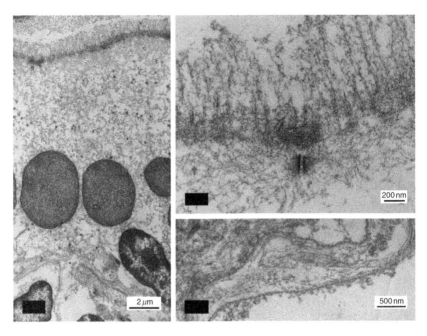

Fig. 10 Electron micrographs of the small intestine pretreated with sodium deoxycholate solution. The mouse small intestine was processed as Procedure I with antimouse laminin antibody, observed with an LSM-10, and then postfixed with glutaraldehyde and osmium tetroxide, dehydrated with a graded series of ethanol, and embedded in EPON 812 through propylene oxide. The ultrathin section was stained with uranium acetate and lead citrate. Original magnification: 4000× (A), 20,000× (B), and 8000× (C). The tissue and cellular architecture is well preserved, although the pretreatment with sodium deoxycholate was performed on a mildly fixed tissue. Filamentous structures in a cytoplasm, microvilli, and desmosome are observed in the enterocyte (A and B). In the basement membrane zone, filamentous structures are also present and collagenous fibers are found within the lamina propria of the villus (C). Adopted from Lee *et al.* (1997).

preparation. The specimen fixed by this fixative showed little autofluorescence and preserved immunoreactivity against various molecules, including laminin, tenascin, type IV collagen, glial fibrillary acidic protein (GFAP), neurofilaments, neurotransmitters, and peptide hormones. This fixative has a comparatively mild action on tissue, but electron microscopy revealed that basic tissue architectures were well preserved after pretreatment with a sodium deoxycholate solution (Lee *et al.*, 1997) (Fig. 10).

IV. Sectioning

Early fetal organs and even the adult intestinal wall of a mouse can be handled as a whole organ. However, more developed and massive organs have a more complicated structure and are not practical as specimens for CLSM. In a specimen prepared by vascular perfusion with fluorochrome-labeled gelatin, emission and excitation light can reach more than 500 μm. Such thick specimens are not practical when fluorescence staining is administered after fixation.

For making a section 100–500 μm thick, a Sartorius-type microtome system, a sliding microtome equipped with a cryostage, is feasible for sectioning without deformation and fracture of a tissue. A vibratome, which is generally utilized for cutting thick sections, is suited for a well-fixed hardened tissue, but it is difficult to prepare useable sections from a weak-fixed soft tissue. It is also difficult for a cryostat, or a cold microtome, to make a thick section because such sections are liable to fragmentate due to freezing.

Prior to sectioning, the specimen is infiltrated with the cryoprotectant solution, 20% (w/v) sucrose and 10% (v/v) glycerol in phosphate-buffered saline (PBS, pH 7.2) through 5–10% (w/v) sucrose in PBS. The specimen on a cryostage should not be so cold that a slice may break before it thaws. The best sections can be obtained when the section thaws on the knife as soon as it is cut and then transferred to PBS with a slender brush.

V. Pretreatment

Pretreatment of the specimen has two purposes: (1) to remove obstacles in order that tracers, such as fluorochromes and antibodies, can penetrate deeply and evenly into the specimens and (2) to reduce unexpected autofluorescence. Theoretically, the best image would be obtained if structural components and a target molecule are left intact and any other constituents, including the plasma membrane and cytosolic proteins, are removed.

Originally, a detergent was utilized for improving the penetration of antibodies and for reducing nonspecific binding of antibodies in the immunoenzyme method for thick sections. Hartman *et al.* (1972) first introduced Triton X-100, a nonionic detergent, in the diluting buffer for a primary antibody and in the subsequent washing buffer to reduce nonspecific, low-affinity protein–protein binding. Grzanna *et al.* (1978) also added Triton X-100 to the diluting buffer and to the washing buffer. Variations in the concentration of Triton X-100 in the first incubation step produced variations in the intensity of staining, no staining of dopamine-β-hydroxylase-containing processes was observed when Triton was omitted. Dehydration, clearing, and rehydration of the whole mount preparations before incubation with antibodies were attempted by Costa *et al.* (1980) and the results were compared with those obtained by freezing and thawing or by using Triton X-100 in the incubation medium. As they reported, freezing and thawing or the use of Triton X-100 in the incubation media did not improve penetration and did not reduce the background fluorescence; indeed the positive nerve fibers appeared disrupted and fragmented. Attempts made to facilitate penetration by allowing the whole mount preparations to air dry or by freezing and thawing or by using detergents failed to solve the problem. The process of dehydrating, clearing, and rehydrating whole mount preparations before incubating with the antibodies improved the penetration of the antibodies dramatically. This treatment seems to have extracted some of the tissue and cell barriers to penetration of large protein molecules. Franklin and Martin (1981) modified a procedure described by Grzanna *et al.* (1978) and incubated tissue slices in PBS with 0.4% Triton X-100 for 15 min at room temperature as a pretreatment. They also tried to improve the permeation of antibodies with absolute methanol at -20 to $-30\,°C$ and then brought to PBS through an ethanol series at $4\,°C$. They found that there was a slight difference between tissue that had been permeabilized by absolute methanol at $-20\,°C$ and by Triton X-100 treatment; the latter tissue seemed to give a slightly weaker fluorescence than the former. Therefore, they selected methanol permeation, although it entails more processing steps. Somogyi and Takagi (1982) simply stated that penetration is enhanced greatly by the freeze–thaw treatment and although this affects preservation adversely, the fine structural details are still fairly good.

As mentioned earlier, the method and effect of pretreatment differ from researcher to researcher. These differences are derived from the diversity in target organs, antigens, the method for fixation, the specificity of antibodies, etc. Therefore, we cannot select the method of choice for any organ and any antigen.

In our early trials, we had performed pepsin treatment on a 4% paraformaldehyde-fixed specimen for the three-dimensional observation of laminin (Fig. 9). The enzymatic digestion by proteinase, including pepsin, is usually carried out prior to immunostaining to restore antigenicity (Finle and Petrusz, 1982; Hashimoto and Hoshino, 1988). However, in the enzymatic digestion of a whole mount specimen or a thick slice, the enzyme acts on the surface of the specimen and gradually on the deeper parts. The surface of the specimen may be digested fully before its deeper part is digested optimally. In case of weak digestion, the surface of the specimen may show intense immunoreactivity whereas little immunoreaction will occur in its deeper part. As the enzymatic digestion proceeds, immunoreaction in the deeper parts becomes intense whereas that on the surface is diminished. Moreover, enzymatic digestion is affected by various conditions, such as fixation of the tissue, pH, ionic strength, and temperature of the enzyme solution, enzyme concentration, and reaction time. Therefore, the enzymatic pretreatment is not suitable for a thick specimen except that the treatment is needed to restore specific antigenicity.

After changing the fixative to 0.5% paraformaldehyde and 15% (v/v) saturated picric acid solution, we attempted pretreatment with a detergent instead of pepsin because the detergent had been utilized to isolate a basement membrane. Von Bruchhausen and Merker (1967) introduced sodium deoxycholate to purify the basement membranes of rat renal cortex for the first time. They found the fine structure of the isolated basement membrane is identical to that of the corresponding structure in the renal cortex. Welling and Grantham (1972) isolated the basement membrane by perfusing renal tubules with sodium deoxycholate or Triton X-100 and stated that several observations support the view that deoxycholate does not alter the physical properties of the basement membrane appreciably. They also mentioned that prolonged incubation of a basement membrane in deoxycholate had no effect on hydraulic conductivity and that a basement membrane isolated in sodium deoxycholate appeared morphologically indistinguishable from untreated membranes when examined by electron microscopy and stained positively with the periodic acid Schiff reagent, indicating preservation of the carbohydrate moieties in the membrane. They had difficulty obtaining consistently clean basement membranes using Triton X-100, although the resulting basement membranes in successful studies appeared identical to those obtained with sodium deoxycholate. Meezan *et al.* (1975) reported a method for the isolation of ultrastructurally and chemically pure basement membranes from kidney glomeruli and tubules and brain and retinal microvessels by the sequential solubilization of cell membranes, intracellular protein, and plasma proteins by Triton X-100 and sodium deoxycholate. They mentioned that direct treatment of the pure organ subfractions with sodium deoxycholate was adequate for the preparation of basement membranes from a few milligrams of tissue, but with larger amounts of tissue, graded solubilization and deoxyribonuclease treatment were necessary to avoid the formation of a gel-like aggregate of DNA with the basement membrane, which was then difficult to disperse.

These previous reports utilized sodium deoxycholate to isolate pure basement membranes from an unfixed tissue and indicated that the ultrastructural and physical properties of the isolated basement membranes are identical with those seen in the intact tissue. We have attempted to apply the pretreatment with Triton X-100 or sodium deoxycholate on a fixed tissue. Pretreatment with sodium deoxycholate reduced autofluorescence from a mouse liver and renal cortex fixed by a phosphate-buffered mixture of 0.5% paraformaldehyde and 15% saturated picric acid solution, 2% paraformaldehyde in sodium phosphate buffer, or 4% paraformaldehyde in sodium phosphate buffer; but that with Triton X-100 had a slight effect. The transparency of the light ranging from 400 to 800 nm in a 100 μm slice of liver and renal cortex fixed with the mixture of paraformaldehyde and picric acid was improved by pretreatment with sodium deoxycholate, but not with Triton X-100. Light and electron microscopic examination of a tissue fixed with a mixture of paraformaldehyde and picric acid and incubated in sodium deoxycholate indicated that the tissue and cellular architecture was not altered; membranous structures and cytosolic macromolecules in a cell were lost (Lee *et al.*, 1997) (Fig. 10). The fluorescence immunostaining of the tissue for laminin and tenascin showed intense fluorescence even deep in the tissue. Prolonged pretreatment with sodium deoxycholate caused no effect on the antigenicity of laminin and tenascin or on the histoarchitecture.

We now routinely utilize a 3% sodium deoxycholate solution as a pretreatment. However, sodium deoxycholate cannot be used in all cases because a target molecule may be extracted by this reagent. Various kinds of detergents are now available commercially. Each detergent has different properties, such as critical micelle concentration, and exerts different actions on lipids and proteins. If a detergent is not successful with a molecule, it is advisable to try another kind of detergent with different electrical properties.

VI. Fluorescent Staining

The most widely used fluorescence staining for CLSM would be the fluorochrome-labeled antibody method. Considering the purpose of three-dimensional analysis, we should select the most specific staining method and avoid the possibility of unexpected background fluorescence. Conventional staining methods, including PAS reaction and Feulgen reactions with fluorescent Schiff-type reagents, may be performed. However, these conventional methods are likely to label so many structures that the three-dimensional image becomes complicated in analyzing their distribution.

On the topic of fluorescent staining, there are two points to which attention should be paid: (1) uniformity in staining and (2) the prevention of nonspecific reaction. To attain uniform staining of a specimen with a specific antibody, the antibody must be able to penetrate evenly and deep into it. Pretreatment of the specimen is needed for this purpose. The concentration of the antibody and the duration of incubation influence the results deeply and must be determined carefully by the researcher. In our experiment, a more desirable result was obtained at a lower concentration and a longer incubation time than those used for staining of the conventional section. In the case of the high concentration of the antibody, the antibody bound with the antigen situated at the surface area of the specimen may aggregate and interfere with the unbound antibody in penetration into a deeper area of the specimen. The surface area of the specimen may show intense immunofluorescence, while the deeper area emits fluorescence weakly.

The specificity of the primary antibody is indisputably important for staining of the specimen as well. In addition, care must be taken regarding the specificity of the secondary fluorochrome-labeled antibody, as the weak-fixed and pretreated specimen is likely to show fluorescence by nonspecific binding of antibodies, especially by the secondary antibody. We usually utilize a fluorochrome-labeled antibody that has been purified by affinity chromatography and absorbed by the serum of a target animal. Some antibodies, however, showed a nonspecific reaction (Fig. 11).

VII. Vascular Cast with Fluorochrome-Labeled Gelatin

A. Procedures II and III

In a series of three-dimensional investigations of fetal organs with the antilaminin antibody, it was difficult to distinguish the basement membrane of vessels from that of the epithelium. In developing organs, vascular nets seem to play a crucial role in morphogenesis (Figs. 12–14 and 18).

Vascular nets in an organ have been investigated by intravascular injection of India ink before. Since the introduction of acrylic monomer as an injection medium by Batson (1955), plastic resins, such as prepolymerized methylmethacrylate and Mercox, have been applied and injection replica of vascular nets were produced (Hodde and Nowell, 1980; Murakami, 1971). After the removal of all

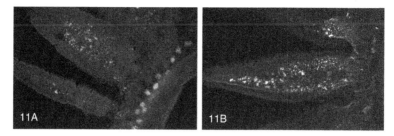

Fig. 11 Examples of nonspecific binding of two kinds of FITC-conjugated antibodies in the small intestinal villi. The mouse small intestine was fixed and pretreated as in Procedure I. The specimen was incubated with FITC-conjugated antirat IgG antibodies without incubation with a primary antibody. Nonspecific fluorescence is found at some cells in the lamina propria and the granules of the Paneth cell. The specificity and nonspecific binding of a secondary antibody should be checked carefully in addition to those of a primary antibody. Modified from Lee *et al.* (1997).

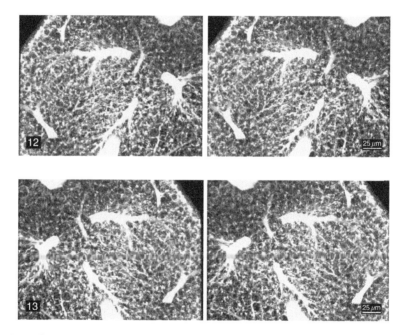

Figs. 12 and 13 Vascular nets in the mouse liver at postnatal day 5. A mouse was perfused with FITC-labeled gelatin and processed as described in Procedure III. The liver was removed, sectioned at 150 μm and observed with an LSM-310. Three-dimensional images were reconstructed with an LSM-510 from 40 images taken at a 2-μm interval from the front side (Fig. 12) and the far sides (Fig. 13). Sinusoidal capillary nets in the young mouse liver are well observed without leakage of FITC-labeled gelatin. As the three-dimensional reconstruction can be performed from both the front and the far sides and with an arbitrary number of serial images, vascular nets of interest can be examined easily.

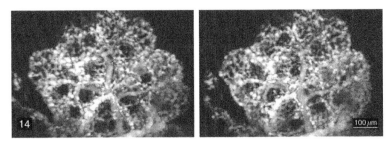

Fig. 14 Vascular nets in a fetal mouse lung at day 15 of gestation. A mouse fetus was perfused with FITC-labeled gelatin as mentioned in Procedure III. The lung was removed and observed with an LSM-10. Three-dimensional images were reconstructed with an LSM-510 from 28 images taken at a 3-μm interval. Numerous vascular nets surrounding the developing bronchia are found.

soft and hard tissues, the injection replica has been principally observed by a scanning electron microscope. However, it is impossible to reveal the interrelationship between vascular nets and the other tissue components because this method requires removing soft tissues for observation.

There are some reports concerning the observation of vascular nets with the light or fluorescence microscopes (Connolly *et al.*, 1988; D'Amato *et al.*, 1993; Holthöfer *et al.*, 1982; Rummelt *et al.*, 1994). In these reports, antibodies against a vascular endothelial cell-specific antigen such as factor VIII and angiotensin-converting enzyme II or lectins specific for a vascular endothelial cell such as *Ulex europaeus* agglutinin I (UAE-I) and *Ricinus communis* agglutinin I (RCA) were utilized to detect endothelial cells. However, some species and strain specificity of these antibodies and lectins may be present and it is not certain whether these tracers react uniformly in each fetal immature to adult mature endothelial cells. Therefore, we attempted to make an injection replica of vascular nets with gelatin labeled with a fluorochrome to examine by CLSM (Hashimoto *et al.*, 1998b,c) (see Appendix). Gelatin has some advantageous characteristics. An aqueous gelatin solution does not solidify in vital warmth. It is solidified not by polymerization, but by cooling. This is particularly favorable for the perfusion of a fetus because no time limit is set for injection as long as the fetus and the gelatin solution are kept warm. There are many reactive residues, including amine in a gelatin molecule. These residues bind covalently with an isothiocyanate derivative of fluorochrome, such as fluorescein isothiocyanate, and are cross-linked easily to each other with fixatives. Once fixed, the solidified gelatin rarely dissolves by warming.

This method is easily applicable to a fetus because no time limit is set for injection and cardiac contractions are maintained during perfusion (Figs. 14 and 18). In addition, a specimen prepared by vascular perfusion of fluorochrome-labeled gelatin can be stained again by the immunofluorescence method (Figs. 17 and 18). This means that the interrelationship between vascular nets and other tissue components can be examined three-dimensionally. This method will aid research on the role of vascular nets.

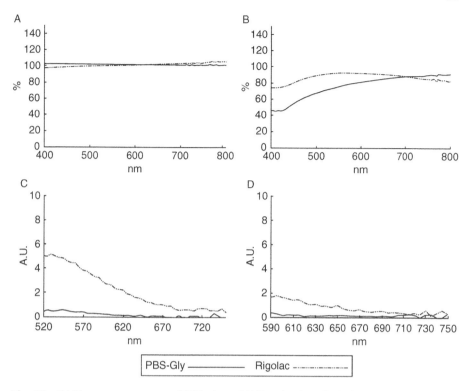

Fig. 15 (A) Transparency spectra of PBS–glycerol (1:9) and polymerized Rigolac, a polyester resin. Rigolac was polymerized as in Procedure IV in a space 170 µm thick between the glass slide and the coverslip. Transparency spectra of the polymerized resin and PBS–glycerol in the space between the glass slide and the coverslip were examined against distilled water with a Carl Zeiss UMSP microspectrophotometer. (B) Transparency spectra of a liver mounted in PBS–glycerol (1:9) and Rigolac. The mouse liver was removed, fixed, and sliced as in Procedure I. The liver slice was mounted in PBS–glycerol or Rigolac. Transparency spectrum of the mounted liver slice was recorded against each mounting media. (C, D) Fluorescence spectra of a liver mounted in PBS–glycerol (1:9) and Rigolac. The liver slice was prepared as in (B). Autofluorescence spectrum of the mounted liver slice was investigated under blue-excitation (C) or green-excitation (D). By mounting in Rigolac, the transparency of the liver slice is improved but autofluorescence of the liver is increased, especially under blue excitation.

VIII. Mounting

The preparation for immunofluorescence observation is usually mounted in an aqueous mounting medium that is a mixture of glycerol and a buffer solution. The most widely used buffer is 0.1 M sodium phosphate buffer (pH 7.4) or 0.01 M sodium phosphate-buffered saline (pH 7.2). In many cases, an antifading agent is added to the mounting medium.

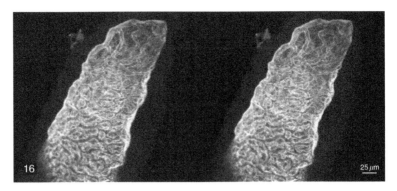

Fig. 16 Stereo pair images of mouse small intestinal villi immunostained for laminin (red) and tenascin (green). The mouse small intestine was processed as Procedure I with rabbit antimouse laminin antibody, RITC-conjugated goat antirabbit IgG antibody, rat antihuman tenascin antibody, and FITC-conjugated goat antirat IgG antibody. Serial red-colored RITC and green-colored FITC images were taken simultaneously at a 1.5-μm interval with an LSM-310 and a Plan NEOFLUOR 40× objective lens (NA 0.75). Three-dimensional images were reconstructed with an LSM-510 from 20 images. By superimposing red-colored RITC and green-colored FITC stereo pair images, the interrelationship of the localization of laminin and tenascin is indicated clearly. (See Plate no. 2 in the Color Plate Section.)

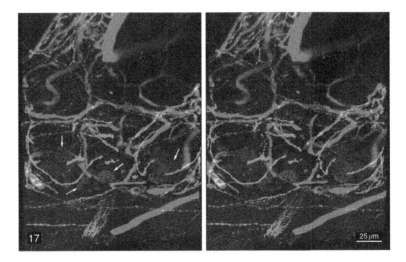

Fig. 17 Stereo pair images of vascular nets and VIP positive nerve fibers at the crypt of the mouse small intestine. The small intestine from a mouse perfused with RITC-labeled gelatin as described in Procedure III was sliced and immunostained for VIP with a rabbit antisynthetic porcine VIP serum (provided by Dr. Yanaihara, Shizuoka College of Pharmacology) and BodipyFL-conjugated goat antirabbit IgG antibody. The specimen was mounted in Rigolac as mentioned in Procedure IV. Serial red-colored RITC and green-colored BodipyFL images were obtained simultaneously at a 0.7-μm interval with an LSM-510 and a 40× objective lens, and stereo pair images were reconstructed from 90 images. The crypt is surrounded with capillary nets at its base and with VIP-positive nerve fibers at various portions. Red fluorescence found in granules of the Paneth cell is nonspecific (arrows). (See Plate no. 3 in the Color Plate Section.)

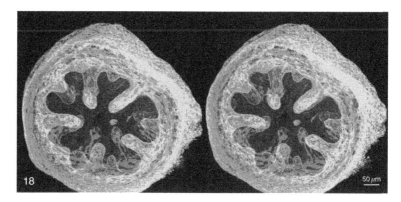

Fig. 18 Stereo pair images of vascular nets and laminin in a mouse small intestine at day 15 of gestation. A mouse fetus was perfused with RITC-labeled gelatin as in Procedure III and the small intestine was obtained. The tissue was sliced at a thickness of 150 μm, pretreated with sodium deoxycholate, and immunostained for laminin with a rabbit antimouse laminin antibody and FITC-conjugated goat antirabbit IgG antibody. Stereo pair images were reconstructed from 55 serial red-colored RITC and green-colored FITC images obtained simultaneously at a 1.99-μm interval with an LSM-510 and a 20× objective lens. In a fetal small intestine, the interrelationship between developing villi and capillary nets is shown clearly. (See Plate no. 4 in the Color Plate Section.)

Fading of fluorescence by continuous excitation is an inevitable problem in fluorescence microscopy, including CLSM. To make a three-dimensional observation, dozens of images must be taken from the same region, although the focal plane is different at each image. Fluorochrome located in a different focal plane is not a little excited by being exposed to a laser light source and quenching of the fluorescence is caused. Several chemicals are reported to prevent fading of the fluorochrome by the addition to mounting media (Böck *et al.*, 1985; Giloh and Sedat, 1982; Johnson *et al.*, 1982). These are sodium azide, sodium iodide, poly-vinylpyrrolidone (PVP), polyvinyl alcohol (PVA), 1,4-di-azobicyclo[2,2,2]octane (DABCO), *p*-phenylenediamine, *n*-propyl gallate, and sodium dithionite. Giloh and Sedat (1982) recommend 0.1–0.25 M of *n*-propyl gallate in glycerol to reduce photobleaching of tetramethylrhodamine and fluorescein. Johnson *et al.* (1982) mention that *p*-phenylenediamine is the most efficient retarding agent and that DABCO, which is an extremely stable compound and is a nonionizing, inexpensive, and readily available, appears to be a suitable alternative. Contrary to these previous reports, Böck *et al.* (1985) state that sodium azide and sodium iodide increase fluorescence intensity and inhibit bleaching but that the other compounds reported previously to exert beneficial effects did not lead to an increase in fluorescence and had questionable effects on bleaching.

We usually use 0.05 M Tris–HCl-buffered saline (pH 8.0) containing 90% (v/v) glycerol and 10 mg/ml of DABCO as a mounting medium. A mixture of glycerol and PBS (9:1) containing *n*-propyl gallate had been used previously, but this mounting medium seemed to induce autofluorescence in a nucleus and the fluorescence of fluorescein was not stable. The Tris–HCl buffer was adopted because the fluorescence intensity of fluorescein derivatives increases and becomes stable as the pH of the solution increases in the physiological range. In our experience, the addition of DABCO reduces the photobleaching of fluorescein and rhodamine, although Böck *et al.* (1985) report that the effect of DABCO on photobleaching is questionable. When a vulnerable fetal organ is to be mounted, it is desirable to increase gradually the concentration of glycerol (30%, 50%, and 70%) in order not to cause shrinkage or deformation of the specimen. This mounting medium is stable over several months. The specimen mounted in this mounting media can be stored for a few weeks at 4 °C, but longer storage increases autofluorescence in cytoplasm and reduces the specific fluorescence of fluorescein and rhodamine.

For mounting a whole organ or a thick section, spacers should be placed between the glass slide and the coverslip in order not to crush it. In our laboratories, two glass strips made from a coverslip are attached to a glass slide as a spacer and a specimen is mounted between the strips. We can obtain coverslips having thicknesses ranging from 0.13 to 0.60 mm. The thickness of the spacer should be approximately equal to that of the specimen.

As mentioned earlier, the specimen for fluorescence microscopy is usually mounted in an aqueous mounting medium. The specimen in the medium is pinched lightly between the glasses and is not anchored firmly. This brings up a problem that the specimen may shift when the coverslip is pushed by an oil-immersion objective lens to take serial images. The shift of specimen causes the deviation of images, resulting in a deformation of the three-dimensional image. A specimen mounted in an aqueous mounting medium increases background fluorescence gradually, even if it is returned to PBS. For these reasons, we have attempted to mount the specimen in a resin to make a semipermanent preparation. The tissue section is mounted conventionally in a balsam or a synthetic styrene resin dissolved in an organic solvent such as xylene through dehydration with a graded series of ethanol. These resins are hardened by the evaporation of solvent. If a whole mount specimen or a thick section is mounted in these resins, it will take a long time to harden completely and shrinkage of the resins during hardening may cause deformation. Franklin and Martin (1981) attempted to embed immunofluorescence-stained tissue slices in hydroxypropyl methacrylate to obtain high-resolution staining of antigens in 1-μm tissue sections. However, 170-μm-thick polymerized hydroxypropyl methacrylate resin showed autofluorescence as well as epoxy resin, EPON 812, in our experiment. An unsaturated polyester resin, Rigolac, which is one of the acrylic resins and is utilized as an embedding medium for transmission electron microscopy, was found to have a high transparency of visible light and no autofluorescence (Fig. 15A). The refractive index to the polymerized polyester resin (a mixture of Rigolac 2004 and Rigolac 70F (7:3)) is 1.5292 (Sato, personal communication), a little higher than that of glass. The transparency of the light through a slice of liver mounted in the polyester resin is higher than that mounted in glycerol/PBS (9:1) (Fig. 15B). When a specimen is mounted in polyester resin between two coverslips spaced with glass strips, the specimen can be observed from both the front and the far side. In the case of observation with an oil-immersion lens, a coverslip is removed from the resin and immersion oil is dropped directly on the resin so that the short working distance of the oil-immersion lens may be compensated. However, polyester resin has some problems. The autofluorescence in the cytoplasm increases especially under blue-excitation (Fig. 15A and D). The fluorescence intensity of fluorochromes conjugated to antibodies may decrease due to polymerization. In the fluorochrome tested, the fluorescence of antibodies labeled with fluorescence isothiocyanate, Oregon Green 488, and tetramethylrhodamine isothiocyanate was reduced, whereas that of BodipyFL was preserved and was observable with a fluorescence microscope and CLSM (Figs. 7, 8, 17, 19 (see color insert), and Fig. 20 (see color insert)). The fluorescence of fluorescein isothiocyanate and rhodamine isothiocyanate conjugated to gelatin was hardly affected. Injection replica of vascular nets by fluorochrome-labeled gelatin was clearly indicated in a specimen 500 μm thick mounted in this polyester resin. The application of resin mounting is still limited, but the specimen mounted in the resin is easy to handle and maintain and would be semipermanently observable if stored in the dark.

IX. Observation

We have been utilizing the CLSMs of Carl Zeiss, LSM-10, LSM-310, LSM-410, and LSM-510. Because the operation of the CLSM depends mainly on the specific control software for it, the comments in this section may not be applicable for every CLSM.

Prior to the observation with CLSM, the specimen should be examined with a conventional fluorescence microscope, especially on the specific and background fluorescence. A confocal image is a computer graphic constructed from tens thousands or millions of pixels whose brightness corresponds with the integrated fluorescence intensity of a limited spectrum in a focal point. If a background fluorescence has a broad or a similar spectrum with a specific fluorescence, it is impossible to distinguish specific fluorescence from background on a confocal image. Although the pseudocolored image may assist it, the real color of specific and background florescence is not reproduced accurately on the pseudocolored image.

The interval between images should be determined from the numerical aperture (NA) of the objective lens used and the size of the pinhole used for obtaining a confocal image. A confocal image is not the image of a mathematical plane, but it includes fluorescence from an area with some thickness. The thickness is conventionally referred to as full-width at half-maximum. The interval between images is principally determined by reference to the value of full-width at half-maximum. If the interval is fairly large, some structures may be lost in the reconstructed three-dimensional image. However, if it is short, so many similar images are collected and it takes a large memory to store and a long time to reconstruct. The latest control software for LSM-510 automatically calculates full-width at half-maximum at the current setting of the object lens, the size of confocal pinhole, and the wavelength of laser beam and suggests half of it as the optimal interval between images.

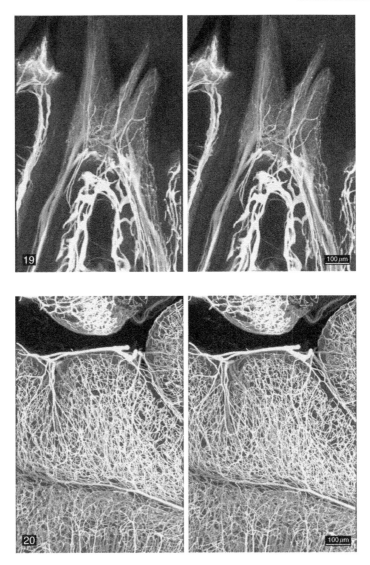

Figs. 19 and 20   Stereo pair images of vascular nets in a tooth (Fig. 19) and cerebellum (Fig. 20) of the mouse. A mouse was perfused with RITC-labeled gelatin and injected 0.11 ml (net) of FITC-labeled gelatin just before the end of the perfusion. The mandible was removed and decalcified with EDTA; the cerebellum was also removed. The specimens were sliced at 500 μm, treated with sodium deoxycholate, and mounted in Rigolac. Stereo pair images were reconstructed from 35 images taken at a 5.54-μm interval (Fig. 19) and from 38 images at a 5.81-μm interval (Fig. 20) with an LSM-510 and a Plan NeoFLUOR 10× objective lens (NA 0.3). The perfusion of the circulatory system from the left ventricle with a fluorochrome-labeled gelatin followed by a small amount (0.1–0.6 ml) of gelatin labeled with another colored fluorochrome showed that arterial vessels are distinguished from the venous vessels by the fluorescence from the gelatin filling them (Hashimoto *et al.*, 1998c). In these figures, venous blood vessels are filled only with the RITC-labeled gelatin and emits red fluorescence, whereas most of RITC-labeled gelatin is extruded by the following FITC-labeled gelatin in the arterial blood vessels,

One of the most important and intrinsic issues found on a confocal image is that the fluorescence intensity from a structure oriented in parallel with the optical axis becomes high and that from a structure perpendicular to the optical axis becomes relatively low (van der Voort and Brakenhoff, 1989). That is, when a tubular structure emits a fluorescence of uniform intensity, confocal images show high intensity on the sidewalls and low intensity on the top and bottom walls (refer to Figs. 1 and 2). This phenomenon is more conspicuous at low magnification than at high magnification because the NA of the lens determines full-width at half-maximum or the thickness of a confocal image.

Three-dimensional images reconstructed with a CLSM will afford meritorious information on biological system to researchers.

═══════ Appendix

Procedure I. Standard Procedure for Sample Preparation

1. Remove a tissue, slice into small pieces with razor blades, and fix by immersion in a fixative consisting of 0.5% paraformaldehyde and 15% (v/v) saturated picric acid in 0.1 M SPB (pH 7.0) for 2 h at room temperature or overnight at 4 °C. Fixation by perfusion with the fixative is desirable, if possible. After perfusion, the tissue is treated in the same manner as described earlier.

2. Rinse the tissue with 0.1 M SPB (pH 7.4) several times until picric acid is no longer found in the buffer.

3. *Sectioning (optional)*: Immerse the tissue in 0.01 M sodium phosphate-buffered saline (PBS, pH 7.2) containing 20% sucrose and 10% glycerol through PBS containing 5% and 10% sucrose. Freeze the tissue on a cryo-stage and section at 100–200 μm. Collect sections in PBS.

4. In the following steps, a 48-well multiwell plate is used to incubate the specimen under mild agitation. Incubate the specimen in a 3% sodium deoxycholate solution for 4 h at room temperature or overnight at 4 °C. The incubation time may be prolonged, if necessary.

5. Rinse the specimen with distilled water two times and then with PBS for three times 1 h each.

6. Incubate the specimen with 10% preimmune serum appropriate for the secondary antibody, such as normal goat serum, overnight at 4 °C.

---

resulting in green to yellowish fluorescence. Vascular nets in a hard tissue such as tooth and bone can be visualized three-dimensionally after decalcification by our methods. In Fig. 19, arteries running through an accessory canal are seen. (See Plate no. 5 and 6 in the Color Plate Section.)

7. Apply the first antibody, such as antilaminin rabbit serum, for 1 or 2 days at 4 °C. The optimal dilution of the antibody should be determined previously.

8. Rinse the specimen with PBS several times 1–3 h each.

9. Apply the secondary antibody, such as FITC-labeled goat antirabbit IgG antibody, for 1 or 2 days at 4 °C.

10. Rinse the specimen as step 8.

11. Mount the specimen in 0.05 M Tris–HCl-buffered saline (pH 8.0) containing 90% glycerol and 10 mg/ml of DABCO.

- Sodium azide is added to the PBS at a concentration of 0.05%.
- For diluting the serum or antibody, PBS containing 1% bovine serum albumin and 0.1% sodium azide is used.
- For double staining, repeat steps 6–10.

Procedure II. Preparation of Fluorochrome–Labeled Gelatin

1. Swell 8 g of gelatin (from bovine skin, approximately 225 Bloom, No. G9382, Sigma Chemical Co., St. Louis, MO) in 30 ml of distilled water in an incubator at 37 °C.

2. Dissolve the gelatin completely in an incubator at 60 °C after swollen.

3. Adjust the total volume of the gelatin solution to 40 ml (the final gelatin concentration is 20%).

4. Raise the pH of the gelatin solution to 10 by adding a 1 N NaOH solution (it requires about 0.8 ml).

5. Dissolve 20–50 mg of isothiocyanate derivative of fluorochrome (fluorescein isothiocyanate, rhodamine isothiocyanate, or tetramethylrhodamine isothiocyanate) in 1 ml of absolute dimethylsulfoxide (DMSO).

6. Gently pour the fluorochrome solution into the warm gelatin solution. The final weight ratio of fluorochrome to gelatin is 1:160–1:400.

7. Place the mixture in a dark incubator at 37 °C overnight and agitate mildly to progress the labeling of the gelatin with the fluorochrome.

8. Transfer the reacted mixture into a seamless dialyzing tube and dialyze against 0.01 M PBS containing 0.01% sodium azide in a dark incubator at 37 °C.

9. Continue the dialysis for 14–20 days; change the dialyzing buffer every day.

10. End the dialysis when no fluorochrome is found in the dialyzing buffer.

11. Place the dialyzed mixture in a dark refrigerator to gel the mixture.

12. Transfer the gelled mixture to a tightly stoppered glass bottle and store in a dark refrigerator.

Procedure III. Perfusion with Fluorochrome-Labeled Gelatin

1. Dissolve the fluorochrome-labeled gelatin in a warm water bath.
2. Adjust the final concentration of fluorochrome-labeled gelatin approximately 10% with 0.01 M PBS.
3. Load the fluorochrome-labeled gelatin into a plastic syringe equipped with a plastic tube and needle.
4. Keep the syringe with the plastic tube and needle in a warm water bath.
5. Perfusion of an animal:

   (a) Whole body perfusion for a small animal:

   - Anesthetize an animal with intraperitoneal injection of sodium pentobarbital solution.
   - Perform a thoracotomy and expose the heart.
   - Insert the needle into the left ventricle and clamp with a Kocher clamp. Care must be taken to keep the pinpoint of needle in the left ventricle.
   - Sever the right atrium or auricle and drain blood from it.
   - Inject the fluorochrome-labeled gelatin solution slowly.
   - Continue the perfusion until no more blood cells flow.

   (b) Perfusion of a fetus:

   - Kill a pregnant animal by excess inhalation of ether or by cervical dislocation.
   - Perform a laparotomy.
   - Remove fetuses with their fetal membrane intact and transfer to ice-cold PBS.
   - Transfer a fetus to warm PBS and cut off its fetal membrane to expose the fetus, the placenta and its umbilical cord under a dissection microscope.
   - Wait until the umbilical artery and vein expand and the pulsation of the umbilical artery is obvious.
   - Insert a 27 to 30-gauge needle or a glass needle connected to a syringe or microinjector via a silicone tube into the umbilical vein. A micromanipulator may help this process.
   - Sever the umbilical artery.
   - Inject the fluorochrome-labeled gelatin solution slowly and drain blood from the umbilical artery. Care must be taken to keep the heart pulsating. A high pressure of the injection may cause the disruption of capillary in the liver and the peritoneal cavity may fill with the leaked fluorochrome-labeled gelatin solution.
   - After a while the fluorochrome-labeled gelatin solution flows from the umbilical artery with blood, but continue the perfusion until no blood cells flow.

(c) In case the governing artery of the target organ is obvious and a needle can be inserted into it, only the organ can be perfused.

- Heparinization prior to the perfusion with fluorochrome-labeled gelatin solution may be recommended if blood coagulation precludes the perfusion in capillaries.
- Perfusion with a fixative solution before the fluorochrome-labeled gelatin solution may improve tissue fixation.

6. Clamp the base of the heart or the vessels proximal to the inserted point and the severed portion with a Kocher clamp after perfusion and immerse the whole body immediately in ice-cold fixative to solidify the fluorochrome-labeled gelatin. The fixative consists of 0.5% paraformaldehyde and 15% (v/v) of a saturated picric acid solution in sodium phosphate buffer as mentioned in Procedure I.

7. After solidification of the fluorochrome-labeled gelatin, remove the target organ and cut into small pieces and fix further in the same fixative overnight at 4 °C in the dark.

8. Rinse the specimen with ice-cold SPB several times.

9. Section the specimen as stated earlier if necessary.

10. Perform pretreatment with chilled 3% sodium deoxycholate solution.

11. Rinse the specimen twice with chilled distilled water for 1 h each.

12. Perform postfixation with 4% paraformaldehyde in SPB if the fluorochrome-labeled gelatin is fixed imperfectly and dissolves during observation.

13. Mount the specimen as in Procedure I.

- The specimen perfused with the fluorochrome-labeled gelatin can be further stained by the immunofluorescence method. The immunofluorescence staining is performed between steps 11 and 12. Each procedure should be carried out at 4 °C.

Procedure IV. Mounting the Specimen in Polyester Resin

1. Prepare the specimen according to the steps 1–10 in Procedure I or the steps 1–11 in Procedure III.

2. Dehydrate the specimen through a graded series of ethanol to 100% ethanol in the dark.

3. Transfer the specimen to pure styrene monomer twice in the dark.

4. Prepare the polyester resin by mixing Rigolac 70F and Rigolac 2004 (3:7) and dissolve 0.05% of benzoyl methyl ether in the mixture. The resin mixture can be stored at 4 °C in the dark for a few months.

5. Immerse the specimen in the resin mixture in the dark.

6. Mount the specimen in the resin mixture between coverslips spaced with glass strips. Seal each edge of the coverslips with paraffin wax.

7. Polymerize the resin mixture under UV (360 nm) irradiation for 30 min at room temperature.

Acknowledgments

We thank Dr. D. C. Herbert, The University of Texas Health Science Center at San Antonio, Texas, USA, for his help during the preparation of this manuscript. We also thank Mr. Fumiyoshi Ishidate (Carl Zeiss Co., Ltd.) for his technical advice and Carl Zeiss Co., Ltd., for providing us with opportunities to utilize the LSM-310 and LSM-410 confocal laser scanning microscopes. The original work described in this manuscript was supported in part by a grant-in-aid from the Ministry of Education, Science, Sports and Culture of Japan (Nos. 08670038 and 09670032), the Special Coordination Funds of the Science and Technology Agency of the Japanese Government, the CASIO Science Promotion Foundation, and the Foundation for Advancement of International Science.

References

Batson, V. (1955). Corrosion specimens prepared with a new material. *Anat. Rec.* **121,** 425.

Belford, D. A., Gole, G. A., and Rush, R. A. (1987). Localization of laminin to retinal vessels of the rat and mouse using whole mounts. *Invest. Ophthalmol. Vis. Sci.* **28,** 1761–1766.

Böck, G., Hilchenbach, M., Schauenstein, K., and Wick, G. (1985). Photometric analysis of antifading reagents for immunofluorescence with laser and conventional illumination sources. *J. Histochem. Cytochem.* **33,** 699–705.

Connolly, S. E., Hores, T. A., Smith, L. E. H., and D'Amore, P. A. (1988). Characterization of vascular development in the mouse retina. *Microvasc. Res.* **36,** 275–290.

Costa, M., Buffa, R., Furness, J. B., and Solcia, E. (1980). Immunohistochemical localization of polypeptides in peripheral autonomic nerves using whole mount preparations. *Histochemistry* **65,** 157–165.

D'Amato, R., Wesolowski, E., and Simth, L. E. H. (1993). Microscopic visualization of the retina by angiography with high-molecular weight fluorescein-labeled dextrans in the mouse. *Microvasc. Res.* **46,** 135–142.

Dhenain, M., Ruffins, S. W., and Jacobs, R. E. (2001). Three-dimensional digital mouse atlas using high-resolution MRI. *Dev. Biol.* **232,** 458–470.

Finle, J. C. W., and Petrusz, P. (1982). The use of proteolytic enzymes for improved localization of tissue antigens with immunocytochemistry. *In* "Techniques in Immunocytochemistry," (G. R. Bullock, and P. Petrusz, eds.), Vol. 1, p. 239. Academic Press, London.

Franklin, R. M., and Martin, M. T. (1981). Pre embedding immunohistochemistry as an approach to high resolution antigen localization at the light microscopic level. *Histochemistry* **72,** 173–190.

Giloh, H., and Sedat, J. W. (1982). Fluorescence microscopy: Reduced photobleaching of rhodamine and fluorescein protein conjugates by n-propyl gallate. *Science* **217,** 1252–1255.

Grzanna, R., Molliver, M. E., and Coyle, J. T. (1978). Visualization of central noradrenergic neurons in thick sections by the unlabeled antibody method: A transmitter-specific Golgi image. *Proc. Natl. Acad. Sci. USA* **75,** 2502–2506.

Hartman, B. K., Zide, D., and Udenfriend, S. (1972). The use of dopamine β-hydroxylase as a marker for the central noradrenergic nervous system in rat brain. *Proc. Natl. Acad. Sci. USA* **69,** 2722–2726.

Hashimoto, H., and Hoshino, K. (1988). Immunohistochemical localization of laminin in fetal mouse lungs. *Acta Histochem. Cytochem.* **21,** 125–136.

Hashimoto, H., and Kusakabe, M. (1997). Three dimensional distribution of extracellular matrix in the mouse small intestinal villi. Laminin and Tenascin. *Connect. Tissue Res.* **36,** 63–71.

Hashimoto, H., Ishikawa, H., and Kusakabe, M. (1998a). Three dimensional analysis of the developing pituitary gland in the mouse. *Dev. Dyn.* **212**, 157–166.

Hashimoto, H., Ishikawa, H., and Kusakabe, M. (1998b). Simultaneous observation of capillary nets and tenascin in intestinal villi. *Anat. Rec.* **250**, 488–492.

Hashimoto, H., Ishikawa, H., and Kusakabe, M. (1998c). Three dimensional investigation of vascular nets by fluorochrome-labeled angiography. *Microvasc. Res.* **55**, 179–183.

Hashimoto, H., Ishikawa, H., and Kusakabe, M. (1999). Development of vascular networks during the morphogenesis of intestinal villi in the fetal mouse. *Acta Anat. Nippon* **74**, 567–576.

Hashimoto, H., Kusakabe, M., and Ishikawa, H. (2009). A novel method for three-dimensional observation of the vascular networks in the whole mouse brain. *Microsc. Res. Tech.* **71**, 51–59.

Hodde, K. C., and Nowell, J. A. (1980). *In* "SEM of Micro-corrosion Casts," pp. 89–106. SEM Inc., AMF O'Hare, Chicago, ILSEM/1980/part II.

Holthöfer, H., Virtanen, I., Kariniemi, A. L., Hormia, M., Linder, E., and Miettinen, A. (1982). *Ulex europaeus* I lectin as a marker for vascular endothelium in human tissues. *Lab. Invest.* **47**, 60–66.

Ito, T., Ohi, S., Tachibana, T., Takahara, M., Hirabayashi, T., Ishikawa, H., Kusakabe, M., and Hashimoto, H. (2007). Development of the mucosal vascular system in the distal colon of the fetal mouse. *Anat. Rec.* **291**, 65–73.

Johnson, G. D., Davidson, R. S., McNamee, K. C., Russel, G., Goodwin, D., and Holborow, E. J. (1982). Fading of immunofluorescence during microscopy: A study of the phenomenon and its remedy. *J. Immunol. Methods* **55**, 231–242.

Larsson, L. I., Goltermann, N., De Magistris, L., Rehfeld, J. F., and Schwartz, T. W. (1979). Somatostatin cell processes as pathways for paracrine secretion. *Science* **205**, 1393–1394.

Lee, L., Hashimoto, H., and Kusakabe, M. (1997). A note on the preparation of whole mount samples suitable for observation with the confocal laser scanning microscope. *Acta Histochem.* **99**, 101–109.

Meezan, E., Hjelle, J. T., Brendel, K., and Carlson, E. (1975). A simple, versatile, nondisruptive method for the isolation of morphologically and chemically pure basement membranes from several tissues. *Life Sci.* **17**, 1721–1732.

Murakami, T. (1971). Application of the SEM to the study of the fine distribution of the blood vessels. *Arch. Histol. Jpn.* **33**, 179–198.

Pearse, A. G. E. (1980). "Histochemistry Theoretical and Applied Volume one," p. 123. Churchill Livingstone, Edinburgh, London, Melbourne and New York.

Rummelt, V., Gradner, L. M. G., Folberg, R., Beck, S., Knosp, B., Moninger, T. O., and Moore, K. C. (1994). Three dimensional relationships between tumor cells and microcirculation with double cyanine immunolabeling, laser scanning confocal microscopy, and computer-assisted reconstruction: An alternative to case corrosion preparations. *J. Histochem. Cytochem.* **42**, 681–686.

Sato, personal communication. Showa Kobunshi Co. Ltd.

Sjölund, K., Sandén, G., Håkanson, R., and Sundler, F. (1983). Endocrine cells in human intestine: An immunocytochemical study. *Gastroenterology* **85**, 1120–1130.

Somogyi, P., and Takagi, H. (1982). A note on the use of picric acid-paraformaldehyde-glutaraldehyde fixative for correlated light and electron microscopic immunocytochemistry. *Neuroscience* **7**, 1779–1783.

Stefani, M., de Martino, C., and Zamboni, L. (1967). Fixation of ejaculated spermatozoa for electron-microscopy. *Nature* **216**, 173–174.

van der Voort, H. T. M., and Brakenhoff, G. J. (1989). 3-D image formation in high-aperture fluorescence confocal microscopy: A numerical analysis. *J. Micorsc.* **158**, 43–54.

Von Bruchhausen, F., and Merker, H. J. (1967). Morphologischer und chemischer isolierter Basal-membranen aus der Nierenrinde der Ratte. *Histochemie* **8**, 90–108.

Welling, L. W., and Grantham, J. J. (1972). Physical properties of isolated perfused renal tubules and tubular basement membranes. *J. Clin. Invest.* **51**, 1063–1075.

CHAPTER 7

Use of Confocal Microscopy to Investigate Cell Structure and Function

Ghassan Bkaily and Danielle Jacques

Department of Anatomy and Cell Biology
Faculty of Medicine, University of Sherbrooke
Sherbrooke, Quebec, Canada

105
DOI: 10.1016/B978-0-12-384658-7.00007-2

I. Introduction

Since writing this chapter in 1999, growing progress in confocal microscopy highly contributed to understanding of physiology and pharmacology as well as intracellular signaling. Furthermore, the recent progress in confocal microscopy permitted opening of a new field of intracellular organelle physiology and pharmacology. Recently, this progress leads to a new concept that the nucleus is a cell within a cell (Bkaily *et al.*, 2006, 2009). The scientific community realized the importance of confocal microscopy and real three-dimensional (3D) reconstruction of the cell, thus the confocal microscopy became a routine technique in many laboratories across the world. This promoted development of more stable and specific fluorescent probes as well as antibodies. All these confocal microscopy related materials became commercially available. The combination of krypton–argon and UV lasers as well as the development of a new line of confocal microscope, the two photons, permitted the scientists to use all ranges of excitation. These two generations of confocal microscopes permitted labeling of two, even three and four targets on and within the cell, in fixed and live preparations. However, any misuse of this outstanding technique due to insufficient knowledge in the field of biological confocal microscopy, image processing, and biophysical properties of fluorescent probes may bring forth invalid results. Therefore, it is highly important that scientist who desires to use the technique of confocal microscopy and real 3D image analysis and measurement should be aware of the science needed to obtain the best reliable results in the field.

Although recent progress in this technique sheds light on the so-called "black box" of working cells, it did open a new field and concept of cell physiology and pharmacology. Despite the recent progress in the field, several questions are sometimes a matter of debate in the literature, such as: Are we actually measuring cytosolic and nuclear free ion? Are all the different ion-specific dyes homogeneously distributed throughout the cell including the nucleus? What is the contribution of intracellular compartments to the cytosolic ionic homeostasis? At what subcellular level does a particular drug produce its effect? How do the ionic channels and receptors of the nuclear membranes contribute to the overall cell function and nuclear signaling? Does cell-to-cell contact play a role in cell function and its response to drugs? What are the effects of photobleaching and/or leak of the fluorescent probe?

Without the high-performance hardware and software systems as well as proper use of confocal microscopy, these questions as well as many others cannot be answered. Today, most laboratories dealing with molecular and cell biology, pharmacology, biophysics, biochemistry, and so on are or in the process of being equipped with this powerful scientific tool. However, to better optimize this technique, scientists need to be familiar with the science behind this technique as well as the basic approaches and limitations of confocal microscopy. This chapter discusses sample preparations to be used, labeling of structures and functional probes, parameter settings, and the development of specific ligand probes, as well as measurements and limitations of the technique. For further information, the reader should refer to other volumes in *Methods in Enzymology* as well as to other key references in the literature (Bkaily *et al.*, 1997b; Paddock, 1996; Rizzuto *et al.*, 1998).

II. Preparations Used in Confocal Microscopy

To date, several cell types have been used routinely for confocal microscopy studies, including cells from heart, vascular smooth muscle (VSM), vascular endothelium, nerve, bone, T lymphocytes, liver, *Xenopus* oocytes, fibroblasts, sea urchin eggs, and basophilic leukemia cells. Most of the work has been done using freshly isolated and/or cultured single cells either in primary culture or from cell lines. There is no doubt that the choice of preparation is crucial for obtaining satisfactory results. For example, to obtain sufficient details from 3D reconstructions, the following criteria should be taken into account: (1) isolated cells should be plated on glass coverslip (plastic should be avoided); (2) cells should be attached to avoid displacement during serial sectioning; (3) serial sectioning should be taken at rest or when the steady-state effect of a compound is reached; (4) to eliminate electrical and chemical coupling between cells, cells should be grown at low density; and (5) cells should be relatively thick and not flattened to allow a minimum of 75–150 serial sections (according to cell type). This could be done by using the cells as soon as they are attached and/or by lowering serum concentration. (6) When cell lines or proliferative cells are used, it is recommended to replace the culture medium overnight by a medium free of serum and growth factors prior to starting the experiments. This will allow the cells to settle in a latent phase of mitosis and will increase the accessibility of receptors and channels for agonist and antagonist actions by washing out the effects of compounds present in the culture medium. (7) Using specific fluorescent markers, the origin, purity, and phenotype of the cells should be checked continually. Because ions and other free functional components of the cells are not distributed homogeneously and because the increase or decrease of the probed free component seems to be faster than computer sampling and acquisition, it is difficult to assess these fast-occurring phenomena at the 3D level. Hence, large rapid scans of the area under interest could be done. (8) Direct access to the nuclei can be attained by using compounds that perforate the cell membrane such as low concentration of triton or ionomycin (Bkaily *et al.*, 1997a,b, 2004, 2006, 2009). This technique permits access to the nucleus without breaking the cytoskeleton and other intracellular components. Furthermore, nuclei can be freed from the cytosol by mechanical or combined enzymatic–mechanical methods (Bkaily *et al.*, 2004). The isolated nuclei are useful for studying not only nuclear membranes proteins presence and function without any cell membrane and cytosolic contribution, but it also permits the delineation and studying of the nuclear envelope, by labeling the nucleoplasm with Syto 11 (Bkaily *et al.*, 2004).

III. Settings of Confocal Microscope

There is no doubt that one of the major elements in confocal microscopy is the setting of optimal parameter conditions. This should be determined prior to initiating any serious experiments and should be captured and maintained for each specific fluorescent probe. This will allow a realistic comparison between results obtained using the same fluorescent probe. Table I shows typical probe data and optical settings for a Molecular Dynamics (Sunnyvale, CA and BIO-RAD (MRC1024) (Thrnwood, NY)) Multi Probe confocal argon laser scanning (CSLM) system equipped with a Nikon epifluorescence inverted microscope and a $60\times$ (1.4 NA) Nikon Oil Plan achromat objective. These optical settings are also valid with other confocal microscopes. When the argon or krypton-argon laser is used, the 488-nm (9.0–12.0 mV) or 514-nm laser line (15–20 mV) is directed to the sample via the 510- or 535-nm primary dichroic filter and attenuated with a 1–3% neutral density filter to reduce photobleaching of the fluorescent probe. The pinhole size could be set at 50 nm for structural determination and 100 nm for ion-specific dyes. In most cases, the image size is set at 512×512 pixels with pixel size of 0.08 µm for small ovoid shaped cells and 0.034 µm for elongated cells. In order to optimize and validate fluorescence intensity measurements, laser line intensity, photometric gain, photomultiplier tube (PMT) settings, and filter attenuation should be determined and kept rigorously constant throughout the duration of the experimental procedures.

IV. Organelle Fluorescent Probes

Several commercially available dyes permit the localization and delimitation of several organelles and membranes such as the plasma membrane (Di-8-ANEPPS), mitochondrial membrane (MitoFluor Green), endoplasmic reticulum (ER) $DiOC_6(3)$, Golgi apparatus (NBDC6-ceramide), and the nucleus (live nucleic acid Syto stains). Once the cultured cell type is available, these probes should be used initially in order to be familiar with their distribution and localization. Once this is done, these structure delimitation probes could be used at the end of each experiment. Other dyes can also be used to delimit organelle structure such as mitochondrial (JC-1) and plasma membrane potential dyes. The choice of the structural delimiting probe depends on the excitation and emission of the fluorescent probe used to study the functional element of the cells. Double and triple labeling of functional fluorescent probes can be performed if the two or three probes are excited and emit at different wavelengths. Also, certain functional probes such as membrane potential-sensitive dyes could be used simultaneously both as delimiting (structural) and functional probes in conjunction with other functional probes. Furthermore, plasma membrane potential-sensitive probes could be used to study cell coupling as well as voltage distribution and cell shortening (Bkaily *et al.*, 2009).

Table I
Probe Data Table and Optical Settings for Argon-Based CLSM

Probe/dye	Application	Ex_{max} (nm)	Em_{max} (nm)	Argon laser line (nm)	Primary beam splitter (nm)	Secondary beam splitter (nm)	Detector 1 or barrier[a] filter (nm)	Detector 2 filter
Fluo-3, AM	Ca^{2+} ion indicator	464	526	488	510	None	510	None
Fura Red, AM	Ca^{2+} ion indicator	458	600	488	535	None	595	None
Calcium Orange, AM	Ca^{2+} ion indicator	550	567	514	535	None	570	None
Sodium Green	Na^{2+} ion indicator	507	532	488	510	None	510	None
Amiloride	Na^+/H^+ pump inhibitor	488	420	488	510	None	510	None
SNARF-1, AM	pH indicator	488–530	Dual 587/636	488/514	535	595 DRLP	600	540 DF30
$DiOC_6(3)$	Endoplasmic reticulum dye	484	501	488	510	None	510	None
MitoFluor Green	Mitochondrion probe	490	516	488	510	None	510	None
BODIPY FL C_5-ceramide	Golgi dye	505	511	488	510	None	510	None
JC-1	Mitochondrial membrane potential dye	490	Dual 527/590	488	510	565 DRLP	570	530 DF30
Di-8-ANEPPS	Cell membrane	481	605	488	510	None	>570	None
Syto 11	Live nucleic acid stain	500–510	525–531	488	510	None	510	None
Angiotensin II-FITC	Angiotensin-binding sites	494	520	488	510	None	510	None
Ang II-TRITC[b]	Angiotensin-binding sites	ND	ND	514	535	None	535	None
PYY-FITC	PYY-binding sites	ND	ND	488	510	None	510	None

[a] If the secondary beam splitter is absent, then value indicated applies to barrier filter.
[b] Custom-prepared.

V. Volume Rendering and Nuclear Calcium, Sodium, and pH Measurements

Once the scanned images are obtained, they can be transferred onto a workstation similar to a Silicon Graphics workstation equipped with Imagespace analysis and volume workbench software modules (Molecular Dynamics). Reconstruction of 3D images of ion-selective dyes are then performed on unfiltered serial sections and are represented as closest intensity projections for calcium, sodium, or pH distribution and look-through extended focus projections for both nucleus and calcium (or Na^+ or pH) colocalization studies. 3D reconstructions of cell organelles can be performed on Gaussian- or median-filtered serial sections. The look-through method depicts a specimen as if it were transparent while being in focus throughout its depth. It is useful for studying both surface and interior features and their relative positions. Closest intensity projections, however, produce a stack of opaque images depicting actual voxel intensities. Calcium, sodium, or pH (or any functional dye) images can be represented as pseudocolored representations according to an intensity scale of 0–255 with lowest intensity in black and highest intensity in white. As a rule, pseudoscales should be included with all images illustrating pseudocolored representations of intensity levels.

Measurement of uptake of calcium, sodium, or any free element of the cell within the nucleus can be performed on both 2D images (individual sections) and 3D reconstructs (section series). The nuclear area following Syto 11 staining can be isolated from the rest of the cell by setting a lower intensity threshold filter to confine relevant pixels. A 3D binary image series of the nuclear volume can then be generated for each cell using the exact same x, y, and z set planes as those used, for example, during calcium ion uptake (Fig. 1A, C, and D). By applying these binary image patterns of the nucleus to the same cell initially labeled for calcium, Na^+, or pH (the binary image serves as a "cookie cutter"), a new 3D projection is created depicting fluorescence intensity levels exclusively within the nucleus. Hence, by "removing" the nucleus from the surrounding cytoplasm, we are then able to measure calcium, Na^+, or pH intensity values in the entire nuclear volume and express it in $microm^3$ while eliminating the possible contribution of the perinuclear space to the measurements.

VI. Rapid Scan Imaging

Rapid line scan imaging can be used to monitor temporal oscillations of a functional probe such as Ca^{2+} dye during the spontaneous contraction or cytosolic, nuclear, and mitochondrial Ca^{2+} oscillation and/or release. Contracting cells or cytosolic and nuclear Ca^{2+} oscillations can be recorded easily at a rate of 330 ms/ scan (3 scans/s) for a total of up to 1000 frames. Each frame, which consists of 32 lines/scan (512 pixels) and a pixel size of 0.42–0.64 μm, enables the visualization of

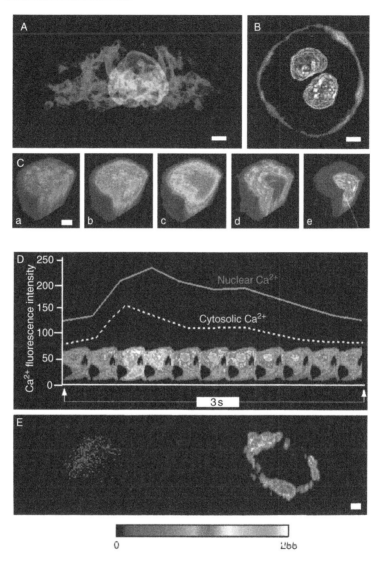

Fig. 1 (A) MitoFluor Green-stained mitochondria (green) followed by Syto 11 staining of the nucleus (red) in an embryonic chick heart myocyte. Stained mitochondria form a ribbon-like mass surrounding the nucleus. Images were generated from 40 serial sections (0.08 μm) at a step size of 0.49 μm. Bar: 2 μm. (B) Di-8-ANEPPS potentiometric dye labeling of plasma membrane potential in an embryonic chick myocyte (pseudocolored) followed by Syto 11 staining of the bilobed nucleus (gray scale). Transmembrane voltage staining appears distributed unevenly around the cell surface with no internalization after 30 min. Bar: 2 μm. (C) Sagittal and cross-sectional view of a 3D-reconstructed perforated plasma membrane ventricular myocyte illustrating intracellular calcium distribution in response to increasing concentrations of extranuclear calcium. Coverslips were mounted in the bath chamber and, after loading with 13.5 μM Fluo-3 AM, cells were bathed in a buffered saline solution containing 100 nM

the entire cell throughout the contractile process with good spatial and temporal resolution. However, if faster acquisition times are needed, line scanning can be performed at a rate of up to 10 ms per image using the Multi Probe CSLM (Molecular Dynamics or BIO-RAD). Figure 1D shows an example.

VII. Coculture of Cells

The reconstitution of interaction between two cell types is an excellent preparation for confocal microscopic studies. For example, single vascular endothelial and smooth muscle cells can be used in coculture conditions.

Because vascular endothelial cells grow less rapidly in culture than VSM cells, vascular endothelial cells are first cultured in the middle of a sterile coverslip using a custom-made watertight compartment 9 mm in diameter. Once the endothelial cells have attached and reached confluence (or subconfluence), the compartment can be removed, leaving a confined area of endothelial cells on the glass lamella. Then, VSM cells can be added in an area adjacent to that of the endothelial cells. On reaching confluence, the VSM cells establish side-to-side and/or overlapping contact with endothelial cells. Once this is achieved, the cocultures can be loaded with Fluo-3 (or other fluorescent probes) for 3D Ca^{2+} measurement using confocal microscopy. An alternate culture method can also be used. This method consists of plating a drop of high-density endothelial cells in the middle of the lamella and

---

of Ca^{2+}. Following rapid perforation of the cell with ionomycin (1–2 min), the cells were then washed quickly and stabilized for 5 min in buffer solution without Ca^{2+}. Image acquisitions were performed at the following settings: 488-nm excitation and 510-nm emission wavelengths, laser power at 9 mV, 3% attenuation filter, and PMT set at 700. Z-axis serial sections (0.08 μm) were performed at a step size of 0.9 μm. Nine identical vertical serial series were obtained 2 min after the incremental addition of 100 nM (A), 700 nM (B), and 1500 nM (C) extranuclear Ca^{2+}. Images are shown as pseudocolored representations according to an intensity scale from 0 to 255 (bottom). At the end of the last concentration of the ion, mitochondria were stained with MitoFluor Green (D) followed by Syto 11 staining of the nucleus (E) for 8–10 min and scanned under identical conditions. The reconstructed images depict a gradual increase of $[Ca^{2+}]_i$ within the mitochondria and the nucleus with increasing concentrations of external calcium. Bar: 2 μm. (D) Time-lapse rapid scans and graphic representations of cytosolic and nuclear-free Ca^{2+} variations of spontaneously contracting 10-day-old chick embryonic ventricular myocytes loaded with 13.5 μM Fluo-3 AM. Whole cell images were generated continuously every 330 ms for 3 min at a resolution of 512 × 512 pixels and 0.17 μm pixel size. Total time span: 3 s. Images are shown as pseudocolored representations according to an intensity scale from 0 to 255. (E) *Left*: Cultured left ventricular endocardial endothelial cells were incubated with 10^{-7} M fluorescein-conjugated human PYY following a 20-min preincubation with a 100-fold excess of cold human PYY. The image represents a sequential rendering of 12 optical sections (step size: 0.52 μm; pixel size: 0.08 μm). Only very little diffuse nonspecific labeling is observed. *Right*: Three-dimensional reconstruction of a labeled ventricular cultured endocardial endothelial cell following a 2-h incubation with fluorescein-conjugated human PYY at room temperature. The image represents a slightly angled 3D rendering of a cell generated from eight optical sections (step size: 0.6 μm; pixel: 0.17 μm) and viewed as a look-through projection. Note the distribution of PYY labeling along the circumference of the cell. (See Plate no.7 in the Color Plate Section.)

allowing the cells to attach overnight. The next day, the culture dish is washed gently and fresh culture medium containing a high density of VSM cells is added to the culture dish containing the attached endothelial cells. This simple technique allows for a better study of the development of cell-to-cell contact between vascular endothelial and smooth muscle cells as a function of time and can be used within an hour or more. Although this technique has a disadvantage in that cell contacts are generated at random, it may, however, more adequately represent a nonhomogeneous organizational contact between two different cell types, such as what we observed in certain pathological conditions and healing processes following surgery or vascular bypass. Also, an interaction between cells of the immune system and endothelial system or at any other cell type could be done in a similar fashion with the exception that the immune cells in suspension should be loaded with the dye, collected by centrifugation, and then added to the attached preloaded monolayer cells.

VIII. Loading of Fluorescent Probes for Confocal Microscopy

A. Cytosolic and Nuclear Ca^{2+} Measurements by Confocal Microscopy: Fluo-3 and Fluo-4

Freshly cultured single cells and cells in primary cultures are excellent preparations for studying cytosolic and nuclear-free Ca^{2+} using 3D confocal microscopy. Several visible wavelength Ca^{2+} dyes were initially tested to select the ion indicator best suited for laser-based confocal microscopy. Of the various Ca^{2+} dyes available, we tested Fluo-3, Fluo-4, Fura Red, Calcium Green-l, and Calcium Orange (Molecular Probes, Eugene, OR). Calcium Orange was the least successful with respect to loading and calcium fluorescence. The dye was very difficult to load in all cell preparations tested regardless of temperature, time of incubation, or dye concentration. Moreover, the addition of Pluronic F-127, a dispersing agent used to aid solubilization, did not improve loading of the dye. Calcium Green-1 loaded reasonably well and displayed good baseline fluorescence. However, in most preparations tested, very little or no Ca^{2+} response was observed in the presence of a number of pharmacological agents known to increase intracellular Ca^{2+}. Fura Red, a long wavelength analog of Fura 2 with a high emission spectrum, was tested equally. Loading was performed at room temperature for 45 min at a concentration of 20 μM. At baseline calcium levels, labeling was relatively uniform in the cytoplasm and nucleus in heart cells but appeared slightly higher in the nucleus of vascular endothelial and smooth muscle cells. Probe fluorescence appeared relatively stable and showed little photobleaching during our experiments. However, one of the singular characteristics of Fura Red is that its fluorescence emission decreases on Ca^{2+} binding in contrast to other indicators where emission intensity increases with Ca^{2+} binding. This particular property of Fura Red allows its use in combinations with Fluo-3 for dual emission ratiometric measurements, thus enabling the expression of fluorescence intensity in terms of Ca^{2+} concentration.

There are some drawbacks; however, this dye, being much weaker than other visible wavelength calcium indicators, must be loaded at higher concentrations. Moreover, the intracellular distribution of both indicators must be identical for ratiometric measurements to be valid. Finally, in some of our experiments involving dual labeling with Fura Red and Fluo-3, both fluorescent dyes increased in response to depolarization of the cell membrane or pharmacological stimulation, making Fura Red unreliable, at least with some cell preparations. Among all Ca^{2+} dyes tested, Fluo-3 and Fluo-4 provided the best overall loading and fluorescence features in all our cell preparations. Both Fluo-3 and Fluo-4 was loaded easily, and was distributed homogeneously within the cytosol and the nucleus, with low photobleaching at a laser attenuation setting of 3%. More importantly, calcium responses to both electrical and pharmacological stimuli were comparable to those using Fura-2 in classical Ca^{2+} imaging studies. Hence, this dye was used in the majority of our calcium experiments with the confocal microscope in intact (Fig. 1D) and perforated plasma membrane (Fig. 1C) of many cell types (Bkaily *et al.*, 1997b).

Single cells from heart, aortic VSM, T lymphocytes, osteoblasts, or endothelium can be cultured on 25-mm-diameter glass coverslips that fit a 1-ml bath chamber. The cells are then washed three times with 2 ml of a balanced Tyrode's salt solution containing 5 mM HEPES, 136 mM NaCl, 2.7 mM KCl, 1 mM $MgCl_2$, 1.9 mM $CaCl_2$, and 5.6 mM glucose, buffered to pH 7.4 with Tris base, and supplemented with 0.1% bovine serum albumin (BSA). The osmolarity of the buffer solution with or without BSA should be adjusted with sucrose to 310 mOsm. The Fluo-3/AM or Fluo-4/AM probe (Molecular Probes) or other Ca^{2+} probes should be diluted in Tyrode–BSA from frozen 1 mM stocks in dimethyl sulfoxide (DMSO) to a final concentration of 6.5 or 13.6 μM. Cells are then loaded by placing the coverslips cell side down on a 50–100 μl drop of diluted probe on a sheet of parafilm stretched over a glass plate and incubated for 45–60 min in the dark at room temperature. The inverted coverslip method offers considerable savings with regard to the amount of probe used, especially at high concentrations. More importantly, the use of smaller aliquots of concentrated stock solution needed to prepare the final probe concentration reduces the percentage of DMSO content in the incubation medium. We found that cell blebbing and photobleaching were encountered commonly with final DMSO concentrations greater than 0.6%. Loading should also be performed in an humidified environment when inverting coverslips on less than 10 μl to avoid evaporation problems. After the loading period, coverslips should be recovered carefully and the cells washed twice with Tyrode–BSA buffer and twice in Tyrode buffer alone. Loaded cells should then be left to hydrolyze for an additional 15 min to ensure complete hydrolysis of acetoxymethyl ester groups. Optical settings for Fluo-3/AM are shown in Table I. To the contrary of what was claimed by one group, Fluo-3, as well as Fluo-4, is homogeneously distributed in the cytosol and the nucleus. The mobility and loading of these probes does not depend on the viscosity of the nucleus, but probably on species differences (Bkaily *et al.*, 2003). At least in our experience, loading with all ionic probes described in this chapter showed homogeneous distribution of the dye in both the cytosol and the nucleus of all cell types including human, rabbit, rat, hamster VSM cells; human, rabbit, rat, hamster vascular endothelial cells; human, rabbit, mouse endocardial endothelium; human, chick, rat, hamster, mouse cardiomyocytes; and human and rat hepatocytes.

B. Cytosolic and Nuclear Na$^+$ Measurement by Confocal Microscopy: Sodium Green/AM

Sodium Green/AM is the only argon laser excitable Na$^+$ indicator available commercially. This sodium ion probe exhibits a 41-fold selectivity for Na$^+$ over K$^+$ and, on binding sodium, increases its fluorescence emission intensity. Loading of Sodium Green into heart and vascular cells is difficult unless coupled with Pluronic F-127 to increase aqueous solubility. When well loaded, the dye is relatively stable, although slightly more subject to photobleaching than Fluo-3. This can be minimized by setting the laser attenuation filter down to 1%.

Single cells can be loaded with 13.5 μM Sodium Green (Molecular Probes) similarly to that described earlier for Fluo-3 with one modification: the Na$^+$-sensitive dye should be loaded in the presence of Pluronic F-127 (20%, w/v). When used alone, Sodium Green loaded poorly into cells. However, when reconstituted initially in equal volumes of DMSO and Pluronic acid and then diluted into the loading buffer (final concentration of 0.1% detergent); the probe offered good fluorescence intensity. Moreover, it was found that the loading temperatures should not exceed 19–29 °C as loading tended to be inconsistent and often compartmentalized. Optical settings for Sodium Green/AM are given in Table I.

After loading the ion probe for 30 min at 19 °C, and if heart cells are to be used, one can see that the nuclear Na$^+$ level is similar to that in the cytosol. However, if resting aortic vascular endothelial and smooth muscle cells are to be used, one can see that the level of cytosolic Na$^+$ is slightly greater in the immediate cytosolic zone surrounding the nucleus. No information is available concerning a possible role of Na$^+$ in nuclear function or whether the nucleus may play a role in cytosolic Na$^+$ buffering. The mechanism by which Na$^+$ crosses the nuclear membrane, as well as its physiological role in the nucleus, if any, is not known. However, it has been demonstrated that confocal microscopy can be highly useful in determining the physiological role of Na$^+$ and the modulation of this ion at the cytosolic and nuclear levels as well as in subcellular organelles (Bkaily et al., 1997a).

C. Intracellular pH Measurement by Confocal Microscopy: Carboxy SNARF-1/AM

Several long wavelength pH fluorescent dyes have been developed and are available commercially such as BCECF, SNARF-1, SNARL, DMNERF, CL-NERF, and HPTS. The use of an argon-based confocal microscope limits the use of all these pH indicators. This limitation can be overcome by modifying the laser capabilities into a multiline argon/UV laser or an argon/krypton laser.

Carboxy SNARF-1, however, is well suited for all types of laser instrumentation and is excited efficiently by both 488- and 514-nm argon laser lines. The emission spectrum of the dye undergoes a pH-dependent wavelength shift, thus allowing the ratio of fluorescence intensities at two emission wavelengths, typically 580 and 640 nm, and hence can be used for more accurate determinations of pH. This pH dye has been applied extensively in many cell types (Chacon *et al.*, 1994; Dunn *et al.*, 1994; Lemasters *et al.*, 1993). The dye has a pK_a of approximately 7.5, making it well suited to cytoplasmic pH measurements (Dunn *et al.*, 1994). It is reported to be less susceptible to photobleaching and exhibits a good pH-dependent shift of its fluorescence emission spectrum (Dunn *et al.*, 1994). Using confocal microscopy, studies on single cells loaded with SNARF-1 suggest that subsarcolemmal and nuclear areas have a pH value near 7.2, but for regions corresponding to the distribution of mitochondria, the pH is 7.8–8.2 (Chacon *et al.*, 1994; Opitz *et al.*, 1994). This raises the possibility that cytosolic pH is not homogeneous and is a function of the location of organelles within the cytosol. It is also possible that pH_i measurements near the plasma (or nuclear and mitochondrial) membrane may lead to erroneously low pH_i values due to the pronounced spectral contribution of the enriched protonated indicator component at its associated apparent alkaline pK_i shift (Opitz *et al.*, 1994). Moreover, pH_i measurement with H^+-sensitive dye may depend on the cell type used, isolation, and/or culture conditions, as well as the degree of metabolic activity. To evaluate intracellular pH changes, cells can be loaded with the ratiometric pH indicator Carboxy SNARF-1/AM, using the inverted coverslip method. Cells should be incubated in darkness for 30 min at ambient temperature with 5 µM of freshly prepared indicator dye in Tyrode–BSA buffer from 1 mM stock solutions reconstituted in DMSO. Following the loading period, cells should be washed and left to hydrolyze for 15 min as described for the Ca^{2+} dyes. Coverslips are then placed in a 25-mm (1 ml) bath chamber for visualization with the confocal microscope. Optical settings for SNARF-1/AM are given in Table I. The use of SNARF-1 dye in chick embryo heart cells revealed a sometimes heterogeneous, sometimes homogeneous pH_i distribution in the cytosol and the nucleus. In some instances, labeling appeared to be more intense at the point of attachment of the cell to the coverslip than at other cell periphery. In heart and vascular cells, the indicator responded well to changes in extracellular pH from 6.5 to 8.5. The dye reached an isosbestic point at an external pH of 6.0 where no difference could be seen between acidic and basic measurements. Also, the distribution of pH levels in heart cells is relatively homogeneous within the cytosol and the nucleus. However, in human vascular aortic smooth muscle cells, the distribution of pH level in the cytosol is less homogeneous and appears to be much more basic within and in the immediate vicinity of the nucleus as compared to cytosolic pH. This was particularly evident when lowering extracellular pH. We should caution that expressing pH_i by quantitative values may lead to significant errors in comparison to qualitative (fluorescence intensity) or ratiometric measurements (Opitz *et al.*, 1994). Care should also be taken even when expressing pH_i in ratiometric terms because of possible intracellular redistribution of the indicator between cytosol and lipophilic cell compartments (Opitz *et al.*, 1994).

D. Endoplasmic Reticulum: Carbocyanine $DiOC_6(3)$ for Endoplasmic Reticulum Staining

Several short-chain carbocyanine dyes such as $DiOC_6(3)$ and $DiOC_5(3)$ and long-chain carbocyanines such as $DIIC_{18}(3)$ have been widely used to visualize the endoplasmic/sarcoplasmic reticulum (ER/SR) (Terasaki *et al.*, 1984; Wadkins and Houghton, 1995). These probes pass through the plasma membrane easily, and while some carbocyanine probes have been reported to stain several other intracellular membranes, such as Golgi and mitochondria, ER membranes are labeled preferentially when used at higher concentrations and are easily distinguishable by their characteristic morphology.

Sarcoplasmic/endoplasmic reticulum can be visualized using the short-chain carbocyanine $DiOC_6(3)$ dye from Molecular Probes. Cells grown on coverslips can be placed in a 25-mm bath chamber and bathed in 1 ml of Tyrode buffer solution. Freshly prepared ER dye should be added at a final concentration of 50 nM obtained after serial dilutions in Tyrode–BSA loading medium from 10 mM stock solutions in DMSO. Labeling is very rapid (within 3–5 min). In some instances, ER staining can be followed by labeling of the nucleus with Syto 11 as described later. This stepwise dual-labeling technique enables the localization of both cell organelles using the same excitation wavelengths. Optical settings for $DiOC_6(3)$ are given in Table I.

The $DiOC_5$ and $DiOC_6$ probes were tested in isolated vascular endothelial and smooth muscle cells as well as in embryonic heart myocytes. The SR can be recognized easily by its filament-like morphology. In 3D reconstructions, the SR appears as leaf-like undulations in heart cells or tubular structures in VSM cells. $DiOC_6$ staining in these cell preparations, when used at concentrations of 10 mM, is more consistent and more intense than that of its equivalent $DiOC_6$. This ER staining can also be used jointly with plasma membrane markers or nuclear stains to evaluate intracellular distribution of the organelle or its spatial relationship with the nucleus. Unfortunately, the use of these carbocyanine probes, apart from being toxic at high concentrations, cannot be used simultaneously (in double-labeling experiments) with standard long wavelength Ca^{2+} or Na^+ indicators such as Fluo-3, Calcium Green, or Sodium Green because of similar excitatory and emission wavelengths (488 nm excitation, 500–530 nm emission range). If one wishes to associate cell function or ionic responses with ER membranes using these particular probes, one approach is to first complete calcium or sodium experiments and then label the organelle with $DiOC_6$ stain to visualize ER localization. Several washes in low calcium buffer to reduce Ca^{2+} content followed by ER staining with 10–50 mM $DiOC_6$ stain produces a signal sufficiently strong, allowing to reduce the gain (2 to 1×), attenuation filter (3–1%), and/or PMT voltage to levels below Ca^{2+} fluorescence detection. The latter staining, being much stronger than the ionic markers, thus enables subsequent correlation between structure and function.

E. Mitochondria: MitoFluor/AM

The new MitoFluor mitochondrion-selective probes developed by scientists at Molecular Probes are novel mitochondrial stains that appear to accumulate preferentially in the organelle regardless of its membrane potential (Molecular Probes Handbook). A 1 mM MitoFluor stock solution in DMSO should be freshly diluted to a final concentration of 50–100 nM in Tyrode–BSA buffer. Labeling can be performed directly in the bath chamber at room temperature in darkness. Staining becomes evident within 5 min after addition of the probe. Excess dye in the surrounding medium should be washed out after 5 min. As in the case of ER/SR staining, mitochondrial staining can be followed by nuclear labeling with Syto 11 and Fig. 1A shows an example. Optical settings for MitoFluor Green are given in Table I.

Figure 1 demonstrates an example of mitochondrial labeling in embryonic chick ventricular myocytes. As evidenced in Fig. 1, fetal heart cells are rich in mitochondria surrounding the nucleus. These organelles may contribute substantially to total measured intracellular Ca^{2+} and their contribution to cytosolic Ca^{2+} buffering should be taken into consideration.

F. Measurement of Membrane Potential Using Confocal Microscopy: Di-8-ANEPPS and JC-1 Probes

Since the late 1960s, the measurement of fluctuations in membrane potential using fluorescent indicators and the search for specific voltage-sensitive dyes with an acceptable signal-to-noise ratio has sparked the development of several commercially available dyes such as thiazole orange, di-0-Cn(3), tetramethylrhodamine ethyl ester, bisaxonal, RH2g2, Di-8-ANEPPS, and the mitochondrial membrane dye JC-1 (Rohr and Salzberg, 1994; Smith, 1990). The Di-4-ANEPPS and Di-8-ANEPPS indicators belong to the class of fast potentiometric dyes of the styryl type. This family of indicators was found to be extremely effective optically and less phototoxic transducers of membrane potential (Rohr and Salzberg, 1994).

Optical imaging of plasma membrane potential can be performed in cells using the fast voltage-dependent potentiometric dye, Di-8-ANEPPS (Molecular Probes). After washings in Tyrode–BSA buffer, coverslip-plated cells are then loaded with the freshly prepared styryl dye (13.5 μM final concentration in Tyrode–BSA buffer) from DMSO-reconstituted stock solutions. Loading of the dye should be performed at room temperature in darkness. Labeling is very rapid, and strong fluorescence intensity is usually obtained within 3–5 min. Figure 1B and C shows examples.

The monomeric mitochondrial voltage-sensitive dye JC-1 can also be used to evaluate intracellular membrane potential responses to high extracellular KCl depolarization. Evaporated stocks of cationic JC-1 (Molecular Probes) can be reconstituted in 100% ethanol and further diluted to a working solution of 10 μg/ml in

HEPES–NaCl buffer. Loading should be performed directly into the experimental bath chamber at room temperature in darkness. Good fluorescence is achieved within 10 min after addition of the probe. Optical settings for Di-8-ANEPPS and JC-1 are given in Table I.

Di-8-ANEPPS is a hydrophobic compound that seems to anchor in the cell membrane and is more stable than other membrane voltage dyes with less leakage (Rohr and Salzberg, 1994). This dye was reported to offer better time recordings of transmembrane voltage changes (about 30 min) in cultured neonatal rat heart cells than other fluorescent voltage probes, before leakage into the cytosol. In experiments with embryonic chick heart single cells as well as human adult aortic vascular endothelial and VSM cells, Di-8-ANEPPS staining is stable for up to 1 h with only a diffuse halo of staining observed in a few cells after 30–60 min, but never in the perinuclear region. The quality of Di-8-ANEPPS membrane potential staining could be influenced by several factors, including cell type as well as species origin (Rohr and Salzberg, 1994). For example, the distribution of transsarcolemmal membrane potential in heart, endothelium, and VSM cells did not appear homogeneous. This irregular distribution of membrane potential in single cells could be due to scattered protein distribution on the sarcolemmal membrane, as was suggested for ion channels. This nonhomogeneity of labeling can be more apparent when single cells are depolarized with high extracellular K^+. Sustained depolarization of the human VSM cell with 30 mM $[K]_o$ rapidly (10 s) induced a nonhomogeneous depolarization of the cell membrane accompanied by cell contracture. This latter contracture remained as long as the cell membrane can be depolarized with 30 mM $[K]_o$.

J-aggregate formation has also been used to visualize mitochondria in a variety of cells (Rohr and Salzberg, 1994; Smiley et al., 1991). The mitochondrial voltage-sensitive probe JC-1, a specific energy potential-dependent mitochondrial dye, was reported to display a fairly narrow red peak that was sensitive to a variety of mitochondrial membrane potential modulating agents (Rohr and Salzberg, 1994). In preparations such as embryonic chick heart cells as well as in adult human aortic vascular endothelial and smooth muscle cells, mitochondrial membrane potentials are well labeled with the JC-1 indicator. Epifluorescence visualization reveals a decrease in green fluorescence (monomer) compared to orange-red fluorescence (JC-1 aggregate) on plasma membrane depolarization. The JC-1 probe does not appear to label the nuclear membrane.

G. Measurement of Receptor Density and Distribution Using Confocal Microscopy

Using 3D confocal microscopy and a Ca^{2+} channel fluorescent probe, as well as an Ang II fluorescent probe, showed that R-type Ca^{2+} channels as well as Ang II receptors are present on sarcolemmal as well as nuclear membranes of several cell types (Bkaily et al., 1997a,b; Paddock, 1996).

We also developed FITC- and BODIPY-conjugated ET-1 as well as Ang II fluorescent probes. Using the ET-1 fluorescent probe, ETA receptors were found to be localized at the plasma and nuclear membranes in nonhomogeneous cluster-like patterns.

An Ang II fluorescent probe was also developed in our laboratory. This probe was found to be more specific and more effective than the one available commercially. Also, the newly developed PYY fluorescent probe was found to be more effective than those available commercially and Fig. 1E shows an example. The optical settings for these probes are given in Table I.

For the time being, many commercially available receptors labeling dye tested did not show a satisfactory labeling of receptors. To develop an agonist or an antagonist-coupled fluorescent probe, the following procedures are recommended: (1) choose a ligand structure that permits attachment of a fluorescent probe without affecting the known active site of the structure; (2) choose a fluorescent probe that is highly stable and does not bind to any structure of the cells; (3) the fluorescent probe alone should be completely washable; (4) no single probe can be coupled with all different ligand types; (5) once the ligand is coupled to the fluorescent probe, the complex must be highly purified in order to isolate only the complex free of dye or ligand; (6) the ligand–dye complex should be tested in isolated cells or preferentially in tissue tension experiments in order to ensure that the ligand is still active and is as specific and powerful as the ligand alone; (7) ensure that the effect of the complex is washable and is antagonized by a well-known antagonist as well as being displaced by the "cold" ligand alone but not with the dye alone; and (8) the stability of the complex should be determined in solution as a function of time as well as a function of storage. It is highly recommended to use the ligand–fluorescent probe complex as soon as it is purified (i.e., within 2 days) and avoid storage if possible. The development of ligand–fluorescent probes, although sometimes difficult, is a much worthwhile venue.

IX. Conclusion

Confocal microscopy imaging studies in single cells and tissue sections confirm the importance of this noninvasive technique in the study of cell structure and function as well as the modulation of working living cells by various constituents of cell membranes, organelles, and cytosol.

The use of nonratiometric dyes does not allow adequate expression of intracellular Ca^{2+} concentrations in absolute values. Thus, care must also be taken when expressing results as either intensity or even ratio values. One approach in addressing these difficulties is by using different probes for the same ion in order to ensure that the results do not contain an artificial component due to loading artifacts. A complementary approach is by selecting cells that exhibit fluorescence intensities within fixed values and by determining the R_{min} and R_{max} values of these probes. In addition, in instances where chromophores are conjugated to a particular hormone or drug, steps including washout, use of cold or unlabeled hormones/drugs, and testing of free dye should be performed. Our results also reveal that confocal microscopy can be used successfully in coculture studies.

Photobleaching of the fluorophore can become a major obstacle in certain instances where the sample must be exposed to higher excitation light because of weak signal emission. In the case of mounted material, agents such as propyl gallate (Chang, 1994), Vectashield (Vector Laboratories, Burlingame, CA), and SlowFade (Molecular Probes) have been used to reduce light-induced photo-damage. In our laboratory, 0.1% phenylenediamine (Sigma, St. Louis, MO) in glycerol–PBS (9.1) and Vectashield are used routinely for mounted slide work. Protecting live material from photodynamic damage, however, is more difficult. Additives such as Oxyrase, ascorbic acid, and high doses of vitamin E have been described as having some antifading properties (Mikhailov and Gundersen, 1995).

With respect to differences in results that may be found between one laboratory and another, they may be due to (1) cell type and origin, (2) experimental conditions, (3) use of 3D measurements, and (4) precise determination of nuclear volume and localization.

Thus, confocal microscopy is extremely powerful in studying contraction, voltage, and chemical coupling between various cell types. One of the limitations, however, is the inability to record serial sections during quick cell responses such as those occurring during cell contraction. Rapid scans can only be performed on single vertical planes, thus limiting access to possible spatial differences in cellular response patterns. However, present results using fluorescent ion and organelle probes illustrate that the 3D study of cell response and function is not only feasible but also important in evaluating cell responses to external stimuli. Moreover, confocal microscopy is a highly useful tool in the assessment of cell pathology, whether it be in single cells or tissue sections. Site-selection probes such as receptor, protein, and second messenger probes, organelle probes, and nuclear stains all provide important indicators not only for the determination of actual structure and location of cell components but also for the study of subcellular distribution and movement of various ions and molecules in working living cells.

For the time being, fluorescence confocal microscopy cannot localize and quantify bound ionic elements of the cell. It is hoped that new technology will soon permit us to determine not only free ions but also bound and/or compartmentalized ions.

Finally, with the help of 3D imaging and volume-rendering capabilities, confocal microscopy constitutes a powerful state-of-the-art technique in the continuing investigation of cell structure and function in normal and pathological conditions.

Acknowledgments

This work was supported by CIHR, NSERC, and Quebec Heart Foundation to Dr. Ghassan Bkaily.

References

Bkaily, G., D'Odéans-Juste, P., Pothier, P., Calixto, J. B., and Yunes, R. (1997a). *Drug Dev. Res.* **42,** 211.

Bkaily, G., Pothier, P., D'Orléans-Juste, P., Simaan, M., Jacques, D., Jaalouk, D., Belzile, F., Hassan, G., Boutin, C., Haddad, G., and Neugebauer, W. (1997b). *Mol. Cell. Biochem.* **172,** 171.

Bkaily, G., Choufani, S., Sader, S., Jacques, D., D'Orléans-Juste, P., Nader, M., Kurban, G., and Kamal, M. (2003). *Can. J. Physiol. Pharmacol.* **81,** 654.

Bkaily, G., Nader, M., Avedanian, L., Jacques, D., Perrault, C., Abdel-Samad, D., D'Orléans-Juste, P., Gobeil, F., and Hazzouri, K. M. (2004). *Can. J. Physiol. Pharmacol.* **82,** 805.

Bkaily, G., Nader, M., Avedanian, L., Choufani, S., Jacques, D., D'Orléans-Juste, P., Gobeil, F., Chemtob, S., and Al-Khoury, J. (2006). *Can. J. Physiol. Pharmacol.* **84,** 431.

Bkaily, G., Avedanian, L., and Jacques, D. (2009). *Can. J. Physiol. Pharmacol.* **87,** 108.

Chacon, E., Reece, J. M., Nieminen, A. L., Zahrebelski, G., Herman, B., and Lemasters, J. J. (1994). *Biophys. J.* **66,** 942.

Chang, H. (1994). *J. Immunol. Methods* **176,** 235.

Dunn, K. W., Mayor, S., Myers, J. N., and Maxfield, F. R. (1994). *FASEB J.* **8,** 573.

Lemasters, J. J., Chacon, E., Zahrebelski, G., Reece, J. M., and Nieminen, A. L. (1993). *In* "Optical Microscopy: Emerging Methods and Applications," (B. Herman, and J. J. Lemasters, eds.), p. 339. Academic Press, New York.

Mikhailov, A. V., and Gundersen, G. G. (1995). *Cell Motil. Cytoskeleton* **32,** 173.

Opitz, N., Merten, E., and Acker, H. (1994). *Pflug. Arch.* **427,** 332.

Paddock, S. W. (1996). *Proc. Soc. Exp. Med.* **213,** 24.

Rizzuto, R., Carrington, W., and Tuft, R. A. (1998). *Trends Cell Biol.* **8,** 288.

Rohr, S., and Salzberg, B. M. (1994). *Biophys. J.* **67,** 1301.

Smiley, S. T., Reers, M., Mottola-Hartshorn, C., Lin, M., Chen, A., Smith, T. W., Steele, G. D., and Chen, L. B. (1991). *Proc. Natl. Acad. Sci. USA* **88,** 3671.

Smith, J. C. (1990). *Biochim. Biophys. Acta* **1016,** 1.

Terasaki, M., Song, J., Wong, J. R., Weiss, M. J., and Chen, L. B. (1984). *Cell* **38,** 101.

Wadkins, R. M., and Houghton, P. J. (1995). *Biochemistry* **34,** 3858.

CHAPTER 8

Combining Laser Scanning Confocal Microscopy and Electron Microscopy

Xuejun Sun

Department of Oncology
Cross Cancer Institute
University of Alberta
Edmonton, Alberta, Canada

I. Update

Combining light microscopy (LM) with transmission electron microscopy (TEM) details is becoming an important tool for biomedical research. The versatility of LM with large variety of (genetically encoded) markers allows observation of cellular processes in living status. Transmission electron microscope offers resolution down to nanometer range with detailed view of cellular architecture. Although the "super-resolution" LM developed in recent years (e.g., STED, PALM, STORM, structured illumination, etc.; for a review, see Huang *et al.*, 2009) offers light microscopic resolution down to as low as 20 nm level, this resolution is still far worse than TEM can offer and, most importantly, the LM super-resolution is only achieved with limited cellular architecture context as all these techniques are based on fluorescence and it is limited to one or few labels. On the other hand, TEM offers details view of cellular architecture with much more information than a typical light microscopic image offers. Association of LM information with TEM details provides extremely valuable insights into the functions of the labeled molecules. This combination of LM and TEM is, however, challenging because of mainly two obstacles: (a) the incompatibility of the labels for two kinds of microscopy and (b) the incompatibility of the tissue processing methods for LM and TEM. Beside the traditional photooxidation method which is widely used in correlative LM–TEM microscopy, the first obstacle is, to some degree, solved with the development of FluoroNanogold (Powell *et al.*, 1997) and quantum dots (Chan and Nie, 1998). These markers are both fluorescence and electron dense which allows the observations of the very same label at both LM and TEM level. Both kinds of markers retain the fluorescence even with TEM tissue processing. Additionally, quantum dots with different emission wavelengths are of different morphology and sizes which allows investigating more than one type of molecules in a specimen at TEM level (Giepmans *et al.*, 2005). However, the incompatibility of the tissue processing for TEM and LM still remains the main obstacle for the successful application of such correlative LM–TEM technique. One approach is to use cryofixation and freeze-substitution which is compatible with both LM and TEM (for a review, see McDonald, 2009). This requires specialized equipments and skills. For routine correlative LM and TEM applications, tissue processing is still usually achieved by finding an acceptable compromising point for both LM and TEM. These involve typically some combination of lowering level of glutaraldehyde in the fixative, a reduction in postfixation with osmication, etc. The methods summarized in this chapter have not been changed since its first publications. However, it is still applicable as some guidelines for combining LM and TEM tissue processing.

II. Introduction

Laser scanning confocal microscopy (confocal microscopy) provides an exciting new tool for biological research. For the first time, it is possible to obtain clear optical sections of relatively thick specimens without physically sectioning the tissue. Additionally, because of the high z-axis resolution of preregistered images it provides, confocal microscopy enables rapid three-dimensional (3D) reconstruction of specimens. The combination of confocal microscopy with powerful computer image-processing techniques gives unprecedented possibilities for visualizing and measuring stained cells in three dimensions (Chen *et al.*, 1995). The technology has gained wide acceptance in biological research.

Although it is true that confocal microscopy provides better resolution than conventional light microscopy, it is still fundamentally light microscopy, limited in resolution by the wavelength of the light source and the numerical aperture of the lens. This is at best about 150 nm with current optics. In many fields of biological research, one needs subcellular resolution better than 150 nm. For example, the establishment of an association of a gene product with a type of organelle leads to important insights into the role of the molecule in the cell. For subcellular study, electron microscopy (EM) is still an indispensable tool for the unambiguous identification of organelles. EM provides a resolution that cannot be matched by any other imaging tool currently available. EM, however, suffers from a lack of 3D information. Because of the poor penetrating power of electrons, observation with EM is often limited to a fraction (nanometer range) of the cellular structure. High-voltage EM provides an alternative that yields both high resolution and some degree of 3D information. It is, however, not readily accessible to most biologists and even with high-voltage EM, specimens no thicker than 1 μm can be observed. Three-dimensional reconstruction is still required to have a 3D view of cells.

Neuroanatomy provides an extreme situation in which 3D information is particularly important. Neurons often are large and have complicated 3D structure. The intrinsic 3D morphology reflects the connections of neurons within a neuronal network. Yet, ultimately, the discrete synaptic connections, and the neurotransmitter(s) used by them, rule the exact roles a particular neuron plays within the network. Study of stained neurons at the level of light microscopy often pinpoints particular loci of interest. Whether there are synaptic connections in the pinpointed areas provides necessary information for understanding the neuron and the circuit. Assessment of the 3D morphology of neurons requires light microscopic study whereas synaptic information is available only through ultrastructural study. Therefore, many investigators have attempted to combine EM with LM (DeFelipe and Fairén, 1993; Hama *et al.*, 1994; Stevens *et al.*, 1980; Ware and LoPresti, 1975). All currently available methods require a certain degree of 3D reconstruction, which, even with advanced computer imaging technology, is still labor-intensive and sometimes impossible for large neurons.

The combination of confocal microscopy and EM is a promising avenue for biologists attempting to solve this problem of 2D versus 3D. Confocal microscopy can provide a valuable overview and 3D information. EM provides ultrastructural detail. A combination of these two technologies is, however, not straightforward because the two technologies require very different tissue processing procedures and staining. First, for confocal microscopy, a specimen must be stained with a fluorescent marker, which is not necessarily electron dense and therefore not suitable for EM study. Second, for confocal observations, the tissue must be translucent, whereas for EM, heavy metals are used that are opaque to visible light. Third, for EM observation, tissue is usually embedded in plastic. This often introduces a large amount of tissue shrinkage, autofluorescence and distortion, which render the correlation of light microscopic information with ultrastructural data difficult. A combination of confocal microscopy and EM has been attempted using the reflection mode of imaging with the confocal microscope (Deitch *et al.*, 1990). With this method, a cell labeled with an electron-dense marker can be imaged in the confocal microscope by bouncing light off of it. This method, however, is not widely applied due to the generally poor quality of reflection in confocal images.

We have been studying the neuronal connections within insect central nervous systems by combining standard confocal microscopy with EM (Sun *et al.*, 1995, 1997, 1998; Tolbert *et al.*, 1996). The two methods described in this chapter rely on a very specific interaction between biotin and avidin. Although the methods were developed primarily for neuroanatomy with biotin-stained neurons, it can be adapted to other tissues to localize various substances within cell using biotinylated antibodies or DNA/RNA probes, as shown in this chapter with an antibody against serotonin, a neuromodulator in insects (Mercer *et al.*, 1995, 1996).

III. Experimental Procedures

A. Overview of Methods

The methods require the introduction of biotin or biotin derivatives into cells. The biotin can subsequently be detected using a fluorescent tag for confocal microscopy and an electron-dense tag for EM observation. Confocal microscopy can provide a guiding tool for thin sectioning to facilitate the correlation of confocal information with EM results. Both methods involve these basic steps of introduction of biotin, introduction of appropriate tags for the biotin, confocal observation, and EM observation.

Both methods have been applied on insect central nervous tissue. Adult central nervous systems of two insect species were used: the moth *Manduca sexta* (Lepidoptera, Sphingidae) and the white-eyed mutant housefly *Musca domestica*

(Diptera, Muscidae). The detailed rearing procedures have been described previously (for moth, see Bell and Joachim, 1976; for fly, see Fröhlich and Meinertzhagen, 1982).

IV. Method 1: Immunogold Fluorescence Method

This method takes advantage of the finding that immunogold reagents do not bind to all the available biotin in biotin-stained cells (Sun *et al.*, 1995). Therefore, the biotin-stained neurons can be incubated with immunogold-conjugated avidin first and then the remaining biotin molecules can be tagged with fluorescent avidin.

A. Cell Staining with Biotin

Biotin was introduced into cells by one of the following three methods.

1. Intracellular Labeling of Neurons

Biotin derivatives, Neurobiotin (Vector Labs, Inc., Burlingame, CA) or biocytin (Molecular Probes, Inc., Eugene, OR), are injected intracellularly into neurons by electrophoresis through a micropipette. For this, an isolated head preparation is used. To gain access to the brain, the mouth parts, frontal cuticle, and muscle systems of the head are removed surgically, and the head is cut off the body and secured with insect pins in a recording chamber where it is superfused continuously with saline solution (149.9 mM NaCl, 3 mM KCl, 3 mM CaCl$_2$, 25 mM sucrose, and 10 mM TES (*N*-tris[hydroxymethyl]-methyl-2-aminoethanesulfonic acid), pH 6.9). A standard electrophysiological recording setup is used to inject the dye as described previously (Matsumoto and Hildebrand, 1981). The tip of a sharp glass microelectrode is filled with either 4% Neurobiotin in 1 M KCl or 4% biocytin in a solution of 0.5 M KCl in 0.05 M Tris buffer (pH 7.4). The shaft of the electrode is filled with 2 M KCl solution. After a stable impalement of a neuron or a period of intracellular recording, a depolarizing (in the case of Neurobiotin, usually 2–10 nA) or hyperpolarizing (in the case of biocytin, 2–10 nA) current is passed through the electrode for a period of 5 30 min to inject label into the cell.

2. Back-Filling of Neurons

Biotin derivatives can also be introduced by back-filling (Sun *et al.*, 1995, 1998), a staining technique widely applied in other animals (reviewed by McDonald, 1992). Back-filling is accomplished by cutting surgically accessible axons of the neurons to be labeled. The cut end of the nerve is then exposed to distilled water for approximately 1 min to open the damaged neurites. A drop of biotin solution (a few crystals of Neurobiotin in distilled water, estimated concentration <0.2%) is placed over the cut area. The whole animal is then placed in a humidified chamber for 5–20 min for the label to be picked up by the damaged cells. The incubation time depends on the distance that the biotin has to travel. Ten minutes is sufficient for about 2 mm of diffusion. Because biotin travels rather quickly within dendrites, there is no need for additional diffusion time. The area of the cut is flushed with saline after staining to wash out the excess biotin. The brain is then dissected out and processed as described later.

3. Immunocytochemical Staining by Biotinylated Antibody

Biotin can also be introduced as common biotin-conjugated probes, such as biotinylated secondary antibodies. We tested this method with a biotinylated secondary antibody. The detail procedure for this type of application is described later in a separate section.

B. Tissue Processing

1. Fixation

The fixation requirements for confocal microscopy and EM are very different. Confocal applications do not tolerate glutaraldehyde fixation because the high autofluorescence introduced by glutaraldehyde will potentially interfere with the fluorescent label. For EM study, a combination of paraformaldehyde and glutaraldehyde is the most common chemical fixation procedure. We find that lowering the concentration of glutaraldehyde to 0.15–0.40% is compatible with most confocal applications while maintaining adequate tissue preservation for EM. Additionally, using long-emission wavelength fluorophores such as Cy3 reduces the problem associated with autofluorescence.

For EM, the fixative solution must be fresh. Immediately after injection of biotin, the brain is dissected out of the head and fixed at 4 °C overnight in freshly made fixative solution comprising 4% paraformaldehyde, 0.32% glutaraldehyde, and 0.2% saturated picric acid in 0.1 M sodium phosphate buffer (pH 7.4). Picric acid is optional. It has been suggested that picric acid stabilizes membranes and precipitates proteins for EM observation (Zomboni and de Martino, 1967). It introduces, however, some additional degree of autofluorescence. We do not find a noticeable difference in preservation of ultrastructure in preparations processed with or without picric acid.

Fixation time can vary, depending on tissue size. We routinely fix moth brains overnight at 4 °C for convenience (approximate size of the brain is 2 × 3 mm). However, fixation times as short as 30 min can be applied at room temperature.

2. Tissue Sectioning

It is necessary to cut the tissue into thin slabs because (1) the immunogold reagent has a limited penetration power and (2) the lenses used for confocal microscopy have a short working distance. After fixation, the brain is washed briefly with sodium phosphate-buffered saline (PBS; 0.1 M, pH 7.4) and embedded in 7% agarose (low melting point agarose; gelling temperature of 2% (w/v) solution: 26–30 °C; Sigma, St. Louis, MO). The brains are then sectioned at 50–80 μm with a Vibratome (series 1000; Technical Products International, Inc., St. Louis, MO). The floating sections are collected in PBS.

3. Permeabilization and Incubation with Immunogold–Conjugated Streptavidin

A common problem in immunogold staining before embedment is the lack of penetration of the gold-conjugated reagent into the specimen because of the relatively large size of the gold particles. To enhance the penetration of the immunogold-conjugated streptavidin, we treat the Vibratome sections with either 0.1% Triton X-100 (J. T. Baker, Inc., Phillipsburg, NJ) in PBS for a period of 10 min or a series of ethanol solutions (30%, 50%, 70%, 50%, and 30% ethanol) for 5–10 min at each step. Optimal penetration is achieved with such permeabilization and using streptavidin conjugated to small (1.4 nm) gold particles. We have attempted to use smaller gold particles (0.8 nm, Nanoprobes Inc., Stony Brook, NY) to enhance the penetration of the reagent, but the 0.8-nm gold particle-conjugated streptavidin penetrated even less than the 1.4-nm gold.

After permeabilization, the sections are incubated with a solution of 1.4-nm gold-conjugated streptavidin (Nanoprobes Inc.) diluted 1:50–1:100 in PBS for 2–4 days on a shaker in an 8 °C cold room. Then 1:200 Cy3-conjugated streptavidin (Jackson ImmunoResearch Labs, Inc., West Grove, PA) is added to the same solution, and the sections are incubated for an additional 8–12 h in the cold room on the shaker. Following this procedure, the biotin-injected neuron is labeled reliably by gold-conjugated and fluorescently tagged streptavidin up to 40 μm into the tissue and could be detected both by confocal microscopy (fluorescence) and by EM (gold particles) (Fig. 1).

C. Confocal Observation

Stained slices of brain are mounted on slides in saline or glycerol-based mounting medium (Vectashield, Vector Labs, Inc.). Care must be taken to avoid drying the tissue for potential ultrastructural damage. We normally use nail polish to seal the slide.

We use a Bio-Rad MRC 600 confocal microscopy (Bio-Rad, Cambridge, MA) mounted on a Nikon Optiphot-2 microscope and a Zeiss LSM 410 mounted on a Zeiss Axiovert 100 M microscope, both equipped with a krypton/argon laser light source for confocal observations. The tissue is usually imaged with a 20× (NA 0.75) Nikon objective lens (Bio-Rad) or with a 25× (NA 0.8) multi-immersion Plan Neofluar objective lens (Zeiss). A multi-immersion lens is highly suitable for such a study to minimize the effects of refractive index mismatch with its adjustable collar. Serial optical sections, usually at intervals of 1–2 μm, are imaged through with its adjustable collar the depth of the labeled neurons and saved as 3D image stacks. Maximum intensity projection of this image stack onto a single 2D plane

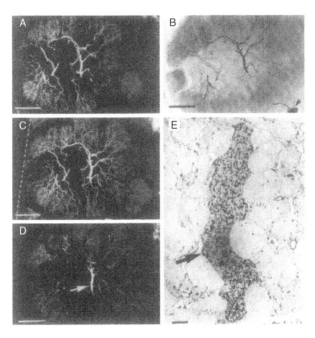

Fig. 1 Example of examination of a Neurobiotin-injected neuron using the immunogold fluorescent method to combine confocal and EM. (A) Intracellular staining of a local interneuron in the antennal lobe of the moth *Manduca sexta*. The neuron was imaged in saline, before plastic embedding, at 3-μm intervals for a total of 18 optical sections. The neuron ramifies extensively throughout the neuropile (antennal lobe). The soma is out of the field of view. (B) Bright-field image of the same neuron after silver enhancement and plastic embedding shows enhanced immunogold staining. The cell body is indicated with an arrowhead. (C) Confocal image of the Epon block demonstrates that the fluorescence is retained. Dashed line indicates the edge of the block. Comparison with confocal image obtained before plastic embedding (A) shows that most of the small processes remain detectable. (D) Single optical section from the 3D image stack at the level of the ultrathin section taken in (E). Arrow indicates the corresponding position of the process. (E) Thin section of the same neuron at EM level showing the ultrastructure of the labeled process, which contains numerous gold particles. Bars: (A–D) 100 μm, (E) 2 μm. Adapted, with permission, from Sun *et al.* (1995).

generates a 2D image of the reconstructed labeled neuron. The image stack is also studied with appropriate software to visualize the neuron in three dimensions. Areas of interest are defined within the stack for TEM study.

There is no evidence of ultrastructural damage by laser radiation on fixed tissue. However, to avoid potential ultrastructural damage by strong laser radiation with the confocal microscope, we try to minimize both the scanning time and the laser intensity by choosing an acceptable fast scan speed and a low laser power setting. This usually involves a trade-off of image quality, imaging duration, and laser intensity.

D. Silver Enhancement of Gold Particles

The 1.4-nm gold particles, too small to allow easy observation under the EM, are enlarged with a silver enhancement solution (Biocell Research Labs, from Ted Pella, Inc., Redding, CA) to facilitate detection in EM. The silver enhancement procedure is modified from the protocol suggested by the manufacture of the kit. In brief, after confocal observation, the selected sections are washed three times (10 min each) with PBS to remove the mounting medium. They are then washed in deionized water (two times at 3 min each) and incubated in the silver enhancement solution on ice in the dark for 8 min to allow diffusion of the solution into the sections. The container containing the sections is subsequently transferred to a water bath at room temperature for another 8 min of reaction in the dark. Timing of the reaction is important and care must be taken to avoid a self-nucleation reaction. The silver enhancement reaction is stopped by several washes with deionized water (three times at 3 min each). The sections are then transferred to sodium phosphate buffer (0.1 M, pH 7.4) for further processing for EM study.

E. Postfixation for EM Study

For EM, it is necessary to stabilize cell membranes by postfixation and then to embed in plastic. A common postfixative for EM is osmium tetroxide. As it was suggested that osmium tetroxide oxidizes silver precipitates of the silver-enhanced gold particles (Basbaum, 1989), tissues are lightly osmicated at room temperature (0.1% osmium tetroxide in phosphate buffer for 5–10 min). The postfixation process is stopped by replacing the solution with phosphate buffer. This procedure stabilizes the cell membranes enough for plastic embedding while retaining enough silver precipitate on the gold particles for EM observation. Osmium tetroxide and its vapor are highly toxic. This postfixation step must be carried out under a well-ventilated fume hood and the waste solution disposed of properly.

F. Plastic Embedding

The postfixed brain slices are dehydrated through a graded series of increasing concentrations of ethanol (50%, 70%, 95%, and twice 100% for 10 min at each step), followed by two changes of 10 min each in propylene oxide (Electron Microscopy Sciences, Fort Washington, PA). The sections are then placed in a 1:1 mixture of Epon/Araldite and propylene oxide for 30 min. The tissue is infiltrated with pure fresh Epon/Araldite overnight on a rotatory shaker. To flat embed the sections, sections are sandwiched between two Aclar sheets (Ted Pella, Inc.) in Epon/Araldite, and this assembly is further sandwiched between two microscope slides with two paper clips to hold the whole assembly together. The whole assembly is then put into a 60 °C oven overnight for polymerization.

One Aclar sheet can be peeled off easily and the sections can be examined with bright-field microscopy (Fig. 1B) or with the confocal microscope. Selected sections are cut out with a razor blade and glued to blank Epon/Araldite blocks with fast cyanoacrylate adhesive (Electron Microscopy Science, Inc.) for microtomy.

G. Confocal Microscopy as Guide for Microtomy

We found that a certain degree of Cy3 fluorescence is retained after processing tissue as described earlier. Therefore, it is possible to use a confocal microscope as a guide for microtomy. For viewing the block on the confocal microscope, the block is trimmed with a glass knife to obtain a smooth surface for confocal observation. It is then held in a specially fabricated brass slide with a hole in the middle. In this way, the block can be imaged in either an inverted or an upright position depending on the configuration of the microscope. These arrangements allow, moreover, a slight tilt of the surface of the block, if the surface of the block is not totally parallel to the image plane, which is a critical prerequisite for determination of the depth of the area of interest. The confocal microscope stage and the z-axis motor are calibrated and the reproducibility of both checked carefully so as to provide accurate depth information. The surface of the block is defined easily by its strong reflecting signal, giving the starting depth of 0 μm. The area of interest can be relocalized with confocal microscopy by reference to the initial set of confocal images. Through the readout of the z-axis motor, the actual depth of the area can be obtained for guiding microtomy. To ease the effort of relocalizing the fiber of interest under EM, the block is trimmed as small as possible.

H. Electron Microscopic Study

After examination with the confocal microscope, the Epon blocks containing labeled neurons are thin sectioned either at regular intervals by alternating thin sectioning and semithin sectioning or at preselected depths of interest in the tissue defined by the confocal study. Sections are cut with a diamond knife on a MT6000-XL microtome (RMC Inc., Tucson, AZ) or on a Reichert-Jung Ultracut microtome (Leica, Deerfield, IL). To facilitate localization of the chosen depth, thin sections are collected so that the ribbon on each grid contains a defined thickness of tissue. Because of the refractive index mismatch and inaccuracies of the microscopic stage, the depth information obtained using the confocal microscope may not be precisely accurate. Additionally, estimates of section thickness on the microtomes will incorporate some error. Therefore, additional thin sections are collected before and after the supposed depth of interest to ensure that the area of interest is contained in the thin sections collected. Thin sections are picked up on Formvar-coated slot grids or on thin-bar copper grids and poststained with saturated aqueous uranyl acetate solution and with lead citrate for 10 min each.

The thin sections are examined with a JEOL JEM-1200EX transmission EM or with a Philips 201C EM. The labeled processes are selectively photographed, usually at magnifications between 5000× and 15,000× for the identification of synapses.

I. Correlation of Confocal Microscopy and EM

In the majority of cases, depth information and the shape of the labeled process in the sections enable easy correlation of the confocal overview with EM detail. However, this correlation must be made with care, as a confocal optical section at the resolution of the lenses used in these experiments may contain the same amount of tissue as is included in as many as 40 thin sections. Moreover, localizing small, labeled profiles in a relative large area under EM is not an easy task. This can be eased by trimming the bock as small as possible to limit the area of view for EM.

One example using the immunogold fluorescence method to combine confocal microscopy and EM is presented in Fig. 1. A neuron in the antennal lobe of the moth brain is stained intracellularly with Neurobiotin and viewed with confocal microscopy before plastic embedding (Fig. 1A) and again in the plastic block (Fig. 1C). The transmitted bright-field microscopy (Fig. 1B) shows the immunogold labeling at the light microscopic level. An electron micrograph (Fig. 1D) of the neuron is shown. Use of this method to investigate the properties of insect neurons has been published in separate papers (Sun *et al.*, 1997).

V. Method 2: Avidin–Biotin Complex Method

This method uses a commercially available kit (avidin–biotin–peroxidase complex (ABC) from Vector Labs, Inc.) to convert some of the fluorescence signal in a fluorescently stained cell into an electron-dense form using 3,3′-diaminobenzidine (DAB) as substrate. It is based on the finding that avidin conjugated to a fluorescent tag can be bridged again with the ABC complex (Sun *et al.*, 1998). Then the peroxidase in the ABC complex can be used to convert DAB to an electron-dense reaction product for EM viewing.

A. Introduction of Biotin

Introduction of biotin to cell is accomplished exactly as the immunogold fluorescence method described earlier.

B. Fixation and Tissue Sectioning

Fixation is carried out as described earlier for the immunogold fluorescence method.

Because the ABC complex can penetrate much more deeply into the tissue than the immunogold reagent, it is possible to investigate much thicker tissue (up to 150 μm) with this method than with the immunogold fluorescence method. Nevertheless, the maximum thickness of the slices is also limited by the working distance of the objective used for confocal imaging. The 25× multi-immersion lenses used in these experiments has a working distance of 130 μm. Therefore, tissue is usually sectioned on the Vibratome at 100 μm as described earlier and the sections are collected in PBS.

C. Detection of Labeled Neurons with Fluorescent Marker

Labeled cells are rendered fluorescent by incubating the floating sections in a fluorescently conjugated avidin (or streptavidin). A variety of different fluorophores conjugated to avidin or streptavidin are available commercially (e.g., Invitrogen). We mostly use Cy3 for its photostability and brightness. Because streptavidin or avidin penetrates into tissue much more readily than immunogold particles, the incubation time is much shorter (1–3 h) than in the immunogold fluorescence method. Moreover, permeabilization is not necessary in this case. We mostly incubate the sections in a Cy3-conjugated streptavidin (Jackson Immuno Research Labs, Inc.) solution (0.5 μg/ml of PBS) for 3 h to overnight (for convenience). The sections are then washed with PBS and mounted without dehydration in Vectashield under a cover glass sealed with nail polish. Again, for good tissue preservation, care must be taken not to let the tissue dry in this process.

D. Confocal Microscopy

Confocal microscopy is carried out as described in the immunogold method described previously. However, because the fluorescence signal is no longer detectable after ABC conversion, all confocal examinations must be carried out before the ABC conversion, unlike the immunogold method where some fluorescence is retained in the plastic block.

E. Avidin–Biotin–Peroxidase Complex Conversion

After the confocal microscopic observation, the sections are transferred to PBS and incubated in an avidin–biotin–peroxidase complex (ABC Elite, Vector Labs, Inc.) solution for 2 h to overnight in a cold room (8 °C) on a shaker. The concentration used is suggested by the kit supplier. The preparations are then washed with PBS, incubated with DAB (0.25% in PBS) for 15 min, and finally reacted with DAB in the presence of H_2O_2 (0.01% in DAB solution) for 5–10 min. Because the reaction is somewhat light sensitive, it is preferable to carry it out under darkness. It is, however, necessary to check the reaction periodically under a dissecting microscope and to stop the reaction before the background stain rises.

The reaction is stopped with several washes of PBS. After such a reaction, the labeled neurons become dark and the fluorescent label is no longer visible.

F. Postfixation and Plastic Embedding

Postfixation with osmium tetroxide would enhance DAB staining. However, one of the common problems with DAB staining for EM is masking of the internal structure of the labeled cell. We found that a good balance between overall tissue preservation and maintenance of visibility of internal structure is achieved by osmicating tissue slices in 0.5% osmium tetroxide in sodium phosphate buffer (0.1 M, pH 7.4) for 15 min with the 100 μm brain slices. The sections are then dehydrated and embedded in plastic as described earlier for the immunogold method. The plastic sheet containing the labeled neurons is initially examined at the light microscopic level. Even though the image quality is somewhat impaired by the thickness of the slices and the density of osmication, the filled cells can be identified clearly under bright-field microscopy. Selected plastic-embedded Vibratome sections containing the labeled neuron are cut out with a razor blade and glued to a blank block as described earlier for thin sectioning.

G. Defining Depth of Areas of Interest with Confocal Microscopy

Because the fluorescence signal is no longer visible after plastic embedding with this method, locating the area of interest in plastic poses a particular challenge. This is overcome by using either the reflection mode of the confocal microscope or the transmitted light detector of the confocal microscope if it is so equipped. Due to the fact it is impossible to set up the Kohler illumination with a plastic block sample on a light microscopy stage, the transmitted images are greatly impaired. Nevertheless, although neither of the methods gives images of high enough quality to allow another 3D reconstruction of the stained cell, both methods allow easy comparison with the original confocal images for identification of the area of interest. Then the depth of that area can be defined from the readout of the z-axis motor of confocal and microtomy can be conducted as described in the immunogold fluorescence method.

H. EM Study

Once the depth of a particular region of interest is defined, the plastic block is thin sectioned and examined by EM as described earlier for the immunogold fluorescence method. An example of a Neurobiotin-stained neuron visualized using this method is presented in Fig. 2. Preservation of ultrastructure is good judging by the appearance of cell membranes and well-preserved synaptic vesicles (Fig. 2D and E).

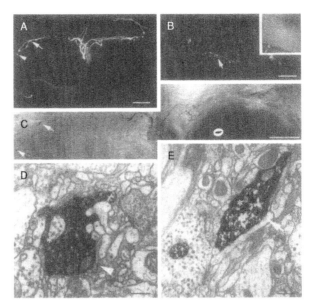

Fig. 2 Example of the use of the ABC conversion method on a Neurobiotin-labeled neuron. (A) Reconstruction of a Neurobiotin-labeled neuron from the fly *Musca domestica*. Neuron was imaged in a 100-μm Vibratome slice. Optical sections were obtained every 2 μm over a depth of 80 μm. Selected varicosities (arrow and arrowhead) indicate areas of interest (varicosities) for further EM investigation. (B) Using the reflecting mode of the confocal microscope, it is possible to locate areas of interest (arrow) in the plastic block to determine the depth of the region (24.5 μm). (Inset) The strong reflecting image of the surface of the block (0 μm). (C) Partial reconstruction from transmitted detector images of confocal indicates that the stain is restricted to the labeled neuron only. Arrow and arrowhead indicate the same regions of the neurons in (A). (D, E) Electron micrographs of the same neurons in the arrowhead (D) and arrow (E) regions in (A–C). Asterisks indicate locations of possible synaptic connections. The quality of tissue preservation is good as indicated by well-preserved cell membrane and synaptic vesicles. Bars: (A) 50 μm, (B) 20 μm, (C) 50 μm, (D, E) 0.5 μm. Adapted, with permission, from Sun *et al.* (1998).

VI. Application of the Method to Immunocytochemistry

The biotin–avidin interaction is well exploited in many fields of biological research because of the specific and strong interaction between biotin and avidin. The methods described earlier, in theory, should be applicable with any biotinylated probe.

As an example, we use an antibody against serotonin, a neurotransmitter/modulator, to demonstrate the use of the method for immunocytochemistry. A requirement of the method is that the antigen must survive use of the small amount of glutaraldehyde that is required for adequate tissue preservation for EM study. This is the case for this specific antibody against serotonin (Sun *et al.*, 1993).

Serotonin immunocytochemical staining is carried out as described previously (Sun *et al.*, 1993) except that a biotinylated secondary antibody is used to introduce

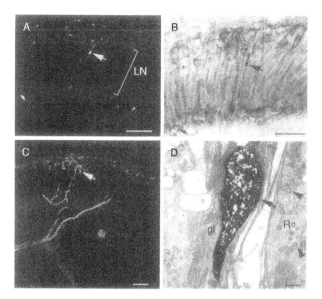

Fig. 3 Extension of the methods to immunocytochemical labeling using labeling with antiserotonin antibody as an example. (A) Single optical section of a serotonin-immunoreactive fiber (which belongs to neuron LBO5HT; Nässel, 1987) in the optical path (lamina, LN) of the fly brain. Arrow indicates a varicosity of interest. (B) Transmitted light image of the same fiber after DAB reaction and plastic embedding. (C) Reconstruction of the entire optical stack (40 optical sections at 1.2-μm intervals) shows the overall view of the neurons in the Vibratome slice. (D) Electron micrograph of the varicosity in (B) shows that it contains numerous darkly labeled vesicles and is surrounded by glia (gl). A neighboring photoreceptor terminal (R), characterized by capitate projections (arrowheads), shows good tissue preservation. Bars: (A–C) 25 μm, (D) 0.5 μm. Adapted, with permission, from Sun *et al.* (1998).

biotin to the labeled neurons. In brief, tissue is fixed and sectioned on a Vibratome as described previously. Then, the tissue is incubated in primary antibody (rabbit antiserotonin, Incstar Corp., Stillwater, MN) in PBS overnight. It is subsequently washed in PBS and incubated with biotinylated goat antirabbit antibody (used at the concentration suggested by the supplier (1:200), Vector Labs, Inc.) for 3 h. The staining is detected with Cy3-conjugated streptavidin for confocal microscopy. Then, the DAB reaction is carried out as described earlier for the ABC complex. An example of such staining is shown in Fig. 3.

VII. Discussion and Conclusion

The two methods described in this chapter are complementary. The immuno-gold method provides better detection than the ABC method because the internal structure is not masked by the stain. Additionally, immunogold particles do not diffuse like the DAB reaction product does. Moreover, the immunogold method provides the possibility to observe the tissue with confocal microscopy after embedding in plastic. Therefore, it avoids the problem of discrepancies caused by tissue shrinkage after plastic embedding when correlating 3D information with ultrastructural details. The immunogold fluorescence method, however, is slow due to the long incubation time required for the immunogold reagent. Additionally, the immunogold method limits section thickness to about 80 µm, which creates additional problems for large cells. The ABC complex method, however, is rapid. It can be carried out in a single day if necessary. Moreover, the ABC method allows much thicker sections to be studied. In summary, the ABC method is suitable for routine applications, whereas the immunogold fluorescence method should be used when better tissue preservation and detection are required.

A. Limitations of Methods

The combination of confocal microscopy and EM allows the investigator to correlate 3D information with ultrastructural characteristics with relative ease. However, identifying the same structure under EM and confocal still requires a high degree of skill. This problem is compounded by the fact that confocal optical sections are much thicker than a single thin section. The problem may be minimized by performing a small 3D reconstruction from serial thin sections.

Both confocal and EM methods require the introduction of biotin. This may not be applicable in tissues with a high level of endogenous biotin.

We used Cy3 almost exclusively as the fluorophore of choice for confocal observation because of its brightness and photostability. Additionally, Cy3 retains a fair degree of fluorescence with the immunogold fluorescence method, allowing confocal microscopic study of the labeled cell in plastic. However, studying plastic sections with confocal microscopy poses other challenges. Opacity of the tissue after osmication made it particularly difficult to image deep into the sections. Moreover, mismatches in refractive indexes between plastic and tissue introduce additional inaccuracies for depth determination.

B. Adaptation to Other Systems

Major obstacles of combining confocal microscopy and EM are the incompatibilities of the markers used for confocal microscopy and for EM. This chapter describes methods of simultaneously or sequentially labeling insect neurons with fluorescence for confocal microscopy and with electron-dense labels for EM. The methods should be adaptable to other systems if certain facts are kept in mind. The requirements of tissue fixation for confocal microscopy are much less demanding than those for EM. For EM, however, there is no universal, ideal fixative for every tissue type. Often, a trial-and-error test is needed to find the best fixation procedure for a specific tissue. Finding the best fixative procedure for EM is a good starting point for developing a combined confocal–EM method for the tissue. By systematically modifying the concentration and/or the time of fixation as we did, one should be able to achieve optimum balance between ultrastructural preservation and good signal-to-noise ratios for fluorescent and electron-dense labels.

Similarly, different types of tissues and different antigens (for immunocyto-chemical labeling) will require different methods and amounts of permeabilization for the access of gold particles. For example, some antigens will not survive dehydration, so alcohol cannot be used as the permeabilization agent. Other methods such as freeze–thawing (van de Pol, 1986) to permeabilize the tissue may be better.

In summary, we have reviewed two methods we developed for combining the advantages of confocal and EM. Although other confocal and EM methods have been described (Deitch *et al.*, 1990), the methods described here are simpler and less tedious. We have proven them to be practical and useful for the combination of confocal microscopy and EM for neurobiological research.

Acknowledgments

The work was carried out in the Laboratories of Drs. Leslie P. Tolbert and John G. Hildebrand at ARL Division of Neurobiology, the University of Arizona, and in the laboratory of Dr. Ian A. Meinertzhagen at the Department of Psychology, Dalhousie University, Canada. The author thanks them for their support. I also thank Dr. L. P. Tolbert for critical reading of the manuscript and suggestions. Additionally, the author thanks Patty Jasma, A. A. Osman, Lesley Varney, and Maggie Klonowska for technical assistance. The work was supported by an award from the University of Arizona, Center for Insect Science to X. J. S., grants to I. A. M. (NIH EY-03592, NSERC OPG 0000065), to L. P. T. and J. G. H. (NIH NS-28495), and to J. G. H. (NIH AI-23253).

References

Basbaum, A. I. (1989). *J. Histochem. Cytochem.* **37,** 1811.

Bell, R. A., and Joachim, F. A. (1976). *Ann. Entomol. Soc. Am.* **69,** 365.

Chan, W. C., and Nie, S. (1998). *Science* **281,** 2016–2018.

Chen, H., Swedlow, J. R., Grote, M., Sedat, J. W., and Agard, D. A. (1995). The collection, processing, and display of digital three-dimensional images of biological specimens. *In* "Handbook of Biological Confocal Microscopy," (J. B. Pawley, ed.)., 2nd edn., **197,** Plenum Press, New York.

DeFelipe, J., and Fairén, A. (1993). *J. Histochem. Cytochem.* **41,** 769.

Deitch, J. S., Smith, K. L., Swann, J. W., and Turner, J. N. (1990). *J. Microsc.* **160,** 265.

Fröhlich, A., and Meinertzhagen, I. A. (1982). *J. Neurocytol.* **11,** 159.

Giepmans, B. N., Deerinck, T. J., Smarr, B. L., Jones, Y. Z., and Ellisman, M. H. (2005). *Nat. Methods* **2**(10), 743–749.

Hama, K., Arii, T., and Kosaka, T. (1994). *Microsc. Res. Tech.* **29,** 357.

Huang, B., Bates, M., and Zhuang, X. (2009). *Annu. Rev. Biochem.* **78,** 993–1016.

Matsumoto, S. G., and Hildebrand, J. G. (1981). *Proc. R. Soc. Lond. B* **213,** 249.

McDonald, A. J. (1992). *Neuroreport* **3,** 821.

McDonald, K. L. (2009). *J. Microsc.* **235**(3), 273–281.

Mercer, A. R., Hayashi, J. H., and Hildebrand, J. G. (1995). *J. Exp. Biol.* **198,** 613.

Mercer, A. R., Kloppenburg, P., and Hildebrand, J. G. (1996). *J. Comp. Physiol. A* **178,** 21.

Nässel, D. R. (1987). *Prog. Neurobiol.* **30,** 1.

Powell, R. D., Halsey, C. M., Spector, D. L., Kaurin, S. L., McCann, J., and Hainfeld, J. F. (1997). *J. Histochem. Cytochem.* **45**(7), 947–956.

Stevens, J. K., Davis, T. L., Friedman, N., and Sterling, P. (1980). *Brain Res.* **2,** 265.

Sun, X. J., Tolbert, L. P., and Hildebrand, J. G. (1993). *J. Comp. Neurol.* **338,** 1.

Sun, X. J., Tolbert, L. P., and Hildebrand, J. G. (1995). *J. Histochem. Cytochem.* **43,** 329.

Sun, X. J., Tolbert, L. P., and Hildebrand, J. G. (1997). *J. Comp. Neurol.* **379,** 2.

Sun, X. J., Tolbert, L. P., Hildebrand, J. G., and Meinertzhagen, I. A. (1998). *J. Histochem. Cytochem.* **46,** 263.

Tolbert, L. P., Sun, X. J., and Hildebrand, J. G. (1996). *J. Neurosci. Methods* **69,** 25.

van de Pol, A. N. (1986). *J. Neurosci.* **6,** 877.

Ware, R. W., and LoPresti, V. (1975). *Int. Rev. Cytol.* **40,** 325.

Zomboni, M. G., and de Martino, C. (1967). *J. Cell Biol.* **35,** 148A.

PART II

Functional Approaches

CHAPTER 9

Volume Measurements in Confocal Microscopy

Carlos B. Mantilla, Y. S. Prakash, and Gary C. Sieck

Departments of Anesthesiology
and Physiology & Biomedical Engineering
Mayo Clinic, Rochester, Minnesota, USA

I. Introduction

The confocal microscope is a convenient tool for obtaining spatially registered two-dimensional (2D) optical sections of three-dimensional (3D) objects (for reviews on confocal microscopy, see Conchello and Lichtman, 2005; Matsumoto, 2002; Pawley, 2006a,b,c; Schild, 1996; Stevens *et al.*, 1994). The ability of the

TECHNIQUES IN CONFOCAL MICROSCOPY
143
DOI: 10.1016/B978-0-12-384658-7.00009-6

confocal microscope to eliminate out-of-focus information is dependent on the presence of an aperture limit before the detector, which determines the thickness of the optical plane imaged for a given objective lens. By restricting the aperture of the detector, out-of-focus light is unable to pass the restricted aperture and, thus does not reach the detector, limiting the signal to the focal plane (Brakenhoff *et al.*, 1989; Inoue, 2006; Wilson, 1989; Wilson and Sheppard, 1984). Image-processing software can then be used to reconstruct a series of optical sections in 3D and thereby make direct volume and surface area measurements of an object of interest (Prakash *et al.*, 1993; Stevens *et al.*, 1994; Turner *et al.*, 1996). When using the confocal technique, a match between optical section thickness and the spacing between optical sections establishes optimal sampling, that is, the least amount of over- or undersampling. However, a perfect match between optical section thickness and stepper motor control of the Z-axis is rarely achieved. Moreover, the small optical section thickness obtained when using objective lenses with high numerical aperture (NA) and magnification necessitates collecting a large number of optical sections. This large data set makes subsequent 3D reconstruction both computer and labor intensive. Accordingly, stereological methods may be applied to reconstruct 3D images from a set of serial optical sections, for example, the Cavalieri principle (Geinisman *et al.*, 1997; Gundersen and Jensen, 1987; Prakash *et al.*, 1994). These stereological methods combined with confocal microscopy greatly lessen the analytical burden of large image sets and offer an unbiased approach to 3D rendering without any necessary assumptions about object shape, size, or orientation.

Recent advances in the speed and sensitivity of image acquisition of confocal microscopy have made it an ideal tool for qualitative and quantitative assessment of cellular dynamics in 3D. Until recently, limitations in detector sensitivity and the need for frame averaging, as well as computer hardware and software limitations had allowed enhanced 3D acquisition only at the expense of considerable reductions in image acquisition rates, especially when obtaining repetitive 3D images of cellular events over time (4D). Accordingly, realistic and reliable 4D confocal microscopy had been possible only under conditions where cellular dynamics were slow enough to be repetitively captured through the volume of a cell. However, recently developed confocal systems can achieve high sensitivity, faster-than-video rates of acquisition in one optical plane and, by combining this high-speed 2D imaging with rapid 3D optical sectioning, are capable of relatively rapid 4D acquisitions. Certainly, new investigations on a variety of cellular dynamics across a wide range of spatial and temporal response patterns can be expected in the near future.

The goal of this chapter is to focus on the techniques and limitations of using confocal microscopy for volume measurements. Practical solutions are offered for calibration and estimation of errors, particularly in the Z-axis. Recent developments in superresolution microscopy are also reviewed. Finally, specific examples are provided for the use of confocal microscopy in volume measurements.

II. Selection of Optics in Confocal Microscopy

The selection of the objective lens is particularly important when subsequent analysis of the optical sections includes 3D reconstruction of an object. It is beyond the scope of this chapter to provide a complete description of objective lenses (for a review, see Keller, 2006). However, several issues regarding the selection of the objective lens have real significance, especially when imaging cellular structures requires minimal loss of transmitted signal (e.g., fluorescence), minimal distortion and maximal signal-to-noise ratio throughout the image field. A basic understanding of the microscope optics design is needed to optimally match the properties of the selected optical components.

When using confocal microscopy for volume measurements, NA is an important determinant in the selection of the objective lens. Generally, the higher the NA, the greater the XY spatial resolution and the thinner the optical section (see below). Another important consideration is the working distance of the objective lens, defined as the distance from the outermost part of the lens to the deepest focal plane in the specimen that can be achieved without compression of the specimen. In general, the working distance is shorter with high-NA lenses, although it varies significantly among lenses and manufacturers. For example, an Olympus UV 40X/ 0.85 NA has a working distance of 0.25 mm, whereas an Olympus UV 40X/1.00 NA only 0.16 mm. When focusing deep into a thick specimen, even lenses with excellent correction properties lose their optimal performance and images of lesser intensity and greater blur are obtained (Inoue, 2006). In many cases, specimen thickness can be optimized to the working distance of the objective lens.

Currently, there is no *one* fully standard, independent test to evaluate the performance of an objective lens. However, manufacturer-provided test procedures, such as the use of fluorescently labeled microspheres and microscopic grids, allow empirical assessment of lens performance under various experimental conditions. Most currently available lenses are near *diffraction limited* (i.e., the corrective properties of the objective lens have been adjusted to match the theoretical limits imposed by medium refractive index and wavelength of light in use) (Keller, 2006), at least in the center of the field. However, long-term use of laser illumination may deteriorate lens performance. In addition, lens aberrations can introduce deviations from the "theoretical" diffraction-limited image. Therefore, regular empirical assessment of lens performance is highly recommended.

Lens aberrations can be grouped as either monochromatic (wavelength independent) or chromatic. Monochromatic aberrations include spherical aberrations, and usually can be easily avoided by the use of lenses with appropriate corrective optics (see manufacturer brochures for details). In particular, dry- or water-immersion lenses of high-NA become very sensitive to imaging conditions, such as the coverslip thickness and unmatched refractive indices (Egner and Hell, 2006). In addition, when focusing deep into a specimen, oil-immersion lenses will no longer have an adequate spherical correction, with consequent loss of signal and resolution. In general, factors that increase spherical aberration also worsen other monochromatic aberrations, particularly when imaging off-axis. Curvature of field is especially important for spatial reconstruction of specimens imaged from thick materials, because of the narrow depth-of-field of the confocal image and the dish-shaped distortion introduced by noncorrected objectives (nonplan) (Sandison *et al.*, 1995). Reconstruction of the 3D stack of optical sections could include correction for the curvature of the field off-axis, but would make calculations more complex.

Chromatic aberrations arise from the fact that the refractive index (η) of every optical medium is dependent on the wavelength of incident light (dispersion), thus affecting the lens focal length for each wavelength (λ). If the objective lens is not properly corrected for the specific light wavelengths in use, the image along the Z-axis at one wavelength will be different from that at another. Obviously, this is particularly important for double or triple-fluorescence labeling of 3D objects, where registration along the Z-axis is important, for example, for colocalization of subcellular structures. Achromat objective lenses include optical materials of linear dispersion, and are usually corrected for only two wavelengths. Apochromat or semiapochromat lenses have excellent correction for chromatic aberrations at three or more wavelengths, but require glass of nonlinear dispersion, that unfortunately might be unsuitable for UV imaging (because of more rapid changes in the η at this end of the spectra) (Keller, 2006). Overall, the best results, in terms of fluorescence transmission and image resolution, are obtained when the emission and excitation wavelengths have similar corrections (Sandison *et al.*, 1995).

In conclusion, a basic understanding of the major components of the optical pathway of the objective lens in confocal systems can help improve on image acquisition and resolution. Most importantly, the adequacy of individual components should be fully evaluated before using any confocal system for specific imaging protocols.

III. Optical Sectioning Using Confocal Microscopy

The confocal microscope, by virtue of its restricted aperture, offers a narrow depth-of-field, reducing problems associated with light scattering. The significant reduction in out-of-focus information and the inherent spatial registration of sequential optical sections make the confocal microscope an excellent tool for direct analysis of 3D structures (Brakenhoff *et al.*, 1989; Inoue, 2006; Stevens *et al.*, 1994; Wilson, 1989; Wilson and Sheppard, 1984). Thin optical sections of physically thick specimens can thus be obtained with complete spatial registration. Furthermore, attenuation of out-of-focus information reduces the need for various image processing and enhancement techniques commonly used with standard light microscopy to improve the quality of images by mathematically removing out-of-focus information *post hoc* (Hiraoka *et al.*, 1990; Shaw, 1994, 2006).

The strength of confocal imaging thus lies in optimal sampling with optical sections and in the faithful reproduction of shape and size along the Z-axis. It is, therefore, important to validate the confocal system in terms of the fidelity of optical sectioning and with regard to the distortions introduced particularly along the Z-axis.

In addition, an important consideration is the potential for photobleaching during optical sectioning, especially when image intensity is also a measured parameter and is critical in determining which parts of the image belong to an object and which do not. In particular, investigators commonly rely on threshold definitions in the calculation of object volume, and bleaching will render different estimations of the same object (or others along the illuminated column). An entire column of the specimen (along the Z-axis) is illuminated by the confocal laser when acquiring images of a single plane, therefore, bleaching will occur outside of the focal plane (Pawley, 2006a,b,c). Fortunately, the rate at which bleaching occurs is inversely proportional to the square of the distance from the focal plane. Nonetheless, as with all fluorescent imaging, limiting the exposure of the specimen to incident light only to that necessary for image acquisition will allow maximal preservation of the sample.

The selection of the most appropriate confocal imaging system for a particular application involves making several choices which may impact spatial and temporal resolution (i.e., the ability to distinguish two points of an object or two temporally separate events). It is well beyond the scope of this chapter to describe the theoretical aspects of confocal imaging and the advantages and limitations of currently available systems (for reviews, see Conchello and Lichtman, 2005; Pawley, 2006a,b,c; Prakash *et al.*, 1999). Nonetheless, a general understanding of the basic differences among the various available confocal systems is required for selection of the optimal system for each particular study. Therefore, a brief description of the general classes of confocal systems is included with emphasis on their suitability for volume measurements.

The confocal principle is based on the introduction of an aperture limit in the light illumination pathway and a corresponding one in the transmitted light path such that only information from a single optical plane is obtained (by convention, Z-axis resolution) (Conchello and Lichtman, 2005; Matsumoto, 2002; Schild, 1996). The use of laser illumination permits coherent illumination of high intensity allowing for better correction of possible chromatic aberration (see Section II). The illuminated specimen reflects light or emits fluorescence, which is registered by a sensitive detector, usually, a photomultiplier tube (PMT) or a cooled charge-coupled device (CCD) (Art, 2006). The signal over time is then digitized and stored in a computer (Pawley, 2006a,b,c).

Scanning microscopes successively illuminate finite sections of the object, therefore increasing the lateral resolution (XY resolution) by reducing the light scattered by neighboring parts of the specimen (Pawley, 2006a,b,c). Scanning introduces a temporal delay in acquisition of an image, especially when using a *pinhole* (circular) aperture. Different scanning methods are available, including stage and beam scanning systems, albeit with limited scan rates (Inoue, 2006; Schild, 1996). To improve temporal acquisition, the tandem scanning microscope, spinning-disk, and acousto-optical modulator systems have been developed (Toomre and Pawley, 2006). A different solution is the use of a rectangular *slit* aperture to produce a confocal image (Wilson and Hewlett, 1990). The slit design is ideally suited for rapid acquisition rates with a reduced signal-to-noise ratio, although it suffers from reduced axial (Z-axis) resolution (Sheppard *et al.*, 2006). Therefore, depending on the fluorescence and intensity distribution in the specimen, the required XY and Z-axis resolutions, and even perhaps on the required temporal resolution, a judicious choice must be made to allow for 3D measurements.

A. *XY* Axis Resolution

In light microscopy, the NA of the objective lens and the wavelength of the incident light (λ) determine the XY (horizontal plane, parallel to the stage by convention) spatial resolution as given by Eq. (1) (Conchello and Lichtman, 2005; Inoue, 2006):

$$XY \text{ spatial resolution} = 0.61 \times \frac{\lambda}{\text{NA}} \tag{1}$$

Currently available confocal systems and objective lenses approximate the theoretical diffraction limit (0.2 μm; described by Abbe, 1884) using visible light sources. However, aberrations in the optical path usually limit actual XY spatial resolution to ∼0.3 μm.

Available confocal imaging systems digitize the optical image within an optical section into picture elements (or pixels) (Pawley, 2006a,b,c). The pixel dimension along any axis is given by the ratio of the total length scanned to the total number of pixels along that axis. In accordance with the practical limit of objective lenses, the finest useful pixel dimension is obviously ∼0.3 μm (pixel resolution). However, it must always be kept in mind that pixel dimension need not necessarily match optical resolution (see Pawley, 2006a,b,c for a detailed discussion). For example, when using low-magnification lenses such as a 10× or 20×, the scanned areas are large, but given the fixed total number of pixels (typically 256–1024), the pixel resolution is likely to be worse than the optical resolution, and there is considerable lost data. With lenses of increased magnification, pixel resolution may approximate the optical resolution, optimizing the digitization process. Magnifications beyond this point result in empty magnification (oversampling), where pixels of decreased dimensions do not provide additional morphological information. The actual minimal resolved area is not changed, but it is now represented by a greater number of pixels. However, under certain conditions, this oversampling may actually be used advantageously by limiting the scanned area but maintaining the total number of pixels (hardware zoom), particularly when imaging objects containing periodic data, for example, Z lines in skeletal or cardiac muscle sarcomeres.

The optimal spatial resolution for acquisition is dictated by the Nyquist theorem, such that sampling should be at least twice the spatial frequency of the feature being imaged within an object (Pawley, 2006a,b,c). In practice, this means obtaining images where the pixel resolution is adjusted to be less than half the dimension of any periodicity within the specimen (Centonze and Pawley, 2006). For example, Z lines are typically separated by 2.5 μm in skeletal muscle fibers. Thus, the optimal spatial resolution of acquisition needs to be at least 1.25 μm. These spatial acquisition limits do not apply only to periodic features within objects, but dictate the ability to resolve differences between features (two-point discrimination). Thus, in the example provided above, a 10× objective lens with a 0.5 NA is barely sufficient to resolve Z line periodicity using visible light (XY resolution of ∼0.62 μm at 515 nm). However, the Nyquist theorem only establishes the minimal requirements for spatial resolution. Oversampling offers distinct advantages particularly in resolving finer features. In the example above, using the 10× objective lens permits a hardware zoom of ∼2× with no loss of spatial resolution. In contrast, neuronal dendrites are frequently 1.0 μm in diameter and bundled with an intervening space of ∼1.0 μm. Thus, to discriminate between two adjacent dendrites, the spatial resolution has to be at least 0.5 μm. Using a 20× objective lens with a 0.7 NA, the XY spatial resolution is ∼0.44 μm, which is barely sufficient, but would not allow for hardware zoom. In contrast, a 60× objective lens with a 1.3 NA provides a XY spatial resolution ∼0.24 μm, approximating the theoretical limit of light microscopy. The use of the 60× objective lens in this specific example also offers the advantage of applying a 2× hardware zoom, which will resolve finer details in dendritic structure.

B. Z-Axis Resolution

The major source of error in volume measurements using confocal microscopy is related to a mismatch between optical section thickness (Z-axis resolution) and the sampling interval along the Z-axis (step size). Optical section thickness depends on the NA of the objective lens, the wavelength of incident light, the size of the confocal aperture and the refractive index (η) of the surrounding medium and tissue, as given by Eq. (2) (Schild, 1996):

$$Z - \text{axis resolution} = \frac{0.45\lambda}{\eta[1 - \cos(\sin^{-1}(\text{NA})/\eta)]} \tag{2}$$

Thus, optical section thickness is inversely related to the NA of the objective lens and directly related to the wavelength of incident light (Toomre and Pawley, 2006). Decreasing aperture size increases Z-axis resolution, but significantly reduces the amount of light reaching the detector with consequent loss of information and worsening of the signal-to-noise ratio (Conchello and Lichtman, 2005; Pawley, 2006a,b,c). Therefore, in selecting the aperture setting, a compromise must be reached between optical section thickness and signal-to-noise ratio. In general, the better the match between the refractive indices of the objective lens, coverslip glass (if any), and the tissue mounting medium, the greater the Z-axis resolution (Egner and Hell, 2006). The theoretical Z-axis resolution can be calculated using Eq. (2); for example, with a $60\times/1.3$ NA oil-immersion objective lens at 515 nm emission wavelength, the theoretical Z-axis resolution is ~0.3 μm. However, several factors such as lens properties (chromatic aberration and flatness of field) and mismatches in refractive indices reduce the actual Z-axis resolution. Therefore, in several practical applications involving volume measurements, the major goal cannot be to achieve the best (theoretical) Z-axis resolution, but to minimize or account for distortions along the Z-axis.

Based on its theoretical determinants (Inoue, 2006; Pawley, 2006a,b,c), the Z-axis resolution is at best twice the XY resolution, and given the number of factors that can affect practical Z-axis resolution, it is difficult to calculate exactly. However, the overall Z-axis distortion can be empirically determined for any particular confocal system (Prakash *et al.*, 1993, 1999). For example, fluorescently labeled latex microspheres of varying diameters embedded in tissue mounting medium can be imaged under different conditions. Microsphere diameters can be measured in both the XY and XZ planes and compared to manufacturer-specified values. The ratio of specified diameter to measured XZ diameter serves as an estimate of overall Z-axis distortion under the specific conditions of tissue preparation, optics and Z-axis step size. Although not the focus of the present discussion, this approach can be utilized to estimate the relative contribution of different factors to the overall distortion. For example, the distortion introduced solely by the mismatch between the refractive indices of embedding medium and the objective lens immersion oil can be estimated by mounting the fluorescent microspheres in the immersion oil rather than tissue mounting medium (Prakash *et al.*, 1993).

The actual Z-axis resolution, under a given experimental condition, can also be determined from the point-spread function (PSF) of the objective lens (for a detailed review, see Egner and Hell, 2006; Hiraoka *et al.*, 1990). Intensity profiles along the Z-axis are obtained when imaging objects of finite size (e.g., 0.2 μm fluorescent beads) which can be averaged to estimate a mean PSF. The full-width half-maximum (FWHM) of the mean profile is then taken to represent the optical section thickness (Fig. 1).

C. Matching Optical Section Thickness to Step Size

The match between optical section thickness and the increment between optical sections, as determined by the stepper motor (step size), determines the optimal sampling conditions in the Z-axis and, ultimately, the reliability of the volume measurements (Prakash *et al.*, 1999). A number of commercially available confocal systems are equipped with stepper motors that control the fine focus knob of the microscope. Most stepper motors have a step size resolution of 0.05–0.20 μm, which is more than sufficient when compared to the Z-axis resolution of light microscope objective lenses. Accordingly, the use of a step size smaller than the optical section thickness results in repeated sampling of more or less the same

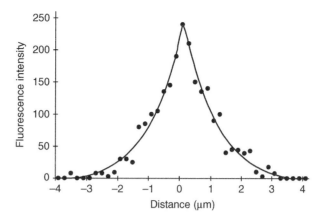

Fig. 1 Empirically determined mean point-spread function (PSF) for a 40×/1.3 oil-immersion objective lens (Nikon Instruments). Fluorescently labeled microspheres (0.2 μm in diameter) were optically sectioned at 0.18 μm step size. The average PSF of the individual microspheres was obtained after alignment along the maximal axis (individual data points), and then fitted in a modified four-parameter Gaussian regression curve (line). The full-width half-maximum was estimated to be 0.8 μm.

optical plane. In contrast, use of a step size larger than the optical section thickness results in an undersampling of the specimen with consequent loss of information. An example of the impact of over- and undersampling in the Z-axis is provided in Fig. 2. Using a Bio-Rad MRC 600 Confocal Imaging System and a 40×/1.3 NA Fluor objective lens (Nikon Instruments), confocal images were obtained of a 10.3 μm fluorescent (488/515 nm) latex microsphere suspended in immersion oil (to minimize distortions due to refractive index mismatches). In the XY plane the measured diameter of the microsphere closely matched the specifications of the manufacturer. At the smallest pinhole aperture size, the step size was adjusted in 0.18 μm increments (stepper motor resolution). As can be seen, the measured diameter of the microsphere in the XZ plane depended on step size. At 0.28 μm step size, microsphere volume was overestimated by ~70% (if no interpolation is performed). In contrast, at 1.54 μm step size, microsphere volume was underestimated by ~45% (without interpolation).

A common mistake is to assume that step size is per manufacturer specification. Accuracy of the stepper motor can be severely affected by backlash and hysteresis. The direction of the focus motor movement determines the size of the step (hysteresis). In addition, changes in direction of the focus motor step can introduce movement of the stage or objective turret (backlash). Hysteresis and backlash in the stepper motor can be estimated by moving the motor from a known initial position to a final position and back, measuring the error in placement. Clearly, the accuracy of optical sectioning is critically dependent on the performance of the Z-axis stepper motor. Errors in the step size introduced by the stepper motor would not directly affect each optical section, but would generate discrepancies

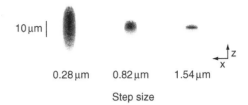

10 μm |

0.28 μm   0.82 μm   1.54 μm

Step size

Fig. 2 2D images of fluorescently labeled latex microspheres (FluoSpheres, Invitrogen) are shown. Measurement of the microsphere diameter on the XY plane closely correlates with the manufacturer specification (10.3 μm). Measured diameters on the XZ plane, on the other hand, show the effect of varying the sampling interval (step size).

between the optical section thickness and the actual step. Furthermore, the specimen thickness may also be incorrectly inferred. As described earlier, this could lead to under- or oversampling and incorrect 3D rendering. Performance of the stepper motor can be empirically determined with the use of large (>10 μm diameter) fluorescent microspheres immersed in oil. The number of steps needed to optically section through the microsphere provides an estimate of actual step size (Prakash *et al.*, 1993).

IV. Estimation of Errors in Confocal Volume Measurements

A. Errors Introduced by Tissue Compression

Tissue preparation and mounting are important considerations in volume measurements using confocal microscopy (Bacallao *et al.*, 2006). Most importantly, the possibility of tissue compression during optical sectioning should be evaluated, especially with high NA lenses with short working distances. One simple approach is to determine the number of optical sections required to scan through a sample of known thickness, for example, a 50-μm thick section of densely stained tissue. In cryosections, sample thickness can be independently calibrated using optical densitometry (Blanco *et al.*, 1991).

B. Errors Introduced by Z-axis Distortion

The greatest source of error in confocal volume measurements is introduced by Z-axis distortion due to a number of factors (see above). Although the contribution of each of these factors could be independently evaluated, the most practical approach is to empirically determine the overall distortion in the Z-axis by assessing confocal images of fluorescent microspheres of known diameter (Prakash *et al.*, 1993). Even under optimized conditions matching optical section thickness and step size, some Z-axis distortion will be present. For example, in the illustration shown in Fig. 2, the optical section thickness was 0.84 μm as calculated by the

PSF of the 40×/1.3 NA oil-immersion objective lens. At step sizes of 0.82 and 1.00 μm, the *XZ* diameter of the 10.3 μm fluorescent (488/515 nm) latex microsphere suspended in immersion oil was 10.6 and 9.5 μm, respectively. In contrast, in the *XY* plane the measured diameter of the microsphere closely matched the specifications of the manufacturer. The differences in measured diameter of the microsphere in the *XZ* plane reflected *Z*-axis distortion. These types of measurements can be repeated for microspheres located at different depths within the working distance of the objective lens. Generally, *Z*-axis distortion worsens at the distal part of the working distance of the objective lens, that is, deeper in the specimen. For example, if fluorescent microspheres were imbedded in mounting medium, rather than immersion oil, and imaged, the *Z*-axis distortion is more pronounced and approximates more closely tissue conditions. In this case, we have estimated an ∼16% overestimation of microsphere volume (Prakash *et al.*, 1993). Such volume measurement errors attributed to *Z*-axis distortion are fairly consistent, and therefore they can be appropriately corrected.

C. Error Introduced by Multicolor Imaging

An additional error can be introduced in volume measurements of multicolor images because different light wavelengths are focused by the objective lens at different planes (dispersion). Therefore, image registration becomes a problem when using different fluorescent indicators, for example, the use of double- or triple-labeling to colocalize subcellular structures. To avoid this problem, the effects of dispersion introduced with each objective lens should be established. This can be accomplished by obtaining *XZ* sections of double- or triple-labeled fluorescent microspheres (e.g., MultiSpeck microspheres with excitation/emission wavelengths of 505/515 and 560/580 nm; Invitrogen). In the example shown in Fig. 3, a two-channel Bio-Rad MRC 600 Confocal System was used to image 4 μm microspheres (MultiSpeck) at two different wavelengths (excitation light: 488 and 568 nm, emission light: 515 and 590 nm, respectively). Registration error can be estimated by the *XZ* depth of microsphere at each wavelength. In addition, the

Fig. 3 Double-labeled fluorescent microspheres (4 μm in diameter; MultiSpeck, Invitrogen) were imaged using a two-channel Bio-Rad MRC 600 Confocal System with an Olympus 40×/1.3 NA objective lens. Correct alignment of the images at the two different wavelengths (emission: 488 and 568 nm; excitation: 515 and 590 nm) provides confirmation of the adequacy of instrument selection and calibration.

difference between the *XY* and *XZ* diameters can be used as an index of the overall distortion along the *Z*-axis for each individual wavelength (Aravamudan *et al.*, 2006; Mantilla *et al.*, 2008; Prakash *et al.*, 1995; Verheul *et al.*, 2004).

V. Practical Limitations of Using Confocal Microscopy for Volume Measurements

A. Size of the Data Set

As with other imaging technologies, improvements in confocal imaging have been accompanied by increased data content. Recent developments in computer technology have considerably expanded the capabilities of data acquisition hardware and software at much reduced cost. However, investigators are still faced with the immense task of probing through large numbers of optical sections. Furthermore, given the 2D nature of computer displays, manipulation and presentation of a set of optical sections can be a daunting task (Lucas *et al.*, 1996; Pawley, 2006a,b,c; Sabri *et al.*, 1997).

Several unbiased, stereological techniques have been used to estimate 3D variables (such as volume) from a limited set of 2D images. These techniques make no assumptions of object shape or orientation; therefore, they are applicable to any situation where random samples are selected (Howell *et al.*, 2002; Mayhew, 1991; Muhlfeld *et al.*, 2010; Royet, 1991; Weibel *et al.*, 2007). The Cavalieri principle is an example of such an unbiased stereological technique that has been verified extensively (Geinisman *et al.*, 1997; Gundersen and Jensen, 1987; Michel and Cruz, 1988; Prakash *et al.*, 1993, 1994). The tissue sectioning protocol of the Cavalieri principle systematically obtains parallel sections of 3D objects. Starting with a random section at one end of the object, uniformly spaced sections are sampled for measurement. Areas from regions of interest are measured in each section, and multiplied by the section spacing to obtain an estimate of volume for these areas. The uniformly spaced sections of fixed thickness are then used to estimate object volume by linearly interpolating between sampled sections. The method of optical sectioning in confocal microscopy also obtains a series of parallel sections of a 3D object; however, unlike the Cavalieri principle, the spacing between sections can be zero with complete confocal optical sectioning, that is, no interpolation is performed. The optical sectioning technique of confocal microscopy can thus be considered a limiting case of the Cavalieri principle, and therefore, the Cavalieri protocol can be applied to digitized images from a confocal system (Prakash *et al.*, 1993). In addition, in the Cavalieri principle, there is no need to match section thickness to the spacing between sections. Therefore, by interpolating between adjacent sections, the Cavalieri principle offers the advantage of reliable estimation of volume from a reduced number of confocal sections. Application of such an interpolation technique to a large set of confocal optical sections reduces data size and makes computations and measurements easier.

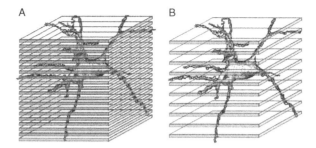

Fig. 4 Application of the Cavalieri principle to estimate object volumes from confocal image sets. Starting with a randomly determined section within the first n sections, every nth successive optical section is selected. A phrenic motoneuron was optically sliced, and the complete stack of optical sections is seen in (A). The Cavalieri principle allows manipulation of a much reduced set of optical sections (B), with minimal loss of information.

In the example shown in Fig. 4, a combination of confocal microscopy and the Cavalieri principle was applied to measurements of phrenic motoneuron somal volumes. In rats, the phrenic motoneuron pool was retrogradely labeled by injection of cholera toxin B fragment into the diaphragm muscle 2 days prior to tissue collection (Prakash *et al.*, 1994). The spinal cord was removed, fixed, and immunostained with Cy5-conjugated antibody to cholera toxin B fragment. Optical sections of the rat cervical spinal cord at 0.6 μm were obtained using a Bio-Rad MRC 600 laser confocal system mounted on an Olympus (BH2) upright microscope (with an Olympus ApoUV 40×/NA 1.3 oil-immersion lens). The long axis of phrenic motoneurons ranges from 30 to 50 μm. Step size was set at different sampling intervals, ranging from 0.6 to 3.0 μm. Thus, the number of sampled optical sections containing segments of a phrenic motoneuron varied from 10 to 80, depending on the sampling interval. Previous studies have suggested that reliable volume estimates can be obtained from approximately 10 sections using the Cavalieri protocol (Royet, 1991). There was no difference in the mean values of phrenic motoneuron somal volumes across the different step sizes (Fig. 5). However, it should be stressed that the size of the set of optical sections using a 3.0-μm step size was only 20% of that using a 0.6-μm step size.

B. Time Constraints in Data Acquisition

Spatial resolution is dependent only on the inherent properties of the optics involved. However, image acquisition results in interdependency between spatial and temporal resolution. Depending on the confocal system, there are temporal limitations in image acquisition in both the XY and XZ planes. The introduction of acousto-optical control has markedly enhanced temporal resolution in the XY plane. Confocal systems are currently available and have full frame (1024 × 1024 and up to 4096 × 4096 pixels) acquisition at video rates (30 frames/s) and limited

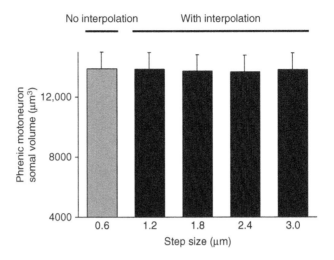

Fig. 5 Bar graph of motoneuron somal volume estimates using different step size intervals with the Cavalieri principle. Noninterpolated confocal measurements correspond to a step size of 0.6 μm. The bar represents mean somal volumes and the error bar the standard error. There is no difference in the mean somal volumes or variance when imaging 20% of the image set (step size 3.0 μm) using the Cavalieri principle.

frame acquisition at up to 1000 frames/s. Of course, with the faster acquisition rates, pixel dwell time is reduced and either laser intensity must be increased or frame averaging must be applied to provide adequate signal-to-noise ratio.

Even with the improvements in frame acquisition rate, the main limitation for volume-repeated volume measurements in time (4D analysis) is control of the Z-axis. With control of the movement of the microscope stage, the mass of the stage introduces significant inertia, causing marked delays for optical sectioning. For example, under unloaded conditions (with motor not attached to the focusing knob of the microscope, or attached to the microscope but with no lenses or specimen stage), the focus motor of an Noran Odyssey XL Confocal Imaging System can achieve a single step in ∼20 ms. However, as with most other focus motors, the Odyssey system also has extremely fine step resolutions (50 nm). Since optical section thickness is optimal at ∼0.6–1.0 μm, several single motor steps are required. Therefore, ∼400 ms is required to achieve a 0.8-μm step size. In addition, delays in software control and communication with the focus motor may prolong this time even further. Although it is difficult to directly determine the temporal resolution for an imaging system during 3D volume measurements, an average time per image can be obtained by measuring the time taken for collecting a set of optical sections. For example, collecting a set of 0.8-μm optical sections through a 10-μm distance requires ∼6 s using the Odyssey system, representing a ∼500-ms delay for each individual step. The time for image acquisition also contributes to the overall delay in 4D imaging, but compared to the delay introduced by Z-axis control, this delay is relatively insignificant. For example, at video rate (30 frames/s) frame acquisition requires ~33 ms. Increasing the acquisition rate to 480 frames/s saves only ~30 ms.

The temporal resolution of 4D volume measurements may be improved by using a faster stepper motor. However, faster motors are also likely to suffer from oscillation artifacts due to the rapid starts and stops ("ringing"), particularly if massive stages are controlled. Therefore, considerable delay may be introduced waiting for the oscillations to dampen. An alternative approach is the use of piezoelectric control of the microscope objective lens. For example, when using piezoelectric control of the objective lens on a Noran Odyssey Confocal System at 480 frames/s and 1 μm step size, 10 optical slices can be obtained in ~67 ms. The major cause of this delay results from the fact that image acquisition is noncontinuous. An optical section is obtained at a set depth, then after the next Z-axis position is reached, another optical section is obtained. This incremental adjustment of the Z-axis position is time consuming because of the mechanical linkages involved. Improvements in 4D acquisition rate could be achieved if Z-axis control was continuous and image acquisition was synchronized to the beginning and end of objective lens movement. The advantage of this approach is that the interaction between image acquisition and Z-axis control is minimal. However, the image set would contain a motion-blurred representation of the object, but it would be possible to deblur the images using mathematical transforms that are currently being used to correct motion artifacts in MRI and CT images (Russ, 2007).

C. Biological Applications of 4D Analysis

Obviously, many intracellular events occur in space, for example, within localized compartments of the cell. These events can be repeatedly sampled during the collection of a set of 3D images. As mentioned above, the main limitation currently is the control of the Z-axis. In the collection of sequential sets of 3D optical sections, if there is no time delay between sequential 3D sets, the time elapsed before a given optical plane is sampled again is simply the total time taken to collect one full stack of 3D images. Under these conditions, the 4D resolution obviously depends on the number of 3D sections. When the different factors contributing to time delays in sequential 3D image acquisition are considered (see above), it may be quite difficult to attain a 3D temporal resolution better than ~600 ms. For example, we repeatedly imaged a 3D set of optical sections of an isolated porcine airway smooth muscle cell. With the Noran Odyssey XL Confocal System, 10 optical sections with a 1-μm step size can be obtained at 15 frames/s, that is, every 667 ms. The Nyquist theorem also applies to 4D analysis of biological intracellular events; therefore, only cellular events occurring with a frequency of every 1334 ms or more can be reliably detected. From previous studies (Prakash et al., 1999), the temporal profile of Ca^{2+} sparks in a porcine airway smooth muscle showed a ~200 ms peak within a very localized area of the cell. Therefore, these imaging conditions can fortuitously detect the occurrence of

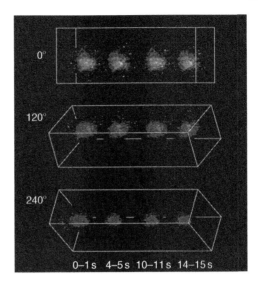

Fig. 6 Rotational views of temporal changes in the 3D volume distribution of Ca^{2+} within a spark in a TSM cell. Such rotational views of 3D reconstructions allow representation of 4D data sets such that temporal changes in all three spatial dimensions can be evaluated. In this example, the volume distribution of the Ca^{2+} spark does not appear to change considerably over time.

sparks in airway smooth muscle cells and would not allow the characterization of the spark (Fig. 6).

In addition to the study of localized cellular events, 4D confocal imaging holds promise for the study of events where the primary result is the actual change in morphology (i.e., shape and/or size) (Paddock, 1996; Petroll *et al.*, 1996; Sabri *et al.*, 1997). In particular, studies relating to cell division can benefit from relatively rapid 3D image acquisition rates, whereas studies of cell differentiation might require more prolonged intervals between sampling. For example, regulation of cell volume in living chondrocytes has been studied in response to osmotic stress or trauma (Bush *et al.*, 2005; Errington *et al.*, 1997). These and other applications are of great interest in the biological sciences.

D. Limitations in Image Processing and Display

Innovative solutions for the limitations imposed by the rate of data storage include resizing of the image to increase the rate of data acquisition from a smaller frame. However, this can result in important loss of spatial information. Obviously, more research has to focus on methods to improve data collection. Fortunately, the impressive improvements in video hardware and software and the ever-faster computer display terminals are likely to greatly reduce problems relating to rapid display. Furthermore, access to extremely rapid, inexpensive data storage devices will also alleviate some of the data storage problems. For example, with new video display technology and computer hardware, single images of reasonable size (e.g., 1024 × 1024 pixels) can be easily displayed and saved within 10–30 ms. However, because of the huge amounts of image data that can be stored with 4D confocal imaging, it is essential that the rate of image acquisition be chosen based on the process of interest, for example, 4D registration of intracellular dynamics or changes in cellular morphology.

After the image set has been collected, the display of the optical section information should allow easy, accessible interpretation of the data content (Lucas *et al.*, 1996; Pawley, 2006a,b,c). For 3D objects, this usually means a 2D representation, using either a computer display or paper. Most image-processing software packages with capabilities for 3D reconstruction employ special algorithms such as voxel gradient shading, depth gradient shading, surface shading, and maximum intensity projections (see Russ, 2007 for a review). Using these algorithms, the 3D reconstructions can be viewed at any desired angle and resectioned if necessary to view the interior of the specimen. The processing of these large amounts of information, imported into image-processing packages such as ANALYZE, VoxelView, and NIH Image, still represents a very time- and effort-consuming task. While the visualization environment for 3D data sets appears to be well under control, the adequate representation of 4D data is still an unresolved issue, although some impressive options exist (Russ, 2007). One option is to display a 2D view of the 3D reconstruction at one defined angle. The time sequence of 3D reconstructions at a certain angle would appear as a movie (when the successive images are played back as video frames). However, when displaying only the view from only one angle, the representation has eliminated the 4D character of the data and made it only 3D (2D spatial over time). This can be obviated by the simultaneous display of multiple angles of view, which would prove very valuable in particular for the spatial analysis of intracellular dynamics over time. In addition, rotational sets can be constructed for successive 3D data sets, where the entire 4D data set rotates as the time sequence is displayed. Obviously, the choice of an appropriate visualization procedure will depend on the desired point to emphasize. For example, to show the temporal changes in the volume occupied by an object (or even compartments within the object), temporal sequences of 3D reconstructions at one or multiple angles would yield the most information in a user-friendly manner. The impressive improvements that are constantly being made in computer technology will surely help in this regard, and evidently, constant attention to optimal data acquisition could limit the amount of data collected to that necessary to address the question(s) of interest.

E. Optical Imaging at Superresolution

Recent developments in optical sectioning microscopy have allowed breaking the laws of physics (and the diffraction limit), effectively reducing the PSF of conventional optics by an order of magnitude. A detailed review of the multiple techniques, their properties, and limitations is beyond the scope of this review.

Two main families of methods have evolved to substantially improve resolution in optical microscopy. In the first (comprising 4Pi and I^5M), two opposing lenses are used to coherently illuminate a sample resulting in constructive interference of the counter-propagating wavefronts that greatly reduce the size of the axial focal plane (Bewersdorf *et al.*, 2006). However, the so-called "sidelobes" above and below the focal plane produce ghost images that must be removed *post hoc* with additional image processing. In fact, using most currently available lenses, applications must be limited to samples with minimal thickness. Although using confocal or multiphoton microscopy helps overcome some of these limitations, the applicability to thick sections will require additional technological developments. Thus, current biological applications are generally limited to single cell layers.

The second family of superresolution methods is based on reversible saturable optical fluorescence transitions (RESOLFT) (Hell *et al.*, 2006). The first introduction of these systems addressed resolution in the XY axis by using scanning stimulated emission depletion (STED) microscopy in a two-laser apparatus. In STED, one laser is used to provide high-intensity light pulses that stimulate emission by a fluorophore until it returns to its ground state. Synchronized pulses from the second laser are used to illuminate the sample such that the excitation spot (diffraction-limited) and the STED spot (with a central minima) overlap, thus creating a confined region of excited fluorophore molecules below the diffraction limit. STED requires specific fluorophore characteristics (e.g., resulting in efficient stimulated emission depletion yet minimal bleaching) and thus, new fluorophores with custom-designed applications are under development.

Although still in their infancy, these new methods for superresolution microscopy offer enormous potential for subcellular and near-molecular characterization of biological specimens with nanometer resolution. The application of these super-resolution imaging techniques to volumetric measurements will require further development so that thicker samples may be used; although recent developments are very promising (Juette *et al.*, 2008; Schmidt *et al.*, 2008). Understanding concepts described within this chapter (e.g., size and shape of the PSF) will be important to quantitative analyses of subcellular structure in 3D.

Acknowledgments

This work is supported by grants AR051173, HL096750, HL074309, and HL088029 from the National Institutes of Health and the Mayo Foundation.

References

Abbe, E. (1884). *J. R. Microsc. Soc.* **4**, 348.
Aravamudan, B., Mantilla, C. B., Zhan, W. Z., and Sieck, G. C. (2006). *J. Appl. Physiol.* **100**, 1617.
Art, J. (2006). *In* "Handbook of Biological Confocal Microscopy," (J. B. Pawley, ed.), 3rd edn., p. 251. Springer Science+Business Media, LLC, New York.

Bacallao, R., Sohrab, S., and Phillips, C. (2006). *In* "Handbook of Biological Confocal Microscopy," (J. B. Pawley, ed.), 3rd edn., p. 368. Springer Science+Business Media, LLC, New York.

Bewersdorf, J., Egner, A., and Hell, S. W. (2006). *In* "Handbook of Biological Confocal Microscopy," (J. B. Pawley, ed.), 3rd edn., p. 561. Springer Science+Business Media, LLC, New York.

Blanco, C. E., Fournier, M., and Sieck, G. C. (1991). *Histochem. J.* **23**, 366.

Brakenhoff, G. J., van der Voort, H. T., van Spronsen, E. A., and Nanninga, N. (1989). *J. Microsc.* **153** (Pt 2), 151.

Bush, P. G., Hodkinson, P. D., Hamilton, G. L., and Hall, A. C. (2005). *Osteoarthr. Cartilage* **13**, 54.

Centonze, V., and Pawley, J. B. (2006). *In* "Handbook of Biological Confocal Microscopy," (J. B. Pawley, ed.), 3rd edn., p. 627. Springer Science+Business Media, LLC, New York.

Conchello, J. A., and Lichtman, J. W. (2005). *Nat. Methods* **2**, 920.

Egner, A., and Hell, S. W. (2006). *In* "Handbook of Biological Confocal Microscopy," (J. B. Pawley, ed.), 3rd edn., p. 404. Springer Science+Business Media, LLC, New York.

Errington, R. J., Fricker, M. D., Wood, J. L., Hall, A. C., and White, N. S. (1997). *Am. J. Physiol.* **272**, C1040.

Geinisman, Y., Gundersen, H. J., van der Zee, E., and West, M. J. (1997). *J. Neurocytol.* **25**, 805.

Gundersen, H. J., and Jensen, E. B. (1987). *J. Microsc.* **147**, 229.

Hell, S. W., Willig, K. I., Dyba, M., Jakobs, S., Kastrup, L., and Westphal, V. (2006). *In* "Handbook of Biological Confocal Microscopy," (J. B. Pawley, ed.), 3rd edn., p. 571. Springer Science+Business Media, LLC, New York.

Hiraoka, Y., Sedat, J. W., and Agard, D. A. (1990). *Biophys. J.* **57**, 325.

Howell, K., Hopkins, N., and McLoughlin, P. (2002). *Exp. Physiol.* **87**, 747.

Inoue, S. (2006). *In* "Handbook of Biological Confocal Microscopy," (J. B. Pawley, ed.), 3rd edn., p. 1. Springer Science+Business Media, LLC, New York.

Juette, M. F., Gould, T. J., Lessard, M. D., Mlodzianoski, M. J., Nagpure, B. S., Bennett, B. T., Hess, S. T., and Bewersdorf, J. (2008). *Nat. Methods* **5**, 527.

Keller, H. E. (2006). *In* "Handbook of Biological Confocal Microscopy," (J. B. Pawley, ed.), 3rd edn., p. 145. Springer Science+Business Media, LLC, New York.

Lucas, L., Gilbert, N., Ploton, D., and Bonnet, N. (1996). *J. Microsc.* **181**(Pt 3), 238.

Mantilla, C. B., Sill, R. V., Aravamudan, B., Zhan, W. Z., and Sieck, G. C. (2008). *J. Appl. Physiol.* **104**, 787.

Matsumoto, B. (ed.) (2002). "Cell Biological Applications of Confocal Microscopy," Elsevier/Academic Press, Amsterdam, Boston.

Mayhew, T. M. (1991). *Exp. Physiol.* **76**, 639.

Michel, R. P., and Cruz, O. L. M. (1988). *J. Microsc.* **150**, 117.

Muhlfeld, C., Nyengaard, J. R., and Mayhew, T. M. (2010). *Cardiovasc. Pathol.* **19**, 65–82.

Paddock, S. W. (1996). *Proc. Soc. Exp. Biol. Med.* **213**, 24.

Pawley, J. B. (ed.) (2006). "Handbook of Biological Confocal Microscopy," 3rd edn. Springer Science+Business Media, LLC, New York.

Pawley, J. B. (2006b). *In* "Handbook of Biological Confocal Microscopy," (J. B. Pawley, ed.), 3rd edn., p. 20. Springer Science+Business Media, LLC, New York.

Pawley, J. B. (2006c). *In* "Handbook of Biological Confocal Microscopy," (J. B. Pawley, ed.), 3rd edn., p. 59. Springer Science+Business Media, LLC, New York.

Petroll, W. M., Jester, J. V., and Cavanagh, H. D. (1996). *Int. Rev. Exp. Pathol.* **36**, 93.

Prakash, Y. S., Smithson, K. G., and Sieck, G. C. (1993). *Neuroimage* **1**, 95.

Prakash, Y. S., Smithson, K. G., and Sieck, G. C. (1994). *Neuroimage* **1**, 325.

Prakash, Y. S., Smithson, K. G., and Sieck, G. C. (1995). *J. Neurocytol.* **24**, 225.

Prakash, Y. S., Kannan, M. S., and Sieck, G. C. (1999). *In* "Fluorescent and Luminescent Probes for Biological Activity," (W. T. Mason, ed.), 2nd edn., p. 316. Academic Press, London.

Royet, J. P. (1991). *Prog. Neurobiol.* **37**, 433.

Russ, J. C. (2007). The Image Processing Handbook, 5th edn. CRC Press, Boca Raton, FL.

Sabri, S., Richelme, F., Pierres, A., Benoliel, A., and Bongrand, P. (1997). *J. Immunol. Methods* **208**, 1.

Sandison, D. R., Williams, R. M., Wells, K. S., Strickler, J., and Webb, W. W. (1995). *In* "Handbook of Biological Confocal Microscopy," (J. B. Pawley, ed.), 2nd edn., p. 39. Plenum Press, New York.

Schild, D. (1996). *Cell Calcium* **19,** 281.

Schmidt, R., Wurm, C. A., Jakobs, S., Engelhardt, J., Egner, A., and Hell, S. W. (2008). *Nat. Methods* **5,** 539.

Shaw, P. (1994). *Histochem. J.* **26,** 687.

Shaw, P. J. (2006). *In* "Handbook of Biological Confocal Microscopy," (J. B. Pawley, ed.), 3rd edn., p. 453. Springer Science+Business Media, LLC, New York.

Sheppard, C. J. R., Gan, X., Gu, M., and Roy, M. (2006). *In* "Handbook of Biological Confocal Microscopy," (J. B. Pawley, ed.), 3rd edn., p. 442. Springer Science+Business Media, LLC, New York.

Stevens, J. K., Mills, L. R., and Trogadis, J. E. (eds.) (1994). "Three-Dimensional Confocal Microscopy: Volume Investigation of Biological Specimens.," Academic Press, San Diego, CA.

Toomre, D., and Pawley, J. B. (2006). *In* "Handbook of Biological Confocal Microscopy," (J. B. Pawley, ed.), 3rd edn., p. 221. Springer Science+Business Media, LLC, New York.

Turner, J. N., Swann, J. W., Szarowski, D. H., Smith, K. L., Shain, W., Carpenter, D. O., and Fejtl, M. (1996). *Int. Rev. Exp. Pathol.* **36,** 53.

Verheul, A. J., Mantilla, C. B., Zhan, W. Z., Bernal, M., Dekhuijzen, P. N., and Sieck, G. C. (2004). *J. Appl. Physiol.* **97,** 1715.

Weibel, E. R., Hsia, C. C., and Ochs, M. (2007). *J. Appl. Physiol.* **102,** 459.

Wilson, T. (1989). *J. Microsc.* **153**(Pt 2), 161.

Wilson, T., and Hewlett, S. J. (1990). *J. Microsc.* **160**(Pt 2), 115.

Wilson, T., and Sheppard, C. J. R. (1984). Theory and Practice of Scanning Optical Microscopy. Academic Press, London.

CHAPTER 10

Quantitation of Phagocytosis by Confocal Microscopy

George F. Babcock★,† and Chad T. Robinson, MS★,†

★Department of Surgery and Cell Biology
University of Cincinnati College of Medicine
Cincinnati, Ohio, USA

†Shriners Hospitals for Children
Cincinnati Burns Institute
Cincinnati, Ohio, USA

I. Update

Recent years have seen an increase in the availability of affordable, high-resolution digital camera systems (e.g., ORDCA, Hamamatsu; AxioCam HR, Zeiss). Pairing these systems with an increase in availability of image analysis software and fluorescent reagents has allowed for better intravital analysis. Also examining the velocity and magnitude of oxidant production allows for a better evaluation of phagocytic efficiency (Fig. 1).

DOI: 10.1016/B978-0-12-384658-7.00010-2

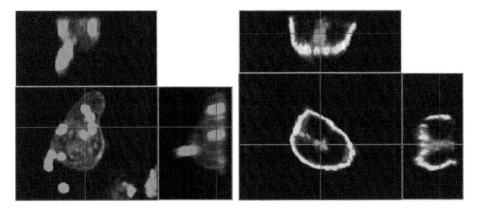

Fig. 1 Time series of human neutrophils phagocytosing and oxidizing *S. aureus*. Fluorescence overlaid with differential interference contrast (DIC). A multitrack, dual channel, two sensitivity approach is utilized to visualize early punctuate structures verses later whole cell signal using a Zeiss LSM510. This allows one to set up a live cell time lapse experiment and not change settings during capture.

At a molecular rather than physiological level, phagocytosis is increasingly being characterized and quantified by the colocalization of a tagged proteins of interest within the phagolysosome. This phagolysosome phenotyping shows promise for intervention with less collateral effects.

Analysis of phagocytosis by indirect means is increasingly being accomplished by tracking the production of oxygen radicals. This has been accomplished in the past with such reagents as dihydroxyrhodamine (DHR). More recently, specialize systems, such as fc oxyburst composed of bovine serum albumin (BSA) that has been covalently linked to dichlorodihydrofluorescein (H2DCF) and then complexed with a purified rabbit polyclonal anti-BSA antibody, have been developed. When these immune complexes bind to Fc receptors, the nonfluorescent molecules are internalized within the phagovacuole and subsequently oxidized to green-fluorescent 2′,7′-dichlorofluorescein. Similarly, pH or calcium-sensitive reagents complexed with particles or microbes are also more widely available.

- Helpful hints:

  Properly aligne the microscope (Kohler illumination, fluorescent alignment, etc.).

  Understand multitrack capture.

  Filter set and dichroic mirror schemes can be evaluated using manufacturers web site.

  Check fluorescent pair compatibility with such programs as Invitrogen's SpectraViewer before use.

  Pair microscope with more quantifiable population averages such as flow cytometry (e.g., ox burst in suspension) or westerns (e.g., immunoprecipitation to verify fret or colocalization).

- Useful web sites:

  www.microscopyu.com (general fluorescent microscopy)

  www.olympusfluoview.com/resources/index (general fluorescent microscopy)

  www.msa.microscopy.org/Ask-A-Microscopist (troubleshooting)

  www.probes.invitrogen.com/resources/spectraviewer (fluorescent pair evaluation)

  www.ibidi.com (specialized slides)

  www.invitrogen.com/site/us/en/home/References/Molecular-Probes-The-Hand-book.html (reagents).

II. Quantification of Phagocytosis by Confocal Microscopy

Phagocytosis has traditionally been defined as being mediated by leukocytes and functioning as the first line of defense against infection. These phagocytic leukocytes, which include neutrophils, monocytes, and tissue macrophages, are considered professional phagocytic cells (Bjerknes, 1998). It is now clear that many other cell types are also capable of phagocytosing various types of particles.

The phagocytic process is the engulfment of particles, which involves the expansion of the plasma membrane to engulf the particle. When performing assays, it is important to separate the events by which the plasma membrane is extended (phagocytosis) from the events that involve the inversion of the plasma membrane (endocytosis). These two processes are often difficult to separate experimentally. Because the endocytic process is only capable of capturing smaller particles, this difference in particle size can be used to differentiate the two processes.

Phagocytic cells are capable of engulfing particles of less than 1 μm in size to large particles up to many micrometers in size. In fact, there have been reports of cells phagocytosing other cell types, which are larger actual phagocytosing cells. Phagocytosis may be divided conceptually into a two-step process (Klebanoff and Clark, 1978; Silverstein et al., 1977). The first step is the attachment of the particle to receptors on the plasma membrane of the phagocytotic cell. The second step involves internalization of the bound particle. Problems associated with the quantification of phagocytosis are related to the nature of this two-step process. The number of particles or relative amount of material phagocytosed must be quantified. If the cell type being examined is a professional phagocyte, during and following phagocytosis, the cells produce large amounts of reactive oxygen intermediates (Babior, 1984). In conjunction with digestive enzymes discharged into the phagosomes, these compounds may result in the quenching of some fluorochromes and/or the destruction of the phagocytosed particle itself. This is further complicated by the fact that the phagocytosed particles, which must be quantified, are intracellular; so in addition to quenching or destruction of the particles, intracellular organelles and/or granules may obscure the particle. An additional complication relates to the fact that particles bind to the outside of cells before being internalized. This external binding, which may be the stronger of the two signals, must be separated from internalization. Confocal microscopy allows the user to focus on various planes within the cell, which permits the investigator to separate truly internalized particles from externally bound particles.

III. Opsonins and Receptors

Some particles in their native form may be ingested by cells without further treatment. However, the phagocytosis of most microorganisms and many other types of particles requires or is enhanced greatly by serum factors termed opsonins (Klebanoff and Clark, 1978; Silverstein *et al.*, 1977). Opsonins form a bridge between the particle and the phagocytic cell (Brown, 1995). The major opsonins include immunoglobulins of the IgG class and complements components of the C3 class. These opsonins bind to several important receptors on phagocytic cells and initiate adherence and uptake of the opsonin-coated particle (Table I). Even in model systems where opsonins are not required, the assay benefits from their inclusion.

IV. Methods Used to Measure Phagocytosis

Many procedures and variations of these procedures have been used to monitor phagocytosis. Some of the common methods are quantitative whereas others are qualitative. Almost any particle of the proper size can be used to measure phagocytosis. However, most of the published phagocytosis procedures have utilized bacteria, yeast, red blood cells, or beads (polystyrene). Particles used for measuring phagocytosis are usually labeled with some type of reagent, which allows for easier detection and more sensitivity. Particles have been labeled with opaque dyes, fluorescent dyes, and radionuclides such as ^{3}H or ^{57}Cr among others (Bjerknes *et al.*, 1989; Goodell *et al.*, 1978; Scheetz *et al.*, 1976). In some cases,

Table I
Opsonin Binding Receptors on Phagocytic Cells

Receptor	Cluster designation	Ligand	Phagocytic cell expression
FcγI	CD64	IgG$_1$, IgG$_3$	Monocytes, neutrophils
FcγII	CD32	IgG$_1$, IgG$_2$, IgG$_3$, IgG$_4$	Monocytes, neutrophils
FcγIII	CD16	IgG$_1$, IgG$_3$	Monocytes, neutrophils, natural killer cells
CR1	CD35	C3	Monocytes, neutrophils
CR3	CD11b/CD18	IC3b	Monocytes, neutrophils, natural killer cells
Various lectin receptors	None	Mannose among other carbohydrates	Most cell types

the quantification of phagocytosis has been accomplished using viable organisms and determining the number of internalized colony-forming units (cfu) (Amoscato *et al.*, 1983).

Confocal microscopy and flow cytometry have become the methods of choice for analyzing phagocytosis, as these methods can provide quantifiable data in a relative short time compared to most other methods. Although confocal microscopy can be utilized for phagocytosis with a particle, which can be visualized by ordinary light microscopy, cells or particles labeled with fluorescent dyes are the most popular. By exciting the fluorescent particle with the appropriate wavelength of light, the particles become much easier to detect and quantify.

V. Choosing Proper Particle for Phagocytosis

Each investigator must decide which type of particle is optimal for their particular model. Some of the properties, as well as the pros and cons, for each of several types of particles are discussed in Table II. It should be noted that Table II is not all inclusive and represents only a partial list.

Table II
Particles Used for Phagocytic Assays

Particle	Detection	Advantages	Disadvantages
Polystyrene beads	Can be observed without further labeling; can be purchased with broad spectrum fluorochrome incorporated into bead; specific fluorochromes can be bound covalently to bead surface	Can be obtained in many sizes; incorporated fluorochromes are not quenched easily; resistant to destruction with most cellular enzymes; specific ligands can be attached to surface of beads to examine receptor binding	Nonviable, nonbiological particle; multiple receptor–ligand interaction; difficult to examine; may bind inherently to lectin receptors
Bacteria	Can be labeled with variety of fluorochromes	Can be used to measure microbiocidal activity and phagocytosis	Difficult to see without labeling due to small size fluorochrome labeling may change adherence and phagocytosis; can be destroyed by professional phagocytes and fluorescence quenches
Yeast	Can be labeled with variety of fluorochromes; can be observed without labeling (externally)	Same as for bacteria	Same as for bacteria; difficult to observe internally with labeling
Red blood cells	Can be observed without labeling; autofluorescent at certain excitation wavelengths	Difficult to quantify by confocal microscopy other than by manual counting	Can be destroyed by cellular enzymes, making difficult to detect; rat@ large particle

VI. Fluorochrome Labeling of Bacteria

Latex particles can be purchased with fluorescent labels incorporated either into the particle or on the surface of the bead. However, the investigator must label the bacteria or yeast. Many dyes can be used to label bacteria; however, fluorescein is the most commonly used. Although it is not necessarily the best reagent to use for this purpose, it is reasonably priced and easy to use. Two methods are commonly employed to label bacteria or yeast with fluorescein. The simplest is to add fluorescein isothiocyanate (FITC) directly to the culture medium (White-Owen *et al.*, 1992). To 100 ml of the appropriate culture medium (trypticase soy broth, brain heart infusion, etc.), 25 µg of FITC is added and stirred gently. All remaining procedures should be performed in reduced lighting, especially fluorescent lighting. The medium is then inoculated with the organism (bacteria or yeast) and incubated with shaking for 18 h at 37 °C. Bacteria are then collected by centrifugation at $2000 \times g$ for 20 min at 4 °C. The organisms should be washed three times in a large volume of phosphate-buffered saline (PBS, 250 ml), recollected, and resuspended in Hanks' balanced salt solution (HBSS) to the appropriate concentration. Organisms prepared in this manner appear to retain their normal growth and attachment properties. Organisms can be frozen without further processing and stored in the dark at −70 °C for at least 6 months. Labeling by this method produces bright and generally stable fluorescence. However, some release of fluorescein over time has been noted. If nonviable organisms are desired, they can be inactivated by heating at 60 °C for 30–60 min following the labeling and washing procedure. Chemical treatment such as formaldehyde has also been used. It should be noted that inactivating bacteria, especially by chemical means, may change the binding properties of the bacteria.

A second labeling method couples the ε-amino group of lysine on the fluorochrome to the bacterial proteins through a thiourea bond (Goding, 1986). Bacteria are washed twice and resuspended in 2 ml of carbonate/bicarbonate buffer (pH 9.5, 8.6 g Na_2CO_3 and 17.2 g $NaHCO_3$ in 1 liter of H_2O). The FITC is dissolved in dimethyl sulfoxide at 10 mg/ml. Two hundred microliters of this solution is added to the bacterial suspension slowly under constant stirring. The mixture should be rotated end over end for 90 min at room temperature. Bacteria should then be washed three times in 100 ml of PBS, aliquoted, and frozen as described earlier. Fluorescein bound by this procedure is very stable if used or stored in the dark. However, the properties of some strains of bacteria and yeast are altered.

VII. Equipment

The minimum equipment required to quantify phagocytosis is a confocal microscope. However, quantification using only confocal microscopy would be tedious at best. An epifluorescence attachment allows for the use of fluorochrome-labeled particles. A confocal microscope with an epifluorescence attachment should be considered the practical minimum. Excitation of fluorochromes with laser light and collection of light emissions using photomultiplier tubes make quantification rather simple (Rodeberg *et al.*, 1997). Other desirable features include laser scanning to automate the process, a temperature-controlled stage for "real time" phagocytosis measurements, and a color charge-coupled device video camera. In addition, an integrated computer with software to analyze the fluorescent signals or the color intensity is also desirable. If adherent cells are used, an inverted confocal microscope is usually necessary.

VIII. Separation of External from Internal Particles

The use of confocal microscopy by itself allows the investigator to separate surface-adherent particles from phagocytosed particles. Although this procedure can be performed with unlabeled particles, fluorescence labeling improves the ability to detect the phagocytosed particle (Fig. 2). For quantification, the individual fluorescent particles detected by fluorescence microscopy must be observed in several planes below the cell surface and then must be counted manually. Particles in a minimum of 100 cells (preferably 200) should be counted. A more efficient method of separating internal particles from external particles is to mark the external particles in a manner such that they can be separated from the internal ones. The methods used most commonly can be divided into two groups. One group of methods involves quenching or resonance energy transfer of the fluorochrome on the external particle to a different emission wavelength, which can be separated from the emission wavelength of the internal particle. Dyes such as ethidium bromide or trypan blue have commonly been used for this purpose (Bassoe, 1993; Fattorossi *et al.*, 1989). These dyes are added immediately prior to reading the assay at final concentrations of 3 mg/ml for trypan blue and 50 μg/ml for ethidium bromide. Another quenching method is the use of Immuno-Lyse (Coulter Corp., Hialeah, FL), which can be used when phagocytic assays are performed on whole blood (White-Owen *et al.*, 1992). This reagent quenches the fluorescence of the external particles and lyses red blood cells.

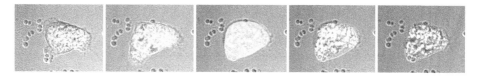

Fig. 2 Orthogonal view created from z-stacks of adherent neutrophils phagocytosing *S. aureus*. (A) Bacteria labeled with BacLight RED can be seen inside and outside the body of the cell. (B) Plasma membrane labeled with Wheat germ agglutinin Alexa488 surrounding fully phagocytosed *S. aureus*. (See Plate no.8 in the Color Plate Section.)

The second set of methods is to use an antibody, which reacts with the particles being used in the assay (Hess *et al.*, 1997). Particles, which are phagocytosed, are isolated and do not react with the antibody, whereas external particles bind the antibody. The antibody can be fluorochrome labeled directly or a fluorochrome-labeled anti-immunoglobulin can be used to detect the primary antibody. It is important to choose an antibody that reacts only with the external particle and that is labeled with a fluorochrome that produces an emission wavelength that can be separated from the fluorochrome associated with the particle. The choice of the fluorochrome for the antibody depends on not only the fluorochrome chosen for the particle but also the wavelengths available for excitation and the filters available for emission. A list of commonly used fluorochromes appears in Table III.

IX. Quantification of Phagocytosis by Confocal Microscopy

Many cell types besides the "professional phagocytes" can phagocytose particles. However, the same procedures can be used to quantify phagocytosis for all cell types. Adjustments must be made to the general procedure depending on the cell type being used. Adherent cells in particular require slight modifications in basic procedures. These modifications are mentioned at the appropriate points in the protocol. The incubation period required for maximal phagocytosis varies greatly, depending on the type of cells being analyzed. This difference can be from minutes to many hours, especially for "nonprofessional" phagocytes such as endothelial cells, which may require especially long incubation periods for maximal phagocytosis to occur. Each investigator must perform a time course

Table III
Common Fluorochromes Useful in Quantifying Phagocytosis

Fluorochrome	Excitation[a] (common wavelengths used)	Emission[a] (common wavelengths collected)
Fluorescein	450–490[b], 488[c]	520–530
Alexa488	450–490[b], 488[c]	520–530
Phycoerythrin	450–490[b], 488[c]	560–590
Rhodamine	510–560[b], 514[c]	610–620
Cy3	510–560[b], 514[c]	610–620
Texas Red	540–580[b], 568[d]	600–660
Cy5	633[e]	675
Allophycocyanin	633[e]	675
Cy7	752[d]	800

[a]Excitation and emission wavelengths used most commonly by typical laser-driven or epifluorescence microscopes. The optimal wavelengths are often different.
[b]Excitation with a mercury arc lamp.
[c]Excitation with an argon ion laser.
[d]Excitation with a krypton ion laser.
[e]Excitation with a helium–neon laser.

experiment to determine when optimal phagocytosis occurs in their particular system. The following procedure is written for neutrophils, which are nonadherent cells that phagocytose particles rapidly. Differences in the technical aspects of the procedures, which must be adjusted when using adherent cells, are stated. The particles used as the phagocytic reagent in this protocol are fluorescein-labeled *Staphylococcus aureus*.

Neutrophils are obtained from whole blood. Phagocytosis can be performed directly on whole blood or on isolated neutrophils. If whole blood is used the neutrophils must be separated from other leukocytes either visually or by the use of a specific marker. In the following protocol, isolated neutrophils are used.

Sufficient numbers of cells can be obtained from 5 ml of blood using ethylene-diaminetetraacetic acid (EDTA) or heparin as the anticoagulant. The blood is separated into neutrophils on mononuclear leukocytes by centrifugation on neutrophil isolation media (Babcock, 1993). Five milliliters of blood is layered over 5 ml of modified Ficoll–Hypaque solution in 15-ml conical centrifuge tubes. The sample is centrifuged at $550 \times g$ for 30 min at 25 °C. The neutrophil band should be collected. This band is the second band going from top to bottom.

The neutrophils are washed twice by centrifugation and resuspended in 5-ml volumes of HBSS. The washed cells are then resuspended to 5×10^6 cells/ml in HBSS or cell culture medium. Cell culture medium is preferred and RPMI 1640 works well. One hundred microliters of this cell suspension is added to 12×75-mm round-bottom polystyrene tubes. Ten microliters of serum as a source of opsonin is added to each tube. Autologous serum should be used when possible, but the sera of certain species are low in opsonic ability and should probably be avoided. If adherent cells are used, they are plated in microwell tissue culture plates (12-well) or Lab-Tek chamber slides (4-well) and grown to confluency. Before use, 100 µl of serum as a source of opsonin is added. *Note*: If fetal bovine serum (FBS) is used, the medium must still be supplemented with opsonins as FBS is usually deficient in opsonic activity. The assay should be performed in triplicate with three tubes or wells being utilized for *each* time point.

The labeled particles, in this case fluorescein-labeled *S. aureus*, are added to each well. The *S. aureus* should be resuspended in culture medium or HBSS to a total volume of 10 µl (100 µl for the adherent cell cultures). Several ratios of particles/phagocyte should be used to determine the optimal ratio. Generally, the optimal concentration ranges from a ratio 2–3:1 particles/phagocytic cell for endothelial cells to 25:1 or even more for neutrophils. In this particular procedure, the 10:1 ratio has been determined to be optimal. The bacteria should be sonicated for 5 s in a bath-type sonicator to break up clumps immediately before using.

Because neutrophils phagocytose bacteria rapidly, time points (incubation times at 37 °C) of 0, 2, 5, 10, 15, 20, and 30 min are used. Following the incubation period, 10 µl of cytochalasin D (5 µM final concentration) is added to the sample, which is chilled immediately to 4 °C. For the time-zero point, 10 µl of cytochalasin D is added to the sample before the addition of *S. aureus* to prevent any phagocytosis. This time point allows for an examination of binding prior to phagocytosis.

Following the proper incubation period and the treatments listed previously, all the samples are washed twice in cold HBSS and resuspended in 200 µl of HBSS. In adherent cultures, enough HBSS is added to just cover the monolayers.

A monoclonal antistaphylococcal antibody or a quenching agent (ethidium bromide or trypan blue) is then added. This step is important to separate internal from external particles. Following a 30-min incubation at 4 °C, the cells are washed once in HBSS and a fluorochrome-labeled goat antimouse antibody is added. Following an additional 30-min incubation at 4 °C, the cells are washed twice and fixed with 1% (w/v) paraformaldehyde.

Controls should include duplicate cultures for each time point, which are incubated at 4 °C instead of 37 °C. Additional controls to monitor nonspecific antibody binding should include tubes containing immunoglobulins, which do not react with either the cells or the bacteria (particles) and are of the same isotype and labeled with the same fluorochrome as the specific antibody. The cells are then held at 4 °C in the dark until analyzed. They can be stored for several days. If a quenching reagent is used, then a slightly different procedure is followed. After the addition of the quenching agent, the cells should be analyzed within a short period of time. These samples should be held at 4 °C in the dark until analysis by confocal microscopy.

X. Data Analysis and Interpretation

Phagocytosis can be quantified in a number of different ways, depending on the type of data collected. The simplest method is to determine the percentage of cells positive for phagocytosis. Any cells containing internal particles are considered phagocytosed. In the example given earlier, green particles would indicate positive phagocytosis whereas green/orange particles are external and not indicative of phagocytosis. Determining the percentage of cells positive for phagocytosis is usually performed even if other quantification methods are used.

A second method involves determining the number of bacteria phagocytosed per cell. Collecting these data is very labor-intensive as each phagocytosed particle must be counted individually. To minimize variations, it is recommended that the particles be counted in at least 200 cells. Data are then expressed as the percentage of cells phagocytosing a mean number of particles.

Most investigators find it easier to determine the relative phagocytosis per cell rather than the actual number of particles per cell. This method requires the computerization of data for analysis. Fluorescence can be collected either by the use of photomultiplier tubes or by a video camera. Data from photomultiplier tubes can be channelized according to intensity and a mean and/or median fluorescence can be obtained. Data captured by the video camera can be contoured according to pixel numbers. The intensity of fluorescence within the pixel contours can then be calculated. Simply stated, the greater the fluorescence intensity per cell, the more phagocytosis per cell. In the protocol given earlier, only the green fluorescence, which represents phagocytosis, is analyzed. Dual-labeled cells, green/

orange in the protocol described earlier, are not analyzed as this represents binding. The green fluorescence of cells incubated consecutively with bacteria at 4 °C and then stained with the appropriate antibody (same as for the experiment) represents the background or nonspecific fluorescence control, as the cells should not phagocytose at 4 °C. Only green fluorescence signals exceeding this value should be included or alternatively subtracted from the experimental values. The isotypic antibody control is used to ensure that the dual fluorescence is indeed specific for organisms (particles) on the outside of cells and is not caused by the nonspecific binding of immunoglobulin molecules.

XI. Summary

Confocal microscopy is an excellent tool to quantify phagocytosis. Depending on the particle used, phagocytosis can be determined by the simple manual counting of internalized particles. If a fluorescence probe is utilized, an analysis of fluorescence intensity can be used for quantification. The basic procedure can be altered in a number of areas to conform with the scientific needs of the investigator. This includes the use of different particles, cell types, fluorescence dyes, and even the degree of sophistication of the instrumentation. The major pitfall encountered when trying to quantify phagocytosis is the inability to separate external from internal (phagocytosed) particles. If not determined properly, data will be erroneous, usually indicating a much higher degree of phagocytosis than actually occurred.

References

Amoscato, A. A., Davies, P. J., and Babcock, G. F. (1983). *Ann. N. Y. Acad. Sci.* **419,** 114.

Babcock, G. F. (1993). *In* "Handbook of Flow Cytometry," (J. P. Robinson, ed.), p. 22. Wiley-Liss, New York.

Babior, B. M. (1984). *Blood* **64,** 959.

Bassoe, C. F. (1993). *In* "Handbook of Flow Cytometry," (J. P. Robinson, ed.), p. 177. Wiley-Liss, New York.

Bjerknes, R. (1998). *In* "Phagocyte Function," (J. P. Robinson, and G. F. Babcock, eds.), p. 187. Wiley-Liss, New York.

Bjerknes, R. C., Bassoe, C. F., Sjursen, H., Laerum, O. D., and Solberg, C. O. (1989). *Rev. Infect. Dis.* **11,** 16.

Brown, E. J. (1995). *BioEssays* **17,** 109.

Fattorossi, A. R., Nisini, R., Pizzolo, J. G., and D'Amelio, R. (1989). *Cytometry* **10,** 320.

Goding, J. W. (1986). Monoclonal Antibodies: Principles and Practice. Academic Press, London.

Goodell, E. M., Bilgin, S., and Carchman, R. A. (1978). *Exp. Cell Res.* **114,** 57.

Hess, K. L., Babcock, G. F., Askew, D. S., and Cook-Mills, J. M. (1997). *Cytometry* **27,** 145.

Klebanoff, S. J., and Clark, R. A. (1978). The Neutrophil: Function and Clinical Disorders. North-Holland, Amsterdam.

Rodeberg, D. A., Morris, R. E., and Babcock, G. F. (1997). *Infect. Immun.* **65,** 4747.

Scheetz, M. E., Thomas, L. J., Allemenos, D. K., and Schinitsky, M. R. (1976). *Immunol. Commun.* **5,** 189.

Silverstein, S. C., Steinman, R. M., and Cohn, Z. A. (1977). *Annu. Rev. Biochem.* **46,** 669.

White-Owen, C., Alexander, J. W., Sramkoski, R. M., and Babcock, G. F. (1992). **30,** p. 2071.

CHAPTER 11

Receptor–Ligand Internalization

Guido Orlandini, ★ **Rita Gatti,** ★ **Gian Carlo Gazzola,** ★
Alberico Borghetti, ★ **and Nicoletta Ronda**[†]

★Department of Experimental Medicine
Via Volturno 39, Parma, Italy

[†]Department of Clinical Medicine
Nephrology and Health Sciences Via Gramsci 14
University of Parma, parma, Italy

I. Introductory Update

The study of ligand–receptor internalization process has been greatly facilitated in last decades by the improvement of morphologic techniques such as confocal microscopy. In particular, the methodology to exploit confocal imaging on living cells described in our paper is still a valuable tool to obtain detailed information on membrane binding, internalization, subcellular localization and co-localization of ligands, and on the timing of the process.

Following our first applications, we studied the internalization of pathologic anti-fibroblast antibodies (AFA) by living fibroblasts, obtaining very interesting details on cell membrane invaginations and tubular structures involved in the process (Ronda

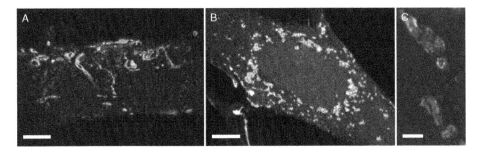

Fig. 1 (A) Initial internalization of IgG (IgG–FITC), purified by patients with AFA in a human fibroblast, scale bar; (B) Subsequent phase with perinuclear localization of IgG–FITC, scale bar 5 μm; (C) Detail of a fibroblast cytoplasm with IgG–FITC enclosed in caveolae (see text), scale bar 2 μm. (See Plate no.9 in the Color Plate Section.)

et al., 2002). Figure 1A shows fluoresceinated IgG (IgG–FITC), purified by patients with AFA, in the first phase of internalization by a living fibroblast, starting from specialized zones of the cell membrane. Following the process, signal moves to the perinuclear area (Fig. 1B), and finally fades after a few hours with aspects consistent with ligand recycling on the cell membrane. We were also able to show that internalized IgG are included in caveolae, known to be involved in receptor–ligand internalization processes. This goal was achieved by an *in situ* fixation technique following the visualization of IgG–FITC internalization in a living fibroblast. After the acquisition of the first image (Fig. 1C, IgG–FITC, green signal) medium was substituted with a fixative (methanol) and then with a monoclonal anticaveolin antibody followed by a Texas Red-coupled secondary antibody (Fig. 1C, red-signal). Our device allows to maintain the exact observation field (no shift along *x–y–z* axis) and to perfectly merge the two images evidencing the topographical relationship between the studied molecules. In more recent studies, we applied our methodology to show specific inhibition by normal human IgG on one of the pathways involved in the internalization of oxidized LDL in endothelial cells (Ronda *et al.*, 2003), to visualize different patterns of internalization of various type of liposomes in living cells, and to demonstrate the exposition of the autoantigen heat shock protein60, functioning as a receptor for specific autoantibodies, on the cell surface of endothelial cells exposed to atherogenic stimuli (Ronda *et al.*, 2009).

Technology improvement and evolution of confocal microscopes since 1999 have led us to change our instrument and slightly modify cell culture devices, although the general principles described in the chapter are still the same. The confocal system currently employed by our group is the LSM 510 Meta scan head integrated with the Axiovert 200 M inverted microscope (Carl Zeiss, Jena, Germany) and endowed with 3 different lasers. It allows multitrack acquisition of many fluorescent signals at the same time, namely through consecutive and independent optical pathways, with an excellent control of the cross talk and covering the whole visible spectrum. Moreover it can perform a spectral analysis of a given fluorescent probe and "acknowledge" its emission spectrum that can be detected and identified in following experiments. Such mode allows to split probes with overlapping spectra with a nominal resolution up to 7 nm at the emission peak. Reckoning with living cells, another major advantage of the new systems is the very low amount of energy delivered to the sample during image acquisition. Moreover this new system is provided with a microincubator that ensures extremely stable and measured micro-environmental conditions (temperature, CO_2 concentration and humidity (Kit Cell Observer, Carl Zeiss, Jena, Germany). Eventually we have designed a new flow chamber to obtain time lapse images of cells grown on opaque biomaterials. This could allow to study ligand receptor interaction in cells grown on special supports mimicking more physiological, *in vivo* conditions. We have exploited this methods to study osteoblast grown on different titanium surfaces (Gatti *et al.*, 2008).

II. Introduction

The interaction between biologically active compounds and target cells has been studied extensively by various quantitative and qualitative approaches as it is the crucial first step in the chain of events leading to the final effect. Additionally it is a suitable process to be studied for pharmacological purposes (Lodish *et al.*, 1995a). The visualization of receptor–ligand binding and internalization has been studied by immunocytochemistry techniques for both light and electron microscopy on fixed, permeabilized tissues or cells. The information achievable by these methods is limited and static, as the localization observed in fixed samples might not correspond to the actual binding site in living cells and the time course of the interaction cannot be evaluated with satisfactory precision (Jackson and Blythe, 1993; Monaghan *et al.*, 1993).

A technique is now available that allows the visualization of the interaction between directly fluoresceinated ligands and living cells and enables one to follow their possible internalization. In living adherent cells, such a methodological approach was hindered previously by the relatively poor resolution of the conventional fluorescence microscope and the unfavorable signal-to-noise ratio. As a consequence, the visual information has not been exhaustive and subcellular localization has been limited to main cell compartments (Lodish *et al.*, 1995a).

More recently, confocal imaging has provided new insights in the observation of fluorescent specimens. The virtual absence of out-of-focus blurring allows a much better definition of probe localization at the subcellular level together with the possibility of exploiting the three-dimensional reconstruction capability of most confocal systems.

We have coupled a self-constructed flow chamber to an inverted confocal scanning laser microscope that allows long-term observation of adherent cells under controlled microenvironmental conditions (Dall'Asta *et al.*, 1997). This method not only provides images of intact, nonfixed cells, but also allows one to change culture conditions and to observe living cell responses directly or to perform two-step staining to identify subcellular structures involved in the observed processes.

The reliability of this procedure has been verified on a well-known model of receptor–ligand internalization, that of insulin and insulin receptor. The same technique has also been applied to studying the interactions between natural human antibodies, circulating in healthy subjects, and living human endothelial cells (ECs), fibroblasts, and proximal tubular epithelial cells (PTECs) (Ronda *et al.*, 1997).

III. Materials

A. Flow Chamber

Despite the ever-increasing number of available fluorescent probes for living cells acting as vital, almost real-time indicators for a series of functional parameters, several technical problems arise in trying to exploit fully confocal laser scanning microscopy of viable cell monolayers. This kind of system is extremely sensitive to changes of the focal plane, the distance between lens and specimen must be very short, and high numerical aperture lenses are mandatory for optimal resolution. Moreover, three major general requisites must be satisfied to achieve improvement over the techniques previously available: (1) perturbation of culture conditions must be kept at a minimum; (2) the system should allow one to extend observation as long as desired; and (3) the relevant stimuli, used in the experimental protocol, should be applied easily and then washed out easily.

The device described in this chapter is simple and inexpensive but fulfills all of the just described criteria. It was developed to fit a Multiprobe 2001-Molecular Dynamics computer scanning laser microscope whose optical "conventional" side is based on a Nikon Diaphot inverted microscope (Sunnyvale, CA).

The upper portion of the flow chamber is a transparent polyacetate block (Figs. 2 and 3). According to the type of experiment to be carried out a hollow, whose depth can vary from 0.1 to 0.5 mm, has been milled on the bottom. When the cover slide (5 cm × 2.5 cm) bearing the cell culture is applied to the bottom of the block, the chamber is completed, with the cover slide and the block serving as the floor and the vault, respectively. According to the depth of the hollow the volume of the flow chamber ranges between 60 and 300 μl. Two vertical tunnels (diameter 4 mm) opening on the upper face of the block allow the introduction of small sylastic catheters (crossing the lid) for inflow and outflow of medium. The inlet and outlet tubes are usually connected to a micropipette and a vacuum apparatus, respectively, but a peristaltic pump for continuous replacement can also be used.

Various sealers, provided they are nontoxic to cells, can be adopted in order to secure the cover slide to the polyacetate block and to prevent medium leakage. In our experience the best results can be achieved with silicon vacuum grease or, better, with a special silicon-based liquid gasket for a high-performance engine (Motorsil D, Arexons, Cernusco, Milano, Italy). In any case, a thin, narrow layer of the sealer must be applied around all the block edges to avoid untoward spreading into the chamber once the block itself is pushed on the cover slide.

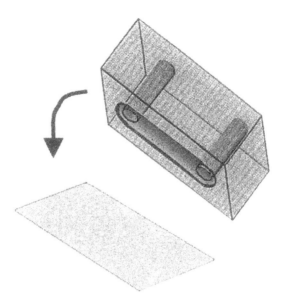

Fig. 2 Schematic representation of the flow chamber. The bottom face of the polyacetate block has to be sealed to the cover slide bearing the cell monolayer.

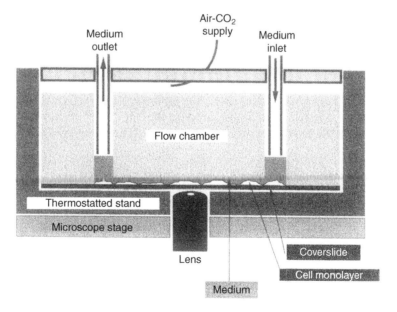

Fig. 3 Schematic cutaway view of the flow chamber lodged in the aluminum stand on the microscope stage.

Fig. 4 View of the stand in place. Note the small tubes for medium substitution and air–CO_2 supply. The cable from the aluminum stand is connected to the temperature control unit. The clear Perspex lid allows field lighting from the above condenser. Microscope stage translation (for field selection) carries about the whole system.

The flow chamber is lodged in a thermostatted stand (Fig. 4), a cylindrical aluminum block (diameter 14 cm × 2 cm) whose base has been opportunely mill finished so as to adhere tightly to the round opening in the microscope stage. A parallelepipedal niche (5 cm × 2.5 cm × 1.4 cm) is carved within the stand in order to contain precisely the flow chamber. The bottom of the niche shows an ellipsoidal slit (3 cm × 1 cm) devoted to objective apposition, wide enough to allow the observation of a large area of the culture by stage translation. The flow cell niche is closed tightly by a Perspex lid with three openings for a medium inlet and outlet and for atmosphere conditioning, respectively. The pressure of the lid also secures the two parts of the system fixed in place. Temperature in the flow chamber is controlled by a resistance coil embedded in the stand. A thermorelay probe is located at the inner surface of the slot; this configuration grants the widest surface for contact between the radiating element and the flow chamber.

B. Confocal Microscope

The confocal scanning laser microscope employed is a Molecular Dynamics Multiprobe 2001 (inverted) equipped with an argon ion laser. Samples are observed through either a 60× or a 100× oil-immersion objective (Nikon PlanApo, NA 1.4), allowing vertical resolution around 1 μm, and the step size is set accordingly. The confocal aperture (pinhole) is set at 50 and 100 μm for 60× and 100× lenses, respectively. For practical purposes, the first setting is chosen whenever possible, as it provides the best vertical resolution (0.8 μm) but, due to the lower brightness, it requires optimal signal intensity.

Because of its high quantum yield, fluorescein can be detected with a very low excitation power, which is crucial when imaging living cells. Settings exceeding

1 mW produce phototoxic events in the culture (such as apoptosis or cell detachment) that are probably related to local energy delivery and consequent heating. Another advantage of using low excitation power is that photobleaching and background noise are negligible.

In postfixation experiments with two tracers (see later), a secondary beam splitter is placed after the pinhole aperture and before the barrier filters. Signals from fluorescein and the second chromophore, usually tetramethylrhodamine isothiocyanate (TRITC) or Texas Red, are then acquired concurrently by two different photomultipliers that can be set independently to correct the unevenness in the quantum yield of the fluorophores.

Once the apparatus is ready for image acquisition, it is important that the field of observation is chosen in bright-field microscopy. This allows one to take time in selecting the best field according to the experimental requirements without delivering useless or even noxious high power.

Fresh medium or the relevant study solutions can be replaced without shifting along the x, y, or z axis. At each experimental step, a section series is acquired along the whole thickness of the cell (for ECs, Sections 5 and 6). Complete scanning yields information about the whole cell, allows one to choose the most representative section, and provides the material for accurate three-dimensional reconstruction.

Whenever it is important to give a global representation of an internalization process, the image series are smoothed with a Gaussian $3 \times 3 \times 3$ kernel filter and three-dimensional reconstruction is performed according to a maximum intensity algorithm. In other words, for all pixels of a given x–y coordinate in the series, the one with the highest intensity is chosen for final image rendering. Image processing is performed on a Silicon Graphics Personal Iris workstation (Image Space Software, Molecular Dynamics).

C. Preparation of Ligand–Fluorescein Conjugates

Human insulin (Sigma, St Louis, MO), purified normal human IgG (pooled normal IgG, Sandoglobulin, Sandoz, Basel, Switzerland), and IgG purified from single healthy donors (Ronda *et al.*, 1997) are coupled to fluorescein isothiocyanate (FITC) (Johnstone and Thorpe, 1982) purified by chromatography on a Sephadex G-25 column, and dialyzed extensively against phosphate-buffered saline (PBS) using a 3500-kDa cutoff membrane (Spectra/Por, Spectrum Medical Industries, Inc., Los Angeles, CA) to allow total elimination of free FITC, as detected by spectrophotometric analysis of the dialysis medium at 495 nm. Fluoresceinated ligands are finally dialyzed in Dulbecco's modified Eagle's medium (DMEM) containing L-glutamine, 50 U/ml penicillin, and 50 mg/ml streptomycin, and filtered with 0.22-μm filters into sterile tubes. The FITC to protein ratio is 6.3 for IgG and 5.5 for insulin, and final protein concentrations are 8 and 6 mg/ml, respectively. The ligand concentrations used, as indicated in the description of each experiment, are obtained by diluting ligand solutions with DMEM with the addition of L-glutamine, penicillin–streptomycin, and fetal calf serum at a final concentration of 5% (v/v). The presence of FCS does not modify ligand–cell interactions, as demonstrated by comparison with serum-free experiments, but is associated with better preserved cell morphology and greater adhesion to the cover slide, especially in the case of ECs. Aliquots of the last dialysis medium of ligand–FITC are filtered and saved for incubating with cells at least 2 h before the beginning of the experiments to further exclude contamination of the ligand–FITC solutions by free FITC (see later). When indicated, possible effects due to lipopolysaccharide (LPS) contamination of the ligand solutions are excluded by repeating the experiments with the addition of 5 μg/ml polymyxin B to all incubation media. Fluoresceinated ligands are kept sterile at 4 °C without preservatives for up to 2 months.

IV. Cell Culture

Human umbilical cord vein ECs (Oravec *et al.*, 1995), human fibroblasts (Gazzola *et al.*, 1980), and human PTEC (Detrisac *et al.*, 1984) at the first passage are grown to subconfluence without attachment factors (to reduce background signal) on a glass cover slide fitting the flow chamber. To obtain good cell density within 24 h, ECs (which proliferate slowly on glass and in the absence of attachment and growth factors) are seeded carefully, placing 0.5 ml of cell suspension (10^5 cells/ml) on the cover slide, letting the cells adhere for 4 h in the incubator, and finally adding the proper amount of culture medium for incubation. All the experiments are performed 24 h after seeding the cells.

V. Membrane Binding of Ligands

The system described allows one to observe membrane binding of a ligand to living cells. In fact, we have been able to show that purified normal IgG binds to the cell membrane of cultured living fibroblasts and is not internalized (Fig. 5). Before starting a specific experiment, it is advisable to obtain a basal image of the cells that had been incubated for 2 h with the dialysis medium saved after the final dialysis of IgG–FITC in DMEM. The complete absence of a fluorescent signal excludes free FITC contamination of the fluoresceinated ligand. Such a preliminary test should be performed before every experiment described in this chapter. We then incubate fibroblasts in the confocal flow chamber for 30 min with IgG–FITC at 2 mg/ml in standard culture conditions, wash them with culture medium, and observe the cells at 10- to 15-min intervals for 2 h. A fluorescent signal is detectable with IgG diluted out to 0.5 mg/ml.

We did not observe IgG–FITC binding to PTEC under the same conditions, even using an IgG concentration of 8 mg/ml. In contrast, normal IgG entered living EC within minutes (see later). To show membrane binding of IgG to ECs, we have inhibited cell energy-dependent processes by setting the flow chamber

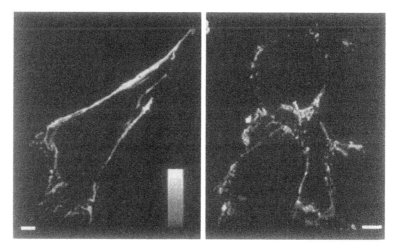

Fig. 5 IgG–FITC membrane binding to a living fibroblast (left-hand side) and to ECs (right-hand side). Bars: 10 μm (left) and 5 μm (right). Reproduced with permission from Ronda *et al.* (1997).

temperature to 27 °C (the lowest temperature tolerated by EC without cell damage in our system) and observed cells after 5 min of incubation with 2 mg/ml of IgG–FITC (Fig. 5).

VI. Receptor–Ligand Internalization

Whenever a signal from the ligand under investigation is detected, it is necessary to check its localization within the cell and it is advisable to identify the nature of the structure or compartment involved. Moreover, it is important to monitor the morphological evolution of the process. To achieve these goals, it is useful to counterstain the cells. This is possible, with no need to change the field of observation, by flushing 1 μM calcein AM through the flow chamber medium. This neutral dye is converted by intracellular esterases into a fluorescent anionic compound, optimally excited at 488 nm with a peak emission around 520 nm, which stains the nucleus and the cytoplasm (nucleus/cytoplasm signal intensity ratio, 2: 1), except for cationic compartments. Calcein allows one to check cell vitality as the fluorescent signal immediately disappears in the presence of membrane damage, despite residual cell esterase activity (Moore *et al.*, 1990). Because the calcein signal often overwhelms that emitted by the internalized ligand (see later), once counterstaining is carried out, possible subsequent changes in the distribution pattern of the internalized ligand cannot be visualized. Therefore, it is possible to program the counterstaining at various times of incubation in different samples or to track the relevant changes in a single microscopic field and to delay counterstaining until appropriate. To verify the actual intracellular nature of the signal, when the raw and the counterstained images of the same field are overlapped digitally, usually the intensity of the latter must be reduced evenly. However, comparison of the two separate images also can provide additional useful information, as will be illustrated later.

As an example of the information achievable related to morphology, timing, and specificity (i.e., receptor involvement) of the internalization of a ligand, we first describe the visualization of receptor–ligand internalization in a well-known model (insulin and ECs) and then in the case of a previously unknown ligand–cell interaction (IgG in ECs).

Confocal observation of living ECs incubated with insulin–FITC allows the direct visualization of insulin internalization, the morphology and timing of which are consistent with previous knowledge of the process. Insulin binding and internalization are almost immediate, with a fluorescent cytoplasmic fibrillar signal evident after only 2 min of incubation with 1 mg/ml insulin–FITC followed by washing with culture medium (Fig. 6A).

The initial fibrillar aspect evolves rapidly, and after 10 min the fluorescence is distributed almost entirely in cytoplasmic granules, some of which are larger bodies of 2–8 μm in diameter (Fig. 6B).

The signal from intracellular insulin–FITC is relatively weak as compared to that of calcein and is no longer appreciable after counterstaining (Fig. 6C and D). For overlapping images, it is necessary to attenuate the calcein signal evenly by 25–35% to show insulin localization together with cytoplasmic staining (Fig. 6E). This is particularly evident from the comparison between Fig. 6B and D. For the acquisition of the latter, the sensitivity of the photomultiplier, which was appropriate for calcein, was too weak to detect insulin–FITC. Therefore the large granules containing insulin and excluding calcein appear as negative bodies. The preservation from calcein loading indicates that these are acidic compartments, likely to correspond to late endosomes involved in insulin catabolism (Carpentier, 1994) (Fig. 6D).

As stated earlier, the incubation of living ECs with IgG–FITC at 2 mg/ml is followed by the internalization of IgG, detectable after 10–15 min of incubation and most evident after 20–30 min. For experiments excluding the presence of contaminants other than IgG that could be responsible for the intracellular fluorescence, see Ronda *et al.* (1997). The intracellular localization of fluorescence can be demonstrated first through a vertical section of a protrusion of a cell (Fig. 7) and by calcein loading, as shown earlier, but without a need to reduce counterstain intensity, as the IgG–FITC signal is higher. The fluorescence pattern is that of a fibrillar network, particularly abundant in peripheral areas of the cytoplasm and in protrusions that apparently connect adjacent cells (Fig. 7, top). After 1 h, most of the fluorescence is localized in cytoplasmic granules and it is reduced greatly after 2 h. The time course, morphology, and inhibition by the low temperature of the IgG internalization process in ECs are consistent with a receptor-mediated mechanism rather than with pinocytosis, which is a slow process, with poor quantitative efficiency, leading to a nonspecific uptake of extracellular medium. The demonstration that pinocytosis was not responsible for our observation came from the lack of internalization of fluoresceinated IgG fragments by ECs under the same

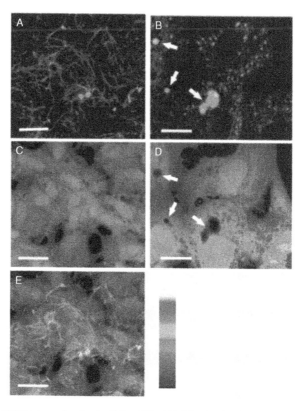

Fig. 6 Insulin–FITC internalization by living ECs. (A) After 2 min of incubation, the insulin–FITC signal appears as filaments. (C) The same field as in (A) after counterstaining with calcein. Cytoplasm and nuclei are now evident. Acquisition parameters fit for calcein, with a higher signal than that of insulin–FITC, do not allow the visualization of intracellular insulin. (E) Overlapping of (A) and (C) shows the actual intracellular localization of insulin. The image shown in (A) was added to that in (C), the intensity of which had been reduced by 30%. (B) In a distinct experiment, the image was acquired after 10 min of incubation with insulin–FITC; the distribution pattern is now mostly granular. Insulin–FITC also accumulates in large bodies, like the one in this field (arrow). (D) Counterstaining demonstrates that granules are cytoplasmic and correspond to structures inaccessible to calcein. Note in particular the large body in this field (arrow); the insulin FITC it contains is not detectable at the low sensitivity of this acquisition setting (see text). Bar (A, C, and E): 20 μm; Bar (B and D): 10 μm. For gray-scale palette, see Fig. 5.

conditions. PTECs, whose nonspecific reabsorption of proteins from preurine is well known, show cytoplasmic granules of IgG after only 48 h of incubation with IgG–FITC or fluoresceinated IgG fragments.

As compared to insulin–FITC, the process of internalization appears morphologically similar, but insulin internalization is faster and, as noted earlier, requires a higher power of excitation and magnification of the signal. Such a difference in fluorescence intensity is likely to be due mainly to the smaller number of FITC

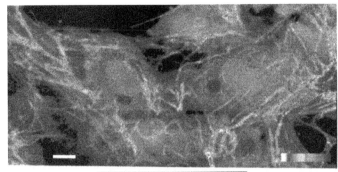

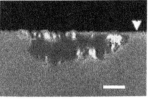

Fig. 7 IgG–FITC internalization by living ECs. (Top) After 15 min of incubation, the IgG–FITC signal was distributed with a cytoplasmic fibrillar pattern. Following calcein addition (which shows cell bodies and demonstrates cell membrane integrity), there was no need to change the acquisition setting or to perform image overlapping because of the high intensity of the IgG–FITC signal. (Bottom) Vertical section of a cytoplasmic protrusion showing intracellular IgG–FITC in a cell not counterstained with calcein. Remnants of extracellular IgG–FITC in culture medium provide contrast for the cell. Arrowhead: cover slide. Bars: 5 μm. Reproduced with permission from Ronda *et al.* (1997).

molecules per molecule of the ligand in the case of insulin (5600 molecular weight) as compared to IgG (150,000), rather than determined by differences in cell receptor number or affinity.

VII. Two-Step Staining for Identification of Subcellular Structures

The fibrillar pattern of cytoplasmic fluorescence shortly following internalization of IgG–FITC and insulin–FITC, together with the well-known function of microtubules in molecular/vesicular intracellular trafficking, suggests the possibility that microtubules are involved in the receptor–ligand internalization systems. We have thus designed an experimental procedure to demonstrate the possible overlapping of signals from internalized fluoresceinated ligands and microtubules stained with a different probe. As a preliminary test we ensured that the cytoplasmic IgG–FITC signal remains unmodified after cell fixation with methanol. We then induced internalization of IgG–FITC in ECs in the confocal flow chamber as described, turned off the thermostat, fixed the cells by flushing 100% methanol at 4 °C through the chamber for 2 min, and washed the cells with PBS at room temperature. We then performed indirect immunofluorescence at room temperature

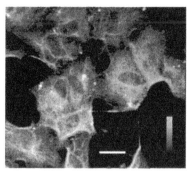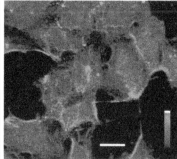

Fig. 8 Simultaneous visualization of EC microtubules and internalized IgG–FITC. Following internalization of IgG–FITC, ECs were fixed and microtubules were stained by indirect immunofluorescence using a rhodamine TRITC-conjugated secondary antibody. Images were then acquired simultaneously with a double-channel system (see text) to show the localization of microtubules (left) and IgG–FITC (right). All IgG–FITC signals correspond to some of the microtubular network. Bars: 10 μm.

using a mouse monoclonal anti-α-tubulin antibody (Sigma) and an antimouse IgG TRITC (λ_{ex} = 552 nm; λ_{em} = 570 nm)-conjugated antibody (Sigma). Images were then acquired using a double-channel system placing a secondary beam splitter (565 nm) after the pinhole and 535-nm ($\pm$15) bandpass and 570-nm long-pass barrier filters before two separate photomultipliers. After acquisition, barrier filters were inverted to check for contamination of fluorescein image by TRITC; indeed the absolute negativity of the field excluded such a possibility. The actual purity of each signal was also enhanced using the "separation enhancement" routine of the software, which subtracts a chosen percentage of the signal of each channel from the other one.

The images obtained show a perfect correspondence between internalized ligand–FITC localization and some of the microtubular filaments (Fig. 8). It is known that microtubules, intermediate filaments, and parts of the endoplasmic reticulum often colocalize (Lodish *et al.*, 1995b), but the actual involvement of microtubules in the internalization of IgG has been demonstrated by the total inhibition of the process obtained by pretreating ECs for 20 min with 100 μg/ml colchicine before incubation with IgG–FITC (Ronda *et al.*, 1997).

Acknowledgments

This work was funded by the Department of Clinical Medicine, Nephrology, and Health Sciences and partly by CNR target project "Biotechnology." The confocal apparatus is a facility of the Centro Interfacoltà Misure of the University of Parma.

References

Carpentier, J. (1994). *Diabetologia* **37**, S117.
Dall'Asta, V., Gatti, R., Orlandini, G., Rossi, P. A., Rotoli, B. M., Sala, R., Bussolati, O., and Gazzola, G. C. (1997). *Exp. Cell Res.* **231**, 260.

Detrisac, C. J., Sens, M. A., Garvin, A. J., Spicer, S. S., and Sens, D. A. (1984). *Kidney Int.* **25,** 383.

Gatti, R., Orlandini, G., Uggeri, J., Belletti, S., Galli, C., Raspanti, M., Scandroglio, R., and Guizzardi, S. (2008). *Micron* **39**(2), 137–143.

Gazzola, G., Dall'Asta, V., and Guidotti, G. (1980). *J. Biol. Chem.* **255,** 929.

Jackson, P., and Blythe, D. (1993). *In* "Immunocytochemistry," (J. E. Beesley, ed.), p. 22. Oxford University Press, Oxford.

Johnstone, A., and Thorpe, R. (1982). "Immunochemistry in Practice", **258**. Blackwell, Oxford.

Lodish, H., Baltimore, D., Berk, A., Zipursky, S. L., Matsudaira, P., and Darnell, J. (1995a). *In* "Molecular Cell Biology," 3rd edn. (J. Darnell, ed.), chapter 16, Scientific American Books, New York.

Lodish, H., Baltimore, D., Berk, A., Zipursky, S. L., Matsudaira, P., and Darnell, J. (1995b). *In* "Molecular Cell Biology," (J. Darnell, ed.), chapter 23, Scientific American Books, New York.

Monaghan, P., Robertson, D., and Beesley, E. J. (1993). *In* "Immunocytochemistry," (J. E. Beesley, ed.), p. 47. Oxford University. Press, Oxford.

Moore, P., MacCoubrey, I., and Haughland, R. (1990). *J. Cell Biol.* **111,** 58.

Oravec, S., Ronda, N., Carayon, A., Milliez, J., Kazatchkine, M. D., and Hornych, A. (1995). *Nephrol. Dialys. Transplant.* **10,** 796.

Ronda, N., Gatti, R., Orlandini, G., and Borghetti, A. (1997). *Clin. Exp. Immunol.* **109**(1), 211.

Ronda, N., Gatti, R., Giacosa, R., Raschi, E., Testoni, C., Meroni, P. L., Buzio, C., and Orlandini, G. (2002). *Arthitis Rheum.* **46**(6), 1595–1601.

Ronda, N., Bernini, F., Giacosa, R., Gatti, R., Baldini, N., Buzio, C., and Orlandini, G. (2003). *Clin. Exp. Immunol.* **133,** 219–226.

Ronda, N., Poti, F., Orlandini, G., Gatti, R., Ardissino, D., Regolisti, G., Cabassi, A., and Fiaccadori, E. (2009). World Congress of Nephrology, Milan, Italy .

CHAPTER 12

Quantitative Imaging of Metabolism by Two-Photon Excitation Microscopy

David W. Piston and Susan M. Knobel

Department of Molecular Physiology and Biophysics
Vanderbilt University Medical School
Nashville, Tennessee, USA

I. Introduction

Many of the important biological discoveries made with fluorescence microscopy have resulted from experiments on fixed samples. Unlike other experimental approaches, such as electron microscopy, fluorescence microscopy offers the possibility of working with living specimen. Since the early 1980s, significant developments have allowed fluorescence microscopy assays of processes in living cells (e.g., Ca^{2+}, membrane potential, vesicular transport). Most of these techniques depend on the addition of an extrinsic fluorescence reporter, which can introduce difficulties in the interpretation of results. Recently introduced fluorescent reporters based on the green fluorescent protein (GFP) have the potential to alleviate some of these problems, but considerable work remains before these will be of general use

189
DOI: 10.1016/B978-0-12-384658-7.00012-6

(Miyawaki *et al.*, 1997). Still, any external indicator dye may alter the process under observation. Instead of using extrinsic probes, we have used intrinsic cellular fluorescence, which offers several advantages for the investigation of cellular metabolism. Because these autofluorescent compounds are natural constituents of every cell, there is no problem of uniform loading of the dye. In addition, these probes can be active participants in cellular processes. However, intrinsic fluorophores are typically not as bright or photostable as artificial probes, so they are more difficult to measure in the microscope. This chapter describes the use of the naturally occurring reduced nicotinamide adenine dinucleotide (phosphate) [NAD(P)H] as a monitor of cellular metabolism. To image these UV-absorbing fluorophores in living cells, we have utilized two-photon excitation microscopy, which minimizes the photodamage associated with NAD(P)H imaging.

This chapter describes the use of NAD(P)H as a metabolic indicator and discusses the two-photon excitation microscopy methods that we use to image its activity. It also details the instrument that we use for these experiments, with emphasis on the important design criteria for this demanding application. Finally, this chapter presents the application of two-photon excitation imaging of NAD(P)H to assay glucose-stimulated metabolism in both pancreatic and muscle cells.

II. Use of NAD(P)H Autofluorescence for Metabolic Imaging

Fluorescence from naturally occurring NAD(P)H can be used as an indicator of cellular respiration and therefore as an intrinsic probe to study cellular metabolism (Chance and Thorell, 1959). Under normal conditions, roughly 25% of the reduced pyridine nucleotides are phosphorylated. Because this amount varies among different cells and because the spectra of NADH and NADPH are indistinguishable in our imaging experiments, we simply refer to both as NAD(P)H. NAD(P)H fluorescence is normally excited with light of ~360 nm and emits in the 400- to 500-nm region. Because the fluorescence yield of reduced forms [NAD(P)H] is significantly greater than for oxidized forms [$NAD^+(P)$], fluorescence intensity can be used to monitor the cellular redox state. Measurement of the $NAD(P)H/NAD^+(P)$ ratio has been developed into a noninvasive optical method to monitor cellular respiration, called redox fluorometry (Chance and Lieberman, 1978), which has been widely used (French *et al.*, 1998; Ince *et al.*, 1992). Biochemical experiments have shown good qualitative agreement between the rise in NAD(P)H levels and the changes in fluorescence intensity measured on the addition of cyanide (Masters *et al.*, 1989).

Because NAD(P)H has a small absorption cross section and a low quantum yield, it is difficult to measure and has the potential to cause considerable photodamage. Furthermore, it absorbs in the UV, which is also more biologically damaging than visible or infrared light. Thus, a researcher interested in imaging metabolic dynamics would never choose NAD(P)H as the fluorophore, except that it is an active participant in cellular metabolic events. This means that it can be used without perturbation of the events under study. Despite the difficulties of imaging this dim, UV-absorbing fluorophore, two-dimensional images of the fluorescence intensity from NAD(P)H have been obtained with a conventional fluorescence microscope from several cellular systems (Coremans *et al.*, 1997; Eng *et al.*, 1989; Halangk and Kunz, 1991; Pralong *et al.*, 1990). These images have also been used to assess metabolic dynamics as a function of pharmacological or electrical stimulation of the cells, but the experiments are limited to whole cell measurements (e.g., of isolated cardiac myocytes) or surface imaging (e.g., of an intact perfused heart).

To measure events with subcellular resolution or within a single cell in an intact tissue accurately, it is necessary to use an optical sectioning microscope. While use of the confocal microscope for optical sectioning is well established, confocal imaging of autofluorescence from living cells is problematic. Unfortunately, UV confocal microscopy is degraded by optical system problems, especially chromatic aberration between UV excitation and visible fluorescence. The introduction of commercial UV confocal microscopes has permitted confocal imaging of cellular autofluorescence dynamics (Masters *et al.*, 1993; Nieminen *et al.*, 1997). However, these UV confocal observations were still limited severely by photobleaching of the autofluorescence and could not be performed in thick tissues. Many of the limitations of confocal microscopy for the imaging of NAD(P)H in living tissue can be overcome using two-photon excitation microscopy (Bennett *et al.*, 1996; Piston *et al.*, 1995, 1999).

III. Two-Photon Excitation Microscopy

A. Background and Concepts

The effective sensitivity of fluorescence microscopy measurements is often limited by out-of-focus flare. This limitation is reduced greatly in a confocal microscope, where the out-of-focus background is rejected by a confocal pinhole to produce thin (<1 μm), unblurred optical sections from within thick samples. A new alternative to confocal microscopy is two-photon excitation microscopy, which excels at imaging of living cells (Denk *et al*, 1990). Two-photon excitation arises from the simultaneous absorption of two photons in a single quantized event, which is dependent on the square of the excitation intensity. Because the energy of a photon is inversely proportional to its wavelength, the two photons should be about twice the wavelength required for single-photon excitation. For example, NAD(P)H that normally absorbs ultraviolet light ($\sim$350 nm) can also be excited by two red photons ($\sim$700 nm). In the case of fluorescence, the emission after two-photon excitation is the same as would be generated in a typical biological fluorescence experiment (Denk *et al.*, 1995). To obtain sufficient two-photon absorption events for an imaging application, very high laser powers are required. These powers are achieved practically using mode-locked (pulsed) lasers, where the power during the peak of the pulse is high enough to generate significant two-photon excitation, but the average laser power is fairly low (<10 mW, just slightly greater than what is used in confocal microscopy).

The application of two-photon excitation to laser scanning microscopy is very powerful. In the microscope, two-photon excitation microscopy is made possible not only by concentrating the photons in time (by using the pulses from a mode-locked laser) but also by crowding the photons spatially (by focusing in the microscope). As a laser beam is focused in the microscope, the only place where the photons are crowded enough to generate an appreciable amount of two-photon excitation is at the focal point. The localization of excitation yields many advantageous effects (Denk *et al.*, 1995). Most importantly, the use of two-photon excitation minimizes photobleaching and photodamage—the ultimate limiting factors in fluorescence microscopy of living cells and tissues. In addition, it is not necessary to use a pinhole to obtain optical sectioning, so flexible detection geometries can be used. For instance, it is now possible to develop a high efficiency direct detection scheme (where the fluorescence does not pass back through the scanning system as it must in a confocal microscope) as described later. Further details about two-photon excitation microscopy are presented elsewhere (for an introductory review, see Piston, 1999; for a more advanced and comprehensive description, see Denk *et al.*, 1995).

Two-photon excitation microscopy is the only currently available method capable of yielding high-resolution images of NADH autofluorescence throughout an extended sample such as the pancreatic islet (Bennett *et al.*, 1996; Piston *et al.*, 1999). The two-photon technique also allows for increased signal detection, and that translates into reduced photobleaching, which in turn leads to a better effective subcellular resolution. Most importantly, though, is the reduction in photodamage, which preserves sample viability and permits extended time-course measurements from living cells. Because the time scale of metabolic events in mammalian tissue may be up to several hours, increased sample viability is paramount. Even for less demanding studies, however, minimization of cellular photodamage is always a worthwhile goal.

B. Two-Photon Excitation Microscope Optimized for Quantitative NAD(P)H Imaging

Because NAD(P)H has a small absorption cross section and a low quantum yield, it is very important to minimize photodamage and to maximize the fluorescence collection. The use of two-photon excitation minimizes photodamage because of its inherent three-dimensional localization. Although there is still a chance of photodamage associated with NAD(P)H excitation in the focal plane, this type of photodamage does not occur out of the focus. To optimize the fluorescence collection system for NAD(P)H, modifications must be made to the optical system. Because normal fluorescence is red shifted and the wavelengths used in confocal microscopy are in the visible, most commercial laser scanning microscopes are designed to increase fluorescence collection in the red wavelengths. This usually means that they are not optically efficient in the deep blue/near-UV range, and use of one of these systems for two-photon excitation requires the addition of an external nondescanned detector. Furthermore, even UV confocal systems (which are more efficient optically in the NAD(P)H fluorescence wavelengths) usually have red-enhanced optics and photomultiplier tubes (PMTs) as well. For two-photon excitation imaging, it is necessary not only to optimize collection, but also to reject the red excitation light, which may be up to 10,000-fold greater than the excitation in a normal confocal system. Thus, to obtain sufficient excitation light rejection, a barrier filter used with two-photon excitation must have four orders of magnitude better rejection in the excitation band than a barrier filter used with confocal microscopy. Part of this rejection can be achieved by replacing the red-enhanced PMTs with red-blind PMTs. Inexpensive, bialkali photocathode PMTs are near ideal for this purpose because they offer high quantum efficiencies in the blue, but are insensitive to wavelengths above 650 nm. The system described in this chapter uses a nondescanned detection pathway with minimal optics and the correct PMTs to maximize the collection of NAD(P)H autofluorescence. Even though this system is entirely custom-built, this type of external detection pathway can be added easily to any confocal microscope system.

A schematic diagram of the two-photon excitation laser scanning microscope is shown in Fig. 1. To produce illumination for two-photon excitation, an all-lines argon ion laser (Coherent Innova 310, Santa Clara, CA) is used to pump a Coherent Mira mode-locked Ti:sapphire femtosecond laser using the X wave mirror set (the mirror set allows tuning from 690 to 960 nm, but for NAD(P)H

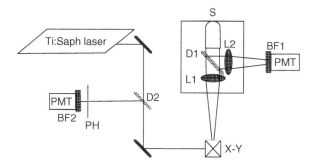

Fig. 1 A schematic diagram of the two-photon laser scanning microscope with optimized nondescanned detection, plus confocal detection. The incoming laser light is raster scanned (X–Y scan mirrors) and is focused onto the sample (S) through the tube lens (L1) and the objective lens. Fluorescence returns down the same path until it is reflected to the detection system by D1 (550 DCLP). The emitted signal is then refocused by a transfer lens (L2) so that the back aperture of the objective is conjugate to the front face of the PMT. A custom 700- to 720-nm blocked low-pass filter (BF1, Chroma Technology) is used to collect the NAD(P)H fluorescence. For confocal detection, D1 is removed and D2 is used. The scanning is done from below the microscope (bottom port of the Zeiss Axiovert 135TV), which allows for access to the stage from both sides for perfusion, temperature control, or microinjection experiments.

we use ~705 nm). The output of the Mira laser at this wavelength is a pulse train of ~150-fs pulses, running at 78 MHz, with an average output power of ~300 mW. For our NAD(P)H imaging experiments, the laser power must be attenuated considerably because we only require ~3 mW at the focal plane. This attenuation can be done with a variable neutral density filter (such as the Newport Corp. 925B, Irvine, CA) or with a low dispersion Pockel's cell modulator designed specifically for ultrafast laser applications (FastPulse Technology 5026, Saddlebrook, NJ). The pulses can be chirped negatively by extracavity prisms to maintain transform-limited pulses at the sample, which will give the highest two-photon excitation. However, we find dispersion compensation to be a highly unnecessary procedure with our optical system (obviously for fiber-coupled systems such compensation is a must). The pulse train is directed onto two orthogonal galvanometer scanners (Cambridge Technology 6350, Cambridge, MA), and the resulting scanned beam is focused onto the sample through the basement port of a Zeiss Axiovert micro-scope. For the objective lens, we use either a $40 \times$ 1.3 NA F-Fluar (to obtain the highest possible signal collection) or a $40 \times$ 1.3 NA plan-Neofluar (when a larger flat field is needed). More highly corrected lenses, such as plan-Apochromats, should not be used for NAD(P)H imaging because they have lower transmission efficiency and the chromatic corrections are not really needed with two-photon excitation of a single fluorophore.

Because the confocal spatial filter is not needed to obtain confocal properties with two-photon excitation, it is best to collect and measure the fluorescence as near to the sample as possible (i.e., before it reaches the scanning mirrors). This eliminates losses at optical surfaces that are not necessary for fluorescence filtering. The entire generated fluorescence signal is collected because no spatial filter is used, nor is it needed because two-photon excitation is confined inherently to the focal volume. Therefore, nondescanned detection allows increased detection efficiency without any loss of three-dimensional discrimination. The fluorescence is split from the reflected signal by a dichroic mirror (with high reflectivity from 380 to 550 nm) to the detection unit that contains a single barrier filter (with high transmittance from 380 to 550 nm, and 10 OD blocking at 700–720 nm) for NAD(P)H autofluorescence. Both of these components are custom made (Chroma Technologies, Brattleboro, VT) for optimal NAD(P)H signal collection. A single transfer lens (antireflection coated from 380 to 550 nm, Melles Griot, Carlsbad, CA) is used to optically map the back aperture of the objective lens to the detection surface of the PMT (Hamamatsu R268, Bridgewater, NJ). Because the PMT is optically conjugate to the back aperture, which is a stationary pivot point of the scanning beam, there should be no effects of any spatial heterogeneities of the photocathode in the resultant image. The use of a bialkali photocathode PMT is best suited for this application. This nondescanned pathway contains only four optical elements (objective lens, dichroic mirror, transfer lens, and barrier filter), and each of these elements is optimized for transmission (or in the case of the dichroic mirror, reflection) of NAD(P)H fluorescence. To obtain the highest gain in the PMT, the supply voltage is set near its maximum (1150 V) throughout the experiments. In our system, the PMT signals are amplified (Hamamatsu C1053), integrated for the pixel duration, and then directed to a frame store card (Data Translation 3852, Marlboro, MA). It should be emphasized that such an external nondescanned detection system can be added easily to any commercial confocal microscope.

IV. Applications to Quantitative Metabolic Imaging

A. General Quantitative Imaging Considerations

The instrument just described allows for extended dynamic studies of many cells simultaneously, thereby permitting observation of the temporal and spatial organization of metabolic activity within intact tissues. To perform quantitative experiments there are some important calibration experiments that must be performed first. These include viability controls, determination of the linearity of detection, and photobleaching controls.

When imaging living cells, sample viability is perhaps the major issue to address. This becomes even more important for NAD(P)H because it is not only a poor fluorophore, but it is also a participant in the metabolic events under investigation. For instance, it may be possible to photoinactivate NAD(P)H by excessive fluorescence excitation, which could alter the cellular redox state. To assay cellular viability in the experiments described here, we check the cellular response to glucose before and after a given laser imaging exposure. Laser irradiation of $\sim$3 mW (average power at the sample) generated signals sufficient for imaging without evidence of cellular damage (i.e., the autofluorescence response to glucose was the same after irradiation as it was before irradiation) even after 1 h of continuous laser scanning irradiation. However, laser irradiation of 5 mW resulted in an immediate slow rise in autofluorescence and led to a significant reduction in the glucose-induced autofluorescence response (Bennett et al., 1996). We further determined that this laser-induced photodamage was due to two-photon excitation, not just the incident red light. To show this, islets were exposed to intense laser illumination that focused into the coverslip (not focused in the islet, so there was no two-photon excited fluorescence generated in the sample). In this case, even 15 mW of unfocused red light did not affect the subsequent glucose-induced NAD(P)H response.

Because laser scanning microscopy uses PMT detectors, which offer a high dynamic range, low noise, and excellent linearity, quantitation of image data is straightforward. However, to make sure that data will be acquired in a linear fashion that will allow quantitation, there are two considerations. First, the analog-to-digital converter that is used to get the PMT signal into the computer must not be saturated (i.e., for a typical 8-bit system, the PMT gain must be set so that all pixel values are <255). Once the correct gain is set, it should not be changed; in our system we always use 1150 V (near the maximum allowable voltage) on the PMT. As described earlier, we are limited to 3 mW of illumination by viability considerations, and with this excitation level we never observe saturated pixels in our NAD(P)H images. Second, the amplifier offset must be set so that the zero pixel value corresponds to the zero fluorescence signal. This can be done using a standard fluorescent sample (we use pure NADH for this) as follows. Image four concentrations (in a ratio of 1:2:3:4) of NADH that each give a good signal but for which the maximum pixel value is <255 at the gain used for cellular imaging. Determine the mean pixel value of each image using the histogram command and plot the mean pixel value versus the concentration. The resulting plot should be a straight line, and the Y intercept will go through zero when the black level is correctly set. If the Y intercept is positive, the offset should be reduced; if the Y intercept is negative, the offset should be increased. Although it may require several trials to achieve the optimum gain/offset combination for a given sample, this calibration only needs to be done once (but it is a good idea to check it occasionally to assure accurate quantitation).

To determine the photobleaching correction, the laser is scanned continuously, and images are generally collected every 10th scan. The whole images are then analyzed as described at the end of this chapter and the results are plotted out. This plot usually shows a decay of fluorescence, which can be fit to an exponential curve. This curve can then be used to correct the results from a time-course experiment. Surprisingly, the photobleaching controls in our two-photon excited NAD(P)H experiments have shown that no bleaching correction is needed. At the excitation levels determined by the viability controls above, there was no measurable photobleaching, even after 1 h of continuous laser scanning of a single optical section. While we cannot be sure why there is no measurable photobleaching, it is likely due to a combination of the low laser intensities used and the excitation localization of the two-photon technique. The small amount of photobleached NAD(P)H may be replaced by natural cellular mechanisms, and as long as the bleaching rate is slow enough, the cells can keep the level of NAD(P)H + NAD$^+$(P) constant during an experiment.

B. Image Analysis and Quantitation

For the quantitation of NAD(P)H autofluorescence data, we perform most of our digital image analysis on Macintosh Power PC computers running NIH Image 1.61 (Bethesda, MD). This free program is good and easy to use and it is available from the NIH via the World Wide Web at http://rsb.info.nih.gov/nih-image/Default. html. For single-cell analysis, we usually use a 25 pixel circular region of interest (ROI) that does not include the cell nucleus. It is also important to use the same ROI for all measurements on that cell (in images acquired at different times or with different glucose concentrations). We also collect data on several 25 pixel ROI that are not within any cells to use as a background standard. These values are averaged and subtracted from the autofluorescence values. Because we have set our acquisition system to have zero offset, this residual background represents the small portion of the excitation light that is detected. The background value in our images has pixel values of 2 or 3, compared with peak autofluorescence pixel values ~100.

C. Quantitative Metabolic Imaging of β Cells in Intact Pancreatic Islets

Insulin secretion from pancreatic β cells is tightly coupled to glucose metabolism. At physiological glucose levels, glucose is phosphorylated by the high K_m hexokinase, glucokinase (GK) (Matschinsky et al., 1998). The kinetic features of GK are quite distinct from the other three mammalian hexokinases: K_m for glucose of ~8 mM, sigmoidal kinetics, and the lack of significant inhibition by glucose 6-phosphate. As glucose signaling proceeds in the β cell, intermediate metabolism results in an increase in the ATP/ADP ratio, which is reflected in a concomitant change in the reduced-to-oxidized NAD(P)H/NAD(P)$^+$ ratio. Thus, the measurement of NAD(P)H autofluorescence is a powerful tool for investigating the glucose response of β cells.

β cells normally exist within the islet of Langerhans, a quasi-spherical micro-organ in the pancreas consisting of ~1000 cells. When β cells are isolated from their natural environment, marked variability in their metabolic responses to glucose has been observed using NAD(P)H autofluorescence as an index of the cellular redox state. These studies have been performed using flow cytometry (Pipeleers et al., 1994) and fluorescence microscopy (Bennett et al., 1996; Pralong et al., 1990). To extend these measurements to the intact islet, we used two-photon excitation microscopy to image glucose-induced NAD(P)H autofluorescence. Using the following protocol for two-photon excitation microscopy to image NAD(P)H levels quantitatively, we have shown that β cells within the islet form a much more uniform population than isolated β cells (Bennett et al., 1996).

1. Islets are isolated from mice by distention of the splenic portion of the pancreas followed by collagenase digestion (Sharp et al., 1973). The splenic portion of the pancreas is preferable because it yields islets that are enriched in β cells (Stefan et al., 1987), which are the focus of our current work.

2. Isolated islets are maintained in petri dishes (Falcon) in culture media consisting of RPMI 1640 (Life Technologies, Inc., Grand Island, NY) with 5 mM glucose (basal), supplemented with 10% fetal bovine serum (FBS; Life Technologies, Inc.) containing 100 units/ml penicillin and 100 μg/ml streptomycin (Life Technologies, Inc.) at 37 °C in an atmosphere of 5% (v/v) CO$_2$. Generally, islets can be maintained in culture dishes for up to 1 week. For longer culture periods, islet viability can be increased by culturing them on extracellular matrix (ECM) (Beattie et al., 1991). However, islets cultured this way should be on ECM-coated coverslips because they cannot be removed from the ECM without damage.

3. Prior to imaging or fixing, islets are attached to the bottom of a MatTek dish (MatTek Corp.) using Cell-Tak (Collaborative Biomedical Products, Bedford, MA). A 0.5-μl drop of Cell-Tak is placed in the center of a MatTek dish and dried for 30 s at 42 °C; the dish is rinsed with Hanks' balanced salt solution (HBSS;

Life Technologies, Inc.) and then 100 μl HBSS is added so that it covers the dried drop of Cell-Tak. We find the MatTek dishes very easy to use for most experiments and they fit into our microscope stage incubator. For experiments that require perfusion of solutions, we can either use these dishes with an input and aspirated outflow or use one of many home-built chambers, which hold a 1-in. diameter coverslip. In the case of chambers, we dry the Cell-Tak onto the coverslip, place the coverslip in the chamber, and then add the islet as described next.

4. Using a dissecting microscope, islets are picked up by a pipette, washed quickly in HBSS, and then placed directly onto the circle of dried Cell-Tak by pipetting through the HBSS. Islets must be washed in serum-free buffer because serum inhibits the attachment to Cell-Tak. We often place two or more islets contiguously in the dish. For many experiments (e.g., comparing islets from normal and transgenic mice), it is extremely useful to perform the imaging of different islets simultaneously. This provides an internal control for each experimental observation.

5. The islet should immediately attach firmly to the Cell-Tak, and then 2 ml of BMHH buffer (125 mM NaCl, 5.7 mM KCl, 2.5 mM CaCl$_2$, 1.2 mM MgCl$_2$, 10 mM HEPES, and 0.1% bovine serum albumin, pH 7.4) with 1 mM glucose is added to the dish. It is important to add the buffer gently to avoid dislodging the islets.

6. The dish containing the islet(s) is then transferred to the microscope stage and allowed to equilibrate for 15 min at 1 mM (basal) glucose. During the imaging experiments, the islets are held at 37 °C using a commercial microincubator (TLC-MI, Adams & List Associates, Westbury, NY). An air stream incubator (Nicholson Precision Instruments, Gaithersburg, MD) is also used to heat the objective to eliminate heat transfer through the glass–oil–objective interface.

7. NAD(P)H autofluorescence measurements can now be made in response to changes in glucose or other treatments. Each image is formed using a 9-s scan. Because we use the external nondescanned detector, all measurements must be made in the dark. NAD(P)H-glucose dose–response images are generally acquired after a 5-min equilibration at each glucose concentration. For extended time-course measurements, we usually take one scan every 10 s; several hundred images can be taken with no noticeable degradation in islet viability.

Figure 2 shows a typical image of β cell NAD(P)H autofluorescence within an intact islet, where signals from both the cytoplasm and mitochondria can be observed. The outlines of single cells are visible as are the nuclei, both of which appear dark. These NAD(P)H imaging experiments simply cannot be performed by confocal microscopy due to the limitations imposed by photobleaching and UV-induced photodamage. In addition, these data open up the possibility of subcellular resolution to investigate differences between cytoplasmic and mito-chondrial metabolism. Figure 3 shows an NAD(P)H-glucose dose–response curve averaged from the responses of 80 cells within four different islets. The apparent K_m of this dose–response is $\sim$8 mM glucose, indicative of the central role of GK in β-cell glucose-stimulated metabolism.

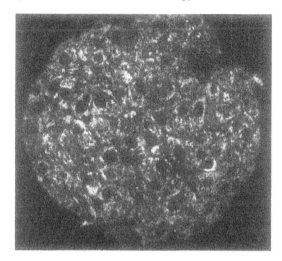

Fig. 2 Optical section of NAD(P)H autofluorescence from an intact islet. The NAD(P)H signal arises from both the cytoplasm and mitochondria, the latter of which sometimes can be seen as punctate bright spots. Cell outlines and nuclei (where there is little or no NAD(P)H) appear dark.

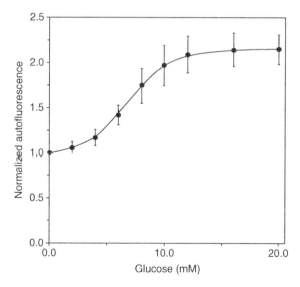

Fig. 3 NAD(P)H autofluorescence glucose dose–response, average of responses from 80 β cells in intact islets. Each NAD(P)H level was measured after a 5-min equilibration with glucose solutions. The inflection point of this curve is consistent with the kinetics of glucose binding to glucokinase.

We have also developed protocols for the indirect immunostaining of glucokinase, insulin, glucagon, somatostatin, and pancreatic polypeptide in intact islets. These immunofluorescence methods allow us to correlate the results of dynamic metabolic imaging with enzyme and hormone levels in the cells. Because the islet is a thick multicellular sample, it is necessary to use low antibody concentrations and long incubation times. Any attempt to rush these steps will lead to brighter immunostaining in cells on the periphery and poor staining of cells in the middle of the islet.

1. Islets used for immunofluorescence staining are fixed in 2 ml of cold (4 °C) 4.0% (v/v) paraformaldehyde (Electron Microscopy Sciences, Ft. Washington, PA) in 10 mM phosphate-buffered saline (PBS), and then incubated at 4 °C for 45 min. This should be added to the dish gently, taking care not to dislodge the islets.

2. The dish is then washed gently three times with cold (4 °C) 10 mM PBS. At this point the sample can be stored at 4 °C in 10 mM PBS for several days before staining.

3. The islets are permeabilized by first pipetting off the PBS and then adding 10 mM PBS with 0.2% Triton X-100; incubate this for 2.5 h at room temperature. All of the staining solutions must be added to and removed carefully from the glass coverslip in the center of the dish with 100 μl volumes for each incubation.

4. Replace the permeabilization solution with the blocking solution (10 mM PBS + 0.2% Triton X-100 with 5% normal donkey serum) and incubate for 2.5 h at room temperature.

5. Replace the blocking solution with fresh antibody dilution buffer (10 mM PBS + 0.2% Triton X-100 + 1.0% bovine serum albumin (Sigma, St. Louis, MO)) and incubate for 20 min at room temperature.

6. This is then replaced with the primary antibody diluted in antibody dilution buffer at a final immunoglobulin G (IgG) concentration of about 10 μg/ml. The dish is then incubated undisturbed overnight at room temperature. This should be placed in a humidified chamber made from a P-100 size petri plate with a circle of damp paper towel placed inside the bottom of the dish. Place the MatTek dish on top of this paper towel carefully.

7. The following day, remove the primary antibody and wash the dish four times carefully with 10 mM PBS + 0.2% Triton X-100 for 20 min each time.

8. Replace the final wash with antibody dilution buffer and incubate for 1 h at room temperature.

9. Replace this with the secondary antibody in antibody dilution buffer, return the dish to the humidified chamber, and incubate overnight undisturbed in the dark at room temperature. Because the secondary antibody is light sensitive, it is helpful to perform these washes with minimal light in the room.

10. The following day, remove the secondary antibody and wash four times, 20 min/wash, with 10 mM PBS + 0.2% Triton X-100.

11. Prior to mounting the sample, rinse it once in 10 mM PBS without Triton X-100. For most sample mounting, we use a few drops of Aqua-Polymount (Polysciences, Inc.). This dries into a solid sample that will last for several months

(at least). The Aqua-Polymount must fill the cutout section in the center of the MatTek disk to the rim, as it will shrink a little as it hardens. For the most demanding high-resolution imaging, we dehydrate the samples and mount them in methyl salicylate, which has the same index of refraction as objective lens immersion oil (Summers *et al.*, 1993).

12. This dish is now ready for analysis by quantitative confocal microscopy and comparison with the two-photon excitation metabolic imaging. It can be stored at room temperature in the dark until imaging is complete.

D. Quantitative Imaging of Muscle Metabolism

Glucokinase is not expressed in peripheral tissue, such as muscle, so in these cells glucose metabolism is mediated by other hexokinases. Because the K_m for glucose of the other hexokinases is much lower than that of GK, this difference can be seen readily in the NAD(P)H responses to glucose. Of the several hexokinase isoforms, hexokinase II (HKII) is thought to be most important in muscle. HKII has been shown by biochemical methods to bind mitochondria (Sui and Wilson, 1997), where it has better access to the ATP needed to phosphorylate glucose and where it is less sensitive to inhibition by glucose 6-phosphate, the reaction product. Further, this mitochondrial binding of HKII is thought to be regulated by insulin. We therefore would like to determine the relationship among HKII distribution, insulin action, and glucose-stimulated metabolism. Redox fluorometry based on NAD(P)H autofluorescence has been used extensively to image metabolism in cardiac muscle (Eng *et al.*, 1989; Ince *et al.*, 1992) and should be quite useful for other muscle types as well. We are using two-photon excitation microscopy to image cultured L6 muscle cells in order to assess the glucose-stimulated changes in NAD(P)H autofluorescence in the entire cell and its mitochondria. For this application, two-photon excitation microscopy offers the unique ability to follow these changes in many individual cells in real time. Similar to the experiments with pancreatic islets, the cells could be fixed and stained immediately after the metabolic experiment, and the precise distribution of HKII can be determined by immunofluorescence. Currently, we are developing methods to image the HKII distribution in living cells using a fusion protein of HKII and the GFP.

Through the two-photon excitation imaging procedure described here, we can derive the kinetics of glucose-dependent changes in mitochondrial and cellular NAD(P)H autofluorescence and attempt to correlate these results with HKII location in the cells. We can also determine whether insulin treatment (either through long-term culture or rapid application) contributes to enhanced cellular glucose metabolism. Once the metabolic effects are well characterized, they can be correlated with the binding of HKII to the mitochondria.

The procedures used are similar to those for islets, but because the cultured cells are not as thick, we can use less input power. We have found that only 2 mW at the sample gives a good signal with no obvious cellular photodamage; even after several minutes of continuous imaging, the morphology of the cell and its NAD(P)H response to glucose are unchanged. The gain and offset of the microscope are the same as used for the islet imaging, and again there was no measurable photobleaching during control experiments.

1. Partially differentiated L6 myotubes (Blau and Webster, 1981) are plated onto coverslip-bottom dishes (MatTek) that have been coated with mouse collagen, type IV (Life Technologies, Inc.), and cultured overnight in RPMI 1640 medium supplemented with 100 nM insulin and 4.5 mM D-glucose. For many cell cultures, it is important to grow them on some kind of matrix to which they will stick. We have used Cell-Tak or collagen in many experiments, but other matrices (e.g., polylysine) may work better for certain cell types.

2. Cells are washed and incubated in insulin- and glucose-free medium for 4 h at 37 °C. It is important to begin with a baseline (in this case zero glucose) that will allow you to ratio the observed changes in autofluorescence.

3. At the start of the experiment, a plate of cells is positioned on the microscope stage of the two-photon instrument, where the cells are maintained at 37 °C. The cells must be focused by eye and then be left to equilibrate to the temperature on the stage for at least 10 min. The temperature-controlled stage fits both MatTek dishes and several custom-built chambers that use 1-in. round coverslips. Failure to equilibrate the sample to the stage will result in motions during the experiments.

4. After the cells have equilibrated on the stage, a single image is taken at zero glucose concentration. As discussed earlier, a reliable baseline is very helpful in data interpretation. Because some cells may have more mitochondria in the optical section than others, the absolute value of the autofluorescence intensity is not of much value.

5. Images are taken 5 min after varying the medium glucose to each concentration. The glucose concentration can be changed by perfusing media with the desired concentration or by adding a small amount of 1 M glucose to the media in order to reach the desired concentration. In either case, we have confirmed that for L6 cells the fluorescence changes are complete within 2–3 min of changing the glucose concentration.

Figure 4 shows the NAD(P)H autofluorescence pattern in several cells at 0.2 mM glucose. The pattern reflects primarily NAD(P)H in mitochondria, which is punctate, but there is also some cytoplasmic distribution of fluorescence. In differentiated cells, the bright mitochondrial fluorescence is evident as columns of mitochondria between muscle fiber striations. With increasing glucose concentration in the medium, an increased fluorescence was apparent, which is shown in Fig. 5 averaged over the same cells at each glucose concentration. A saturable increase in averaged fluorescence is apparent, with an apparent K_m that is in between that of the transporter (6 mM) and the endogenous hexokinase (0.1 mM). The response of various cells was quite homogeneous (i.e., all cells show a similar change). The addition of insulin to the overnight culture medium

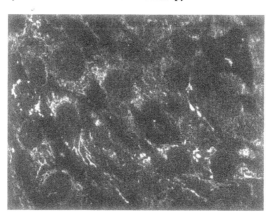

Fig. 4 Optical section of NAD(P)H autofluorescence from a field of L6 myotubes. In this case the NAD(P)H signal arises mostly from the mitochondria, which can be seen as brighter punctate objects. In the more highly differentiated cells, fluorescence is evident in mitochondria collimated between muscle fiber striations. As in Fig. 2, cell nuclei appear dark.

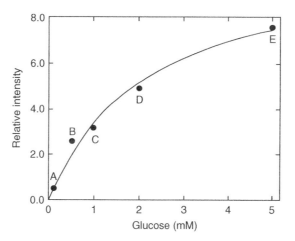

Fig. 5 NAD(P)H autofluorescence glucose dose–response, average of responses from five L6 cells from Fig. 4. Each NAD(P)H level was measured after a 5-min equilibration with glucose solutions. Unlike the dose–response curves from β cells shown in Fig. 3, the inflection point of this curve is consistent with the kinetics of hexokinases other than glucokinase.

in step 1 increased the NAD(P)H autofluorescence signals significantly at all glucose concentrations (including zero glucose). This suggests persistent effects of the hormone on the cellular redox state. Overall, these results show that we can define the kinetics of glucose utilization in real time from single cells using two-photon excitation microscopy.

Acknowledgments

The authors thank Drs. James May, Dick Whitesell, and Daryl Granner for help with the L6 cell experiments and Mr. Guangtao Ying for help in constructing the two-photon excitation microscope. This work has been supported by an NIH grant to D. W. P. (DK53434) and the Vanderbilt Cell Imaging Resource (supported by CA68485 and DK20593).

References

Beattie, G. M., Lappi, D. A., Baird, A., and Hayek, A. (1991). *J. Clin. Endocrinol. Metab.* **73,** 93.

Bennett, B. D., Jetton, T. L., Ying, G., Magnuson, M. A., and Piston, D. W. (1996). *J. Biol. Chem.* **271,** 3647.

Blau, H. M., and Webster, C. (1981). *Proc. Natl. Acad. Sci. USA* **78,** 5623.

Chance, B., and Lieberman, M. (1978). *Exp. Eye Res.* **26,** 111.

Chance, B., and Thorell, B. (1959). *J. Biol. Chem.* **234,** 3044.

Coremans, J. M., Ince, C., Bruining, H. A., and Puppels, G. J. (1997). *Biophys. J.* **72,** 1849.

Denk, W., Strickler, J. H., and Webb, W. W. (1990). *Science* **248,** 73.

Denk, W., Piston, D. W., and Webb, W. W. (1995). In "The Handbook of Biological Confocal Microscopy," 2nd edn. (J. Pawley, ed.), p.445. Plenum, New York.

Eng, J., Lynch, R. M., and Balaban, R. S. (1989). *Biophys. J.* **55,** 621.

French, S. A., Territo, P. R., and Balaban, R. S. (1998). *Am. J. Physiol.* **275,** C900.

Halangk, W., and Kunz, W. S. (1991). *Biochim. Biophys. Acta* **1056,** 273.

Ince, C., Coremans, J. M., and Bruining, H. A. (1992). *Adv. Exp. Med. Biol.* **317,** 277.

Masters, B. R., Ghosh, A. K., Wilson, J., and Matschinsky, F. M. (1989). *Invest. Ophthalmol. Vis. Sci.* **30,** 861.

Masters, B. R., Kriete, A., and Kukulies, J. (1993). *Appl. Opt.* **32,** 592.

Matschinsky, F. M., Glaser, B., and Magnuson, M. A. (1998). *Diabetes* **47,** 307.

Miyawaki, A., Llopis, J., Heim, R., McCaffery, J. M., Adams, J. A., Ikura, M., and Tsien, R. Y. (1997). *Nature* **388,** 882.

Nieminen, A. L., Byrne, A. M., Herman, B., and Lemasters, J. J. (1997). *Am. J. Physiol.* **272,** C1286.

Pipeleers, D., Kiekens, R., Ling, Z., Wilikens, A., and Schuit, F. (1994). *Diabetologia* **37,** S57.

Piston, D. W. (1999). *Trends Cell Biol.* **9,** 66.

Piston, D. W., Masters, B. R., and Webb, W. W. (1995). *J. Microsc.* **178,** 20.

Piston, D. W., Knobel, S. M., Postic, C., Shelton, K. D., and Magnuson, M. A. (1999). *J. Biol. Chem.* **274,** 1000.

Pralong, W. F., Bartley, C., and Wollheim, C. B. (1990). *EMBO J.* **9,** 53.

Sharp, D. W., Kemp, C. B., Knight, M. J., Ballinger, W. F., and Lacy, P. F. (1973). *Transplantation* **16,** 686.

Stefan, Y., Meda, P., Neufeld, M., and Orci, L. (1987). *J. Clin. Invest.* **80,** 175.

Sui, D. X., and Wilson, J. E. (1997). *Arch. Biochem. Biophys.* **345,** 111.

Summers, R. G., Stricker, S. A., and Cameron, R. A. (1993). *In* "Methods in Cell Biology," (B. Matsumoto, ed.), Vol. 38, p. 265. Academic Press, San Diego, CA.

CHAPTER 13

Trafficking of the Androgen Receptor

Virginie Georget, Béatrice Terouanne, Jean-Claude Nicolas, and Charles Sultan

INSERM U439
Pathologie Moleculaire des Recepteurs Nucleaires
Rue de Navacelles Montpellier
France

I. Introduction

The androgen receptor (AR) belongs to the superfamily of nuclear receptors characterized by a common structure and mechanism of action (Evans, 1988; Mangelsdorf *et al.*, 1995). The AR is an androgen-dependent transcription factor composed of four domains: the amino-terminal transcription activation domain; the DNA-binding domain, which interacts with a specific DNA sequence called the androgen-responsive element (ARE); the hinge region, which includes the nuclear localization signal (NLS); and the carboxy-terminal ligand-binding domain (LBD). For its transcriptional activity, the AR needs to bind the androgen, pass the nuclear membrane, and interact as a dimer with the ARE to trigger a cascade of transcriptional events.

The analysis of the subcellular localization of the steroid receptors has usually been performed by immunotechniques (immunocytochemistry or immunohisto-chemistry). It is generally acknowledged that the estrogen receptor (ER) and the progesterone receptor (PR) are predominantly nuclear, with a continuous shuttle between the nucleus and cytoplasm (Guiochon-Mantel *et al.*, 1996). The intracel-lular localization of the mineralocorticoid receptor (MR), the glucocorticoid receptor (GR), and the AR are more controversial (Guiochon-Mantel *et al.*, 1996). According to the immunostaining protocol, these receptors have been described as being either in the cytoplasm or in the nucleus in the absence of ligand, and exclusively in the nucleus after incubation with ligand. These techni-ques require the fixation and permeabilization of cells, which can lead to artifacts in the pattern of subcellular localization. Moreover, the AR can be in different states, that is, associated with the heat shock proteins in an unliganded form or associated with DNA or transcription factors in the liganded form. The accessibil-ity of the epitope to antibodies may vary for these different forms and this could induce artifactual results in the immunostaining. Having considered all the limits of immunocytochemistry, we developed a model using a chimera of AR fused to the green fluorescent protein (GFP). This fluorescent reporter permitted the visualization of the AR in living transfected cells (Georget *et al.*, 1997).

We first verified that the fusion protein (GFP–AR) conserved the functional characteristics of the AR. We demonstrated the advantages of this GFP–AR tool versus immunodetection. The intracellular dynamics of the AR were evaluated and quantified in living cells, which suggested some applications of the GFP–AR model, such as antiandrogen screening and androgen insensitivity study.

II. Plasmid Construction and Functional Properties of GFP–AR

A. Construction

Two different GFPs are used: wild-type GFP contained in pC1–GFP (Clontech, Palo Alto, CA) and GFP-S65T, a red-shifted variant of GFP in which a mutation in the chromophore has been introduced, contained in pC1–S65T-GFP (Clontech); but only GFP-S65T is fused to the AR (Fig. 1). To digest pC1–S65T-GFP with *Xba*I, the vector is transformed in DM1 cells (*dam*-host) (GIBCO-BRL, Cergy-Pontoise, France) to demethylate the *Xba*I site. The AR cDNA is isolated from pCMV5–hAR by *Xma*I–*Xba*I digestion and ligated at the cognate sites of

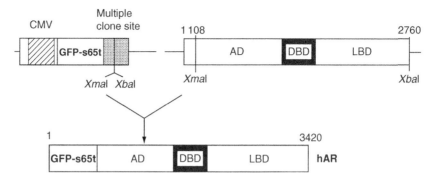

Fig. 1 Construction of GFP expression plasmids and structure of GFP fusion proteins. Shown is the schematic structure of pC1–S65T-GFP, encoding a full-length red-shifted variant of GFP, and the schematic structure of the human AR cDNA with its three domains: the activation domain (AD), the DNA-binding domain (DBD), and the ligand-binding domain (LBD). The amino-terminal truncated hAR cDNA (*Xma*I–*Xba*I) was fused with GFP-S65T.

pC1–S65T-GFP digested with *Xma*I and *Xba*I. In pGFP–AR, GFP-S65T is fused to the amino terminal of a human AR (hAR) that lacks the first 36 amino acid residues. The ligation product is transformed in DH5α *Escherichia coli* cells and positive clones are selected by neomycin antibiotics. Constructions are verified by enzymatic digestion and then sequenced in ligation fragments to verify the correct reading frame.

B. Functional Properties of GFP–AR: Comparison with Androgen Receptor

The protein expression, the androgen-binding characteristics, and the transcriptional activity capacity of the GFP–AR are evaluated by transient transfection experiments and cotransfection assays (Georget *et al.*, 1997). Mammalian cells expressing GFP exhibit stronger fluorescence when grown at 30 °C (Pines, 1995). All experiments are performed at 30 and 37 °C.

1. Materials

The COS-7 cell line is transfected with 10 μg of pCMV5–hAR, pC1–S65T-GFP, or pGFP–AR in dishes with a 10-cm diameter for protein expression detection by Western blot; and with 50 ng of pCMV5–hAR or pGFP–AR with 0.25 μg of pCMV-β-galactosidase per well in 12-well tissue culture dishes for androgen-binding assay in whole cells.

The CV-1 cells are transfected for the transactivation assay with 50 ng of pCMV5–hAR or pGFP–AR with 0.25 μg of pCMV-β-galactosidase and 1 μg of androgen-dependent reporter gene (*p*-tyrosine aminotransferase-thymidine kinase (TAT-tk) regulating the luciferase gene) per well in 12-well tissue culture dishes.

2. Methods

Cells are transfected by the calcium phosphate DNA precipitation method 8 h after trypsinization. The precipitate is removed 12 h later and cells are cultured in Dulbecco's minimal essential medium (DMEM; GIBCO-BRL) supplemented with penicillin (100 units/ml) and streptomycin (100 μg/ml) in the presence or absence of hormones for 30 h at 30 or 37 °C.

a. Western Blot

Cells are solubilized in lysis buffer (160 mM Tris (pH 6.9), 200 mM dithiothreitol (DTT), 4% sodium dodecyl sulfate (SDS), 20% glycerol, 0.004% bromophenol blue) in the presence of protease inhibitors (1 mM phenylmethylsulfonyl fluoride (PMSF), 0.05 mM leupeptin, 0.01 mM pepstatin), boiled for 5 min, and centrifuged at 13,000 $\times$ g for 10 min. The supernatant is subjected to SDS–polyacrylamide gel electrophoresis (SDS–PAGE) and then Western-transferred onto nitrocellulose membranes by electroblotting. The filters are saturated with a 10% milk solution and incubated either with anti-AR antibody (SpO61) (van Laar *et al.*, 1989) diluted 1:2000 or with anti-GFP antiserum (anti-GFP; Clontech) diluted 1:1000. The filters are then incubated in the presence of a peroxidase-conjugated antirabbit IgG diluted 1:5000. Blots are developed using the enhanced chemiluminescence (ECL) detection system (Amersham, Les Ulis, France).

b. Androgen–Binding Assay

The transfected cells are incubated in duplicate for 2 h at 37 °C with various concentrations (0.05–0.3 nM) of synthetic androgen [^{3}H]R1881 (total binding) and duplicate wells are incubated together with 100 nM unlabeled R1881 (nonspecific binding). The cells are then washed with cold phosphate-buffered saline (PBS) and harvested in lysis buffer (25 mM Tris–H_3PO_4 (pH 7.8), 2 mM DTT, 2 mM EDTA, 1% Triton X-100, and 10% glycerol). Aliquots are counted for radioactivity and for β-galactosidase activity assay. After subtraction of nonspecific binding from total binding, the dissociation constant (K_d) and the maximum androgen-binding sites (B_{max}) were derived from the Scatchard representation.

c. Transactivation Assay

CV-1 cells are cultured for 30 h at 30 or 37 °C in serum-free medium with various concentrations of R1881 (3 $\times$ 10^{-12}–10^{-9} M). Transfected cells are lysed as described above. The luciferase activity is measured by the reaction of lysate with the luciferin solution: 270 μM coenzyme A, 470 μM luciferin, 530 μM ATP, 20 mM Tris–H_3PO_4, 1.05 mM $MgCl_2$, 2.7 mM $MgSO_4$, 0.1 mM EDTA, and 33 mM DTT. Luciferase activity is measured in an LKB (Bromma, Sweden) luminometer and expressed in arbitrary units of luminescence.

3. Results and Comments

By immunoblot analysis, the GFP–AR was visualized with an anti-AR antibody and an anti-GFP antibody as a protein of 130 kDa. GFP was detected with an apparent molecular mass of 27 kDa and the AR was detected with a mass of 110 kDa. The GFP and GFP–AR were less expressed than the AR, which could be due to the difference in the vectors containing the AR or GFP and the GFP–AR. The use of the two antibodies revealed that the GFP–AR was sensitive to degradation because in both cases smaller bands were observed. The expression level appeared to be identical for cells cultured at 30 and 37 °C but the GFP–AR was more degraded at 37 °C.

In transient transfected cells, the androgen-binding affinity of the GFP–AR was conserved in comparison with the wild-type AR, and the B_{max} value was slightly reduced. The GFP–AR was able to transactive the androgen-regulated reporter gene only if the cells were cultured at 30 °C. This phenomenon has already been reported for the GFP–GR fusion protein (Ogawa *et al.*, 1995). The GFP–AR demonstrated an induction luciferase activity with a dose dependence identical to that of the AR, but it reached a reduced maximal activity in comparison with the AR. This partial activity of the GFP–AR can be explained by a low expression of the protein previously observed by Western blot, by the deletion of amino acids 1–36 of the amino-terminal cDNA of the AR in GFP–AR, or by a cumbersome conformation of the fusion protein, which is less appropriate than the AR to transactivate reporter gene.

The GFP–AR, however, conserves the essential functional characteristics of the AR, such as the androgen-binding characteristics and transactivation capacity. The GFP tag does not modify the properties of the AR and can thus be used to determine the subcellular localization of the AR.

III. GFP–AR Localization by Fluorescence and by Immunodetection

The subcellular localization of the GFP–AR has been determined by detection of GFP fluorescence and by antibodies conjugated to Texas Red (rhodamine isothiocyanate (RITC) filter) within the same cell. Because the reported immunodetection results of AR localization were controversial, we tested different conditions of cell permeabilization for indirect immunofluorescence.

A. Materials

1. COS-7 cells

COS-7 cells are cultured directly on microscope glass coverslips (20 mm× 20 mm) and transfected using the calcium phosphate method with 1 μg of pGFP–AR, pC1–GFP, or pC1–S65T-GFP, as described above.

2. Solutions

Fixation solution: paraformaldehyde, 4% in PBS

Permeabilization and washing solution:

In PBS: Saponin (0.05%, 0.1%, or 0.3%) or Triton X-100 (0.02%, 0.1%, and 0.3%)–
   Tween 20 (0.05%)

Saturation solution: normal donkey serum, 2.5% diluted in permeabilization
   solution

Antibody dilution solution: 0.5% bovine γ-globulin diluted in permeabilization
   solution.

3. Antibodies

SpO61, rabbit polyclonal anti-AR antibody (van Laar *et al.*, 1989)

Anti-GFP, a rabbit anti-GFP antiserum (Clontech)

Texas Red-conjugated donkey anti-IgG rabbit antibody (Jackson Immunoresearch, West Grove, PA).

4. Microscopes

Conventional direct epifluorescence microscope with a 40× objective and a
fluorescein isothiocyanate (FITC) filter are used for GFP detection and an RITC
filter for immunofluorescence visualization.

B. Methods

The following protocol was written for immunofluorescence in whole COS-7
cells; appropriate modifications can be adapted to other cell types.

Cells are incubated in the presence or absence of 10^{-6} M R1881 for 1 h at 37 °C.
The cells are then placed on ice for 10 min, washed twice with PBS, and fixed for 1 h
at room temperature in 4% paraformaldehyde. The cells are washed twice with
PBS and incubated with various concentrations of saponin or Triton X-100
(to permeabilize the cells) for 30 min at room temperature with gentle agitation.
For saturation of nonspecific sites, the cells are treated for 30 min with saturation
solution, the saponin or Triton X-100 concentration of which corresponds to the
concentration used for permeabilization. The slides are then incubated for 2 h with
either anti-GFP or anti-AR antibody diluted 1:500. After cell washing, the slides
are incubated for 2 h with the secondary antibody diluted 1:100. After another
wash, coverslips are mounted on the slides with fluorescent mounting medium
(Dako, Carpinteria, CA) and observed under a microscope the next day.
To control the specificity of immunostaining, some slides are incubated with just
the primary antibody or the secondary antibody.

C. Results and Comparison of Permeabilization Conditions

The cells transfected with GFP-S65T or its corresponding fusion protein (GFP–AR) were easily identified by GFP fluorescence. The level of GFP fluorescence differed from that of cellular autofluorescence, and the transfected cells corresponded to approximately 10% of the COS-7 cells. Conversely, the fluorescence was not detected in cells expressing wild-type GFP. GFP-S65T was distributed throughout the cytoplasm and nucleus of the COS-7 cells, and this distribution was not modified by addition of androgen. The GFP–AR was localized predominantly in the cytoplasm in the absence of hormone. The intensity of the detected fluorescence depended on the cell depth. The higher perinuclear fluorescence exhibited by some cells was thus due to a thicker cytoplasm around the nucleus rather than to a particular staining pattern. After incubation with 10^{-6} M R1881 for 1 h, the GFP–AR was predominantly nuclear. This revealed the hormone dependence of GFP–AR trafficking.

In immunocytochemistry, the spatial distribution of the AR was strongly affected by the conditions of permeabilization. A pattern of localization identical to GFP fluorescence was obtained when cells were permeabilized with 0.3% saponin or 0.02% Triton X-100 (Fig. 2). With these concentrations, the GFP–AR was cytoplasmic in the absence of hormone and nuclear after hormonal incubation. Control experiments were performed with transfected cells expressing the AR, and the same cellular

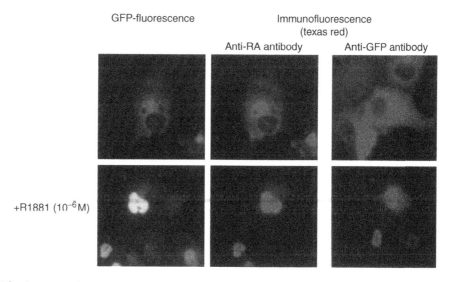

Fig. 2 Comparison of GFP fluorescence and immunofluorescence. The subcellular localization of the GFP–AR in COS-7 cells was analyzed by GFP fluorescence and by immunofluorescence within the same cell. Some cells were incubated with 10^{-6} M R1881 at 37 °C for 1 h before immunostaining. The cells were permeabilized with 0.3% saponin and incubated with either anti-GFP or anti-AR antibody. The same cell was observed with an FITC filter for GFP fluorescence and with an RITC filter for immunodetection of Texas Red-conjugated antibody. The same results were obtained when cells were permeabilized with 0.02% Triton X-100. (See Plate no. 10 in the Color Plate Section.)

distribution was observed. However, as a function of the concentration of permeabilization solution, the spatial distribution differed. Indeed, the cells permeabilized with Triton X-100 (0.1% or 0.3%) exhibited a nuclear localization of immunofluorescence in the absence and in the presence of hormone. A lower concentration of saponin (0.05% or 0.1%) revealed a cytoplasmic immunofluorescence in the absence of hormone, as visualized in the same cell by GFP fluorescence. In the presence of hormone, the GFP–AR was immunodetected in the cytoplasm, whereas the GFP fluorescence was nuclear. This was verified with the anti-GFP and anti-AR antibodies.

Given the results of AR immunodetection, we concluded that GFP was of potentially great utility. Moreover, GFP can be visualized in living cells, which suggests that the localization of the GFP–AR could be studied in terms of its dynamics along with the analysis of the fate of the AR in the same living cell.

IV. Various Fluorescence Microscope Technologies for GFP Analysis

Analysis of the subcellular localization of GFP could be done using various fluorescence microscope technologies such as the confocal scanning microscope, the laser scanning cytometer, and the epifluorescence microscope. Each technique presents some advantages and disadvantages for specific GFP applications.

A. Confocal Scanning Microscope

In confocal scanning microscopes, the object is viewed not as a whole but as a series of points on which both illumination and detection are tightly focused. This avoids the problem often encountered with fluorescence, that is, the swamping of the in-focus fluorescence signal by noise in the form of fluorescence from out-of-focus portions of the object. In some applications, the confocal microscope produces a dramatic increase in the sharpness and contrast of ultimate images. This technology could be applied to GFP to determine a precise localization of GFP fusion protein. However, the confocal scanning microscope is not adapted for kinetics studies. The time required for the analysis of one cell is too long for the study of a rapid process and it can induce substantial photobleaching of the fluorescence. Therefore, this technology is not adapted for dynamics studies.

B. Laser Scanning Cytometer

ACAS 570 (adherent cell analysis and sorting; Meridian, Okemos, MI): The ACAS provides fluorescence measurement of structure, function, and response at the single-cell level. Cells of interest are positioned above the objective and the focused laser beam is pulsed as the stage moves to scan the cell. Data are collected at intervals as small as 0.25 μm.

1. Advantages

- Sensitive, quantitative measurement is obtainable with minimal photobleaching.
- The inverted fluorescence microscope allows maximum flexibility in sample handling, from glass slides to multiwell culture dishes.
- The motorized x–y stage permits storage of cell coordinates for repeated scans of the same cell over time. The ACAS 570 is well adapted for kinetics studies.
- The ACAS 570 permits the measurement of cellular dynamics, such as receptor mobility, by FRAP (fluorescence redistribution after photobleaching) techniques. FRAP techniques involve photobleaching of the fluorescent molecule in discrete subcellular areas with a high-intensity laser pulse, thereby creating nonfluorescent patches. The rate at which the patches are filled by unbleached molecules from surrounding areas is measured by periodic, low-intensity laser scanning.

2. Disadvantages

- The absence of an epifluorescence microscope in parallel with the ACAS 570 leads to some difficulties in detecting rapidly the positive cells in a transient transfection.
- The time required to scan one cell is too long for visualization of rapid processes.
- The low resolution and sharpness do not permit identification of organelles in the cell.
- The ACAS 570 can be used for quantitative kinetics to study only predefined processes.

C. Epifluorescence Microscope

For most GFP applications, epifluorescence microscopes are sufficient. A microscope coupled to a camera is a good system for analysis of GFP, providing an excellent compromise between sufficient sensitivity, good resolution, simplicity of operation, and high rapidity of image acquisition. The images can be quantified with NIH Image software. This microscopic system is well adapted for kinetics study.

Of these three technologies that we have tested, we prefer the epifluorescence microscope and have used it to study GFP–AR kinetics in COS-7 cells.

V. Intracellular Dynamics and Quantification of GFP–AR in Living Cells

In transfected cells, the dynamics of the GFP–AR in the presence of androgen can be evaluated in the same living cell by epifluorescence microscopy and, after quantification of fluorescence, in the various cellular compartments.

A. Quantification of GFP–AR Dynamics

1. Materials

We use an inverted epifluorescence microscope (Diaphot 200; Nikon, Tokyo, Japan) with 10×, 20×, and oil immersion 60× differential interference contrast (DIC) objectives. This microscope is coupled to a CCD camera (Night Owl; EGG, Berthold, Germany) which permits image acquisition and processing with Winlight software (EG&G).

2. Methods

COS-7 cells are cultured in 2×2 cm^2 Lab-Tek chamber slides (Nunc, Naperville, IL) and transfected with 2 μg of pGFP–AR. These chamber slides have the depth of a coverslip and thus the cells can be observed with the oil immersion 60× objective, with good resolution and low fluorescence noise. The medium is replaced with phenol red-free DMEM before microscopy, because the phenol red exhibits a high background fluorescence. The transfected cells, easily and rapidly distinguished, are maintained at 37 °C during the kinetics studies. A single cell is recorded with the CCD camera with a 0.1-s time acquisition in DIC visible light and a 1-s time acquisition in fluorescence. This image corresponds to time zero of the kinetics. The medium is removed and replaced by 1 ml of medium with androgen. The same cell is observed and recorded every 15 min (an observation every 5 min could have damaged the cell after several observations).

All of the images recorded with Winlight software were saved as TIFF images for quantification by NIH Image software. We measured the fluorescence of the nuclear area and the total cellular area using the same surface for each image. Thus, the intensities of pixels were summed within the nuclear and total cellular areas and corrected for background fluorescence. The percentages of nuclear fluorescence compared with the total cellular fluorescence are calculated and represented on graphs as a function of time in minutes.

3. Results and Comments

Cytoplasmic fluorescence was observed without hormone, as previously described, after immunostaining. The cells were then incubated with the natural androgen dihydrotestosterone (DHT). After 15 min of incubation at 37 °C with 10^{-6} M DHT, the cytoplasmic fluorescence diminished, whereas the nuclear fluorescence increased (Fig. 3A). After 60 min, the fluorescence signal was predominantly nuclear (85% of total receptors were nuclear), although a complete translocation was never observed. All of the cells in the chamber exhibited the same kinetics. In addition, we determined the energy dependence of AR nuclear import, as has been previously described for the PR (Guiochon-Mantel *et al.*, 1991). Cells were pre-incubated for 1 h with inhibitors of ATP synthesis (sodium azide, 10 mM) in glucoseminus DMEM supplemented with 6 mM 2-deoxyglucose. With the

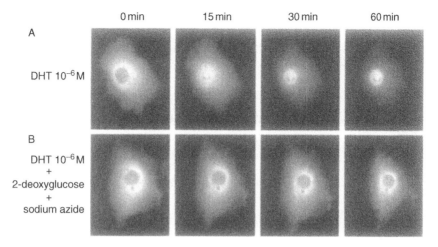

Fig. 3 Dynamics of GFP–AR translocation induced by DHT. COS-7 cells expressing the GFP–AR were analyzed directly by epifluorescence microscopy. Living cells were observed and recorded after 15, 30, and 60 min of incubation with 10^{-6} M DHT alone (A) or with inhibitors of ATP synthesis (B).

epifluorescence microscope, we followed the fate of the AR after addition of 10^{-6} M R1881. While the GFP–AR entered rapidly into the nucleus under normal conditions, the GFP–AR in the presence of the inhibitor of ATP synthesis remained cytoplasmic after 1 h (Fig. 3B). This suggested that AR nuclear transport is an energy-dependent process. We also tested DHT at various concentrations (10^{-9}, 10^{-7}, and 10^{-6} M). The initial rate of nuclear import, evaluated by quantification, was identical with different concentrations (Fig. 4).

GFP is an effective reporter, providing good temporal and spatial resolution in living cells. The GFP–AR model appears to be a new tool for studying AR translocation.

VI. Applications of GFP–AR

A. Screening of Antiandrogens

The antiandrogens are antihormones that repress the action of androgens. The binding of antiandrogen to the AR induces an abnormal conformation of the AR that could block the mechanism of AR action at different steps, such as the dissociation of heat shock proteins, nuclear import, dimerization, and the interaction with DNA or transcription factors (Kuil and Brinkmann, 1996). We hypothesized that the antiandrogen that would be able to block a preliminary step such as nuclear import could be an efficient antiandrogen with a reduced agonist activity. The antiandrogens are classified in two groups: the steroidal antiandrogens (R2956 and cyproterone acetate) and the nonsteroidal antiandrogens (OH-flutamide, inocoterone, bicalutamide or casodex, and nilutamide or anandron).

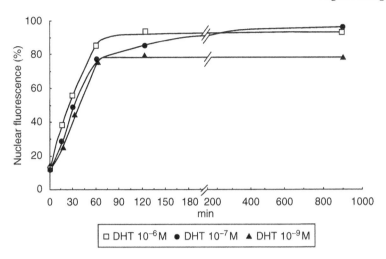

Fig. 4 Quantification of nuclear trafficking of the GFP–AR in the presence of various concentrations of DHT (10^{-6}, 10^{-7}, and 10^{-9} M). The nuclear import of the GFP–AR was analyzed in transfected cells and images were quantified by NIH Image software. For each set of conditions, the intensities of pixels were summed within the individual nuclei and the total cellular areas and corrected for background fluorescence. The percentages of nuclear fluorescence were calculated and pooled for each point.

We tested the ability of each antiandrogen alone to induce AR nuclear import and, in competition with androgen, their ability to block the nuclear translocation induced by androgen. In parallel, we evaluated the agonistic and antagonistic activities of each antiandrogen by cotransfection experiments.

1. Methods

Transcriptional activity is evaluated as previously described with CV-1 cells transfected with the GFP–AR and an androgen-dependent reporter gene. Cells are incubated with various concentrations of antiandrogens in the presence or absence of R1881 at 10^{-10} M, the concentration at which the maximal luciferase induction was obtained. The incubation of antiandrogen alone determined the agonistic activity and, in competition with androgen, the antagonistic property.

Concerning nuclear trafficking, COS-7 cells were transfected with the GFP–AR and incubated with 10^{-6} M antiandrogen alone or in competition with DHT at 10^{-9} M, a concentration at which nuclear import was efficient.

2. Results and Comments

Most of the antiandrogens exhibited agonistic activity (inocoterone, cyproterone acetate, R2956, and OH-flutamide); bicalutamide displayed low agonistic activity and nilutamide had no agonistic activity even at 10^{-6} M. These last two are the most efficient antiandrogens. Only nilutamide (10^{-6} M) did not induce AR

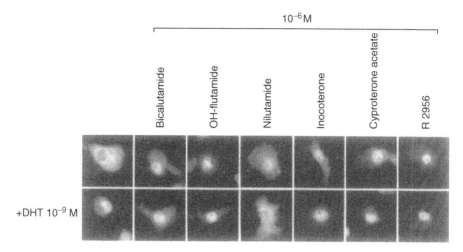

Fig. 5 Nuclear import of the GFP–AR in the presence of antiandrogens. GFP–AR-transfected cells were incubated with various steroidal and nonsteroidal antiandrogens at 10^{-6} M alone or in competition with 10^{-9} M DHT. After 1 h, cells were observed by epifluorescence microscopy.

nuclear transfer and it was able to block the transfer normally induced by androgen (Fig. 5) (Georget and Sultan, 1999). This reflects and explains its antagonistic activity. Regarding bicalutamide, it quite likely blocks the mechanism of AR action in a step that follows translocation.

The GFP–AR is a new tool for screening for new antagonists that are able to block nuclear transport, which would provide more potent antiandrogens in the treatment of endocrine disease. We next applied the GFP–AR model in the study of environmental antiandrogens.

B. Application to Natural Androgen Receptor Mutants Described in Patients with Androgen Insensitivity Syndrome

Defects in the *AR* gene can cause androgen insensitivity syndromes, with a wide spectrum of phenotypes ranging from the complete female phenotype in 46,XY subjects (complete androgen insensitivity syndrome (CAIS), grades 6 and 7) to partial androgen resistance (or partial androgen insensitivity syndrome (PAIS), grades 1–5) (Quigley *et al.*, 1995). Abnormalities of the *AR* gene have been identified in the eight exons of *AR* and can vary from complete or partial gene deletions to, most often, single base mutations (Gottlieb *et al.*, 1998). A single amino acid substitution in the AR protein may thus partially or completely affect the mechanism of AR action. Like many investigators, we have evaluated the activity of these mutants by *in vitro* techniques, previously described, to establish the relationship between the genetic mutation identified in the *AR* gene and the phenotype observed in the patient. This is not always easily done for the partial form of androgen resistance. The GFP–AR model has been applied to AR mutants to evaluate the trafficking of two mutants in the LBD and to determine the usefulness of this method in the investigation of PAIS.

1. Methods

Two mutations detected in patients with PAIS, substitutions R840C (grade 3) (Beitel *et al.*, 1994; Bevan *et al.*, 1996) and G743V (grade 4 or 6) (personal data; and see Nakao *et al.*, 1993), and one mutation detected in CAIS, L707R (Lumbroso *et al.*, 1996), which is used as "negative control," are recreated in the *AR* gene sequence cloned in pCMV5 by site-directed oligonucleotide mutagenesis. The construction of the AR mutants fused to GFP is then obtained in the same manner as for the GFP–AR. The expression level, the androgen-binding characteristics, and the transactivation properties are evaluated as previously described in the section concerning the wild-type GFP–AR. The intracellular dynamics of the GFP–AR mutants were evaluated in transfected COS-7 cells incubated with various concentrations of DHT. For each concentration, real-time kinetics is studied with the epifluorescence microscope and images are quantified with NIH Image software, as described above.

2. Results and Comments

In transient transfection assays, with an identical expression level, GFP–AR-G743V and GFP–AR-R840C displayed decreased androgen-binding affinity compared with the GFP–AR and were fully effective in transactivation, but only at high concentration. This explained the abnormal function of the AR in these patients but did not permit differentiation in terms of the severity of phenotype. Moreover, the androgen-binding characteristics revealed a faster dissociation of androgen for GFP–AR-G743V than for GFP–AR-R840C. As expected, we observed no androgen binding by and no transcriptional activity of GFP–AR-L707R.

The nuclear trafficking of the GFP–AR mutants was evaluated in terms of two parameters: the rate of nuclear transfer and the maximal amount of receptors imported into the nucleus. At 10^{-6} M DHT, the GFP mutants associated with PAIS entered the nucleus in a fashion similar to that of the wild-type GFP–AR, whereas GFP–AR-L707R remained cytoplasmic (Fig. 6). Concerning exclusively the mutants associated with PAIS, the rate and maximal degree of nuclear import at 10^{-7} M DHT were both reduced, even more so for GFP–AR-G743V. The difference between mutants was more pronounced at 10^{-9} M DHT because GFP–AR-G743V entered the nucleus with even slower kinetics (Georget *et al.*, 1998).

We observed that the nuclear transfer capacities of these LBD mutants are in correlation with the severity of the phenotype and, thus, the GFP–AR model could

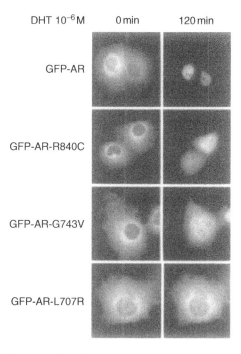

DHT 10⁻⁶ M   0 min   120 min

GFP-AR

GFP-AR-R840C

GFP-AR-G743V

GFP-AR-L707R

Fig. 6 Quantification of the trafficking of the GFP–AR and GFP–AR mutants associated with PAIS (GFP–AR-R840C and GFP–AR-G743V) or CAIS (GFP–AR-L707R). Transfected COS-7 cells were incubated with 10^{-9} M DHT for 2 h. The fusion proteins were detected in live cells using an epifluorescence microscope coupled to a CCD camera.

be a useful complementary tool to understanding the phenotype–genotype relationship of AR function in patients with AIS.

In conclusion, the GFP–AR model has permitted analyses of the dynamics of GFP–AR trafficking in naive cells. Nuclear transfer could also be evaluated in androgen target cells, such as prostatic cells.

References

Beitel, L. K., Kazemi, E. P., Kaufman, M., Lumbroso, R., DiGeorge, A. M., Killinger, D. W., Trifiro, M. A., and Pinsky, L. (1994). *J. Clin. Invest.* **94**, 546.

Bevan, C. L., Brown, B. B., Davies, H. R., Evans, B. A. J., Hughes, I. A., and Patterson, M. N. (1996). *Hum. Mol. Genet.* **5**, 265.

Evans, R. M. (1988). *Science* **240**, 889.

Georget, V., and Sultan, C. (1999). in preparation.

Georget, V., Lobaccaro, J., Térouanne, B., Mangeat, P., Nicolas, J., and Sultan, C. (1997). *Mol. Cell. Endocrinol.* **129**, 17.

Georget, V., Térouanne, B., Lumbroso, S., Nicolas, J. C., and Sultan, C. (1998). *J. Clin. Endocrinol. Metab.* **83**, 3597.

Gottlieb, B., Lehvaslaiho, H., Beitel, L. K., Lumbroso, R., Pinsky, L., and Trifiro, M. (1998). *Nucleic Acids Res.* **26**, 234.

Guiochon-Mantel, A., Lescop, P., Christin-Maitre, S., Loosfelt, H., Perrot-Applanat, M., and Milgrom, E. (1991). *EMBO J.* **10,** 3851.

Guiochon-Mantel, A., Delabre, K., Lescop, P., and Milgrom, E. (1996). *J. Steroid Biochem. Mol. Biol.* **56,** 1.

Kuil, C. W., and Brinkmann, A. O. (1996). *Eur. Urol.* **29**(Suppl. 2), 78.

Lumbroso, S., Lobaccaro, J. M., Georget, V., Leger, J., Poujol, N., Terouanne, B., Evainbrion, D., Czernichow, P., and Sultan, C. (1996). *J. Clin. Endocrinol. Metab.* **81,** 1984.

Mangelsdorf, D., Thummel, C., Beato, M., Herrlich, P., Schütz, G., Umesono, K., Blumberg, B., Kastner, P., Mark, M., Chambon, P., and Evans, R. M. (1995). *Cell* **83,** 835.

Nakao, R., Yanase, T., Sakai, Y., Haji, M., and Nawata, H. (1993). *J. Clin. Endocrinol. Metab.* **77,** 103.

Ogawa, H., Inouye, S., Tsuji, F. I., Yasuda, K., and Umesono, K. (1995). *Proc. Natl. Acad. Sci. USA* **92,** 11899.

Pines, J. (1995). *Trends Genet.* **11,** 326.

Quigley, C. A., Debellis, A., Marschke, K. B., Elawady, M. K., Wilson, E. M., and French, F. S. (1995). *Endocr. Rev.* **16,** 271.

van Laar, J. H., Voorhost-Ogink, M. M., Zegers, N. D., Boersma, W. J. A., Claassen, E., van der Korput, J. A. G. M., Ruizeveld de Winter, J. A., Van der Kwast, T. H., Mulder, E., Trapman, J., and Brinkmann, A. O. (1989). *Mol. Cell. Endocrinol.* **67,** 29.

CHAPTER 14

Confocal Microscopy: Kallikrein Proteases and Kinin Peptide Receptors

Kanti D. Bhoola,[*] Celia J. Snyman,[†] and Carlos D. Figueroa[‡]

[*]Lung Institute of Western Australia
University of Western Australia
Nedlands, Perth Western Australia, Australia

[†]Department of Biochemistry
Genetics and Microbiology
University of KwaZulu-Natal, Pietermariztburg, South Africa

[‡]Institute of Anatomy
Histology and Pathology
Faculty of Medicine, Universidad Austral de Chile, Valdivia, Chile

I. Preamble

The laser scanning confocal microscope is now established as an invaluable tool in developmental biology for improved light microscope imaging of fluorescently labeled cells. The universal application of confocal microscopy in biomedical research has stimulated improvements to the bright-field microscopes as well as the synthesis of novel probes for imaging biological structures and physiological processes. The key feature of confocal microscopy is its ability to produce blur-free images of thick specimens at various depths. In a laser scanning confocal microscope a laser beam passes through a light source aperture and then is focused by an objective lens into a small (ideally diffraction-limited) focal volume within a fluorescent specimen. A mixture of emitted fluorescent light as well as reflected laser light from the illuminated spot is then recollected by the objective lens. A beam splitter separates the light mixture by allowing only the laser light to pass through and reflecting the fluorescent light into the detection apparatus.

Essentially a confocal microscope has a pinhole, that is an efficient aperture system for rejecting out-of-focus fluorescent light. The image comes from a thin section and a small field depth of the sample. By scanning many thin sections, one can build up a very clear three-dimensional (3D) image of the sample. Confocal microscope has slightly better resolution horizontally, as well as vertically than a bright-field microscope. In practice, the best horizontal resolution of a confocal microscope is about 0.2 μm, and the best vertical resolution is about 0.5 μm. As the laser scans over the plane of interest a whole image is obtained pixel by pixel and line by line, while the brightness of a resulting image pixel corresponds to the relative intensity of detected fluorescent light. The beam is scanned across the sample in the horizontal plane using one or more (servo-controlled) oscillating mirrors. This scanning method usually has a low-reaction latency and the scan speed can be varied as slower scans provide a better signal-to-noise ratio resulting in better contrast and higher resolution. After passing a pinhole the fluorescent light is detected by a photodetection device (photomultiplier tube (PMT) or avalanche photodiode) transforming the light signal into an electrical one which is recorded by a computer.

Development of a spinning disk, or Nipkow disk, by Paul Gotlieb of Germany in 1884, had the merit of high-speed scanning, allowing high-frequency frame capturing of events in live cells. Significant drawbacks of this technology, however, are that it allows viewing in only one plane, and alignment of the pinholes for emitted and excitation beams may be problematic. More recently, the Yokogawa CSU-10 disk system was introduced, which provides radically more light, ensures better uniformity and makes possible completely new laser options.

A. What is the Advantage of Using a Confocal Microscope?

Confocal microscopy offers several advantages over conventional optical microscopy, including controllable depth of field, the elimination of image degrading out-of-focus information, and the ability to collect serial optical sections from thick specimens. The key to the confocal approach is the use of spatial filtering to eliminate out-of-focus light or flare in specimens that are thicker than the plane of focus. There has been a tremendous explosion in the popularity of confocal microscopy in recent years, due in part to the relative ease with which extremely high-quality images can be obtained from specimens prepared for conventional optical microscopy, and in its great number of applications in many areas of current research interest.

Successful confocal imaging relies on the most favorable combination of pinhole size, lowest laser power and acquisition settings to achieve optimal image brightness.

B. What's New?

Confocal laser scanning microscopy is a valuable system for obtaining high-resolution images and 3D reconstructions. Techniques for 3D light microscopy of a wide variety of biological specimens are rapidly maturing. The next advances are to improve image contrast, make image resolution as isotropic as possible, and perform quantitative analysis. Special need is to quantify and trace individual structures, and to montage high-resolution fields with the view of producing an associated computer database representation. The microscopy of biological specimens has traditionally been a two-dimensional (2D) imaging method for analyzing what are in reality 3D objects. This has been a major limitation of the application of one of science's most widely used tools. Specific advantages would accrue from the 3D imaging of the data to achieve new capabilities not possible with 2D imaging.

C. Fundamental and Practical Challenges

Multifluorochrome labeling is experiencing a renaissance in current cell structure research. Cutting-edge biophysical technologies including total internal reflection fluorescence (TIRF) microscopy, single-molecule fluorescence, single-channel opening events, fluorescence resonance energy transfer (FRET), fluorescence recovery after photobleaching (FRAP), high-speed exposures, two-photon imaging, fluorescence lifetime imaging (PALM/STORM). These and other tools are becoming increasingly important as they link cellular images to molecular events, a key goal of modern molecular and cellular biology.

D. Current Models of Confocal Microscopes

The essentials of confocal microscopy, namely optical sectioning, multiple lasers in the visible light and ultraviolet ranges, multidimensional imaging, and improved contrast and resolution of images, are provided by all of the models on the market. The following are some of the family of confocal microscopes that are commercially available:

1. *Confocal laser scanning microscopes*: Leica TCS STED, the Leica TCS SMD, and the Leica TCS SP5 X, Leica TCS SP5 II, Olympus FluoView TM FV1000, NORAN OZ CLSM, Zeiss LSM 710.
2. *Spinning-disk (Nipkow disk) confocal microscopes.*
3. *Programmable Array Microscopes (PAM).*

Each of these classes of confocal microscope has particular advantages and disadvantages, most systems are either optimized for resolution or high sensitivity for video capture.

E. Future Expansion

Confocal microscope technology is continuously upgraded to incorporate and evaluate new techniques. Currently, systems are being developed to incorporate a 410-nm laser as an addition to the QLM (488 laser) module and a fast Roper Cascade II/512 camera to one of the current scanning confocal microscopes imaging—but at lower resolution. Cutting edge development of confocal laser scanning microscopy now allows better than video rate (60 frames/s) imaging by using multiple microelectromechanical systems (MEMS)-based scanning mirrors. New developments in this arena is the multiphoton confocal microscope and additionally fluorescence lifetime imaging.

II. Introduction

A. Historical Overview

In the 1970s, a new generation of confocal microscopes was introduced to the scientific world simultaneously in Oxford and Amsterdam. These technical innovations were combined in 1977 and comprised the theoretical background for the geometry of confocal imaging by Wilson and Sheppard (1994), both members of the Oxford group. These authors also pioneered the description of the nonlinear relationship between fight and atoms of an illuminated object and the theory of Raman spectroscopy using laser scanners.

The first convincing practical demonstration of a confocal microscope with improved resolution was performed by Brakenhoff *et al.* (1978) at the University of Amsterdam, using a microscope equipped with high numerical aperture lenses, designed for imaging in the transmission mode. Soon after that, the new technique was ready for practical application in various fields of medicine and biology. Clinical examination techniques in ophthalmology immediately profited from the new technique. With confocal microscopy it became possible to routinely inspect the endothelial cells at the inner corneal side of a patient's eye. For this particular application, Koester designed a confocal slit scanner for which he obtained a US patent in 1979 (USP 4,170,398). Another application was added later, which allowed the 3D imaging of the retina.

Carl Zeiss developed the first prototype of a laser scanning microscope in 1982. However, Wijnaendts van Resandt *et al.* (1985) presented the first successful demonstration of optical sectioning of fluorophore-labeled biological material in 1985, who were also the first to perform a section along the z-axis. This marked the time when all the key technologies necessary for the production and digital recording of 3D data sets became available. Among the prerequisites were a set of perfectly designed optical lenses, high-sensitivity detectors, versatile fluorophore markers, and last but not least, powerful lasers capable of producing coherent light of the correct wavelength required for the desired excitation of a given fluorophore. Furthermore, integrated computer systems were developed for controlling the scanning mechanics and for storing and managing the vast amount of data required for reconstructing 3D images. Van der Voort *et al.* (1985) reported the first description of such a completely integrated version of a confocal microscope. These instruments, now referred to as the first generation and equipped with a laser as the light source, had been designed as object scanners, referring to movement of the object stage, whereas the beam was fixed in place.

The alternative version with a beam scanner was more suitable for the study of living cells, as it uses rapidly scanning mirrors to scan a laser beam across the specimen. This construct only became available in 1985, but it is now a standard feature of the confocal microscope. Currently, there are several different versions of confocal microscopes on the market, one of which is illustrated in Fig. 1.

The trend toward linking even more powerful computer workstations with confocal microscopes is already evident. With increased computational power, it is now possible to view an object from all possible spatial orientations and combine the morphological information with quantitative physiological data such that a full representation of the structures and dynamic changes of a cell during physiological events can be obtained (Engelhard and Knebel, 2010) (see Section I.D).

B. Nature of Confocal Imaging

For more than 200 years, light microscopy has provided biologists with a powerful system to unravel the structure of microorganisms, to analyze cell and tissue structure, and to study the dynamics of cell function. Further, with

Fig. 1 Confocal microscope assembly. The image visualization and capturing system of the confocal microscope comprises a fluorescent microscope and a controlling computer workstation linked to a 3D image analyzer.

conventional light microscopes, researchers have always been limited by the preparation procedures necessary to produce suitable contrast for visualizing structures. With the advent of the electron microscope came the ability to decode structures of biological objects much smaller than those observed by light microscopy. Although light microscopy has played a crucial role in characterizing tissue structure and electron microscopy in providing powerful visuals of cell ultrastructure, the images are static. Research into the functions of nonfixed living cells requires an experimental approach that permits observation of live cells. Living specimens offer unique imaging problems due to their opacity, motion, and photo-absorption. Using noninvasive techniques, advancements in fluorescence microscopy have permitted real-time analysis of free ions, second messengers, retrieval of peptide receptors, and many other molecular events in living cells and molecules on the surface of nonfixed cells or membranes with considerable precision using novel and specific probes and ligands.

The confocal microscope provides optimal operating conditions for the study of living tissue. What distinguishes confocal microscopy from conventional light microscopy is its ability to optically slice tissue sections or ceils. In the confocal microscope, all structures being out of focus are suppressed at image formation, providing images free of out-of-focus information. This is achieved because the object is not illuminated and imaged as a whole at the same time, but in sequence at one focal point after the other, with the optical scanning of each plane. An additional advantage is the ability to create, with increased image resolution, 3D constructs.

C. Principal Elements—Lasers and Fluorescent Probes

1. Lasers and Image Capture

The laser unit is the illumination source, and in general, the fluorescence measured is proportional to the laser power level. Although total *laser output power* is usually regulated, the amount of power in each line of a multiline laser may not be, and may vary widely with time. The *wavelength* affects optical performance, and through the absorption spectrum of the dye, it determines the amount of fluorescence produced. The following components are typically found in a confocal microscope (Fig. 2). (1) The laser light source has the versatility that allows suitable wavelengths to be selected either individually or as a combination. Currently, a number of lasers are available: HeNe 543 nm, HeNe 633 nm, and Ar 458, 488, 514, and 566 nm for the visible light range and the KrAr 351, 364 nm for the ultraviolet range. (2) It should have a scan unit for moving the illuminating beam across the object. (3) Detectors to record the amount of photons coming from the object are usually designed for detecting fluorescence and reflected light. Optionally an additional detector can be installed that would allow viewing of the object in a nonconfocal transmission mode. (4) A central processing unit to control

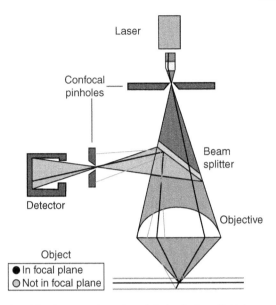

Fig. 2 Confocal optics. After it passes through a pinhole, the laser light beam widens and becomes concentrated by the objective that focuses the beam to a discrete point in a specified focal plane. Photons emitted from the excited fluorophore are directed through a second pinhole for capture by the detector system. Observe that the brown and red lines showing light paths that produce the blurred image in a conventional microscope are not passed through the second pinhole.

the hardware and data storage unit to store and manipulate image data. (5) A scanning system attached to a research microscope.

With these components, a multifunctional imaging system is assembled capable of performing all major techniques of microscopical analysis, namely confocal multiple fluorescence detection and reflection, as well as basic nonconfocal transmission techniques such as absorption, phase, and differential interference contrast. The combination of these imaging techniques and image processing allow the manipulation of all aspects of an image, including the rotation of the object in space. Therefore, a much more detailed view of the specimen is possible. This explains why confocal microscopes have become extremely important in medicine and biological science.

2. Basic Steps

1. Illumination of a spot by confocal microscopes, which achieve high resolution of a selected plane in a specimen, occurs in three basic steps. The light is focused by an objective lens into an hourglass-shaped beam so that the bright beam strikes one spot on a specimen, where it excites a specific fluorophore.

2. Light emitted the excited fluorophore passes through a second pinhole to the detecting device, or photomultiplier. Up to four detectors can be supplied with custom spectral bands simultaneously.

3. In earlier confocals one laser line had to be split and only the light beams with the correct wavelengths was deflected onto the sample. These days, however, confocals are fitted with many laser sources, each with a very specific wavelength (Table I).

4. The pinhole blocks out those rays reflected by illuminated parts of the specimen and excited fluorophores lying above and below the plane of interest.

5. The confocal uses a variable size detection pinhole for one primary reason: since the size of the pinhole is related to the section thickness. (*Actually, this should not be used as an option to generate a brighter image, since it compromises the confocal principal. Rather, use varying gain settings, or change the laser attenuation.*) In order to achieve optimal optical section thickness, one needs to match the pinhole diameter to the diffraction pattern in the intermediate plane (the Airy unit).

6. In order to produce a 2D image, the illuminating light spot is rapidly moved from point to point and in consecutive lines along the object until the entire plane has been scanned. The optical path depicted in Fig. 3 describes the geometry underlying the formation of just a single image spot. Alternatively, in slit scanners, a scan line instead of a spot, is used to scan the image, giving faster results and less negative impact on the sample, for example, in live cells.

7. Another important feature is that serial optical sections may be obtained by moving the focal plane progressively through the specimen, as the objective lens is positioned closer to or further from the sample. To obtain a 3D image of an object, successive planes in a specimen are scanned. It produces a stack of images, each of which is an optical selection; such selections are analogous to images of fine slices cut physically from a specimen.

Table I
Laser and Arc-Discharge Spectral Lines in Bright-Field and Confocal Microscopy

Source	Ultraviolet	Violet	Blue	Green	Yellow	Orange	Red	Near-IR
Argon ion	351, 364	–	457, 477, 488	514	–	–	–	–
Diode	–	405, 440	–	–	–	–	635, 640	650, 685
DPSS	355	430, 442	457, 473	532	561	593	638	660, 671
He–Cd	322, 354	442	–	–	–	–		
Kr–Ar	–	–	488	–	568		647	676
Krypton- ion	–	–	–	–	–	–	647	676, 752
Helium–neon	–	–	–	543	594	612	633	–
Mercury arc	365	405, 436	–	546	579	–	–	–
Xenon arc	–	–	467	–	–	–	–	–
Metal halide	365	435	495	520, 545	575	–	625	685

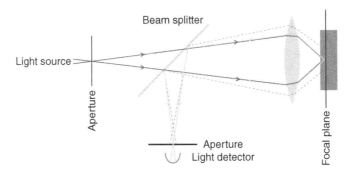

Fig. 3 Confocal microscopic imaging. Detailed organogram of the laser beam light path that complements the diagrammatic representation in Fig. 2.

3. Fluorescent Probes

The laser scanning confocal microscope is routinely used to produce digital images of single-, double-, and triple-labeled fluorescent samples. The use of red, green, and blue (*RGB*) color is most informative for displaying the distribution of up to three fluorescent probes labeling a cell, where any colocalization is observed as a different additive color when the images are colorized and merged into a single three-color image. The combination of colors within a three-color merged image is important for clearly conveying the biological information collected by the confocal microscope. The true emission colors of two of the most commonly used fluorophores, rhodamine, and fluorescein are conveniently red and green, respectively, and overlapping domains of expression are yellow. These are the colors observed by eye in a conventional epifluorescence microscope equipped with the appropriate filter sets for simultaneous double-label imaging, which is now available from most of the microscope manufacturers.

Fluorescent proteins: Blue fluorescent protein (*BFP*), cyan fluorescent protein (*CFP*), green fluorescent protein (*GFP*), yellow fluorescent protein (*YFP*), enhanced yellow fluorescent protein derivative (*EYFP*), kindling fluorescent protein 1 (*KFP1*), red fluorescent protein (*DsRed*).

A large number of fluorescent probes are available that, when incorporated in relatively simple protocols, specifically stain certain cellular organelles and structures (Table II). Among the plethora of available probes are dyes that label nuclei, the Golgi apparatus, the endoplasmic reticulum, and mitochondria, and also dyes such as fluorescently labeled phalloidins that target polymerized actin in cells. Such dyes are very useful in multiple labeling approaches to locate antigens of interest having specific compartments in the cell. Fluorescence *in situ* hybridization (*FISH*) has advantages in resolution and sensitivity of probe detection that are further enhanced when coupled with the LSCM, and is a valuable approach for imaging the distribution of fluorescently labeled DNA and RNA sequences in cells. In addition, brighter fluorescent probes are

Table II
Fluorophores Used Commonly in Biomedical Science

Fluorophore	Absorbance (nm)	Emission		
			Color	Use/function
Aminoactinomycin D	546	647	Yellow	Labels nucleic acids in intact or permeabilized cells
Cy3	550	570	Red	Detection of antigen–antibody complexes in immunohistochemistry
DAPI	358	461	Blue	Labeling and counterstaining of chromosomes
Ethidium bromide	518	605	Brown/red	Detects breaks in nucleic acids
FITC	494	518	Green	Detection of antigen–antibody complexes in immunohistochemistry and FISH
Fluo-3	506	526	Green	Fluorescent indicator for calcium during signal transduction in excitable cells
Cito Tracker Red	578	599	Red	Labels active mitochondria selectively
Texas Red	595	615	Red	Flow cytometry, detection of antigen–antibody complexes in immunohistochemistry, and labeling unpolymerized G-actin
NARF-1	575	635	Red/orange	Label to study cell morphology, adhesion, communication and migration
CYTO 17	621	634	Red	Fluorescent labeling of DNA and RNA in live cells

The range of fluorophores with their absorption and emission wavelengths used in confocal research is listed, together with their use in identifying cellular structures, dynamics of cell function, and intracellular transport of ions.

currently available for LSCM imaging of total DNA in both nuclei and isolated chromosomes.

Fluo-3 and rhod-2 are examples of dyes that have been synthesized to change their fluorescence characteristics in the presence of certain ions such as calcium. New probes have been developed for imaging gene expression, including, for example, the jellyfish GFP, which enables gene expression and protein localization to be observed *in vivo*. The use of GFP has enabled the monitoring of gene expression in a number of different cell types including living mammalian cells, and plants (using the 488 nm line of the excitation laser). Mutants of GFP with spectral variations are available for use in multilabeling experiments, and these have also found use for avoiding interference from autofluorescence in living tissues.

Autofluorescence of tissues occurs naturally in many cell types, and can be a major source of background interference during imaging. For example, chlorophyll in yeast and plant cells fluoresces in the red part of the spectrum. Certain reagents, such as glutaraldehyde fixative, are sources of autofluorescence, which can be reduced by treatment with borohydride. Autofluorescence can sometimes be avoided by using fluorophores that can be excited at wavelengths that are out of the range of natural autofluorescence.

D. Merits of the Confocal Microscope

Confocal microscopy offers several advantages over conventional optical microscopy, including controllable depth of field, the elimination of image degrading out-of-focus information, and the ability to collect serial optical sections from thick specimens. The key to the confocal approach is the use of spatial filtering to eliminate out-of-focus light or flare in specimens that are thicker than the plane of focus. Advantages of the confocal microscope over a conventional microscope and the major improvements offered by the confocal microscope on performance may be summarized as follows: (1) Light rays from outside the focal plane will not be recorded. (2) Defocusing does not create blurring, but gradually cuts out parts of the object as they move away from the focal plane. The practical consequence is that these parts become darker and eventually disappear. The feature is called optical sectioning. (3) True, 3D data sets can be recorded. (4) Scanning the object in the xy direction as well as in the z direction (along the optical axis) allows one to view the object from all sides. (5) Due to the small dimension of the illuminating light spot in the focal plane, stray light is minimized. (6) By image processing many slices can be superimposed, giving an extended focus image that can only be achieved in conventional microscopy by reduction of the aperture and thus sacrificing resolution. (7) It is therefore clear that quantitative measurements are operator dependent because until all intensities above and below the specimen are physically gathered, analysis of the images cannot be performed.

E. Optical Sectioning

Similar to the optics in a conventional microscope, the depth of the focal plane in the confocal microscope is determined in particular by the numerical aperture of the objective used and the diameter of the diaphragm. What is the thickness of an optical section for a given pinhole size? This question is not simple to answer. Optical sections do not have well-defined edges like a section taken from a microtome. In fact, the z-intensity profile is shaped like a bell. The thickness of an optical section is defined as the full-width half-maximum (FWHM) of this bell-shaped profile. The exact profile depends on many factors, such as the correction of the optical components, lens type, refractive index, and matching of indices. However, the most important factor is the specimen itself. Taking all these factors into consideration, the empirical section thickness can deviate from theoretical section thickness by one-third or more. At a wider detection pinhole the confocal effect can be reduced. This means that by widening the pinhole to a maximum width, an image can be achieved that is similar to that observable in a conventional microscope, where blurred parts surrounding a particular object or structure, arising from the out-of-focal plane, contribute to the image.

F. Technical Aspects: Size of Detector Pinhole

Due to diffraction, a point of light in the focal plane is imaged as a bright disk surrounded by increasingly dimmer rings. The bright center is called the Airy disk (G. B. Airy, 1801–1892, British astronomer). The diameter of the Airy disk depends on several optical parameters, such as wavelength, numerical aperture, objective magnification, and additional magnification factors of the system. Setting the size of the detector pinhole to the size of the Airy disk achieves optimal performance. Due to diffraction, setting the detection pinhole diameter less than the diameter of the Airy disk does not result in a thinner optical section.

The performance in optical resolution of a microscope is expressed as FWHM, which is two times the distance from a point-like object at which its luminosity drops by half the maximum light intensity measured in the center of the object. Two point-like objects separated by a distance of one FWHM from each other are just recognizable as two disparate structures, for instance, they can be resolved optically. The FWHM is defined by the wavelength of light used for illuminating the object and by the aperture of the objective lens. Although this is true for conventional as well as confocal microscopes, the actual resolution is still better by a factor of 1.4 in the confocal setup.

This is described by the formula: FWHMlat = 0.4A/NA. The wavelength is designated as A and the numerical aperture (NA) is $n \sin \alpha$, where n is the refractive index of the immersion medium and α is the aperture angle of the entrance lens of the objective. The immersion medium is placed between the objective lens and the object.

It is noteworthy that the axial resolution is much more dependent on the numerical aperture than the lateral one; also, objectives that allow air as the immersion medium have a better axial resolution than those that require liquid media at the small numerical aperture. Axial resolution can be deduced from the following formula: FWHM~x = 0.45/[n(1 − cosa]. The actual values for axial and lateral resolution measured in commercial high-performance lenses come close to the theoretical values within the range of a few percentage points.

When living samples are to be studied, it is usually necessary to mount them in a chamber that provides all of the necessary requirements for life, and that will also allow sufficient access by the objective lens to image the desired area. Objective lenses that are capable of the highest resolution generally are those with the highest magnification and the highest numerical aperture (Table III).

G. Optimal Image Quality

As in other microscopes, the useful wavelength range of light is limited by the design of the objective lens. A very careful correction of chromatic lens aberrations is necessary for high-quality fluorescence images. Demands on the quality of chromatic correction become even more pressing as modern applications of multiple color fluorescence need even wider ranges of chromatic correction in order to

Table III
Objective Lens Parameters and Optical Section Thickness

Objective		Pinhole	
Magnification	NA	Closed (1 mm)	Open (7 mm)
60×	1.40	0.4	1.9
40×	1.30	0.6	3.3
40×	0.55	1.4	4.3
25×	0.80	1.4	7.8
4×	0.20	20.0	100.0

increase the number of fluorophores when applied simultaneously in a single experiment. This in turn is only possible if lasers can be provided that generate the suitable spectrum of excitation wavelengths. To this end, modern mix gas lasers have been introduced that can be tuned to generate lines over the entire spectrum of visible light. However, the intensity for excitation illumination in a microscope should not exceed certain levels because of light saturation phenomena within the object, which can have adverse effects such as a reduction of optical resolution and inaccurate readings in quantitative measurements. Another important parameter in obtaining optimal images of quality is the signal-to-noise ratio, which depends on the type of detector and the design of the scanner. Optical discrimination between image points within and outside the focus is realized best in single beam scanners equipped with small confocal pinholes. Slit scanners and multiple spot scanners produced by a Nipkov-type disk suffer from imperfect optical discrimination and hence reduced contrast. For the detection of fluorescence signals in single beam scanners the signal-to-noise ratio is in practice determined only by the number of photons. This is why confocal imaging systems must be designed for optimal light transmission and maximum yield of photons at the detector, both of which depend heavily on the quality of lenses and the width of the apertures in the light path. Consequently, the signal-to-noise ratio is also the major limiting parameter in determining the scan rate, namely the time necessary to build a single image.

H. Engineering Aspects of Scanners

The following types of scanners may be distinguished.

1. The speed with which an object can be moved relative to the beam in object scanners is limited by mass inertia of the moving parts and by viscosity and surface interactions of the immersion media with lens and object. The advantage, however, is that because of the fixed position of the beam, object scanners require less sophisticated designs for the optical path. Mechanical work in object scanners is performed by galvanometers, piezo crystals, and electromotors fitted with suitable gears. Objective scanners have an instrument design similar to that of object scanners except that the object itself is not moved. However, the obvious

restrictions imposed on the capability to perform high frequency movements have prevented a widespread application of both types of scanners. Object movement is still used but only for the slower movements along the *z*-axis.

2. Beam scanners, however, are sufficiently fast to satisfy the requirements of a confocal system. They operate as single or multiple beams or slit scanners. Between these, the single beam scanners appear to bring together the best combination of features. Although slit scanners have the fastest rates of image buildup, the signal-to-noise ratio is inferior compared to single point scanners because stray light from out-of-focus areas is blocked off less effectively. In addition, slit scanners have a built-in asymmetry in their resolving power. Disk scanners also have disadvantages in that they have a lower fluorescence yield compared to single point scanners because in the latter, pinhole position and geometry can be adapted more easily to the experimental needs. In slit scanners, movements of the beam are realized by mounting polygonal mirrors on a motor axis or by galvanometers. Nipkov disks are rotated by electromotors. A very fast deflection of mirrors is achieved in acoustooptical devices, which, however, are curbed in their performance by the fact that the crystals used in these devices reflect light differently depending on the wavelength. Their use is therefore limited to semiconfocal slit scanners in fluorescence applications.

Generally, the scan rate must be set as a compromise between speed and image quality and this in turn is dictated by the signal-to-noise ratio defined by the statistical profile of the photons collected. The latter is a result of the level of fluorescence output and conditions set by other experimental prerequisites. In principle, a higher fluorescence output can be produced using more powerful lasers; however, the intensity that can be applied maximally is limited by a saturation effect occurring in the process of fluorophore excitation. Scan rates in fluorescence applications, therefore, do not usually exceed one image per second. If image quality is not a limiting factor, images may be captured at the much faster rate of three to five images per second. (See preamble on the Yokogawa CSU-10 disk system.)

1. Confocal Image

Confocal images can typically be recorded at variable pixel densities with the highest resolution at a setting of 1024×1024 pixels. Depending on the size of the object there may be from 16 to several hundred optical sections compiled in a single 3D data set. In addition to the digitized information on the 3D position of an image pixel, the data set contains intensity values for each pixel recorded by 8-bit leading of 256 gray values. Such data sets, therefore, encompass several megabytes of information. This amount of data requires a fairly powerful computer system equipped with large memory storage media. Hence, it is obvious why technical realization of confocal microscopy was historically closely related to the developments in the computer field.

J. Three-Dimensional Constructs of Images

One of the most pressing problems in medical research is cancer. When cancer develops within a patient, cells of the body suffer subtle internal deviations from the normal developmental pattern, leading to more pronounced changes in cell shape and organization of tissues. Monitoring these changes as early as possible and describing the unfolding features of the many types of cancer may be achieved by preparing 3D constructs of confocal images of the morphology of cancer cells. Looking deeper inside the cell, it is becoming more important to study the microarchitecture of the cell nucleus with the aim of understanding the processes underlying cell growth and differentiation. Modern techniques of cell and molecular biology, such as *in situ* hybridization and immunofluorescence, provide the means for the study of spatial order in the nucleus, particularly by visualizing chromosomal domains at interface or in dividing nuclei, when the genome is split and partitioned to the progeny cells. Kinetochores are important structures in the cell nucleus, which are thought to be of instrumental importance during mitosis. In order to quantify structural changes that occur in the course of these processes, complex 3D segmentation must be applied, which eventually leads to the exact determination of kinetochore positions. As far as actual changes in the tissue occur, there is great interest in applying confocal microscopy to monitoring the progression or remission of bone tumors. Particularly important are lesions in the bone caused by the activity of special cells called osteoclasts. These cells are also active during normal bone development, and the disintegrating bone structures caused by them have long been known from electron microscope studies. However, it has been difficult to determine the degree of damage inflicted on the bone. Using the confocal microscope, it is possible to determine the size of the lesion in a bone quantitatively and thus come to a realistic estimation of the scale of damage inflicted on the bone structure.

K. Simultaneous Detection of Separate Fluorophores

A great number of suitable fluorophores are known to differ in their emission and excitation spectra (see Table II). These differences can be used to simultaneously visualize two or more cell structures labeled with specific fluorophore-conjugated markers. Such an application is beginning to gain importance in the localization of several genes in the nuclear genome as well as in studies on the spatial arrangement of the cytoskeleton within the microarchitecture of cells.

L. Value of Confocal Microscopy in Cell Research

Confocal microscopy is a new imaging technique allowing new insight into the morphology and dynamics of cells and tissues. In contrast to the performance of conventional microscopy, larger tissues, or densely packed structures are no longer obstacles in arriving at high-quality images. The capability to produce consecutive optical sections with a thickness of less than 0.5 m enables direct recording of 3D data sets. Suitable programs to determine surface features, volume, near-neighbor analysis of given structures, and views of the total structure from any angle in space can process these data. Studying dynamic aspects of cell activities is facilitated greatly by confocal microscopy in conjunction with suitable fluorophores. New applications that are becoming possible are the quantitative estimation of local ion concentrations and membrane potentials in cells and tissues. Similar experiments using conventional microscopy techniques have produced ambiguous results in the past. The fact that the geometrically defined focal spot can be positioned precisely anywhere in the specimen provides ideal preconditions for a number of additional interesting experiments with the microscope results in the past. The fact that the geometrically defined focal spot can be positioned precisely anywhere in the specimen provides ideal preconditions for a number of additional interesting experiments with the microscope.

Confocal microscopy involves the collection and computation of numerous data related to structure and physiological activity of the cell or tissue under study. Most data, such as multiple fluorophores and transmission signals, fluorescence recovery course, and color and Raman spectroscopy, can be recorded simultaneously. The integration of such information provides an insight that is much closer to reality than has been ever possible to obtain before confocal microscopes became available to researchers in cell and tissue medical research and in material science (Table II).

M. Medical and Biological Applications of Confocal Microscopy

Due to the strength, versatility, and little limitation of the instrument in the choice of objects or preparation techniques, the applications for confocal microscopy are numerous. Material science was one of the first areas that took advantage of the 3D imaging capability in the reflection mode. Because it is not necessary to place the object into a vacuum chamber, such as that used previously for scanning electron microscopy, handling and processing of samples are much less time-consuming.

N. Visualization of Living, Nonfixed Cells

Apart from the immunofluorescence techniques already mentioned, cell structure research depends on the availability of fluorophores that can be used to stain specific cell components in the living cell. The lipophilic dicarbocyanine dye (DiOC6) (Molecular Probes, Inc.) is one that accumulates in the endoplasmic reticulum (ER) and, because it can be monitored over time, provides information on the structural dynamics and functional importance of ER. However, it was also found that immobile, fixed sites exist that determine the location and overall shape of the ER over time. The same dye also stains the mitochondria of cells. Additionally, laser-scanning confocal microscopy has been used in developmental biology.

It has been particularly useful for the visualization of cellular actin together with chromatin in human arrested preimplantation embryos and in unfertilized oocytes obtained by *in vitro* fertilization (Levy *et al.*, 1997).

Spontaneous calcium release (SCR) from the sarcoplasmic reticulum (SR) initiates contractile waves in both uni- and multicellular cardiac preparations. The dynamics of contractile waves, originating from sites in the SR and visualized as propagating bands of shortened sacromeres, have been studied extensively. When loaded with the esterified derivative fluor3-acetoxymethyl ester (Fluo-3-AM, Sigma Chemicals, St. Louis, MO), cell suspensions of isolated ventricular myocytes have been used for intracellular release studies of calcium. With the development of Ca^{2+}-sensitive fluorescent probes, it has been possible to confirm that local changes in intracellular Ca^{2+} levels, described as "Ca^{2+} waves," are associated with the spontaneous contractile activity of myocytes.

O. Medical Science Research Projects

Ethics: Ethical permission for all studies of Bhoola and Colleague has been obtained from the ethics committee of the Faculty of Medicine, University of Natal and the Sir Charles Gairdner Hospital ethics committee.

III. Methods and Procedures

A. Collection and Processing of Neutrophils

Whole blood was obtained by venipuncture through a precision glide 21-gauge Vacutainer needle (Becton-Dickinson Vacutainer Systems, UK) from healthy volunteers and patients presenting with infection. Blood was carefully layered on 3.5 ml Histopaque 1119/1077 (Sigma, UK) and centrifuged at $1500 \times g$ for 30 min at $20\,^{\circ}C$ to form a band. Neutrophils from this layer were harvested with a Pasteur pipette, mixed with an equal volume of phosphate-buffered saline (PBS), centrifuged ($500 \times g$), and the process was repeated twice.

B. Tissue Collection

Brain, blood vessels, and control kidney tissue were obtained at autopsy: Individuals declared dead on arrival as a result of trauma not involving the head or sudden unexplained death (the patients had sustained gunshot wounds to the chest or abdomen, acute myocardial infarction, or carbon monoxide poisoning) were immediately refrigerated and maintained at $4\,^{\circ}C$ in the hospital morgue. The age, sex, cause, and time of injury and death were noted. The ages ranged from 24 to 40 years. At postmortem, performed within 24 h of death, the jugular veins were opened first to allow free drainage of blood from the brain. Immediately following its removal, the brain was divided sagittally along the midline. One-half was suspended in 5% formal saline (41% formaldehyde in 0.9% NaCl, 1:8, v/v) and was dissected after 48 h. Topographical distribution of the various brain areas was established from three neuroanatomical texts (Barr and Kiernan, 1993; Fix, 1974; Roberts *et al.*, 1987) as well as maintaining orientation by marking appropriate areas with India ink (Reeves, London) and 1% Alcian blue (pH 2.5, Sigma, St. Louis, MO). Blood vessel and control kidney samples are fixed similarly in 5% formal saline for 24 h.

1. Brain and Lung Tumor Tissues

Samples of human brain and lung tumor were obtained at surgery.

2. Gastric Tissue

Ten patients with dyspepsia were endoscoped, and two biopsy specimens from each were taken from the fundus and antrum. Lung tissue control tissue was obtained surgically by antrectomy from the antrum of a single patient. Patient selection was restricted to males and nonpregnant females between 18 and 75 years old undergoing routine evaluation for dyspepsia. Patients exhibiting any severe abnormality of the renal, respiratory, or cardiovascular system were excluded. Patients who had undergone recent esophageal or stomach surgery for an ulcer and those who presented with pyloric stenosis, perforation, or arterial bleeding were also excluded.

3. Renal Tissue

Renal tissue samples were obtained by ultrasound-directed needle biopsy from 18 patients with deteriorating renal function.

C. Tissue Processing

All of the tissue sections were fixed routinely in 5% (v/v) buffered formaldehyde-saline for 24 h at room temperature. After fixation the tissue samples were dehydrated in a graded series of ethanol and embedded in paraffin wax.

D. Histology

Sections (4/μm thick) were cut on a rotary microtome (Leica, Germany) and adhered onto glass slides. These are stained with hematoxylin and eosin for histology and grading of disease; Giemsa, to determine bacterial presence; as well as periodic acid-Schiff, Luxol fast blue, methamine silver, Masson's trichrome, and elastic von Gieson. The control tissue showed "normal" histology.

E. Antibodies

1. Tissue Kallikrein (TK) Antibody

Recombinant TK was obtained from Dr. M. Kemme (Institute for Biochemistry, Technical University of Darmstadt, Darmstadt, Germany). The antibody was produced in a goat host using $100/\mu g$ of recombinant TK conjugated to $125/\mu l$ TiterMax adjuvant injected intramuscularly at 4-week intervals over a 4-month period. Isolation and characterization of IgG were performed according to the method described by Johnston and Thorpe (1982). Aliquots ($30/\mu l$) are stored at $-20\,°C$ and reconstituted when required.

2. Kinin B_2 Receptor Antibody

Eight polyclonal antibodies, fully characterized for specificity, had been raised to synthetic peptides of the amino-terminal and loop regions encoded by the rat kinin B_2 receptor cDNA and based on the homology between these regions with the human receptor. However, only four were shown to react strongly with human epithelial cells and neutrophils (Haaseman *et al.*, 1994). Homology between the peptide sequences of intracellular domains one (ID1, LHK) and two (ID2, DRY) and the fourth extracellular domain (ED4, DTL) in the rat and human receptors are 80%, 100%, and 75%, respectively. It was, therefore, decided to perform experiments with a combination of antibodies directed to these regions. The antibodies were kindly donated by Werner Muller-Esterl (Department of Pathobiochemistry, Johannes Gutenberg University of Mainz, Duesbergweg 6, D-6500 Mainz, Germany).

3. Kinin B_1 Receptor Antibody

Fred Hess (Merck Research Laboratories, R80M-213, Rahway, NJ) kindly supplied a specific antibody directed at the C terminus of the peptide of the kinin B_1 receptor (I-S-S-S-H-R-K-E-I-[F-Q-L]-F-W-R-N)-monoclonal antibody

F. Immunocytochemistry

1. Nonfixed Cells

The pellet of neutrophils was resuspended in oxygenated (95% O_2 and 5% CO_2, v/v) PBS containing 5 mM glucose and were then dispensed into microfuge tubes for incubation with kinin-releasing enzymes. Control neutrophils were incubated without the enzyme. Following an incubation period of 20 min, the cells were pelleted ($1000 \times g$ for 10 min at room temperature), spotted onto glass slides and air dried. Cells were washed twice with PBS containing 1% human IgG, 1% bovine serum albumin, and 0.2% (w/v) sodium azide in order to block nonspecific binding of the antisera to Fc receptors and to prevent internalization of the receptors, respectively. Cells were first immunolabeled with the appropriate primary antibody (antibody against kinins, tissue kallikrein, and kininogen fragments or kinin receptors) under humid conditions and then labeled with fluorescein isothiocyanate (FITC)-conjugated specific antispecies secondary antibody for 30 min. For method controls, the primary antibody was replaced by PBS buffer. The slides were mounted with 10% PBS/90% (v/v) glycerol and viewed with a confocal microscope.

2. Fixed Cells

Isolated neutrophils were floated onto glass slides and fixed with 4% (w/v) paraformaldehyde in PBS for 2 min. The labeling procedure was similar to that used for nonfixed cells prior to viewing with the confocal microscope.

3. Fixed Tissue

Sections (3–4/μm thick) were cut from wax-embedded tissue on a rotary microtome (Leica) and adhered onto poly(L-lysine)-coated glass slides. Sections were dewaxed using xylene, rehydrated with increasingly dilute ethanol solutions (100%, 90%, 70%, and 50%) and distilled H_2O as the final rehydrant, and trypsinized with 0.01% trypsin (Sigma) for 3 min. The tissue was then heated at 80 °C in 0.1 M sodium citrate buffer for 10 min in a microwave oven (25% power, Sharp R-4A52) for antigen retrieval and allowed to cool at room temperature for 20 min.

Prior to incubation with the primary antibodies the sections were incubated with 1% (w/v) human immunoglobulin G (IgG) (Sigma) for 20 min to block nonspecific sites. Incubation of the tissue sections with specific, primary antibodies (100/μl), diluted in 1 mM phosphate buffer, pH 7.2, was performed at 4 °C for 18 h. These antibodies comprised (1) goat antihuman recombinant TK IgG diluted 1:1000, (2) rabbit antikinin B_2 receptor peptide IgGs (a combination of the following three antibodies: ED4 DTL, 283; ID1 LKH, 280; and ID2 DRY, 277) diluted 1:200, and (3) rabbit antikinin B1 receptor peptide IgG diluted 1:100 (v/v).

All incubations were carried out in a humidified chamber maintained at 21 °C. Between incubations the sections were washed by submerging in 0.01 M PBS, pH 7.4, for 10 min. Binding of the primary antibody was detected with a sheep antigoat or sheep antirabbit IgG coupled to a fluorescein probe, either FITC (488 nm excitation, nm average emission) or Cy3 fluorophore (516 nm excitation, 552 nm average emission), at room temperature for 30 min.

Replacement of the primary antibody with nonimmune IgG served as a method control. The loss of immunolabeling following preabsorption of the specific primary antibody with its respective antigen, namely (1) recombinant TK, (2) the kinin B_2 receptor peptides (D-T-L-L-R-L-G-V-L-S-G-C for ED4, L-H-K-T-N-C-T-V-A-E for ID1, and D-R-Y-L-A-L-V-K-T-M-S-M-G-R-M for ID2), and (3) the kinin B_1 receptor peptide (I-S-S-S-H-R-K-E-I-[F-Q-L]-F-W-R-N), demonstrated absolute specificity.

The labeled slides were finally mounted with 90% (v/v) glycerolphosphate. Immunofluorescence was observed with a Leitz DM IRB confocal microscope (Leica), and fluorescent emission was analyzed using the Leica TD4 confocal microscopy system (Leica).

G. Image Analysis

Confocal images were typically recorded at a pixel density of 225 × 225 pixels. The gray scale ranges from 0 to 256 was divided into eight equal phases (POLl Look-Up-Table), with each phase having a lower and upper gray-scale value threshold. With image analysis using the Analysis 2.1 Pro system (Soft-Imaging Software GmbH, 1996, Germany), the regions of interest in each image were encircled. This information was used to calculate the number of pixels falling within each phase, as well as the area of the circled areas. Therefore, the mean high intensity of immunolabeling in (*n*) number of cells was quantified in pixel/μm^2.

IV. Images of Tissue Kallikrein (Protease) and Kinin Receptors (Peptide Receptors) Captured by Confocal Microscopy

A. Literature Review

There was a paucity of research on the imaging of receptor populations by confocal microscopy since only a few citations were captured on literature survey. Studies using the confocal microscope had been performed mainly on neutrophils or on cultured cells, and also on some fixed tissue.

Bacterial *N*-formyl peptides chemotactically attract neutrophils to sites of infection. The binding of *N*-formyl peptide to its G-protein-coupled receptor on the neutrophil membrane triggers signal transduction accompanied by the release of second messengers. Using a fluorescent probe, the distribution and possible internalization of the receptor were examined by confocal microscopy. Activated *N*-formyl peptides receptors seem to aggregate and bind to the membrane prior to internalization by a process sensitive to cytochalasin (Johansson *et al.*, 1993).

The neutrophil CD14 receptor for lipopolysaccharide (LPS) regulates the chemotactic response of these cells to gram-negative bacteria. In cells so attracted the surface-expressed CD14 receptor becomes internalized, and this process can be visualized by confocal microscopy (Rodeberg *et al.*, 1997).

Receptors to chemokine proteins are members of the seven transmembrane G-protein-coupled family and play a role in inflammation. The receptor-mediated endocytosis of chemokines was investigated by Solari *et al.* (1997) using both fluorescently coupled chemokine agonists and antagonists. The distribution of the fluorescent label was followed confocally, and analysis of the results indicated that internalization had been triggered by agonists and not by antagonists. The question of the internalization of high-affinity neurotransmitters has been addressed by Grady *et al.* (1996). Receptors for substance P and neurokinin 1 showed internalization and recycling of the receptors. This is clearly a phenomenon common to all peptide receptors following activation of the cell initiation of signal transduction.

V. Visualization of Proteases and Receptors on Nonfixed Cells by Confocal Microscopy

A. Contact-Phase Factors on the Human Neutrophil Membrane

The contact-phase assembly comprises the serine proteases, plasma prekallikrein (PpreK) and tissue prokallikrein (TproK), the endogenous substrates for these enzymes, H- and L-kininogens, and the clotting factors, XI and XII (Bhoola and Fink, 2006; Bhoola *et al.*, 1992). In humans the two kallikreins are different proteins with the common capacity to generate the vasoactive peptides known as kinins. TK, found in the epithelial cells of many organs, was discovered a decade ago in granules of the neutrophil (Figueroa *et al.*, 1989) In contrast, plasma kallikrein is synthesized as a precursor enzyme in hepatocytes together with H-kininogen. Both proteins are secreted into the circulation where they form a complex (Bhoola and Fink, 2006). The proenzyme (PpreK) docks on domain 6 of H-kininogen, in equilibrium with factor XI. On discovery of TK in human neutrophils by Figueroa and Bhoola (1989), these authors developed a concept that components of the kallikrein–kinin system may be linked or spatially orientated on the neutrophil membrane (Figueroa and Bhoola, 1989). The unique optical feature of the confocal laser scanning microscope enabled the very first mapping of the essential proteins that comprise the contact-phase assembly on the external surface of the neutrophil cell membrane (Henderson *et al.*, 1994; Kemme *et al.*, 1999). Additionally, L-kininogen and TproK also appear to have binding sites on the membrane. Specific antibodies to each of these proteins, linked on incubation with FITC-labeled F(ab)2 secondary antibodies, provided a powerful new imaging technique that established that these proteins were bound to the external surface of the neutrophil membrane (Fig. 4, A1). Images of TK mRNA are illustrated in Fig. 4, A2–3. The precise molecular sequences involved in the binding process have not as yet been fully elucidated for some of these proteins.

B. Kinin B2 Receptor on External Surface of the Neutrophil Membrane

The kallikreins form kinin peptides which mediate their cellular actions through the seven transmembrane B_1 and B_2, receptors, which are coupled to heterotrimeric G proteins, including Gaq (the pathway linked to intracellular mobilization of calcium and phosphoinositide turnover) and $Ga_{12,13}$ (the pathway linked to c-Jun kinases). Further, mitogenic kinins formed by hK1 and hKB1 from endogenous proteins called H- and L-kininogens, regulate the growth and proliferation of

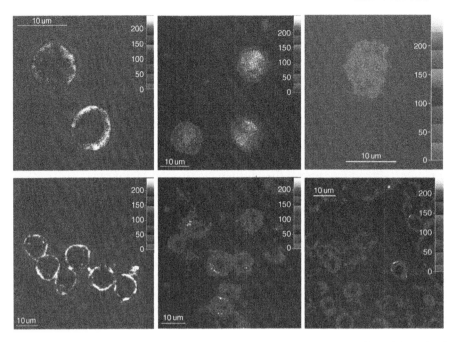

Fig. 4 Confocal images of tissue kallikrein (*in-situ* hybridization) and cleavage of the kinin moiety in the kininogen molecule on the external surface of the neutrophil. **A1**, TK; **A2**, hKLK1 (TK) mRNA antisense; and **A3**, hKLK1 (TK) mRNA sense. **B1**, Kinin moiety in the kininogen molecule on the external surface of the neutrophil membrane; **B2**, loss of kinin moiety from the neutrophil membrane when incubated with bacterial enzymes for 15 min; and **B3**, for 30 min. *Note*: The intensity of fluorescent immunolabeling is indicated by the color bar strip: white, maximal; red, high; yellow, moderate; turquoise, minimal; and blue, nil. (See Plate no. 11 in the Color Plate Section.)

cancer cells through their G-protein coupled B_1 (BR1) and B_2 (BR2) receptors. The monomeric family of kinin antagonists (e.g., [Leu^8] des-Arg^9-BK and HOE 140) inhibit only the Gaq-coupled pathway (Bhoola and Fink, 2006; Blaukat, 2003; Mahabeer and Bhoola, 2000).

The kinin B_2 receptors are constitutive, whereas the cell surface expression of the kinin B_1 receptors is mainly induced. The amino acid sequence of the membrane core and the transmembrane loops of the two receptors were determined from their cDNA profile (reference). The subsequent raising of antibodies to the external and internal loop sequences of the receptor enabled the visualization of the receptors on the cell membrane of epithelial cells. Having established the presence of kinin-releasing enzymes and their substrates on the neutrophil membrane, insight clearly suggested the possible occurrence of autoreceptors to kinins also on the external surface. Experiments were designed, therefore, to examine this possibility. With antibodies raised to its external loops, the kinin B_2 receptor was immunolocalized by confocal microscopy on the external surface of nonfixed human neutrophils (Haasemann *et al.*, 1994; Naidoo *et al.*, 1999).

C. Kinin Moiety Translocated from Kininogen Molecule on the Surface of the Neutrophil Membrane

The kinin nona- and decapeptides are primary mediators of the cellular response to tissue injury. For this reason they are important therapeutic targets. The kinin moiety resides in domain 4 of the kininogen molecule. Once released from kininogen by the enzymatic action of the kallikreins, kinins cause vascular dilatation and, by increasing capillary permeability, promote the traffic of neutrophils into sites of tissue injury and infection. We have, therefore, examined the question of whether kinins are released from the kininogen molecule on the external surface of circulating neutrophils when incubated with bacterial enzymes.

Using confocal microscopy linked to an image processing program, we captured the fluorescence emitted from an FITC-labeled $F(ab)_2$ secondary antibody complexed to a monoclonal primary antibody to bradykinin (Fig. 4, B1). The images provided a qualitative estimate of the amount of antigen visualized on each optical plane. There was an almost complete loss of the kinin moiety in a dose-dependent manner from the surface membrane of the neutrophils incubated with either nagarse or serratiopeptidase (Fig. 4, B2–3) and the kallikreins. Similarly, neutrophils harvested from patients with infections also showed a loss of the peptide, with the segments on either side of the kinin molecule, namely domains 1–3 and 5–6, still attached to the neutrophil membrane. In confirmation electron microscopy studies, immunoreactive kinin, seen as clusters of gold particles on the surface of the membrane, was absent from the circulating neutrophils of patients with systemic infection (Bhoola, 1996; Naidoo et al., 1996a).

VI. Visualization of Proteases and Receptors on Fixed Cells by Confocal Microscopy

A. Tissue Kallikrein in Human Brain

TK is a serine protease that has the capacity to generate kinins that are vasoactive peptides with multiple functions. Except for the localization of TK in human prolactin-secreting adenomas, its apparent deficiency in Alzheimer's disease, and its presence in human cerebrospinal fluid (CSF), the regional distribution of TK in human brain has not been defined. The study was the first attempt at mapping the regional distribution of TK in the human brain (Raidoo et al., 1996). The presence of this enzyme in several areas of the human brain provided a basis for investigating the localization by confocal microscopy and the expression of kinin receptors in the human brain. Further, such a study may aid in the understanding of the pathophysiological role of these receptors and their cell surface orientation to other peptide receptor systems in the human brain.

The regional distribution and the cellular localization of TK in human brains, collected within 24 h of death, were determined by labeling with specific antibody to recombinant TK. Thus far this enzyme has been visualized in neurons of the hypothalamus, thalamus, cerebral gray matter, and reticular areas of the brain stem, as well as in cells of the anterior pituitary and choroid plexus by light and confocal microscopy (Fig. 5, A1). In the central nervous system, a greater amount of TK exists in the proenzyme form and is located preferentially in the neuronal cell bodies and their processes in both the cerebral cortex and the brain stem. The cellular distribution of TK in specific areas suggests a role for TK in the neurons and epithelial cells of the brain. The question whether the functional importance of TK may relate to a particular cell type remains to be elucidated.

B. Kinin Receptors on Human Neurons

Knowledge of the distribution of kinin receptors in the human brain should enhance our understanding of the neurophysiological role of kinins. Furthermore, induction of the kinin B_1 receptor may be important in the pathogenesis of neural diseases. Using polyclonal antibodies directed to specific regions of the B_1- and B_2-kinin receptors and standard immunolabeling techniques, we have determined the localization of these receptors on neurons of specific areas of the human brain by confocal microscopy (Raidoo and Bhoola, 1997). Kinin B_2 receptors were identified in neurons of the brain stem, basal nuclei, cerebral cortex, thalamus, and hypothalamus (Fig. 5, A2). Kinin B_2 receptor immunolabeling was also observed in the endothelial lining of the superior sagittal dural sinus and ependyma of the lateral and third ventricles, whereas kinin B_1 receptors were localized on neurons of the thalamus, spinal cord, and hypothalums. Although the binding of radiolabeled bradykinin to neuronal membranes has been demonstrated, this study provided the first conclusive evidence for the existence of immunoreactive kinin B_1- and the further confirmation of kinin B_2 receptors on human neurons.

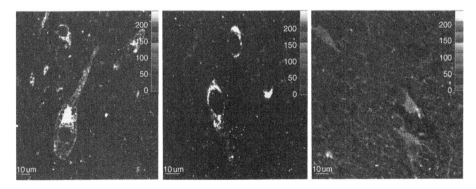

Fig. 5 Confocal images of tissue kallikrein (TK) and kinin receptors in the human brain. **A1**, TK in the cell body and axons of the neurons; **A2**, kinin B2 receptors on neurons; and **A3**, kinin B1 receptors on astrocytic tumor cells. *Note*: The intensity of fluorescent immunolabeling is indicated by the color bar strip: white, maximal; red, high; yellow, moderate; turquoise, minimal; and blue, nil. (See Plate no. 12 in the Color Plate Section.)

The presence of immunoreactive kinin B_2 receptors in neurons of the human brain implies that the receptor-mediated interaction of kinins may occur in the human nervous system and further supports a neuronal or humoral role. The localization of TK in choroidal epithelial cells and the occurrence of immunoreactive kinin B_2 receptors in the ependyma of the lateral and fourth ventricles, as well as in the endothelial lining of the superior sagittal dural sinus, suggest that kinins may regulate the homeostatic balance between the cerebral vasculature and brain parenchyma.

Kinin B1 receptors in the hypothalamus and thalamus may also modulate the cellular actions of kinins in neural tissue. The presence of kinin B_1 receptors on neurons in the substantia gelatinosa and some interneurons of the spinal cord, as well as specific thalamic nuclei in all of the brains studied, suggests a role for kinins in nociception. Our knowledge of the localization of kinin receptors will be valuable, especially since a new generation of potent antagonists could be targeted to specific cellular functions.

C. Identification of Tissue Kallikrein and Kinin Receptors in Astrocytomas of Human Brain

Some of the major diagnostic and prognostic proteinases associated with cancer include the cathepsins (D, B, and L), collagenase, and urokinase-type plasminogen activator. The serine protease TK is present in tumors of the breast (ductal breast cancer cells), lung (Lewis lung tumor cells), stomach (gastric carcinoma cells), and pituitary (pituitary prolactin-secreting adenomas). By degrading components of the extracellular matrix, these enzymes facilitate tumor proliferation and invasion. In addition, the vasodilator effect of the bioactive peptides, bradykinin and kallidin, generated by TK, would increase vascular permeability, thereby enhancing metastasis as well as providing additional nutrients important for tumor growth (Wu *et al.*, 2002).

In order to determine whether TK has a role in the pathogenesis of tumors, its localization was carried out by confocal and electron microscopy and confirmed by *in situ* hybridization. Because kinins act by signal transduction mechanisms linked to kinin B_1 and B_2 receptors, it was necessary to elucidate the occurrence of these receptors in normal human cerebrum and astrocytic tumors.

Our studies have established that TK and kinin B1 receptors are abundant in proliferating tissue, whether of inflammatory or cancer origin. This observation is supported by the demonstration of TK and kinin B_1 receptors in the astrocytes showing malignant transformation (Raidoo *et al.*, 1999) as illustrated in Fig. 5, A3. These findings also suggest that, like other proteases, components of the kallikrein–kinin system may have diagnostic and prognostic relevance and those specific inhibitors of TK, kinin B_1, and B_2 receptors or their gene expression may be of therapeutic value in tumors of the central nervous system.

D. Images of Tissue Kallikrein and Kinin Receptors in Human Blood Vessel and Atheromatous Plaques

Kinins are very potent dilators of arterioles, constrict veins, and contract endothelial cells. The specificity of pharmacological responses to kinin agonists is determined by the kinin receptor. Whereas kinin B2 receptors are stimulated by bradykinin and lysyl-bradykinin (kallidin), kinin B_1 receptors are activated by the desArg analogs of these two peptides.

The first step was to visualize TK as the preferred kinin-releasing enzyme in blood vessels by confocal microscopy. Immunoreactive TK was localized in the cytoplasm of endothelial cells and, with lesser intensity, in the smooth muscle cells of muscular arteries and arterioles (Fig. 6, A1). In atheromatous plaques of larger blood vessels (Fig. 6, A2), the enzyme was present in endothelial cells and foamy macrophages (Raidoo *et al.*, 1997).

Immunoreactive kinin B_1 and B_2 receptors were also observed on endothelial cells and smooth muscle cells of the tunica media of regional blood vessels. In contrast, only kinin B_2 receptors were located on the endothelium of the renal vein. The higher density conical images of these receptors on blood vessels suggest that they may be uniquely placed to homeostatically regulate blood pressure in humans.

E. Tissue Kallikrein in *Helicobacter pylori*-Associated Gastric Ulcer Disease

The association of *H. pylori* (Hp) with ulcer disease is a common form of gastric disorder involving mucosal damage and invasion of the mucosa by polymorphic inflammatory cells with concomitant changes in the epithelial cell structure. Bacteria are thought to adhere by specific junction zones to the epithelial cell surface, resulting in the degeneration of the mucosal layer. The relative status of TK in

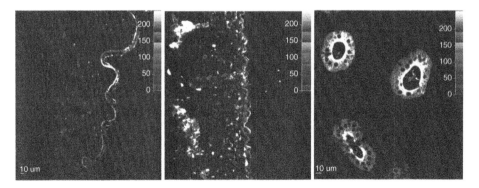

Fig. 6 Confocal images of tissue kallikrein (TK) in human blood vessels and salivary gland. **A1**, TK in normal artery; **A2**, TK in atheromatous plaque; and **A3**, TK in ductal cells of the salivary gland (positive control). *Note*: The intensity of fluorescent immunolabeling is indicated by the color bar strip: white, maximal; red, high; yellow, moderate; turquoise, minimal; and blue, nil. (See Plate no. 13 in the Color Plate Section.)

antral and fundic biopsies, obtained endoscopically from 10 patients suspected of having gastric disorders, was examined by Naidoo *et al.* (1997). For cellular evidence of inflammation, the tissue was stained with hematoxylin and eosin and classified as mild, active, chronic, and chronic active gastritis. The presence of the infective agent Hp was determined by Giemsa staining. For the localization of TK, slide-mounted tissue sections were subjected to peroxidase–antiperoxidase (PAP) and immunofluorescent staining using a goat IgG antibody to recombinant human TK. Confocal microscopy results revealed TK along the luminal border of the deep pyloric glands in antral control tissue removed during partial resection of the stomach (Fig. 7, A1). The surface epithelia and superficial glands showed no labeling. The fundic control tissue showed an absence of TK in the superficial and surface epithelial glands, but was positive in the acid-producing parietal cells. Fundic biopsy specimens showed similar immunoreactivity in these areas. In contrast, in the inflamed pyloric mucosa, there was a shift of TK localization to the basal part of the glandular cells and TK was also expressed in the superficial glands that showed cellular evidence of regeneration (Fig. 7, A2). In fundic biopsies, there was no change observed in the sites of TK localization (similar to control tissue). It was noted that even though infection by Hp could be demonstrated in 8 of the 10 subjects, the inflamed mucosa showed no discernible difference in staining patterns between infected and noninfected tissues.

F. Kinin Receptor Status in Normal and Inflamed Gastric Mucosa

No documented studies have been reported on the occurrence of kinin B_1 and B_2 receptors in the mammalian gastric mucosa. This first study attempted to immunolocalize by confocal microscopy sites of B_1 and B_2 kinin receptors in the human

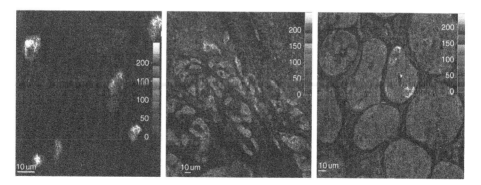

Fig. 7 Confocal images of tissue kallikrein (TK) and kinin B2 receptors in human gastric glands. **A1**, TK in parietal cells of the glands; **A2**, TK in parietal cells and deep glands in gastritis; and **A3**, kinin **B2** receptors on parietal cells and cells of the deep gastric glands. *Note*: The intensity of fluorescent immunolabeling is indicated by the color bar strip: white, maximal; red, high; yellow, moderate; turquoise, minimal; and blue, nil. (See Plate no. 14 in the Color Plate Section.)

(a)

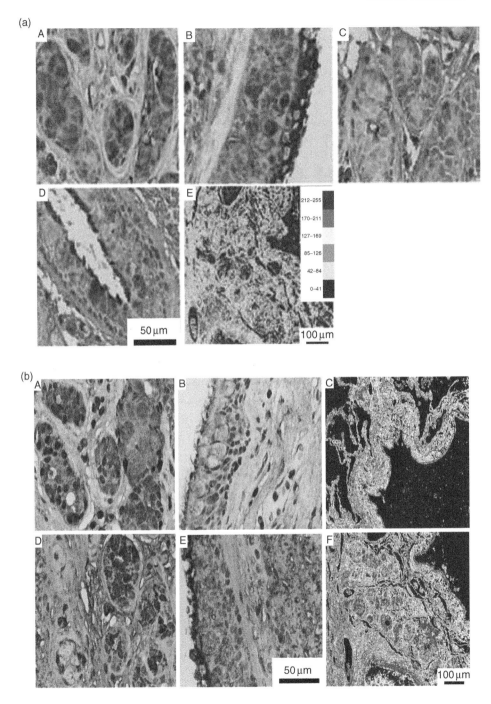

(b)

pyloric gastric mucosa and to evaluate its role in gastritis (Bhoola *et al.*, 1997). Biopsies were obtained from patients with dyspepsia during endoscopic examination of the patient. Diagnosis and grading of the gastritis were performed following histological examination. In gastritis, destruction of the normal mucosal glandular architecture occurs with subsequent regeneration of the epithelial cells.

Kinin B_2 receptors are known to mediate the physiological action of kinins, especially with regard to mucous production. Schachter *et al.* (1986) postulated that TK may play a role in goblet cells by processing mucoproteins. During inflammation destruction of mucous-secreting cells occurs, which would explain the loss of the kinin B_2 receptors.

Because the antrum is affected more by inflammation, the kinin receptor loss is more readily discernible in this region. Inflammation involves the migration of neutrophils, macrophages, and lymphocytes to the antrum. There is transmigration of the neutrophils through the basement membrane of the deep glands, resulting in cryptitis. Additionally, the acute inflammatory cells infiltrate the pyloric glands, resulting in the formation of abscesses in the crypts, with the concomitant loss of epithelial cell membranes. This finding may explain the reduction in the localization of the B_2 kinin receptors (Fig. 7, A3). Later there is a loss of the glands themselves and replacement by fibrous tissue.

The inflammatory process is also characterized by the release of cytokine that causes the induction of the kinin B_1 receptors in active gastritis. Whether the stem cells normally carry the gene for these receptors or whether they are synthesized de novo is open to further study by Molecular Probes. This first study on the identification of kinin receptors on the gastric mucosal cells indicates a role for kinin B_1 receptors in gastritis and may provide a new pathway for treatment of gastritis with kinin receptor antagonists. Follow-up studies after treatment of the inflammation with a combination of kinin B_1 and B_2 receptor antagonists are, therefore, indicated. Further, the occurrence of kinin B_2 receptors on parietal cells points to a novel role for kinin B_2 receptor antagonists in the management of patients.

Fig. 8 **(a)** Localization of tissue prokallikrein and plasma prekallikrein in the normal lung: (A) TproK-submucosal glands, (B) TproK-bronchial epithelium, (C) PPK submucosal glands, (D) PPK labeling in bronchial epithelium, (E) confocal microscopy image of negative control (nonimmune rabbit serum). **(b)** Localization of kinin B_1 and B_2 receptors in the normal lung: (A) B_1-submucosal glands, (B) B_1-bronchial epithelium, (C) B_1 confocal, (D) B_2 submucosal glands, (E) B_2-bronchial epithelium, (F) B_2-confocal. In peroxidase–antiperoxidase immunolabeling, the brown color indicates positive labeling. In confocal images, color scale indicates intensity of fluorescence (pixels/μm^2) as pseudo-colors applied to the gray-scale images: specific labeling depicted as magenta (255–212), red (211–170), yellow (169–127), and pink (126–85), nonspecific labeling, light blue (84–42), no labeling, dark blue (41–0). Scale bar = 50 μm for bright-field and 100 μm for confocal images. *Controls*: primary antibody either omitted or replaced by DAKO nonimmune rabbit serum (Bhoola *et al.*, 2001; Singh *et al.*, 2008). (See Plate no. 15 in the Color Plate Section.)

(a)

(b)

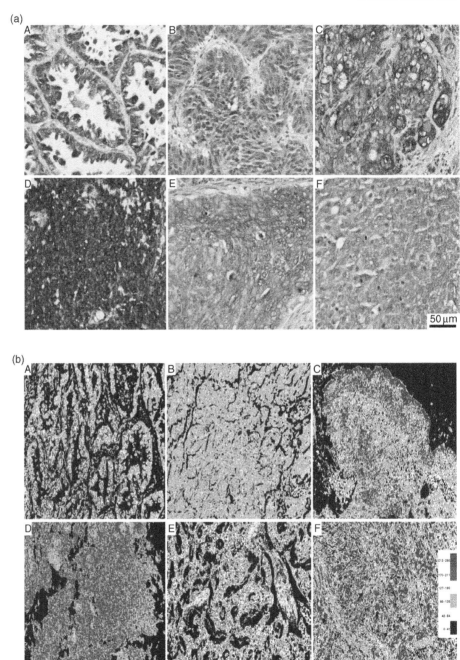

50 µm

100 µm

G. Tissue Kallikrein in Transplant Kidney

Literature survey has established a decrease in the excretion of urinary TK in transplant patients with a further reduction of the enzyme during episodes of acute rejection. The localization of TK in biopsies of the transplant kidney was compared at cellular and subcellular levels to autopsy-derived normal renal tissue (Naidoo *et al.*, 1996b). Renal biopsies from 18 transplant patients with deteriorating renal function were obtained. Immunolabeling for TK, using a polyclonal goat antibody raised against recombinant human TK, was performed following routine enzymatic, confocal immunofluorescence, and electron microscopic techniques. In normal kidney tissue, TK was immunolocalized in the distal connecting tubules (Fig. 8, A1). By comparison, the renal transplant tissue showed a reduction in the intensity of label, but maintained the sites of localization (Fig. 8, A2). Although TK was confined mainly at the luminal side of the cell on electron micrographs, some label was noted along the basolateral membranes. In the transplant kidneys, there was a reduction in the overall number of gold particles counted, which correlated with the decreased intensity observed on confocal immunocytochemistry. In addition, there was a shift to a basolateral orientation of the immunolabel. Edema, tubulitis, and vasculitis characterize acute rejection. Destruction of the tubule cells and leakage of TK into the interstitial tissue space and the resultant effect of the formed kinins on renal capillary vasculature could explain the observed renal parenchymal oedema and transplant rejection.

H. Kinin B2 Receptors in Acute Renal Transplant Rejection

The mechanisms of renal rejection are complex and involve cell-mediated immunological reactions. The question of whether the immediate acute rejection process is initiated by inflammatory molecules such as kinins prompted us to examine the status of kinin receptors in patients undergoing acute transplant rejection (Ramsaroop *et al.*, 1997). In the normal kidney the kinin B_2 receptors observed are confined mainly to the convoluted portion of the distal tubules and the collecting duct (Fig. 8, A3). No immunolabeling at these positions was seen in the transplant kidney, indicating considerable downregulation of the receptors.

Fig. 9 (**a, b**) Localization of tissue prokallikrein in lung carcinomas and mesothelioma. Tumor tissue visualized by bright-field microscopy (**a**) and confocal microcopy (**b**). Notation for the tumors of the lung and pleura: (A) adenocarcinoma, (B) squamous cell carcinoma, (C) large cell carcinoma, (D) small cell carcinoma, (E) lung carcinoid, (F) mesothelioma. In peroxidase–antiperoxidase immunolabeling, the brown color indicates positive labeling. In confocal images, color scale indicates intensity of fluorescence (pixels/μm^2) as pseudo-colors applied to the gray-scale images: specific labeling depicted as magenta (255–212), red (211–170), yellow (169–127), and pink (126–85), nonspecific labeling, light blue (84–42), no labeling, dark blue (41–0). Scale bar = 50 μm for bright-field and 100 μm for confocal images. *Controls:* primary antibody either omitted or replaced by DAKO nonimmune rabbit serum (Bhoola *et al.*, 2001; Singh *et al.*, 2008).

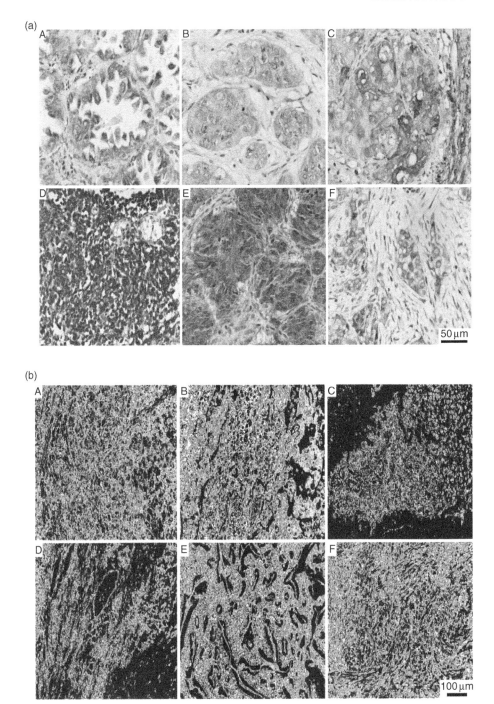

(a)

(b)

50 µm

100 µm

I. Kallikrein–Kinin Cascade Genes and Proteins in Carcinomas of the Lung and Pleura: A Novel Approach for Cancer Research

Lung carcinomas are currently the most commonly diagnosed of all nonskin cancers, both in Australia and throughout the world. Each year in Australia, lung carcinomas are responsible for almost 45,000 years of life lost before the age of 75. Malignant mesotheliomas are strongly associated with exposure to asbestos, with a latent period of 15–40 years between exposure and the onset of symptoms. Survival ranges from 5 to 16 months, with a mortality of 75% within 1 year of diagnosis. Lung carcinomas and mesotheliomas are associated with aberrant gene expression allowing uncontrolled growth and proliferation of cancer cells. The kallikrein–kinin cascade is emerging as a significant signaling pathway in several human cancers and mesotheliomas and is implicated in tumorigenesis, angiogenesis, cell proliferation, and metastasis (Bhoola *et al.*, 2001; Clements *et al.*, 2004).

Tissue (hK1) and plasma (hKB1) kallikreins are serine proteases that act on endogenous, multifunctional, H- and L-kininogens to produce biologically active kinin peptides (Bhoola *et al.*, 1992). The cellular actions of kinin peptides are mediated through B_1 and B_2 receptors (Mahabeer and Bhoola, 2000). Recent reports implicate tissue (*KLK1* to *KLK15*) and plasma (*KLKB1*) kallikreins in the carcinogenic process arising from induction of one or more of the genes. Recent evidence suggests that the serine proteases, *tissue kallikrein (hK1)* and *plasma kallikrein (hKB1)*, hydrolyze macromolecules of the extracellular matrix and thereby regulate *the invasiveness and metastasis* of tumors. In addition, the mitogenic *kinin peptides* regulate the *growth and proliferation of cancer cells* by binding to their respective G-protein coupled B_1 (B_1R) and B_2 (B_2R) kinin receptors.

The synthesis and evaluation of a new family of cytotoxic kinin receptor antagonists has led to a resurgence of interest in the importance of kallikrein–kinin cascade proteins in cancer. The new generation of dimerized kinin receptor antagonists have a unique mode of action causing apotosis of cancer cells by a novel mechanism that involves inhibition of Gaq signaling (e.g., inhibition of intracellular calcium release by bradykinin) and stimulation of the $Ga_{12,13}$ pathway (inhibition of growth and induction of apoptosis) (Chan *et al.*, 2002). Of these

Fig. 10 **(a, b)** Localization of plasma prekallikrein in lung carcinomas and mesotheliomas. Tumor tissue visualized by bright-field microscopy **(a)** and confocal microcopy **(b)**. Notation for the tumors of the lung and pleura: (A) adenocarcinoma, (B) squamous cell carcinoma, (C); large cell carcinoma, (D) small cell carcinoma, (E) lung carcinoid, (F) mesothelioma. In peroxidase–antiperoxidase immunolabeling, the brown color indicates positive labeling. In confocal images, color scale indicates intensity of fluorescence (pixels/μm^2) as pseudo-colors applied to the gray-scale images: specific labeling depicted as magenta (255–212), red (211–170), yellow (169–127), and pink (126–85), nonspecific labeling, light blue (84–42), no labeling, dark blue (41–0). Scale bar = 50 μm for bright-field and 100 μm for confocal images. *Controls:* primary antibody either omitted or replaced by DAKO nonimmune rabbit serum (Bhoola *et al.*, 2001; Singh *et al.*, 2008).

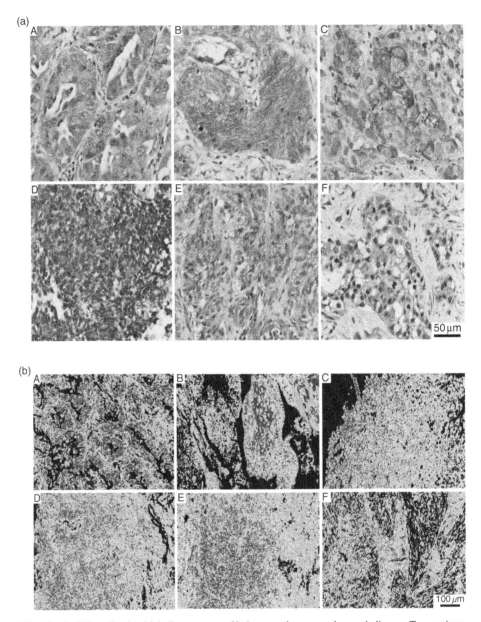

Fig. 11 (**a, b**) Localization kinin B$_1$ receptors of in lung carcinomas and mesotheliomas. Tumor tissue visualized by bright-field microscopy (**a**) and confocal microcopy (**b**). Notation for the tumors of the lung and pleura: (A) adenocarcinoma, (B) squamous cell carcinoma, (C) large cell carcinoma, (D) small cell carcinoma, (E) lung carcinoid, (F) mesothelioma. In peroxidase–antiperoxidase immunolabeling, the brown color indicates positive labeling. In confocal images, color scale indicates intensity of fluorescence (pixels/μm^2) as pseudo-colors applied to the gray-scale images: specific labeling depicted as magenta (255–212), red (211–170), yellow (169–127), and pink (126–85), nonspecific labeling, light blue (84–42), no labeling, dark blue (41–0). Scale bar = 50 μm for bright-field and 100 μm for confocal images. *Controls:* primary antibody either omitted or replaced by DAKO nonimmune rabbit serum (Bhoola *et al.*, 2001; Singh *et al.*, 2008).

(a)

(b)

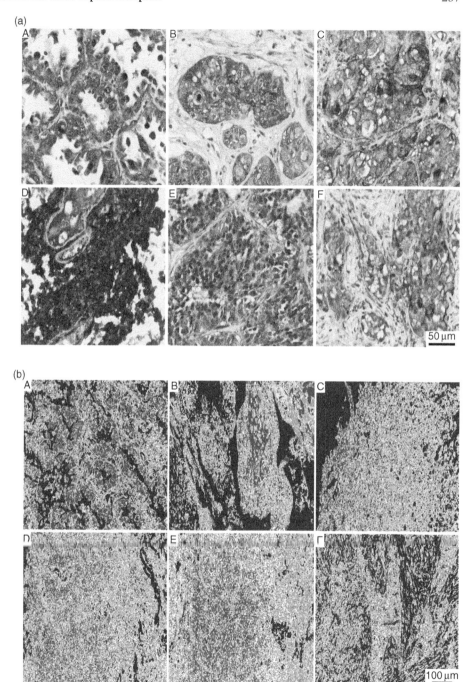

50 μm

100 μm

Fig. 12 **(a, b)** Localization kinin B$_2$ receptors of in lung carcinomas and mesotheliomas. Tumor tissue visualized by bright-field microscopy **(a)** and confocal microcopy **(b)**. Notation for the tumors of the lung and pleura: (A) adenocarcinoma, (B) squamous cell carcinoma, (C) large cell carcinoma, (D) small cell carcinoma, (E) lung carcinoid, (F) mesothelioma. In peroxidase–antiperoxidase immunolabeling, the brown color indicates positive labeling. In confocal images, color scale indicates intensity of fluorescence (pixels/μm^2) as pseudo-colors applied to the gray-scale images: specific labeling depicted as magenta (255–212), red (211–170), yellow (169–127), and pink (126–85), nonspecific labeling, light blue (84–42), no labeling, dark blue (41–0). Scale bar = 50 μm for bright-field and 100 μm for confocal images. *Controls:* primary antibody either omitted or replaced by DAKO nonimmune rabbit serum (Bhoola *et al.*, 2001; Singh *et al.*, 2008).

dimerized kinin antagonists the most potent anticancer agent has been a molecule in which the N-terminal is cross-linked and dimerized with suberimide (B9430) (Stewart *et al.*, 2004). Another potentially powerful inhibitor is Domain 5 of H-kininogen (kininostatin), the endogenous substrate of plasma kallikrein, which inhibits critical steps in the angiogenic process, namely inhibition of new blood vessel formation within the tumor. Proof of concept is provided by the *killing of small cell carcinoma cells* by dimeric *kinin receptor antagonists* and by the *inhibition of hK1-induced tumor cell migration by kallistatin*, a specific inhibitor of hK1.

We have demonstrated increased expression of kallikrein–kinin cascade proteins in several lung carcinoma subtypes (Chee *et al.*, 2008; Singh *et al.*, 2008) and in human pleural mesotheliomas (Chee *et al.*, 2007) (Figs. 8a, b; 9a, b; 10a, b; 11a, b; 12a, b). Increasingly the kallikrein serine proteases are considered to be important biomarkers for endocrine malignancies, in particular prostate (PSA) and ovarian cancers. The significance of the kallikrein–kinin signaling pathway in human lung cancer is demonstrated by both the induction of apoptosis in small cell carcinomas following treatment with dimeric kinin receptor antagonists and the inhibition of tumor cell migration by kallistatin, a specific inhibitor of hK1.

Acknowledgments

We thank the Foundation for Research Development (FRD), Medical Research Council of South Africa (MRC, SA), Cancer Research Association of South Africa (CANSA), and University of Natal Research Fund (NURF) for generous financial support. We thank also the Lung Institute of Western Australia for research funds.

References

Barr, M. L., and Kiernan, J. A. (1993). The Human Nervous System: An Anatomical Viewpoint Lippincott, Philadelphia.

Bhoola, K. D., and Fink, E. (2006). *In* "Kallikrein–Kinin Cascade in Encyclopedia of Respiratory Medicine," pp. 483–5493. Elsevier Ltd.

Bhoola, K. D. (1996). *Immunopharmacology* **33**(1-3), 247–256.

Bhoola, K. D., Figueroa, C. D., and Worthy, K. (1992). *Pharmacol. Rev.* **44**(1), 1–80.

Bhoola, R., Ramsaroop, R., Naidoo, S., Muller-Esterl, W., and Bhoola, K. D. (1997). *Immunopharmacology* **6,** 161.

Bhoola, K., Ramsaroop, R., Plendl, J., Cassim, B., Dlamini, Z., and Naicker, S. (2001). *Biol. Chem.* **382,** 77–89.

Blaukat, A. (2003). *Andrologia* **35,** 17–23.

Brakenhoff, G. J., Blom, P., and Bakker, C. (1978). *Proc. ICO* **11,** 215.

Chan, D., Gera, L., Stewart, J., Helfrich, B., Verella-Garcia, M., Johnson, G., Baron, A., Yang, J., Puck, T., and Bunn, P., Jr. (2002). *Proc. Natl. Acad. Sci. USA* **99**(7), 4608–4613.

Chee, J., Naran, A., Misso, N. L., Thompson, P. J., and Bhoola, K. D. (2008). *Biol. Chem.* **389,** 1225–1233.

Chee, J., Singh, J., Naran, A., Misso, N. L., Thompson, P. J., and Bhoola, K. D. (2007). *Biol. Chem.* **388** (11), 1235–1242.

Engelhard, J., and Knebel, W. (2010). Confocal Spectrum Leica Lasertechnik GmbH, Im Neurnheimer Feld 518, Heidelberg, Germany.

Figueroa, C. D., and Bhoola, K. D. (1989). *In* "The Kallikrein–Kinin System in Health and Disease— Leucocyte Tissue Kallikrein: An Acute Phase Signal for Inflammation," (H. Fritz, J. Schmidt, and G. Dietz, eds.), p. 311. Limbach-Verlag, Braunschweig.

Figueroa, C. D., MacIver, A. G., Dieppe, P., Mackenzie, J. C., and Bhoola, K. D. (1989). *Adv. Exp. Med. Biol. Kinins V* **247B,** 207–210.

Fix, J. D. (1974). Atlas of the Human Brain and Spinal Cord. Waverly Press, Baltimore, Maryland, USA.

Grady, F., Gamp, P. D., Jones, E., Baluk, P., McDonald, D. M., Payan, D. G., and Bunnett, N. W. (1996). *Neuroscience* **75**(4), 1239.

Haasemann, M., Figueroa, C. D., Henderson, L., Grigoriev, S., Abd Alia, S., Gonzales, C. B., Dunia, I., Benedetti, E. L., Hoebeke, J., Jarnagin, K., Cartaud, J., Bhoola, K. D., *et al.* (1994). *Braz. J. Med. Bio. Res.* **27,** 1739–1756.

Henderson, L. M., Figueroa, C. D., Muller-Esterl, W., and Bhoola, K. D. (1994). *Blood* **84,** 474–482.

Johansson, B., Wymann, M. P., Holmgren-Peterson, K., and Magnusson, K. E. (1993). *J. Cell Biol.* **121**(6), 1281.

Johnston, A., and Thorpe, R. (1982). *In* "Immunocytochemistry in Practice," p. 41. Blackwell Scientific, Oxford.

Kemme, M., Podlich, D., Raidoo, D. M., Snyman, C., Naidoo, S., and Bhoola, K. D. (1999). *Biol. Chem.* **380,** 1321–1328.

Levy, R., Benchaib, M., Cordonier, H., Souchier, C., Guerin, J. F., and Czyba, J. C. (1997). *Ital. J. Anat. Embryol.* **102**(3), 141.

Mahabeer, R., and Bhoola, K. D. (2000). *Pharmacol. Ther.* **88,** 77–89.

Naidoo, S., Ramsaroop, R., Naidoo, Y., and Bhoola, K. D. (1996a). *Immunopharmacology* **33,** 157–160.

Naidoo, Y., Naidoo, S., Nadar, R., and Bhoola, K. D. (1996b). *Immunopharmacology* **33,** 387–390.

Naidoo, S., Ramsaroop, R., Bhoola, R., and Bhoola, K. D. (1997). *Immunopharmacology* **36,** 263.

Naidoo, Y., Snyman, C., Raidoo, D. M., Bhoola, K. D., Kemme, M., and Muller-Esterl, W. (1999). *Brit. J. Haematol.* **105**(3), 599–612.

Raidoo, D. M., and Bhoola, K. D. (1997). *J. Neuroimmunol.* **77,** 39.

Raidoo, D. M., Ramsaroop, R., Naidoo, S., and Bhoola, K. D. (1996). *Immunopharmacology* **33**(1-3), 39–47.

Raidoo, D. M., Mahabeer, R., Naidoo, S., and Bhoola, K. D. (1999). *Immunopharmavcology* **43,** 255–263.

Raidoo, D. M., Ramsaroop, R., Naidoo, S., Muller-Esterl, W., and Bhoola, K. D. (1997). *Immunopharrnacology* **36,** 153.

Ramsaroop, R., Naicker, S., Naicker, T., Naidoo, S., and Bhoola, K. D. (1997). *Immunopharmacology* **36,** 255.

Roberts, M., Hanaway, J., and Morest, D. K. (1987). Atlas of the Human Brain in Section. Lea and Febiger, Philadelphia, PA.

Rodeberg, D. A., Morris, R. E., and Babcock, G. F. (1997). *Infect. Immun.* **65**(11), 4747.

Schachter, M., Longridge, D. J., Wheeler, G. D., Metha, J. G., and Uchida, Y. (1986). *J. Histochem. Cytochem.* **34,** 927.

Singh, J., Naran, A., Misso, N. L., Rigby, P., Thompson, P. J., and Bhoola, K. D. (2008). *Int. Immunoparmacol.* **8**(2), 300–306.

Solari, R. I., Offord, R. E., Remy, S., Aubry, J. P., Wells, T. N., Whitehorn, E., Oung, T., and Proudfoot, A. E. (1997). *J. Biol. Chem.* **272**(15), 9617.

Stewart, J. M., Gera, L., Chan, D. C., York, E. J., Stewart, L. T., Simkeviciene, V., and Helfrich, B. (2004). *Chest* **125**(5 Suppl.), 148S.

Van der Voort, H. T. M., Brakenhof, G. J., Valkenburg, J. A. C., and Nanninga, N. (1985). *Scanning* **7,** 66.

Wijnaendts van Resandt, R. W., Marsman, H. J. B., Kaplan, R., Davoust, J., Stelzer, E. H. K., and Stocker, R. (1985). *J. Microsc.* **138,** 29.

Wilson, T., and Sheppard, C. J. R. (1994). Theory and Practice of Scanning Optical Microscopy. Academic Press, London.

Wu, J., Akaike, T., Hayashida, K., Miyamoto, Y., Nakagawa, T., Miyakawa, K., Müller-Esterl, W., and Maeda, H. (2002). Identification of bradykinin receptors in clinical cancer specimens and murine tumor tissues. *Int. J. Cancer* **98,** 29.

CHAPTER 15

Multiphoton Excitation Microscopy, Confocal Microscopy, and Spectroscopy of Living Cells and Tissues; Functional Metabolic Imaging of Human Skin *In Vivo*

Barry R. Masters, Peter T. C. So, Ki Hean Kim, Christof Buehler, and Enrico Gratton

Arlington, Virginia, USA

I. Introduction

Multiphoton excitation microscopy combined with functional metabolic imaging based on NAD(P)H is an important development in cellular imaging. Nonlinear multiphoton excitation processes were predicted by Göppert-Mayer (1931). An example of this nonlinear process is two-photon-induced fluorescence in which the simultaneous absorption of two photons of red light (800 nm) causes the subsequent emission of blue light (Fig. 1).

In practical terms this new technology has been implemented in nonlinear optical microscopes that use pulsed near-infrared light to induce the fluorescence of chromophores (ultraviolet and visible) in cells, tissues, and organs. The rate of a two-photon absorption process is a function of the square of the instantaneous intensity of the excitation light. The microscope objective focuses the light pulses into a diffraction-limited volume. At sufficiently high light intensity absorption and fluorescence can be induced from the fluorophores at a rate suitable for microscopy. Photobleaching and phototoxicity are limited to the same focal region. Outside the focal volume, the nonlinear multiphoton process occurs with a much lower probability due to the reduced photon flux.

Denk *et al.* (1990, 1991) first demonstrated this new type of microscope based on two-photon excitation of molecules by integrating a laser scanning microscope

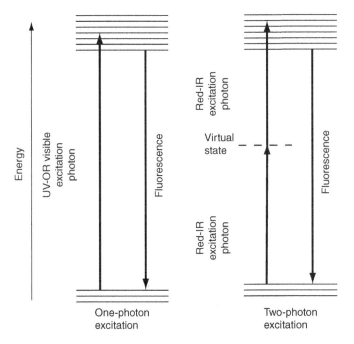

Fig. 1 Jablonski diagram of two-photon process versus single-photon process. For the two-photon excitation process there is a short-lived virtual state shown by a dashed line.

(scanning mirrors, photomultiplier tube detection system) and a mode-locked laser that generates femtosecond pulses of near-infrared light. The pulses of red or near-infrared light (700 nm) were less than 100 fs in duration and the laser repetition rate was about 100 Hz. The high intensity, short pulses of near-infrared light cause significant multiphoton excitations; however, the low average power incident on the sample minimizes cell and tissue damage. The capabilities of these nonlinear microscopes include improved spatial resolution without the use of pinholes or slits for spatial filtering, improved signal strength, deeper penetration into thick, highly scattering tissues, confinement of photobleaching and photodamaging to the focal volume, and improved background discrimination.

In order to correctly interpret the results of applying multiphoton excitation microscopy to three-dimensional functional imaging of thick, highly scattering *in vivo* tissue such as human skin it is necessary to demonstrate the following. First, images of the cells based on the cellular autofluorescence are obtained and these images are correlated with similar images of the same tissue obtained with reflected light confocal microscopy. This use of correlative microscopy is important to demonstrate the histology of the tissue based on images acquired with multiphoton excitation microscopy. Second, the correct interpretation of skin physiology using multiphoton microscopy and spectroscopy requires that the tissue fluorophores responsible for the fluorescence contrast be identified correctly. Because only a few fluorophores in tissues have been studied, the correct identification of the fluorescent species requires a knowledge of its excitation and emission spectra, which requires knowing whether the fluorophores are excited by either two or higher photon processes. The characterization of the multiphoton process can be performed in the following manner. For a two-photon excitation process, it is necessary to show a plot of the quadratic dependence of the fluorescence intensity on the excitation power. A plot of the logarithm of the fluorescence intensity versus the logarithm of the excitation power should have a slope of 2 for a two-photon excitation process. Similarly, a slope of 3 indicates a three-photon process. Finally, for each type of cell or tissue studied, the emission spectra and the lifetime of the fluorescence should be determined to confirm the nature and biochemical state of the fluorophore, for example, NAD(P)H or flavoprotein.

To demonstrate a multiphoton excitation process it is necessary to show the nonlinear dependence of the fluorescence on the laser excitation power. In the human skin system, we have measured the fluorescence intensity as a function of excitation power to determine whether the excitation at 960 nm is due to two-photon excitation or involves higher photon processes (Fig. 2). A 0.5-mm-thick slice of human skin from the second digit of the right hand was excised. The sample was sandwiched between a piece of cover glass and a microscope slide. The fresh sample was imaged immediately. The incident power was controlled by a polarizer. Fluorescence images of the same area were collected at power levels ranging from 1 to 10 mW with 960 nm excitation. Equivalent portions (10×10 pixel areas) of these images were averaged and compared. The slopes of the plots verify multiphoton excitation processes.

How does multiphoton excitation microscopy compare with confocal laser scanning microscopy (Masters, 1996b)? The main limitations of two-photon

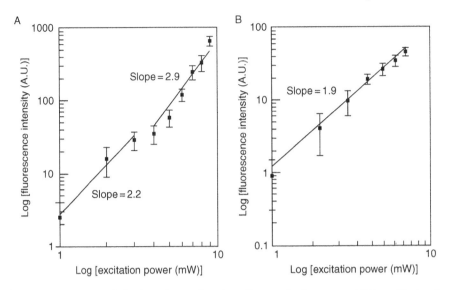

Fig. 2 Fluorescence emission intensity as a function of laser excitation power at 960 nm. The log–log plot is the experimental verification of a multiphoton excitation process; the slopes of 3 and 2 indicate three-photon and two-photon excitation processes, respectively. From Masters *et al.* (1997b).

excitation microscopy are (1) multiphoton excitation microscopy is only suitable for fluorescent imaging; reflected light imaging is not currently available, and (2) the technique is not suitable for imaging highly pigmented cells and tissues that absorb near-infrared light. Multiphoton excitation microscopy has the following important advantages: (1) reduced phototoxicity, (2) reduced photobleaching, (3) increased penetration depth, (4) ability to perform uncaging or photobleaching in a diffraction limited volume, (5) ability to excite fluorophores in the ultraviolet without a ultraviolet laser, (6) the excitation and the fluorescence wavelengths are well separated, and (7) no spatial filter is required.

II. Functional Metabolic Imaging of Cellular Metabolism Based on NAD(P)H Fluorescence

Redox fluorometry is a noninvasive optical method used to monitor the metabolic oxidation–reduction (redox) states of cells, tissues, and organs. It is based on measuring the intrinsic fluorescence of the reduced pyridine nucleotides, NAD(P) H, and the oxidized flavoproteins of cells and tissues (Chance *et al.*, 1979; Masters, 1988, 1990, 1993a; Masters and Chance, 1984, 1993; Masters *et al.*, 1981, 1989). Both the reduced nicotinamide adenine dinucleotide, NADH, and the reduced nicotinamide adenine dinucleotide phosphate, NADPH, are denoted as NAD(P) H. The fluorescence of NAD(P)H is in the range of 400–500 nm with a single-photon absorption in the region of 364 nm. Redox fluorometry is based on the fact that the quantum yield of the fluorescence, and hence the fluorescence intensity, is greater for the reduced form of NAD(P)H and lower for the oxidized form. For flavoproteins, the quantum yield, and hence the fluorescence intensity, is higher for the oxidized form and lower for the reduced form. The reduced pyridine nucleotides are located in both the mitochondria and the cytoplasm. The flavoproteins are uniquely localized in the mitochondria. Fluorescence from oxidized flavoproteins occurs in the region from 520 to 590 nm with a single-photon absorption in the region of 430–500 nm. Fluorescence from the reduced pyridine nucleotides is usually measured in tissue investigations as the measured fluorescence is higher than in the case of the flavoprotein fluorescence. Redox fluorometry has been applied to many physiological studies of cells, tissues, and organs.

Functional imaging of cellular metabolism and oxygen utilization using the intrinsic fluorescence has been studied extensively in various types of cells and tissues. Comprehensive reviews of instrumentation, techniques, and experimental results based on fluorescence measurements of the autofluorescence of cells have been published previously. Specific studies based on redox fluorometry include redox measurements of *in vivo* rabbit cornea based on flavoprotein fluorescence (Fig. 3), monitoring of oxygen tension under the contact lens in live rabbits, chemical analysis of nucleotides and high-energy phosphorous compounds in the various layers of the rabbit cornea, and redox fluorescence imaging of the *in vitro* cornea with ultraviolet confocal fluorescence microscopy.

One important aspect of this technique, which is often overlooked, is the experimental verification of the chemical identities of the fluorophores that contribute to the fluorescence signal. Several experimental techniques are used to verify that the predominate contribution to the measured fluorescence is from the reduced pyridine nucleotides, NAD(P)H. Methods of microanalysis and enzyme cycling have been used to measure the concentrations of NADH and NADPH, the reduced pyridine nucleotides, in the normoxic state and in the anoxic states of the rabbit cornea. This analysis was performed for the epithelial, the stromal region, and the endothelium. Another method (Fig. 3B) measures the time course of the increase of fluorescence following full reduction of the oxidized pyridine nucleotide, NAD^+, following the application of cyanide, which blocks cytochrome oxidase in the mitochondria. An alternative technique to verify NAD(P)H as the source of fluorescence, which is discussed later, is to measure the fluorescence spectrum and the fluorescence lifetime.

III. Ultraviolet Confocal Fluorescence Microscopy Compared with Multiphoton Excitation Microscopy

It is instructive to compare redox imaging of NAD(P)H fluorescence from the cornea with both single-photon confocal microscopy and two-photon excitation microscopy. The cornea is the 400-μm-thick, transparent tissue located on the front surface of the eye.

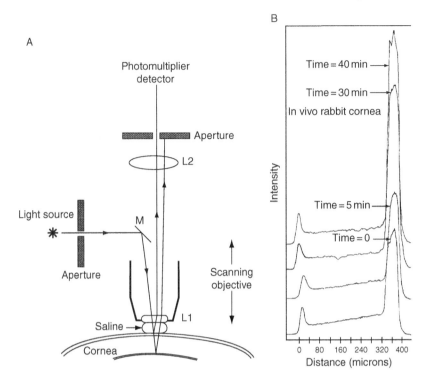

Fig. 3 (A) Schematic diagram of the *z*-scan confocal microscope developed for optically sectioning the cornea *in vivo*. The light source is connected to the confocal microscope with a quartz fiber-optic light guide. Two conjugate slits make the instrument a confocal microscope. One slit is imaged onto the cornea; the second slit is in front of the photon detector. The *z*-scan confocal microscope has a scanning microscope objective that moves on the optical axis under computer control via a piezoelectric driver. This device scans the focal plane through the full thickness of the cornea under computer control. (B) *In vivo* redox metabolic functional imaging of the rabbit cornea. The *z*-scan confocal microscope shown in (A) repeatedly scans the 400-µm-thick cornea and measures the NAD(P)H fluorescence intensity as a function of position in the cornea and shows the time dependence of the intensity of NAD(P)H fluorescence (across the full thickness of the cornea *in vivo*) after the application of a thick PMMA contact lens on the cornea. The small peak on the left side is the corneal endothelium. The larger peak on the right side is from the corneal epithelium. Time is the duration that the contact lens was on the cornea. Successive *z* scans of the cornea are displaced vertically for clarity. Note that during the 40-min period of contact lens wear (blocks oxygen flow into the cornea from the air) the NAD(P)H fluorescence intensity of the epithelium doubled. This effect is reversed completely when the contact lens is removed.

In confocal microscopy studies, an ultraviolet confocal laser scanning fluorescence microscope was used to investigate the *ex vivo* rabbit cornea. The laser scanning confocal microscope used an argon–ion laser with a light output at 364 nm. The excitation wavelength of 364 nm and the emission at 400–500 nm were used to image the fluorescence of the reduced pyridine nucleotides, NAD(P)H. A Zeiss water-immersion microscope objective 25× with a numerical aperture of 0.8 and corrected for the ultraviolet, was used for imaging the fluorescence.

Rabbit eyes were obtained from New Zealand White rabbits. The rabbits were euthanized with an injection into the marginal vein of the ear with ketamine hydrochloride (40 mg/kg) and xylazine (5 mg/kg). The eyes were freed immediately of adhering tissue and swiftly enucleated. A specimen chamber was designed and developed to maintain the eye in a physiological state during the data acquisition period and to gently immobilize it during microscopic imaging. The intact eye was placed in the specimen chamber, perfused with aerated, bicarbonate Ringer's solution, which contained 5 mM glucose and 2 mM calcium, and kept at 37 °C. The specimen chamber preserved the normal corneal morphology for several hours; however, in the absence of the special specimen chamber, cells in the cornea of the *in situ* eye showed time-dependent morphological alterations (Masters, 1993b).

Analysis of the images obtained with ultraviolet confocal laser scanning fluorescence microscopy showed photobleaching of the NAD(P)H during the course of the measurements. The microscope objective was designed for the refractive index of water. However, there are significant variations of the index of refraction across the full 400 μm of corneal thickness. While the average refractive index of the cornea is known, the profile of refractive index is unknown. The microscope objective had chromatic aberrations for excitation and emission wavelengths. These effects resulted in reduced image quality in the ultraviolet confocal microscope.

Two-photon excitation microscopy was used to image cells of the freshly excised rabbit cornea (Masters *et al.*, 1993, 1997c; Piston *et al.*, 1995). The initial experiments used a modified confocal scanning laser microscope (Bio-Rad, Richmond, CA, MRC-600) that was coupled to a Zeiss Universal upright microscope. The epi-illumination collimator lenses were removed. This is an example of how to modify an existing Bio-Rad laser scanning confocal microscope in order to convert it to an instrument for multiphoton excitation microscopy. It is suggested that in order to maximize the sensitivity of the modified instrument, the fluorescence light from the specimen should not be descanned prior to photon detection. Instead, the fluorescence light should be detected with external photon detection, which bypasses the descanning unit. This follows from the physics of the multiphoton excitation process; the optical sectioning capability is on the excitation side and it is derived from the nonlinear optical absorption of the fluorophore.

A Satori ultra fast dye laser from Coherent Corporation was used to generate 150-fs pulses of 705-nm light at a repetition rate of 76 MHz. A dichroic mirror (Omega Optical 550DCSP) was used to direct the excitation light to the specimen. The microscope objective was a Leitz water-immersion objective lens (50×/1.0) with a free working distance of 1.7 mm. The resulting fluorescence from the NAD(P)H in the cornea was collected by the microscope objective and descanned by the confocal microscope scanning mirrors. The confocal aperture was open in order to maximize the fluorescence signal. Other microscope modifications included replacing the internal mirrors with those optimized for 380–650 nm and replacing the existing photomultiplier tube (PMT) with a Thorn-EMI 9924B. The average excitation power at the specimen was 10–15 mW.

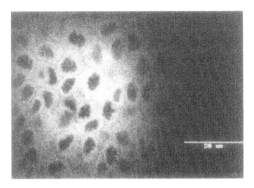

Fig. 4 An example of multiphoton excitation microscopic imaging based on the fluorescence of NAD(P)H. The dark oval regions are cell nuclei of the basal epithelial cells in a freshly excised rabbit cornea. The cell borders are shown as dark lines between adjacent cells. Scale bar: 50 μm. From Piston *et al.* (1995).

NAD(P)H fluorescence was imaged from the superficial cells of the corneal epithelium and from the corneal endothelium (Fig. 4). In comparison to single-photon confocal microscopy there were several advantages. There was minimal photobleaching of the NAD(P)H. Images from the basal epithelial cells showed improved contrast and image quality.

More recently, a specially designed and constructed multiphoton excitation microscope was used to investigate the functional imaging of the *in situ* cornea based on NAD(P)H imaging. The use of emission spectroscopy was used to further validate that the fluorophore was NAD(P)H. The use of external photon detection in this design (described later) permitted multiphoton imaging of the stromal keratocytes in the normoxic cornea. In comparison, the modified Bio-Rad instrument with descanning of the fluorescence prior to photon detection was not capable of imaging the weakly fluorescent stromal keratocyte cells in the absence of complete reduction of the pyridine nucleotides with cyanide treatment, which results in a doubling of the concentration of NAD(P)H. The two-dimensional imaging of the corneal endothelial cells based on NAD(P)H functional redox imaging with multiphoton excitation microscopy and spectroscopy presents a new technique for investigating the heterogeneity of cellular respiration in the corneal cell layers.

IV. Laser Sources for Multiphoton Excitation Microscopy

The appropriate selection of a laser light source is critical in designing a multiphoton excitation microscope. For a one-photon absorption process, the rate of absorption is the product of the one-photon absorption cross section and the average of the photon flux density. For a two-photon absorption process, in which two photons are absorbed simultaneously by the fluorophore, the rate of absorption is given by the product of the two-photon absorption cross section and the average squared photon flux density. In respect for the work of Maria Göppert-Mayer, the units of two-photon absorption cross section are measured in GM (Göppert-Mayer) units. One GM unit is equal to $10–50$ cm^4 s/photon.

To achieve a reasonable probability of the occurrence of a multiphoton excitation process, a pulsed laser source of high peak power is required. A mode-locked dye laser can provide sufficient excitation intensity at the focal volume of the microscope objective. Typical laser pulses are in the red region of the spectrum (630 nm) with a duration of less than 100 fs and a repetition ration of 80 MHz. These pulses are of sufficient intensity to generate fluorescence in a fluorophore that usually absorbs at 315 nm. Alternatively, two photons in the infrared region at 1070 nm could excite a fluorophore that usually absorbs at 535 nm in the visible region. The two photons must interact simultaneously with the molecule. The following quantities are useful for rough calculations. Typically the laser pulse width is 10^{-13} s, the fluorescence decay time is 10^{-9} s, and the pulse separation time is 10^{-8} s. At a laser pulse repetition rate of 100 MHz, there is a laser pulse every 10^{-8} s, or once every 10 ns. Two-photon excitation processes do not require that the two photons absorbed simultaneously be of the same wavelength. Two different wavelengths can be combined by superimposing pulsed light beams of high peak powers. The two wavelengths can be chosen using the following equation:

$$\frac{1}{\lambda_{ab}} = \frac{1}{\lambda_1} + \frac{1}{\lambda_2} \tag{1}$$

where λ_{ab} is the short wavelength of the absorber and λ_1 and λ_2 are the incident beam wavelengths.

The most critical component of a multiphoton microscope is the laser light source. The fluorescence excitation rate is insufficient for scanning microscopy in deep tissue applications unless femtosecond or picosecond pulsed lasers are used. Although the use of a CW infrared laser for two-photon imaging of cell cultures has been achieved, its applicability in deep tissue work is limited. The choice between femtosecond and picosecond light sources for multiphoton excitation remains controversial. As discussed, multiphoton excitation can be achieved with both types of lasers. To generate the same level of fluorescence signal, picosecond laser systems will need a significantly high average input power whereas femtosecond laser systems have a much higher peak power.

Several laser sources are available for multiphoton excitation microscopy (Gratton and vandeVen, 1995; Valdmansis and Fork, 1986; Wokosin et al., 1996; Xu and Webb, 1997; Xu et al., 1996). Important considerations in laser selection for multiphoton excitation microscopy include peak power, pulse width, range of wavelength tunability, cooling requirement, power requirement, and cost. The range of tunability should cover the region of interest for multiphoton excitation processes of typical fluorophores. It is important to note that single-photon and two-photon excitation spectra for many molecules are different. Published two-photon excitation spectra (plots of two-photon excitation cross sections as a function of wavelength) are extremely useful in optimizing the selection of a laser wavelength for a specific molecule. This chapter focuses on multiphoton excitation microscopy based on NAD(P)H functional imaging; however, progress is being made in synthesizing molecules with very large two-photon absorption cross sections.

Titanium–sapphire laser systems are good choices. These systems provide a high average power (1–2 W), high repetition rate (80–100 MHz), and short pulse width (80–150 fs). The titanium–sapphire laser provides a wide tuning range from below 700 nm to above 1000 nm. Titanium–sapphire lasers require pump lasers. The older systems use argon ion lasers, but the new ones use solid-state diode-pumped Nd:YAG lasers. The newer lasers are basically turnkey systems. Ultracompact (the size of a shoe box), single wavelength femtosecond lasers combining diode, Nd:YAG, and titanium–sapphire lasers are becoming available commercially. Furthermore, other single wavelength solid-state systems, such as diode-pumped Nd:YLF lasers and diode-pumped erbium-doped fiber laser systems, have also become available.

V. Pulse Compensation Techniques

For multiphoton excitation processes the number of photon pairs absorbed for each laser pulse is related inversely to the pulse width. The passage of ultrashort pulses from the mode-locked laser through a dielectric medium results in pulse broadening due to group velocity dispersion. This pulse broadening, called pulse dispersion, is a serious problem in multiphoton excitation microscopy because it results in a reduction in the probability of multiphoton excitation. This is observed as a strong reduction in the intensity of the fluorescence signal from the fluorophores in the specimen. An experimental technique called "prechirping" the laser pulses can be used to compensate the laser pulse dispersion. A "prechirp" unit can be constructed with two prisms and mirrors. This system compensates for the group velocity dispersion and causes all of the different wavelengths in each pulse to arrive simultaneously at the specimen after propagating through the microscope optics.

An important paper describes how to measure the group velocity dispersion for a microscope with a variety of microscope objectives and how to "prechirp" the pulses to achieve pulse compensation (Wolleschensky *et al.*, 1997). A Michelson-type interferometric autocorrelator was attached to the microscope, which was used to measure the pulse length of the laser pulses at the specimen. For example, a Coherent MIRA 900-F titanium–sapphire femtosecond laser system (780 nm, 72 MHz, 92 fs) was adapted to the Zeiss confocal laser scanning microscope LSM 410 with a C-APOCHROMAT 40×/1.2 water-immersion microscope objective. The measured pulse length at the specimen was 355 fs. Recompression of the pulse width to 135 fs at the specimen was achieved with the use of a "prechirp" unit consisting of a pair of prisms.

VI. Multiphoton Excitation Microscope

The instrumentation and design of a basic multiphoton microscope have been described in a number of previous publications (Art, 1995; Dong *et al.*, 1995; Kim *et al.*, 1998; Masters *et al.*, 1997b, 1998a,b; So *et al.*, 1995, 1996). In addition to the laser light source, a critical element is a high-throughput microscope system with beam-scanning electronics. High numerical aperture microscope objectives are critical for efficient two-photon excitation and the detection of low-level signals. As compared to a confocal system, chromatic aberration is not a crucial factor as the different color emission light does not have to be descanned. Maximizing infrared transmission decreases the scattering of the high-power excitation light, which interferes in detecting the low-level fluorescence signal. Most implementations of multiphoton scanning microscopes incorporate infrared femtosecond light sources with an existing scanning confocal microscope (Fig. 5). This

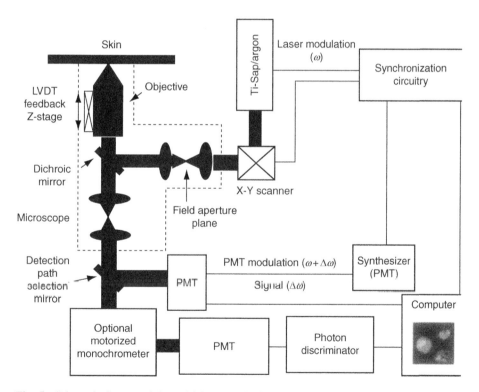

Fig. 5 Schematic diagram of the multiphoton excitation scanning microscope. Two detection beam paths are shown: one for measuring fluorescence lifetimes and the other for measuring the fluorescence emission intensity/spectrum. These two beam paths are selected by rotating a single mirror. PMT, photomultiplier tube; LVDT, linear variable differential transformer; titanium–sapphire/argon, argon ion laser pumped titanium–sapphire laser. From Masters *et al.* (1997b).

implementation is fairly straightforward where only mirrors in the scanning head have to be modified to efficiently reflect infrared laser light. It is also critical to modify the emission beam path such that high loss descanning optics is not used. Because the mechanooptics involved in multiphoton microscopes is much simpler than the confocal system, modifications needed to convert a high-throughput fluorescence microscope for multiphoton scanning are actually quite simple. It mainly involves modification of the excitation light path to incorporate a scan lens and to provide an electronic interface to synchronize the scanner mirror system and the data acquisition electronics.

Properly designed detection electronics is essential for a high-performance multiphoton scanning microscope. Typical scanning systems use PMTs with analog detection circuitry. PMTs have good quantum efficiency (10–25%) in the blue-green region, but poor (less than 1%) quantum efficiency in the red. Although analog circuity does not provide very sensitive detection at very low signal levels, it can handle fairly high-intensity signals without saturation. It is also critical that analog conversion circuitry have sufficient dynamic range in cellular imaging applications where signal strength can vary greatly. Typically, a dynamic range of at least 12 bits is desirable. Another common detection method uses a single-photon-counting scheme that is very sensitive at extremely low light situations and has excellent dynamic range, but suffers from the difficulty of easy saturation at high intensity. Other detectors that are promising for multiphoton microscopes are high-sensitivity single-photon-counting avalanche photodiode detectors, which have excellent quantum efficiency (over 70%) at the red spectra range and very good efficiency (30–50%) in the green range.

The laser used in the microscope constructed at the Laboratory for Fluorescence Dynamics, University of Illinois at Urbana-Champaign, and at the Department of Mechanical Engineering, MIT, is a mode-locked titanium–sapphire laser (Mira 900, Coherent Inc., Palo Alto, CA). Two-photon excitation microscopy at 730 and 960 nm was used. A Glan–Thomson polarizer and a laser pulse picker are used either alone or in combination to control the excitation laser pulse train profile. The beam-expanded laser light is directed into the microscope via a galvanometer-driven x–y scanner (Cambridge Technology, Watertown, MA). Images are generated by raster scanning the x–y mirrors. The excitation light enters the Zeiss Axioskopt microscope (Zeiss Inc., Thornwood, NY) via a modified epiluminescence light path. The scan lens is positioned such that the x–y scanner is at its eye point, whereas the field aperture plane is at its focal point. Because the objectives are infinity corrected, a tube lens is positioned to recollimate the excitation light. The scan lens and the tube lens function together as a beam expander, which overfills the back aperture of the objective lens. The excitation light is reflected by the dichroic mirror to the objective. The dichroic mirrors are custom-made, short-pass filters (Chroma Technology Inc., Brattleboro, VT) that maximize reflection in the infrared and transmission in the blue-green region of the spectrum. The water-immersion objective used in most of these studies is a Zeiss c-Apochromat 40× with numerical aperture of 1.2. The objective axial position is driven by a stepper motor interfaced to a computer. The microscope field of view is about 50–100 mm on a side. The typical image acquisition time is about 2 s.

The fluorescence emission is collected by the same objective and is transmitted through the dichroic mirror along the emission path. An additional barrier filter is needed to further attenuate the scattered excitation light because of the high excitation intensity used. Because two-photon excitation has the advantage that the excitation and emission wavelengths are well separated (by 300–400 nm), suitable short-pass filters such as 2 mm of the BG39 Schott glass filter (CVI Laser, Livermore, CA) eliminate most of the residual scatter with a minimum attenuation of the fluorescence. A descan lens is inserted between the tube lens and the PMT. The descan lens recollimates the excitation. It also ensures that the emission light strikes the PMT at the same position independent of scanner motion. A single-photon-counting signal detection system is implemented. The fluorescence signal at each pixel is detected by a R5600-P PMT (Hamamatsu, Bridgewater, NJ), which is a compact single-photon-counting module with high-quantum efficiency.

The instrumentation for wavelength-resolved spectroscopy measurements in multiphoton microscopy is essentially similar to the technology used in one-photon systems with two exceptions. First, because the two-photon excitation wavelength is red shifted to twice the normal spectra region, there is always good separation between two-photon excitation and emission spectral regions. This wide separation allows the emission intensity to be isolated easily from the high-intensity excitation. More importantly, the overlap between excitation and emission spectra in the one-photon case prevents the shorter wavelength region of the emission spectra to be studied, but this is not a problem in the two-photon case. Second, two-photon excitation at the properly chosen wavelength region can simultaneously excite fluorophores in the near-ultraviolet, blue, green, and red ranges. In the multiple labeling experiments, this capability allows all the fluorescent probes to be excited simultaneously and their individual distributions mapped using wavelength-resolved spectroscopy. For a rough wavelength separation, as in applications where a number of chromophores need to be resolved, different wavelength channels can be constructed using a number of dichroic filters. When finer wavelength resolution is needed to study spectral features, single-point measurement using monochromameters or spectrographs can be performed.

The lifetime-resolved method is another powerful spectroscopy technique that has been integrated with multiphoton microscopy. Lifetime imaging can resolve multiple structures in cells and tissues similar to wavelength-resolved imaging. It can also be used to quantitatively measure cellular metabolite concentration such as calcium using nonratiometric probes such as Calcium Green. The instrumentation involved in lifetime imaging inside a multiphoton microscope is very similar to that of a confocal system. In the confocal system, lifetime measurements have been performed using both time-domain and frequency-domain techniques. Because multiphoton microscopy is still a relatively new approach, only the incorporation of frequency domain techniques has been attempted. Lifetime measurements were performed using frequency-domain heterodyning techniques, which have been described previously (So *et al.*, 1995).

VII. Simultaneous Multiphoton Excitation Microscopy and Reflected Light Confocal Microscopy

There are many experimental studies when it would be of great advantage to have the capability of simultaneous multiphoton excitation microscopic and reflected light confocal imaging. Because the multiphoton excitation microscope is limited to fluorescence imaging, it is very important to have a simultaneous capability to study the morphology of cells and tissues based on reflected light confocal microscopy. In the original two-photon excitation microscope the infrared photons that are reflected by the sample are rejected at the dichroic mirror. If an additional photon detection path is incorporated into the standard multiphoton excitation microscope, these infrared photons scattered from the sample could be collected and detected to form a confocal image (Kim *et al.*, 1998). The reflected infrared photons are collected by the microscope objective and then collimated by the excitation tube lens and scan lens combination. This collimated beam is then descanned by being directed backward through the x–y scanning mirrors. This descanned beam is then focused by a lens through a confocal pinhole aperture onto a photodetector. A polarizer can be used to reduce the specular reflection of the infrared light beam back into the confocal detector path (Fig. 6). Choices of the photodetector include a fast silicon photodiode (PDA50, Thorn Labs, Newton, NJ) or a high-sensitivity avalanche photodiode (Advanced Photonics, Camarillo, CA). Data acquisition of the single-photon-counting circuit for the multiphoton excitation fluorescence signal and the reflected light signal are synchronized by a custom interface circuit. The only effect on the multiphoton excitation microscope is a factor of two loss in the excitation power at the input beam splitter. This loss of power can be compensated easily by increasing the initial power of the laser source.

VIII. Video Rate Multiphoton Excitation Microscopy

There are a number of studies on living cells, tissues, and organisms in which a video rate of data acquisition is required (Bewersdorf *et al.*, 1998; Brakenhoff *et al.*, 1996; Guild and Webb, 1995; Rajadhyaksha *et al.*, 1995). For example, in order to acquire a stack of images of thick human skin *in vivo*, it is necessary to minimize the motion of the skin. This problem is substantially easier if the time of data acquisition can be reduced to the range of tens of milliseconds per frame. One technique to achieve video rate data acquisition is to use a rapidly rotating polygon mirror for scanning on one axis and to use a galvanometer-driven mirror for scanning on the orthogonal axis (Fig. 7). Other promising approaches include line scanning and multifocal scanning microscopes.

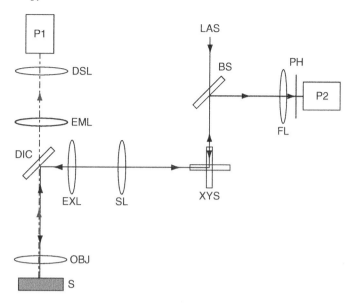

Fig. 6 A schematic of a new multiphoton excitation microscope design showing the new reflected light confocal beam path. S, sample; OBJ, objective; DIC, dichroic mirror; EXL, excitation tube lens; SL, scan lens; XYS, x–y scanner; BS, beam splitter; FL, focusing lens; PH, pinhole aperture; LAS, laser light source; EML, emission tube lens; DSL, descan lens; P1, fluorescence PMT; P2, confocal PMT.

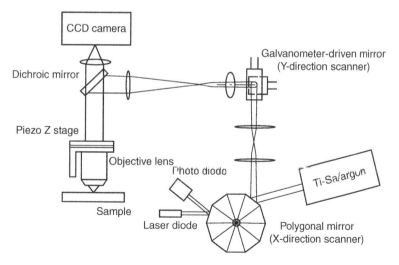

Fig. 7 A schematic of a new multiphoton excitation microscope design showing the system components critical to video rate capability. The 50 facet polygonal mirror can run at a maximum speed of 3000 rpm. A pair of relay lens project the laser spot from the polygonal mirror facets onto the galvanometer-driven x–y scanner. Only the y-axis of the galvanometer scanner is used for raster scanning because the scanning in the x direction is provided by the polygonal mirror. The x-axis galvanometer-driven mirror is used for easy positioning of the scan region.

IX. Multiphoton Excitation Microscopy of *In Vivo* Human Skin: Emission Spectroscopy and Fluorescent Lifetime Measurements

Multiphoton excitation microscopy at 730 and 960 nm was used to image *in vivo* human skin autofluorescence (Fig. 8) (Masters *et al.*, 1997b). The lower surface of the right forearm (of one of the authors) was placed on the microscope stage where an aluminum plate with a 1-cm hole is mounted. The hole is covered by a standard cover glass. The skin was in contact with the cover glass to maintain a mechanically stable surface. The upper portion of the arm rested on a stable platform prevented motion of the arm during the measurements. The measurement time was always less than 10 min. The estimated power incident on the skin was 10–15 mW. The photon flux incident on a diffraction-limited spot on the skin is on the order of 10 MW/cm^2. We observed individual cells within the thickness of the skin at depths from 25 to 75 μm below the skin surface. No cells were observed in the stratum corneum. These results are consistent with studies using reflected light confocal microscopy.

In order to show the three-dimensional distribution of the autofluorescence, we acquired optical sections with the two-photon excitation microscope and formed a three-dimensional visualization across the thickness of the *in vivo* human skin (Figs. 9, 10, and 11) (Masters, 1996a; Masters *et al.*, 1996, 1997a,b).

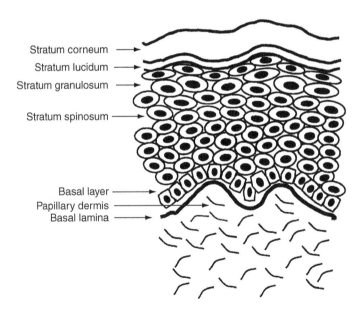

Fig. 8 A vertical section of skin showing the following cell layers from the skin surface to the dermis: stratum corneum, stratum lucidum, stratum granulosum, stratum spinosum, basal cell layer, and the papillary dermis. Individual cells forming the stratum corneum and the thin stratum lucidum are not shown. From Masters *et al.* (1997b).

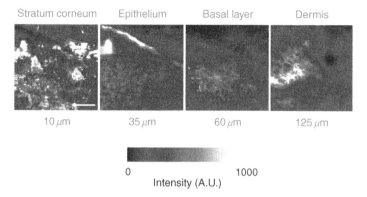

Fig. 9 Multiphoton excitation microscopy of human skin with an excitation wavelength of 780 nm. The stratum corneum is clearly seen on the cell surface. Keratinocytes 15–20 µm in diameter were imaged at an approximate depth of 40–50 µm in the epidermis. Basal cells of about 10 µm were observed at a depth of 60 µm below the skin surface. Punctated fluorescence was observed in the cytoplasm of the larger cells. Similar findings were reported previously. These fluorescent organelles are likely to be mitochondria with a high concentration of NAD(P)H. In the dermis layer between 80 and 150 µm, collagen/elastin fibers can be clearly discerned. Bar: 32 µm.

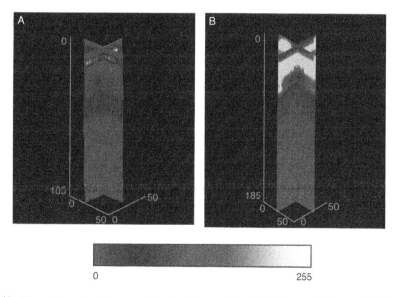

Fig. 10 Three-dimensional images of *in vivo* human skin: (A) 730 nm excitation and (B) 960 nm excitation. *x* and *y* orthogonal slices are shown. The axis dimensions are in micrometers. (A) The top bright layer corresponds to the stratum corneum. The second bright band at a depth of 80–100 µm is the basal cell layer at the top of the papillary dermis. (B) The top bright band extends throughout the stratum corneum. From Masters *et al.* (1997b).

It is important to characterize the source of the fluorescence that is imaged with multiphoton excitation microscopy. Two types of measurements are useful in the characterization of the fluorophore: emission spectroscopy and lifetime measurements. We measured these characteristics at selected points on the skin. Fluorescent spectra were obtained close to the stratum corneum (0–50 μm) and deep inside the dermis (100–150 μm). Measurements were made for both 730- and 960-nm excitation wavelengths corresponding to one-photon excitation wavelengths of about 365 and 480 nm, respectively. The fluorescent lifetimes were measured at

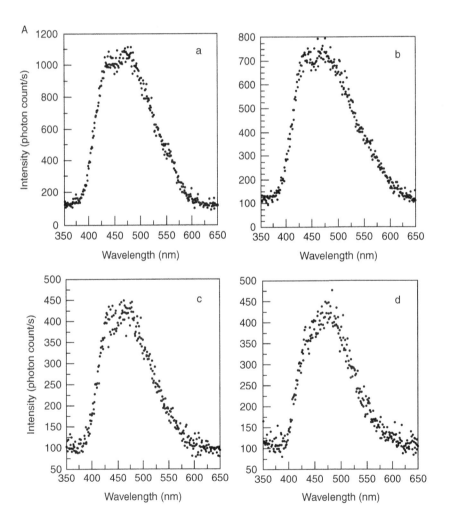

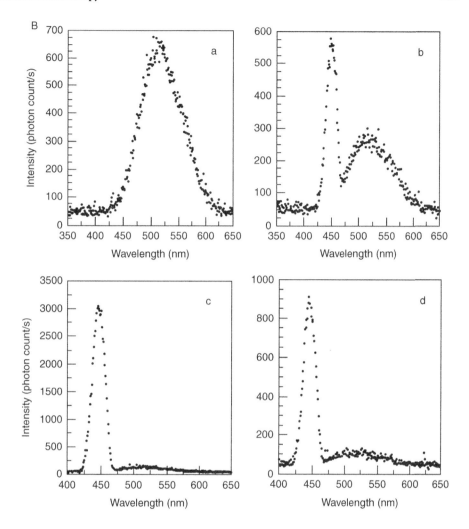

Fig. 11 Point emission spectrum of *in vivo* human skin at 730 nm (A) and 960 nm (B) excitation. Intensity is given in photon counts per second. (a, b) Two distinct points on the surface of *in vivo* human skin between 0 and 50 μm deep. (c, d) Two distinct points at a depth between 100 and 150 μm. From Masters *et al.* (1997b).

selected points on the skin to complement the fluorescent spectral data obtained. The lifetime results support NAD(P)H as the primary source of the autofluorescence at 730 nm excitation. The chromophore composition responsible for the 960-nm excitation is more complicated.

X. Use of Laser Pulse Picker to Mitigate Two-Photon Excitation Damage to Living Specimens

For a single-photon process, fluorescence excitation typically uses continuous wave lasers and requires an average power of about 100 µW. For a two-photon excitation microscope, lasers with 100-MHz and 100-fs pulses are often used and require an average power of about 10 mW. This corresponds to a peak power on the order of 1 kW. For typical cells and tissues, the two-photon photodamage pathways still need to be better determined (König *et al.*, 1996, 1997). Because typical two-photon excitation microscopy requires milliwatt average power and laser pulse trains consisting of nanojoule pulses, severe thermal damage can occur resulting from one-photon excitation. This situation is particularly severe for highly pigmented cells such as melanophores. Furthermore, the high peak power required for two-photon excitation may produce cell and tissue damage through dielectric breakdown mechanisms. For multiphoton excitation microscopy, oxidative photodamage is also generated. The processes that result in photodamage to the specimen are complex. As a comparison between two-photon and confocal methods, less photodamage occurs in the two-photon case as the excitation volume is less than a femtoliter, whereas photodamage in the confocal case occurs throughout the optical path in the specimen.

During a two-photon excitation process, it is important to realize that three-photon excitation can also be generated. Although the emission filters may be selected for fluorescence detection from the two-photon process, there may still be photodamage from the unintentional three-photon excitation process. When intentionally using three-photon excitation for tissue imaging, it is important to realize that three-photon excitation microscopy often requires an order of magnitude higher incident laser power than two-photon excitation in order to achieve similar rates of excitation. It is important to study the associated three-photon cellular damage mechanisms and to identify the damage threshold. Nevertheless, three-photon excitation microscopy provides a useful technique to excite those fluorophores with single-photon absorption bands in ultraviolet and short-ultraviolet wavelengths.

For the successful study of highly pigmented cells such as melanophores, it is critical to reduce thermal damage by reducing power deposition from individual laser pulses as well as reducing heat accumulation from the pulse train. The power management of laser pulses is typically accomplished by using an attenuator to decrease laser pulse energy. With energy sufficiently attenuated such that thermal damage from individual pulses is insignificant, heat accumulation from the pulse train remains a major problem. It is possible to reduce the laser pulse energy further to minimize the effect of heat accumulation. However, given a constant pulse width, reducing the pulse energy decreases fluorescence excitation efficiency quadratically. It is preferable to reduce the laser pulse repetition rate by using a laser pulse picker that maximizes the ratio between the peak power and the average

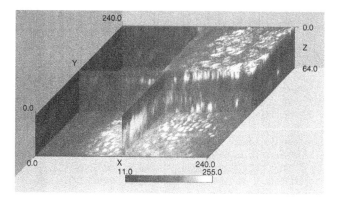

Fig. 12 Three-dimensional image of human skin *in vivo*. A tandem scanning reflected light confocal microscope was used to acquire the optical sections. The human skin was visualized with Dicer software (Spyglass, Inc., Champaign, IL), which permits the display of three-dimensional volume data as a set of orthogonal slices. From Masters *et al.* (1996).

power. This strikes the best compromise between minimizing thermal effects and maintaining fluorescence excitation efficiency. We have accomplished this goal using a laser pulse picker (Masters *et al.*, 1999). If a Glan–Thompson polarizer is used to reduce the light intensity of the laser pulse train, then the laser average power, as well as the laser peak power, is reduced. With an ideal pulse picker, the average light intensity of the laser pulse train can be decreased, not affecting the peak intensity. This is because the pulse picker operates by removing a fraction of the laser pulses, but does not affect the peak power of each laser pulse. Real pulse pickers, which are based on acoustooptical deflectors, result in about 50% instantaneous power loss. We have demonstrated that when the laser power is attenuated with a polarizer, the fluorescence intensity of the fluorophores decreases quadratically with the excitation power. When laser power is attenuated with a laser pulse picker, the fluorescence intensity decreases only linearly with the excitation power. It is concluded that decreasing the pulse repetition rate with a laser pulse picker is the preferred method of choice for mitigating tissue thermal damage when studying living cells and tissues with a multiphoton excitation microscope. In the future, three-dimensional optical biopsy will provide a new diagnostic tool (Fig. 12).

References

Art, J. (1995). Photon detectors for confocal microscopy. *In* "Handbook of Biological Confocal Microscopy," (J. B. Pawley, ed.), p. 183. Plenum Press, New York, NY.

Bewersdorf, J., Pick, R., and Hell, S. W. (1998). *Opt. Lett.* **23**, 655.

Brakenhoff, J., Squier, J., Norris, T., Bliton, A. C., Wade, W. H., and Athey, B. (1996). *J. Microsc.* **181**(Pt 3), 253.

Chance, B., Schoener, B., Oshino, R., Itshak, F., and Nakase, Y. (1979). *J. Biol. Chem.* **254**, 4764.

Denk, W. J., Strickler, J. H., and Webb, W. W. (1990). *Science* **248**, 73.

Denk, W., Strickler, J. P., and Webb, W. W. (1991). U.S. Patent 5,034,613 .

Dong, C. Y., So, P. T. C., French, T., and Gratton, E. (1995). *Biophys. J.* **69**, 2234.

Göppert-Mayer, M. (1931). *Ann. Phys. Lpz.* **9**, 273.

Gratton, E., and vandeVen, M. J. (1995). Laser sources for confocal microscopy. *In* "Handbook of Biological Confocal Microscopy," (J. B. Pawley, ed.), p. 69. Plenum Press, New York, NY.

Guild, B., and Webb, W. W. (1995). *Biophys. J.* **68**, 290a.

Kim, K. H., So, P. T. C., Kochevar, I. E., Masters, B. R., and Gratton, E. (1998). *In* "Functional Imaging and Optical Manipulation of Living Cells," (D. L. Farkas, and B. J. Tromberg, eds.). SPIE, Bellingham, WA.

König, K., So, P. T. C., Mantulin, W. W., Tromberg, B. J., and Gratton, E. (1996). *J. Microsc.* **183**(3), 197.

König, K., So, P. T. C., Mantulin, W. W., and Gratton, E. (1997). *Opt. Lett.* **22**, 135.

Masters, B. R. (1988). *In* "The Cornea: Transactions of the World Congress of the Cornea III," (H. D. Cavanagh, ed.), p. 281. Raven Press, New York, NY.

Masters, B. R. (1990). *In* "Noninvasive Diagnostic Techniques in Ophthalmology," (B. R. Masters, ed.), p. 223. Springer, New York, NY.

Masters, B. R. (1993a). *In* "Medical Optical Tomography: Functional Imaging and Monitoring," (G. Müller, B. Chance, R. Alfano, S. Arridge, J. Beuthan, E. Gratton, M. Kaschke, B. R. Masters, S. Svanberg, and P. van der Zee, eds.), p. 555. The International Society for Optical Engineering, Bellingham, WA.

Masters, B. R. (1993b). *Scan. Microsc.* **7**(2), 645.

Masters, B. R. (1996a). *Bioimages* **4**(1), 13.

Masters, B. R. (1996b). *In* "Selected Papers on Confocal Microscopy," (B. R. Masters, ed.). SPIE, The International Society for Optical Engineering, Bellingham, WA.

Masters, B. R., and Chance, B. (1984). Noninvasive corneal redox fluorometry. *In* "Current Topics in Eye Research," (J. A. Zadunaisky, and H. Davson, eds.), p. 140. Academic Press, London.

Masters, B. R., and Chance, B. (1993). *In* "Fluorescent and Luminescent Probes for Biological Activity," (W. T. Mason, ed.), p. 44. Academic Press, London.

Masters, B. R., Falk, S., and Chance, B. (1981). *Curr. Eye Res.* **1**, 623.

Masters, B. R., Ghosh, A. K., Wilson, J., and Matschinsky, F. M. (1989). *Invest. Ophthalmol. Vis. Sci.* **30**, 861.

Masters, B. R., Kriete, A., and Kukulies, J. (1993). *Appl. Opt.* **32**(4), 592.

Masters, B. R., Gonnord, G., and Corcuff, P. (1996). *J. Microsc.* **185**(3), 329.

Masters, B. R., Aziz, D. J., Gmitro, A. F., Kerr, J. H., O'Grady, T. C., and Goldman, L. (1997a). *J. Biomed. Opt.* **2**(4), 437.

Masters, B. R., So, P. T. C., and Gratton, E. (1997b). *Biophys. J.* **72**, 2405.

Masters, B. R., So, P. T. C., and Gratton, E. (1997c). *Cell Vision* **4**(2), 130.

Masters, B. R., So, P. T. C., and Gratton, E. (1998a). *Ann. N. Y. Acad. Sci.* **838**, 58.

Masters, B. R., So, P. T. C., and Gratton, E. (1998b). *Lasers Med. Sci.* **13**(3), 196.

Masters, B. R., Dong, C. Y., So, P. T. C., Buehler, C., Mantulin, W. M., and Gratton, E. (1999). *J. Micros.* in preparation.

Piston, D. W., Masters, B. R., and Webb, W. W. (1995). *J. Microsc.* **178**, 20.

Rajadhyaksha, M., Grossman, M., Esterowitz, D., Webb, R. H., and Anderson, R. (1995). *J. Invest. Dermatol.* **104**, 946.

So, P. T. C., French, T., Yu, W. M., Berland, K. M., Dong, C. Y., and Gratton, E. (1995). *Bioimaging* **3**, 49.

So, P. T. C., French, T., Yu, W. M., Berland, K. M., Dong, C. Y., and Gratton, E. (1996). *In* "Fluorescence Imaging and Microscopy," (X. F. Wang, and B. Herman, eds.), p. 351. Wiley, New York, NY.

Valdmansis, J. A., and Fork, R. L. (1986). *IEEE J. Quantum Electron.* **QE-22**, 112.

Wokosin, D. L., Centonze, V. E., White, J., Armstrong, D., Robertson, G., and Ferguson, A. I. (1996). *IEEE J. Sel. Top. Quantum Electron* **2**(4), 1051.

Wolleschensky, R., Feurer, T., and Sauerbrey, R. (1997). *Scanning* **19**(3), 150.

Xu, C., and Webb, W. W. (1997). *In* "Nonlinear and Two-Photon-Induced Fluorescence," (J. Lakowicz, ed.), p. 471. Plenum Press, New York, NY.

Xu, C., Williams, R. M., Zipfel, W., and Webb, W. W. (1996). *Bioimaging* **4**, 198.

CHAPTER 16

Video–Rate, Scanning Slit Confocal Microscopy of Living Human Cornea *In Vivo*: Three-Dimensional Confocal Microscopy of the Eye

Barry R. Masters and Matthias Böhnke

Arlington, Virginia, USA

I. Biomicroscopy of the Living Eye from Slit Lamp to Confocal Microscope

The optimal application of confocal microscopy to ocular tissue requires specially designed and optimized instrumentation and an examination technique that takes into account the unique structural and optical properties of the human eye (McCally and Farrell, 1990). This article describes the unique capabilities of a video-rate, scanning slit confocal microscope and its application in the living human cornea. A confocal microscope provides two enhancements compared to a standard light

DOI: 10.1016/B978-0-12-384658-7.00016-3

microscope: enhanced lateral resolution and enhanced axial resolution. The latter property is the basis of its capability to optically section a thick, highly scattering specimen.

The cornea is an avascular, transparent, living optical element in the front portion of the eye. Investigation of the physical basis for the transparency of the cornea and the structural alterations that affect corneal transparency, such as corneal wounds and corneal refractive surgery, are topics of active study. For the designer of a video-rate, clinical confocal microscope the optical problem is how to image a moving, transparent cornea with very little intrinsic contrast.

There is a direct and interesting lineage from the confocal microscope developed by Goldmann, to the development of the specular microscope by Maurice, Koester, and others, to the various types of clinical confocal microscopes (Koester and Roberts, 1990). These instruments represent partial solutions to the problem of how to image thin optical sections from a 500-μm-thick transparent, moving object: the human cornea *in vivo*.

II. Optical Principles of Confocal Microscope

It is useful to illustrate optical principles of the confocal microscope with a simple diagram. A confocal microscope is a type of microscope in which a thick object, such as the cornea, is illuminated with a focused spot of light. The same microscope objective used to illuminate a point, P1, in the object is used to collect the scattered and reflected light from the same point P1. This is illustrated in Fig. 1. For simplicity, we show two separate microscope objectives placed on opposite sides of a thick translucent object. The microscope objective on the left side is used to illuminate a point (P1) in the object, whereas the microscope objective on the

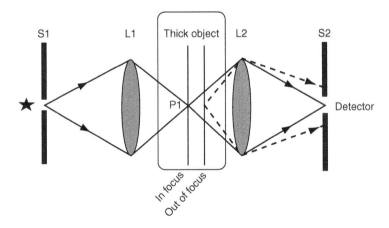

Fig. 1 Diagram illustrating the principle of a confocal imaging system. From Böhnke and Masters (1999b).

right side is used to detect the illuminated point within the object. A point source of light at aperture S1 is focused by lens L1 onto a small spot in the focal plane within the object. The second lens L1 is positioned so that its focus is also in the same focal plane and coincident (confocal) with the spot of illumination. Lens L2 forms an image of the illuminated spot at aperture S2. All of the light that lens L2 collects from the illuminated spot enters the aperture at S2 and reaches the detector.

How does a confocal microscope discriminate against light that is not in the focal plane? For real thick scattering objects, the point of light imaged by lens L1 onto the thick object will focus the light as a spot in the focal plane; however, there will be light of lower intensity in the double cone on both sides of the focal plane. In Fig. 1 the light paths that are from an out-of-focus plane are shown as dotted lines. In this case, some of the scatted light from an out-of-focus plane is collected by lens L2. In this case the collected light is spread out at aperture S2. Only a small amount of the light that is spread out enters aperture S2. Therefore, the detector will detect only a small amount of the light from the out-of-focus planes. This is the physical basis for strong discrimination against out-of-focus light in a confocal microscope. In Fig. 1 the diameter of the apertures is enhanced greatly for visual clarity. Clinical confocal microscopes have a single microscope objective. The single objective is typically used for point illumination of the object and to simultaneously image the point of illumination on the aperture of the light detector (Masters, 1996). The optical principle of the confocal microscope has been simply illustrated for a single point of the object being illuminated and the same point being imaged simultaneously on the detector. Both aperture S1 and aperture S2 are cofocused (confocal) on the same point in the focal plane. This is the origin of the word confocal.

III. Objective Scanning Confocal Microscope

Figure 2 illustrates the mechanical and optical components of a confocal microscope with computer-controlled objective scanning on the z-axis. The device is a confocal microscope because it contains two conjugate slits. The confocal microscope has been used in the vertical mode for work on tissue culture and for studies of *in vitro* eyes. When the confocal microscope is shifted into the horizontal mode, it is used for *in vivo* studies on animals or human subjects. This confocal microscope can be used to measure the following signals: z-axis profiles of fluorescence from NAD(P)H, fluorescence from oxidized flavoproteins, fluorescence from extrinsic probes, that is, mitochondrial or nuclear stains, and simultaneous measurements of back-scattered light. This confocal microscope is used to investigate the metabolism of the cells in the cornea *in vivo*. The unique feature of the confocal microscope is a z-scanning objective. A piezoelectric driver, under computer control, scans the microscope objective along the z-axis; this motion of the objective scans the focal plane across the thickness of the cornea. The z-axis spatial resolution of this instrument system is shown in Fig. 3.

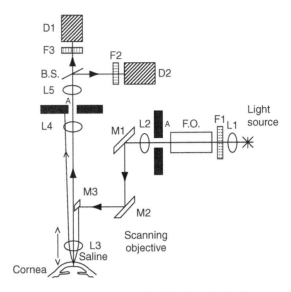

Fig. 2 Schematic diagram of the scanning one-dimensional confocal microscope showing a light ray path. The light source is either a laser or a mercury arc lamp connected to the microscope by a fiber optic. F1 and F2 are narrow band interference filters to isolate the excitation wavelengths. F3 is a narrow band interference filter to isolate the emission light. There are two conjugate slits. M1, M2, and M3 are front surface mirrors, and B.S. is a quartz beam splitter. L3 is the scanning objective 50×, NA 1.00. The piezoelectric driver scans the microscope objective along the optic axis of the eye. This confocal microscope is suitable for use with tissue culture or in the horizontal mode for use with living animals or human subjects. The depth resolution is 6 μm with a 100× microscope objective and 18 μm with a 50× microscope objective. From Masters (1988).

Several design features of this instrument were incorporated into more recent designs of clinical confocal microscopes (Böhnke and Masters, 1999a). The slit confocal microscope has *two adjustable slits* that can be used to optimize the signal-to-noise ratio from the cells in the cornea and the image contrast. At the early stages of its development, it was decided to eliminate the applanating microscope objectives, which were designed for specular microscopes, and to use nonapplanating, high numerical aperture (NA) water-immersion microscope objectives (Leitz 50× and 100× water-immersion microscope objectives). These microscope objectives are used routinely with the more recently developed video-rate, scanning slit confocal microscopes.

IV. Confocal Microscopy of Cornea

A. Tandem Scanning Nipkow Disk Confocal Microscopes

A video-rate, tandem scanning confocal microscope was developed by Petran and coworkers to optically section living neural tissue. Petran acknowledged the contribution of the Nipkow disk, which was invented over 100 years ago by Nipkow

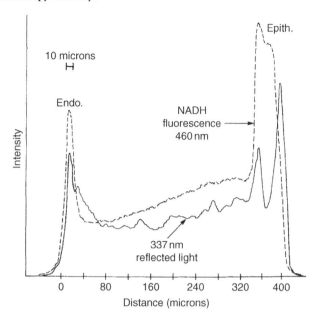

Fig. 3 One-dimensional *in vivo* confocal microscopy of the living eye. An optical section through a rabbit cornea illustrating the range resolution for back-scattered light (solid line) and NAD(P)H fluorescence emission (broken line). The intensity of the back-scattered light is 10 times that of the fluorescence. The tear film is on the right-hand side of the scan and the aqueous humor is on the left-hand side. The ordinate is relative intensity and the abscissa is distance into the cornea. From Masters (1988).

(Masters, 1996). The principle of the tandem scanning confocal microscope is as follows. Sets of conjugate pinholes (20–60 µm in diameter in various instruments) are arranged in several sets of Archimedes spirals. Each pinhole on one side of the disk has an equivalent and conjugate pinhole on the other side of the disk. The light transmission of the different disks varies from 0.25% to 1%. The illumination light passes through a set of pinholes (about 100 at a time) and the reflected light passes through a conjugate set (about 100 at a time) on the other side of the disk. As the disk rotates it causes the illumination to focus on the focal plane of the cornea, and the detected light passes through sets of conjugate pinholes. Both the illumination and the reflected light are scanned in parallel over the sample to generate the two-dimensional image of the focal plane. Because the ratio of the area of the holes to the area of the disk is usually only about 0.25%, only a small fraction of the illumination reaches the sample, and a similar small fraction of the light reflected from the sample passes the disk and reaches the detector (Lemp *et al.*, 1986). The microscope has a very low light throughput. Therefore, the illumination must be very bright (a xenon or mercury arc lamp is usually required). The potential advantages of the tandem scanning confocal microscope include (1) video-rate operation, (2) true color confocal imaging, and (3) the image can be viewed with the eye.

Because of their low light throughput, confocal microscopes based on a tandem scanning Nipkow disk are best suited for reflected light confocal imaging of highly reflecting specimens (i.e., hard tissue such as bone) (Masters, 1990, 1995a; Masters and Kino, 1990; Masters and Paddock, 1990a). Therefore, although the resolution of a pinhole-based confocal microscope is higher than a scanning slit-based confocal microscope (using the same microscope objectives and light source), it is a poor choice for weakly reflecting specimens such as the human cornea *in vivo*. The low intensity of light, reflected from the cornea and passed through the tandem scanning disk with a 0.25% transmission, reaches the intensified video camera results in single video images of marginal image quality and low contrast. This result is predominately due to noise and low-signal level. Thus, there is a need for digital processing (frame averaging) after image acquisition in order to obtain acceptable images.

B. Scanning Slit Confocal Microscopes

An alternative to point scanning is to use a slit of illumination that is scanned over the back focal plane of the microscope objective (Masters, 1995a). The advantage of this optical arrangement is that because many points on the axis of the slit are scanned in parallel, the scanning time is markedly decreased. The microscope operates at video rates. Another very important advantage is that scanning slit confocal microscopes have superior light throughput as compared to point scanning Nipkow disk systems. The disadvantage is that the microscope is truly confocal only in the axis perpendicular to the slit height. In comparison to a pinhole-based confocal microscope, a slit-based confocal microscope provides lower lateral and axial resolution. This comparison is for the same wavelength of illumination and reflected light and the same microscope objective in each case. However, for confocal imaging of weakly reflecting living biological specimens such as the cornea, the trade-off of lower resolution with higher light throughput is acceptable and preferable. Several arrangements have been developed to provide the scanning of the slit of illumination over the specimen and the synchronous descanning of the reflected light from the object.

There are several advantages to scanning slit confocal microscopes. The slit height can be adjusted, which allows the user to vary the thickness of the optical section. What is more important, the user can vary the slit height and therefore control the amount of light that reaches the sample and the amount of reflected light that reaches the detector. This is important for samples that are very transparent and therefore can be imaged with the slit height very small; more opaque samples require that the slit height is increased. The light throughput is much greater for a slit scanning confocal microscope than for a confocal microscope based on the Nipkow disk containing sets of conjugate pinholes.

A confocal microscope has enhanced transverse (x and y coordinates in the plane of the specimen) and axial resolution (z coordinate, which is orthogonal to the plane of the specimen) in comparison with a nonconfocal microscope using the same wavelength of light and the same microscope objective. It is the enhanced axial resolution in a confocal microscope that permits the improved optical sectioning of thick specimens and their three-dimensional reconstruction. The transverse resolution of a confocal microscope is proportional to the NA of the microscope objective. However, the axial resolution is more sensitive to the NA of the microscope objective. Therefore, to obtain the maximum axial resolution, and hence the best degree of optical sectioning, it is necessary to use a microscope objective with a large NA. For work with living cells and tissues, we recommend long working distance, water-immersion microscope objectives with a high NA such as the Leitz $50\times$, NA 1.0.

The advantage of a slit scanning confocal microscope over those based on Nipkow disks containing pinholes is shown in the following example. For cases of weakly reflecting specimens, such as living, unstained cells and tissues, the advantage of the much higher light throughput from the slit scanning systems is crucial for observation. The basal epithelial cells of the normal, *in vivo* human cornea cannot be observed with a tandem scanning confocal microscope. However, corneal basal epithelial cells are always observed *in vivo* in normal human subjects when they are examined with a video-rate, scanning slit *in vivo* confocal microscope. The reason for this discrepancy is that although the tandem scanning confocal microscope has higher axial and transverse resolution, the very low light throughput of the disk does not pass enough reflected light from the specimen to form an image on the detector (in a single video frame) that has a sufficient signal to noise and, therefore, contrast to show an image of the cells.

A video-rate, direct view tandem scanning confocal microscope was developed in the mid-1960s by Petran and coworkers based on a spinning Nipkow disk. The disk contains many sets of conjugate pinholes, one pinhole is on one side of the disk and its conjugate pinhole is on the other side of the disk. A number of confocal microscopes based on the tandem scanning disk have been developed for use in imaging the eye.

Lemp *et al.* (1986) produced a series of studies on the rabbit eye and the *in vivo* human cornea based on the tandem scanning confocal microscope. They used a low-NA-applanating microscope objective developed for specular microscopy. The *in vivo* cornea was flattened by the applanating microscope objective. Disadvantages of the system are high noise in the intensified video camera and scan lines on the single images. Postprocessing and frame averaging reduced the noise and removed the scan lines; however, the instruments were no longer video rate. In general, single frames acquired with low-light-intensified video cameras are noisy; frame averaging is usually required to increase the signal-to-noise ratio. In addition, many of the acquired video frames are blurred because of eye motion; postprocessing is usually required to align and average sequential video frames and to preserve the best nonblurred images. We define video rate as the acquisition and display of usable video frames. If data acquisition is video rate (video) but postprocessing, such as frame averaging, is required to enhance the quality of the images, then the system cannot be called video rate.

A clinical confocal microscope based on a Nipkow disk with an intensified video camera as a detector was developed by the Tandem Scanning Corporation, Inc. It used a higher numerical objective than was used in the first system that Lemp used at Georgetown University. Their later version of the instrument contained an internal focusing lens that varied the depth of focus while the applanating microscope objective was held stationary on the surface of the deformed cornea. The design of an internal focusing lens was first proposed by Masters (1990).

There are several disadvantages to using a tandem scanning confocal microscope in imaging the living cornea. The most important problem is with the tandem scanning disk itself. In order to reduce the cross torque between sets of adjacent pinholes, the pinholes are configured with a separation distance. The result is that only about 1% or less of the area of the disk is used for the pinholes. On the illumination side the incident light is very bright as observed by the patient; however, only about 1% or less of the light passes through the tandem scanning disk and reaches the cornea. The reflected light from the cornea is collected by the microscope objective and only 1% or less of this light (the signal containing the image) is passed through the disk toward the photon detector. Therefore, most of the signal from the cornea is lost and never detected. In the real instrument there are significant additional light losses from optical elements within the microscope and at the photocathode of the video camera. The result is that individual video frames need to be postprocessed after image acquisition, averaged, and digitally enhanced in order to reduce the noise in the final image. The video-rate property of the tandem scanning confocal microscope is lost due to the necessity for digital postprocessing of the images.

The advantage of an applanating microscope objective is that it helps stabilize the cornea against motion along the optical axis of the eye. The eye still undergoes rotatory motion, although it is reduced.

Although in 1990 we advocated the use of applanating objectives, based on our experience in using the wide-field specular microscope with its applanating microscope objective, we have since decided against the use of applanating objectives. Our experience is that applanating microscope objectives, which press against and deform the shape of the cornea in the applanating process, induce artificial alternations in the structure of the cornea. This observation has been confirmed by other groups using a modified, Koester wide-field confocal microscope with applanating objectives. They also observed that the instrument caused flattening-induced corneal bands and ridges (Auran *et al.*, 1994). Therefore, in the development of a video-rate, scanning slit clinical confocal microscope, we introduced the use of nonapplanating, high NA, water-immersion microscope objectives. The front surface of these microscope objectives does not contact the corneal surface directly; between the front surface of the objective and the surface of the cornea is a layer of an index matching gel. This feature is an important distinction between the scanning slit clinical confocal microscope described in this article and the alternative confocal microscopes used by other investigators.

C. Video–Rate, Scanning Slit Clinical Confocal Microscope

A new, video-rate, scanning slit confocal microscope was developed by Thaer for the observation of the *in vivo* human cornea. The image of a slit is scanned over the back focal plane of the microscope objective. The slit width can be varied in order to optimize the balance of optical section thickness and image brightness. The instrument is based on the double-sided mirror, which is used for scanning and descanning. This confocal microscope used a halogen lamp for illuminating the slit. The detector is a video camera that acquires images at video rates. This confocal microscope can image basal epithelial cells and the adjacent wing cells in the living human cornea due to its high light throughput. This design was first developed into a video-rate confocal microscope over 20 years ago. Svishchev designed and constructed a video-rate confocal microscope based on an oscillating two-sided mirror (bilateral scanning) and used this microscope to observe living neural tissue in the reflected light mode (Svishchev, 1969, 1971). Figure 4 illustrates the optical design of the video-rate, scanning slit *in vivo* confocal microscope developed by Dr. A. Thaer. The design consists of two adjustable slits placed in conjugate planes of the confocal microscope. Both scanning of the illumination slit over the back focal plane of the microscope objective and descanning of the reflected light from the object are accomplished with an oscillating two-sided mirror. The optical design of the video-rate, scanning slit *in vivo* confocal microscope developed by Dr. A. Thaer incorporates the double-sided mirror slit scanning system first developed by Svishchev (Masters and Thaer, 1994a; Masters *et al.*, 1994; Wiegand *et al.*, 1995). The double-sided mirror slit scanning system of Svishchev has also been incorporated into other "bilateral scanning" confocal microscopes.

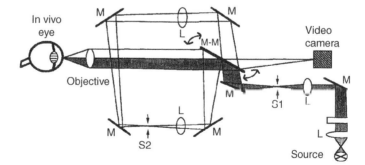

Fig. 4 Optical system of a clinical video-rate, scanning slit confocal microscope. The light source is a halogen lamp. S1, S2, adjustable confocal slits; L, lens. The objective shows the position of the microscope objective. One-half of the objective is used for illumination of the cornea and one-half of the objective is used to collect the light from the cornea. The illumination light path is drawn with black dots, and the collection light path is shown in white. M, fixed front surface mirror; M–M, oscillating double-sided mirror that is used for both scanning and descanning. The video camera is an intensified video camera. From Böhnke and Masters (1999a).

For corneal imaging, a 25×/0.65 NA, a 40×/0.75 NA, or a 50×/1.0 NA water-immersion lens (Leitz, Germany) can be used. The resolution of the different objectives and the thickness of the optical section to some extent are influenced by the light levels used and the geometry and the reflectivity of the structures studied. As a practical guideline, the lateral resolution of the 25× objective is about 1.4 μm, whereas the thickness of the optical sections is about 16 μm. The lateral resolution of the 50× objective is better than 0.8 μm, whereas the thickness of optical section is about 10 μm.

The following design parameters were incorporated into the video-rate, scanning slit confocal microscope (Fig. 5). (1) The use of nonapplanating, high NA water-immersion microscope objectives, Leitz 50× and Leitz 100× microscope objectives. (2) The microscope objective would use a methylcellulose gel to optically couple the tip of the microscope objective to the cornea. There was no applanation or direct

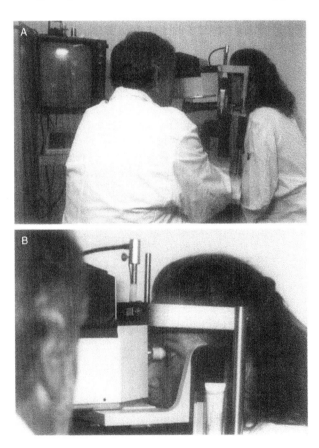

Fig. 5 (A) The video-rate, scanning slit confocal microscope in a clinical setting. (B) Close-up of the instrument showing the nonapplanating water-immersion microscope objective with a drop of index matching gel that contacts the cornea.

physical contact, which deforms the cornea and introduces artificial folds and ridges in the cornea, between the objective and the surface of the cornea. (3) One-half of the NA was used for illumination, and one-half of the NA was used for collection of the reflected and fluorescent light. (4) Optical sectioning in the plane of the cornea was obtained with two sets of conjugate slits. The slit heights are variable and adjustable. (5) An oscillating, two-sided mirror (bilateral scanning) was used for scanning the image of the slit over the back focal plane of the microscope objective and for descanning the reflected and back-scattered light collected by the microscope objective from the focal plane in the specimen. (6) The light source is a 12-V halogen lamp. For fluorescence studies, a mercury arc lamp or a xenon arc lamp can be used. (7) The scanning was synchronized with the readout of an interline CCD camera in order that the full vertical resolution of the intensified CCD camera could be utilized.

The microscope used standard nonapplanating microscope objectives with RMS threads that are interchangeable. Several different microscope objectives can be used, which permits the use of various magnifications and fields of view. Typically a Leitz 50×, 1.0 NA water-immersion objective is used. When a larger field of view is required, a Leitz 25×, 0.6 NA water-immersion microscope objective is used.

An intensified video camera (Proxitronic) with video output to a Sony S-VHS tape recorder is used. The synchronization of the bilateral scanning and the readout of the intensified video camera have been described previously. The PAL video format provides 625 lines. The S-VHS video recorder provides high bandwidth. In parallel with the video recording, there is a video monitor so that the operator can observe the confocal images of the subject's eye in real time. This video-rate, scanning slit confocal microscope does not require any frame averaging for producing the image quality and contrast shown in this article.

What are the advantages of using a scanning slit confocal microscope such as is described and demonstrated in this article? Slit scanning confocal microscopes have a much higher light throughput than confocal microscopes based on a Nipkow disk. This has two consequences. First, the illumination incident on the patient's eye can be much less. This allows for a much longer duration of the use of the confocal microscope on the patient's eye without the severe patient discomfort and high light intensity that is necessary with the use of the confocal microscope based on the Nipkow disk. Second, it is possible to image the low-reflecting layer of wing cells that are immediately adjacent to the basal epithelial cells in the normal human cornea. This layer of wing cells has been imaged, at video rates, as single video frames without the need for any analog or digital image processing using the video-rate, scanning slit confocal microscope. No other video-rate confocal microscope has been able to image these wing cells in normal, *in vivo* human cornea. This new confocal instrument has unique advantages over other confocal systems. The bright, high-contrast confocal images of the wing cells and the basal epithelial cells demonstrate its unique optical characteristics. The low reflectivity of wing and basal cells in normal human cornea presents a low-contrast benchmark test specimen for various types of confocal microscopy. The high rejection of stray light and the narrow depth of field, coupled with the high NA microscope objective

(1.0 NA), results in the ability of the instrument to image these cell layers clearly in live normal human cornea. The clear advantage of slit scanning confocal microscopes for ophthalmic diagnostics and basic eye research is best appreciated when the basal epithelium in the anterior cornea is imaged. The video-rate, scanning slit confocal microscope provides high-contrast, high-resolution images of both wing cell and basal epithelial cells in the normal *in vivo* human eye.

In 1990, it was suggested that an internal lens system would permit focusing at different depths within the cornea; however, in order to keep light losses to a minimum, and therefore maximize the light sensitivity of the z-scanning instrument, it was decided to use a z-scanning microscope objective (Masters, 1990). This capability of a z-scanning confocal has been implemented into the clinical video-rate, scanning slit confocal microscope. In addition to microscopic pictures with a lateral movement of the scanning slit, a confocal measurement of the tissue reflectivity can be performed by an automated z scan through all layers of the cornea. For this procedure, the lateral scan is switched off. Reflectivity, as recorded by a photomultiplier, can be used to measure corneal haze, which may be of special interest in keratorefractive procedures.

V. Optical Low-Coherence Reflectometry to Measure Position within Cornea

Optical low-coherence reflectometry instruments can noninvasively measure distances within the *in vivo* cornea with a precision of less than 1 μm (Böhnke *et al.*, 1998a,b; Masters, 1999). In order to relate the distance that the focal plane of the microscope objective has moved into the thickness of the cornea, it is necessary to know the profile of the refractive index across the thickness of the cornea. This requirement exists for any form of optical pachometry and includes optical low-coherence reflectometry.

The technique is rapid, noncontact, and noninvasive and can easily be incorporated into a variety of ocular imaging instruments. Optical low-coherence reflectometry is an optical imaging technique based on a Michelson interferometer and a low-coherence light source. It can obtain micrometer resolution cross-sectional imaging of biological tissue. The technique is analogous to ultrasound B-mode imaging; the difference is that with optical low-coherence reflectometry, light is used instead of acoustic waves. Low-coherence optical reflectometry is noninvasive in that is operated in a noncontact fashion. It has been used to measure corneal thickness and retinal thickness and for ocular biometry. Optical low-coherence reflectometry has promise as a useful noninvasive technique for biometry of the anterior segment. It has been applied to measure corneal thickness in a clinical setting. This new optical pachometer has been implemented on a slit lamp headrest as well as on an eximer laser system for refractive surgery. Optical low-coherence reflectometry is an ideal noninvasive technique to measure position within the full thickness of the cornea at a precision of 1 μm.

VI. Clinical Examination Technique with Confocal Microscope

The technique of video-rate, scanning slit confocal microscopy of the cornea has been applied at the University Eye Clinic, University of Bern, in about 1000 clinical sessions since 1994. This accumulated clinical experience is factored into the current clinical examination technique, which is presented here in sufficient detail for others to follow (Böhnke and Masters, 1999a). Care must be taken not to mechanically damage the cornea.

The instrument used for our investigations is a video-rate, scanning slit confocal microscope. The microscope is equipped with a halogen lamp for illumination and a slit scan-synchronized, high-sensitivity video camera with adjustable black level suppression. Phototoxicity from the halogen lamp is negligible. Patients have reported this examination to be less disturbing than contact corneal endothelial cell photography with a specular microscope. Complications have not been observed in our patients. The healthy cornea with its low reflectivity requires a high primary light output from the halogen lamp. In cases of corneal opacities such as scar tissue or other deposits, the light source has to be dimmed down to a much lower level.

In clinically normal corneas, the halogen lamp will have to be set to full power to supply enough reflected light from the corneal structures. In pathological corneas with scar tissue or other highly reflective contents, the lamp power has to be reduced by about 50%. With selected filters, which can be inserted into the optical path of the microscope, the spectrum emitted from the halogen lamp can be confined to selected spectra, either improving optical penetration by selecting longer wavelengths or improving image contrast by blocking out longer wavelengths. For studies using fluorescent dyes, a 300-W mercury lamp with an appropriate fluorescence excitation and emission filter set is supplied. Depending on the experience of the investigator, the instrument should be readjusted periodically to give an optimum and homogeneous illumination of the optical section, a perfect alignment of the scanning slits, and the lowest possible levels of stray light, which may degrade the image contrast. The projected confocal slit width, which can be adjusted in some instruments, was selected to be 10 μm, which is a compromise between the best resolution and the best illumination and contrast.

In order to check the homogeneity of the field illumination and also to practice coordination of the microscope movements, a piece of black paper is placed vertically in front of the microscope. The microscopy is equipped with a 10× objective and advanced until the surface structure of the object is visible. The halogen lamp has to be dimmed for this procedure. With this overall weakly reflective object, the technical status of the microscope optics can be checked easily and, if required, also adjusted. By mounting the black paper not strictly *en face* before the objective, the investigator can practice to judge and establish perpendicularly by using extra drives tilting the frontal plane of the microscope.

To investigate the patient's cornea, the instrument objective is brought from its most backward position into optical contact with the corneal tissue by a high viscosity acrylic ocular gel. The S-VHS tape recorder is started when an optimum centration has been achieved, as judged from a well-centrated light reflex on passing the epithelial or endothelial layer. The video-rate *en face* sections from all layers of the cornea are then recorded on videotape. The position of the optical plane in the z-axis is controlled with a manual micrometer drive. Bowman's layer and the corneal endothelium are used as additional reference structures for the z-axis position.

For a detailed analysis, the recorded video sequences are reviewed frame by frame. For the patients' record, selected frames can be printed with a video printer. Pictures for publication can be photographed directly off the monitor. Alternatively, if the appropriate equipment is available, selected frames can be digitized and entered directly into an image file server to be available either online or be exposed to photographic film with a laser film printer. We recommend that the examiner routinely digitize images from all corneal layers, plus specific findings for a given case.

Patients are evaluated routinely by clinical slit lamp biomicroscopy before and after the confocal scanning procedure. Before the examination with the confocal microscope, the patients are informed about the nature of the confocal scanning procedure. The investigator should decide before-hand which objectives are going to be used. For a quick routine examination, the $40\times$ objective offers the best overview type of characteristics. However, this objective will miss some details visible with the $50\times$ objective and does not give a generous overview (and a better link to slit lamp morphology) like the $25\times$ objective. In most of our patients, we try to work with all three objectives and select the one or two most appropriate to image the specific pathology in further follow-up sessions.

One drop of acrylic eye gel is placed on the microscope objective and the instrument is moved to a full backward position. After topical anesthesia with one drop of 0.4% Novesine or any other topical anesthetic, the patient's head is positioned on an adjustable headrest. The confocal microscope with the $40\times$ objective is then placed 1–1.5 mm above the apex of the corneal center. The patient is asked to look into the light so that the optical center of the cornea is aligned with a lateral accuracy of probably less than 1 mm. The microscope is then brought into optical contact with the cornea by manual advancement of the micrometer-controlled z drive. From this point on, all further x–y–z movements of the instrument are controlled from the video-rate picture displayed on the monitor. The image is recorded on an S-VHS tape recorder and can be reviewed in slow motion. Simultaneously, the observer comments on the videotape sound track on position and other findings to supplement the information recorded. Care must be taken not to mechanically damage the cornea.

For manual imaging with the $40\times$ and $50\times$ objective setup, the following pattern is usually employed.

1. Establish optical contact with cornea, focus on basal epithelial layer, centrate on the surface to image parallel sections, and start videotape.

2. Move backward to image superficial epithelial cell layers.

3. Proceed to basal cell layer.

4. Proceed to Bowman's layer and look for subepithelial nerve plexus.

5. When first stromal keratocytes are visible, return to step 2.

6. Repeat steps 2–5 until enough frames of epithelium have been sampled.

7. From Bowman's layer proceed to endothelium at about 0.1–0.5 mm/s with manual micrometer advancement.

8. When endothelium reflex just above the anterior chamber becomes visible, recentrate the microscope if required and then return to the keratocyte layer just above Descemet's membrane.

9. Return to endothelium and anterior chamber, making sure that some frames of the endothelium have been captured.

10. Repeat steps 8 and 9.

11. Return to epithelium and reverse movement as in step 7.

12. Repeat steps 2–11 at least once or more times if required.

If required, the same procedure (steps 1–12) can be carried out with the 50× and 25× microscope objectives. For every eye and microscope objective, a minimum of 0.5- to 1-min good-quality tape recording should be collected. Extended recording times are usually due to a nontrained investigator, which lead to decreased patient comfort and later on to extended tape reviewing times. With sensitive or less cooperative patients, the recording time, however, may be extended to obtain at least 1000 useful optical sections from all corneal layers. Included in this number is usually a minimum of at least optical sections from all z positions of the central corneal stroma. If a specific finding is observed during the examination, for example, in the epithelium, extra recording time is spent on the region of interest.

To investigate noncentral locations on the cornea, the limbus, or even conjunctival areas, the patient's direction of gaze is aided with a fixation light for the contralateral eye. For these special locations, the angle of the frontal plane has to be tilted to achieve perpendicularity to the surface of the tissue studied. Occasionally, a somewhat oblique section may also be interesting, as it gives more information on the thickness of some structures imaged in one optical section. The videotapes should be evaluated in slow motion or single frame mode of the video recorder.

From patient examinations stored on S-VHS tapes, we usually digitize a standard set of frames (from 40× or 50× objective recordings) for every patient, which includes (1) epithelial surface cells, (2) epithelial wing (intermediate) cells, (3) epithelial basal cells, (4) subepithelial nerve plexus/Bowman's layer, (5) first keratocyte layer, (6) anterior stroma keratocytes (10–100 μm below/behind Bowman's layer), (7) intermediate stroma (100–350 μm depth), (8) posterior stroma (350 to >500 μm depth), (9) most posterior keratocyte layers, just anterior to Descemet's membrane, and (10) endothelium (see Figs. 6–11).

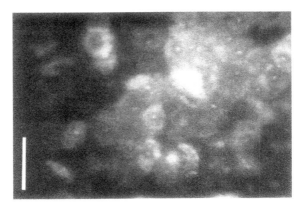

Fig. 6 Confocal section of surface cells in a normal human cornea. Dark and light cells are visible with dark nuclei. In some cells the borders are bright, possibly indicating a loss of contact and the process of desquamation. Bar: 50 μm. From Böhnke and Masters (1999a).

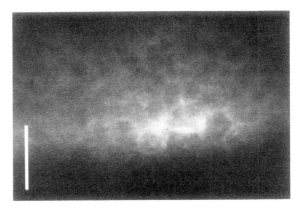

Fig. 7 Confocal section of a normal human cornea at the level of intermediate or wing cells. The orientation of the section is slightly oblique, showing maturation and enlargement of the cells from top to bottom. Bar: 50 μm. From Böhnke and Masters (1999a).

In addition, we digitize a variable number of frames that show specific findings for a given case. The digitized files are given identifying numbers, which clearly link them to a record of the database carrying information about the identity of the patient, the location of the recording, and relevant technical information regarding the microscope. When video images are digitized, we do not apply image enhancement techniques except for a median filter function, which basically makes the image appear smoother and reduces pixel noise. With the advent of digital video recording technology, the video images can now be stored in a digitized format with a unique identifier of every image on a given tape. Thus, the need to digitize selected frames within due time to prevent deterioration of image quality may be

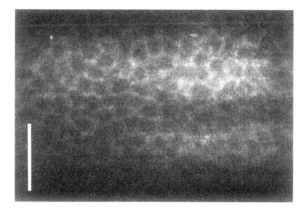

Fig. 8 Confocal section of a normal human cornea at the level of basal cells. Dimly reflective cell borders are shown. Bar: 50 μm. From Böhnke and Masters (1999a).

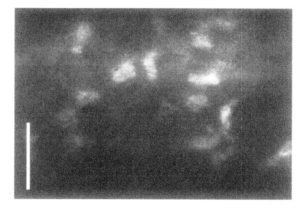

Fig. 9 Confocal section of a normal human cornea at the level of the most anterior keratocyte layer. Bar: 50 μm. From Böhnke and Masters (1999a).

reduced; however, a critical review of a recorded sequence after the recording session is still mandatory.

In contrast to the extended sessions in the early days of the procedure, a central corneal examination of all layers now typically consists of 40–100 s of recording time for every microscope objective used. Considering patient and instrument handling, the total examination time does not exceed 10 min. It should be kept in mind, however, that the observer has to review the tape frame by frame later on and digitize and store selected images. The total time for review, selection, processing and storage of the confocal images can be much more than 1 h per patient. If, however, only one specific aspect has to be answered, the total investigation time can be considerably shorter.

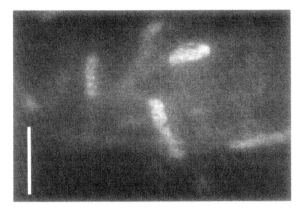

Fig. 10 Confocal section of a normal human cornea at the level of the most posterior keratocyte layer. Keratocyte nuclei with oval shape and cytoplasmic invaginations, which are the most predominant type throughout all tissue layers. A second type of elongated nuclei in the last or the second to last keratocyte layer just before Descemet's membrane is visible. Bar: 50 μm. From Böhnke and Masters (1999a).

Fig. 11 Confocal section of a normal human cornea at the level of the corneal endothelium. Bar: 50 μm From Böhnke and Masters (1999a).

VII. Clinical Findings with Scanning Slit Confocal Microscope

The scanning slit confocal microscope provides *in vivo* details of the normal and pathologic structure of the cornea with a resolution of less than 1 μm (Böhnke and Thaer, 1994; Masters, 1993, 1995b; Masters and Thaer, 1994b, 1995, 1996). The confocal microscope provides *en face* images of the cornea; the plane of the image is orthogonal to the thickness of the cornea. These images are very different and are oriented perpendicular from the typical sections obtained in histopathology in which the tissue is cut along the thickness of the cornea (vertical sections). Some examples from various clinical conditions are given (Böhnke and Masters, 1997; Böhnke *et al.*, 1997; Cadez *et al.*, 1997, 1998; Chiou *et al.*, 1997; Früh *et al.*, 1995; Masters and Böhnke, 1998).

1. *Contact lens-associated changes* (Böhnke and Masters, 1997; Cadez *et al.*, 1998). This is an important example of a new corneal degeneration associated with microdot deposits in the corneal stoma. This clinical study is a seminal example of the efficacy of the scanning slit confocal microscope to image submicrometer cellular alterations in the living human cornea. In patients with long-term contact lens wear, all layers of the cornea stroma show multiple fine particles with high reflectivity. These changes persist for years after discontinuing contact lens wear. These stromal deposits of submicrometer size have not been reported previously. This may be due to the superior resolution and contrast that is achieved routinely with the video-rate, scanning slit confocal microscope. These microdots were only visible when the 50×/1.0 NA water-immersion microscope objective was used (Fig. 12).

2. *Corneal trauma.* After a plant injury from *Dieffenbachia*, confocal microscopy was used to locate the oxalate crystals in all layers of the cornea down to Descemet's membrane. Over time the crystals were found to fragment and dissolve without signs of cell toxicity or inflammatory infiltrate (Fig. 13).

3. *Corneal infection.* Confocal microscopy is an excellent tool used to easily identify a fungal infection in the cornea (Fig. 14).

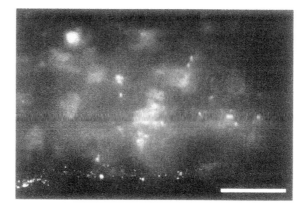

Fig. 12 Confocal section of a normal human cornea after 15 years of soft contact lens wear. Numerous small structures, called microdots, are present in all regions and layers of the corneal stroma. These structures are associated frequently with keratocytes, giving rise to the theory that they consist of lipofuscin granules that are of intracellular origin. Some of these granules may be deposited in the extracellular compartment, where they stay and accumulate over time. Bar: 50 μm. From Böhnke and Masters (1999b).

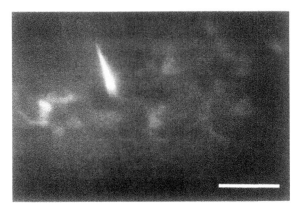

Fig. 13 A corneal injury with oxalate crystals derived from the plant sap of *Dieffenbachia*. Four weeks after the trauma, needle-like highly reflective crystals are visible in all layers of the corneal stroma down to Descemet's membrane. Confocal section of highly reflective oxalate crystal just below the basal epithelial cells, located in Bowman's layer. Bar: 50 μm. From Böhnke and Masters (1999a).

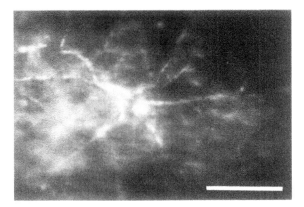

Fig. 14 Confocal microscopic image showing fungal mycelia growing in the stroma. Fungal infection after minor corneal trauma in a contact lens wearer. Bar: 50 μm. From Böhnke and Masters (1999a).

4. *Corneal surgery*. After lamellar cornea surgery, the viability of the keratocyte population, as well as the progress of reinnervation, can be studied. One year postoperatively, normal patterns of keratocytes and nerve fibers are found in the donor lenticule. Following photorefractive keratectomy, various changes even in the deep corneal layers can be found. A new finding is the occurrence of a highly reflective spindle-shaped structure, which in size may correspond to some extracellular deposit oriented along the collagen fibers. These changes persist for years after photorefractive keratectomy and possibly correlate with the amount of tissue ablated and the type of instrument used (Fig. 15).

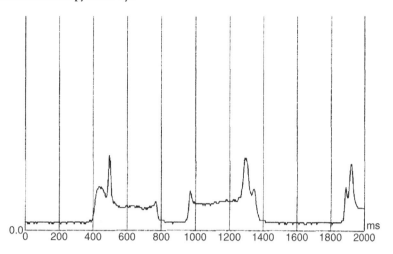

Fig. 15 Repetitive forward–backward confocal z scan of a human cornea with grade 1 haze 1 year after photorefractive keratectomy. The small amount of subepithelial scar tissue causes a significant peak in the stromal light scatter. In each z scan the intensity of scattered light is plotted on the ordinate in arbitrary units, and the distance within the cornea is plotted on the abscissa. The largest peak of scattered light in each scan is due to the reflectivity from the subepithelial region of the cornea. From Böhnke and Masters (1999a).

VIII. Three-Dimensional Confocal Microscopy and Visualization of the Living Eye

Three-dimensional visualization is an important application of confocal micros-copy based on the enhanced z-axis resolution. This feature permits the acquisition of a stack of z sections (optical sections) and the subsequent three-dimensional visualization. The three-dimensional visualization of the living *in situ* eye and studies of the structural changes of the three-dimensional structures over time (four-dimensional visualization) are important applications of multidimensional confocal microscopy (see Figs. 16 and 17). There are many diverse applications of confocal imaging and three-dimensional reconstruction of the living eye. Each application involves its special techniques for both data acquisition and three-dimensional visualization. Examples of the authors' prior work in the three-dimensional visualization of stacked optical sections acquired with a confocal microscope include the three-dimensional visualization of the full thickness of the *ex vivo* rabbit cornea (Masters, 1991; Masters and Farmer, 1993; Masters and Paddock, 1990b). This application required the use of a specially developed specimen chamber that is required to maintain the freshly excised eye in a normal physiological state and with minimal stress on the cornea during data acquisition with a confocal microscope (Masters, 1993). The *in vivo* human fundus and optic nerve were visualized in three dimensions from a stack of optical sections acquired with a modified scanning laser ophthalmoscope (Fitzke and Masters, 1993).

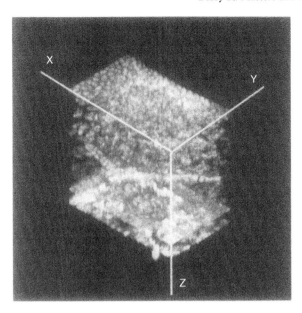

Fig. 16 Three-dimensional visualization of an *in situ* rabbit cornea based on the maximum intensity technique. The single layer of endothelial cells is shown on the top. The epithelium is shown at the bottom. A large bifurcating nerve fiber is shown in the anterior stroma. A stack of 365 confocal reflected light images were used to generate projections. The step size, which is the distance on the *z*-axis between consecutive optical sections, was set equal to the pixel size. In a maximum intensity projection the three-dimensional image is formed from the maximum intensity or the brightest pixel along each line of sight. The visualization was performed with software from Molecular Dynamics, Inc.

The *in vivo* human lens has been visualized in three dimensions from a series of optical slices acquired with a rotating slit confocal microscope (Masters and Senft, 1997).

Special techniques of multidimensional confocal microscopy and visualization of the living eye have been described. As in all techniques of microscopy, the fidelity of the images and their correct interpretation require appropriate instrumentation as well as a proper microscopic technique. Correlative microscopy with both confocal microscopy and electron microscopy has been shown to be critical in the analysis of confocal reflected light images (Masters, 1998a,b,c; Masters *et al.*, 1997). For *ex vivo* studies of living ocular tissue, the maintenance of the normal physiological state and cell and tissue morphology required a specially developed specimen chamber (Masters, 1993). For clinical confocal microscopy of the living human eye, the correct examination techniques are required to achieve the full potential of this important new optical technique. Video-rate, scanning slit confocal microscopes offer many advantages over Nipkow disk (pinhole-based), tandem scanning confocal microscopes for imaging the *in vivo* human cornea. This noninvasive, nonapplanating confocal microscope provides a new tool to image cellular details of the *in vivo* living human eye.

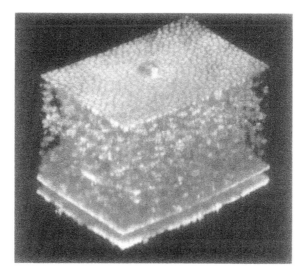

Fig. 17 Three-dimensional visualization of an *in situ* rabbit cornea based on the volume rendering technique. The reconstruction is shown in an isometric view. The single layer of endothelial cells is at the top. The epithelium is shown at the bottom. The full thickness of the central cornea is 400 μm. The volume rendering technique was performed with Voxel View software (Vital Images, Fairfield, IA).

Acknowledgments

This work was supported by a grant (B. R. M.) from the NIH, National Eye Institute, EY-06958. The authors thank Dr. Andreas A. Thaer for collaboration in the development of the clinical confocal microscope. We thank Dr. David Hanzel of Molecular Dynamics for helping us process the confocal images.

References

Auran, J. D., Koester, C. J., Rapaport, R., and Florakis, G. J. (1994). *Scanning* **16**, 182.
Böhnke, M., and Masters, B. R. (1997). *Ophthalmology* **104**, 1887.
Böhnke, M., and Masters, B. R. (1999a). *Progr. Retina Eye Res.* **18**(5), 1.
Böhnke, M., and Masters, B. R. (1999b). *Progr. Retina Eye Res.* **18**(3), 1.
Böhnke, M., and Thaer, A.A (1994). *In* "Bildgebende Verfahren in der Augenheilkunde," (O.-E. Lund, and T. N. Waubke, eds.), p. 47. Ferdinand Enke, Stuttgart, Germany.
Böhnke, M., Schipper, I., and Thaer, A. (1997). *Klin. Monatsbl. Augenheilkd.* **211**, 159.
Böhnke, M., Chavanne, P., Gianotti, R., and Salathe, R. P. (1998a). *J. Refract. Surg.* **14**, 140.
Böhnke, M., Wölti, R., Lindgren, F., Gianotti, R., Bonvin, P., and Salathe, R. P. (1998b). *Klin. Monatsbl. Augenheilkd.* **212**, 367.
Cadez, R., Früh, B., and Böhnke, M. (1997). *In* "Kongress der Deutschsprachigen Gesellschaft für Intraokularlinsen: Implantation und refraktive Chirurgie," (D. Vörösmarthy, ed.), p. 403. Springer, Berlin, Germany.
Cadez, R., Früh, B., and Böhnke, M. (1998). *Klin. Monatsbl. Augenheilkd.* **212**, 257.
Chiou, A. G., Cadez, R., and Böhnke, M. (1997). *Br. J. Ophthalmol.* **81**(2), 168.
Fitzke, F. W., and Masters, B. R. (1993). *Curr. Eye Res.* **12**, 1015.
Früh, B. E., Korner, U., and Böhnke, M. (1995). *Klin. Monatsbl. Augenheilkd.* **206**(5), 317.

Koester, C. J., and Roberts, C. W. (1990). *In* "Noninvasive Diagnostic Techniques in Ophthalmology," (B. R. Masters, ed.), p. 99. Springer, New York, NY.

Lemp, M. A., Dilly, P. N., and Boyde, A. (1986). *Cornea* **4**, 205.

Masters, B. R. (1988). *In* "The Cornea: Transactions of the World Congress on the Cornea III," (H. D. Cavanagh, ed.), p. 281. Raven Press, New York, NY.

Masters, B. R. (1990). *In* "Confocal Microscopy," (T. Wilson, ed.), p. 305. Academic Press, London.

Masters, B. R. (1991). *Mach. Vis. Appl.* **4**, 227.

Masters, B. R. (1993). *Scann. Microsc.* **7**(2), 645.

Masters, B. R. (1995a). *Comments Mol. Cell. Biophys.* **8**(5), 243.

Masters, B. R. (1995b). *Opt. Eng.* **34**(3), 684.

Masters, B. R. (1996). Selected Papers on Confocal Microscopy, Vol. MS131. The International Society for Optical Engineering, Bellingham, WA.

Masters, B. R. (1998a). *Opt. Express* **3**(10), 356. Available from:http://www.opticsexpress.org.

Masters, B. R. (1998b). *Opt. Express* **3**(9), 332. Available from:http://www.opticsexpress.org.

Masters, B. R. (1998c). *Opt. Express* **3**(9), 351. Available from:http://www.opticsexpress.org.

Masters, B. R. (1999). Selected Papers on Optical Low Coherence Tomography. SPIE Press, Bellingham, WA.

Masters, B. R., and Böhnke, M. (1998). *In* "Corneal Disorders: Clinical Diagnosis and Management," (H. Leibowitz, and G. Waring, eds.), p. 123. Saunders, Philadelphia, PA.

Masters, B. R., and Farmer, M. A. (1993). *Comp. Med. Imag. Graph.* **17**(3), 211.

Masters, B. R., and Kino, G. S. (1990). *In* "Noninvasive Diagnostic Techniques in Ophthalmology," (B. R. Masters, ed.), p. 152. Springer, New York, NY.

Masters, B. R., and Paddock, S. (1990a). *J. Microsc.* **158**(2), 267.

Masters, B. R., and Paddock, S. W. (1990b). *Appl. Opt.* **29**, 3816.

Masters, B. R., and Senft, S. L. (1997). *Comp. Med. Imag. Graph.* **21**(3), 145.

Masters, B. R., and Thaer, A. A. (1994a). *Appl. Opt.* **33**(4), 695.

Masters, B. R., and Thaer, A. A. (1994b). *Microsc. Res. Techn.* **29**, 350.

Masters, B. R., and Thaer, A. A. (1995). *Bioimages* **3**(1), 7.

Masters, B. R., and Thaer, A. A. (1996). *Bioimages* **4**(3), 129.

Masters, B. R., Thaer, A. A., and Geyer, O. C. (1994). *In* "Contact Lens Practice," (M. Ruben, and M. Guillon, eds.), p. 390. Chapman & Hall, London.

Masters, B. R., Vrensen, J., Willenkens, B., and Van Marle, J. (1997). *Exp. Eye Res.* **64**(3), 371.

McCally, R. L., and Farrell, R. A. (1990). *In* "Noninvasive Diagnostic Techniques in Ophthalmology," (B. R. Masters, ed.), p. 189. Springer, New York, NY.

Svishchev, G. M. (1969). *Opt. Spectrosc.* **26**(2), 171.

Svishchev, G. M. (1971). *Opt. Spectrosc.* **30**, 188.

Wiegand, W., Thaer, A. A., Kroll, P., Geyer, O. C., and Garcia, A. J. (1995). *Ophthalmology* **102**, 568.

CHAPTER 17

In Vivo Imaging of Mammalian Central Nervous System Neurons with the In Vivo Confocal Neuroimaging (ICON) Method

Sylvia Prilloff,★ Petra Henrich-Noack,★ Ralf Engelmann,★,†
and Bernhard A. Sabel★

★Medical Faculty
Institute of Medical Psychology
Otto-von-Guericke University of Magdeburg, Magdeburg
Germany, email: Bernhard.Sabel@med.ovgu.de

†Carl Zeiss MicroImaging GmbH (Jena)
Product Management BioSciences
Laser Scanning Microscopy

I. Update

With "*in vivo* confocal neuroimaging" (ICON) we use the eye as a window to the brain to obtain live images of cellular physiology and pathophysiology in the central nervous system (CNS). Since the first publication of the ICON technique (Sabel *et al.*, 1997), several lines of research have developed as anticipated, but some unforeseen applications have emerged, too.

TECHNIQUES IN CONFOCAL MICROSCOPY
307
DOI: 10.1016/B978-0-12-384658-7.00017-5

Several research projects have been completed with imaging the dynamics of retinal ganglion cell (RGC) morphology with ICON (Prilloff *et al.*, 2010). Fluorescent dyes injected into the superior colliculus are retrogradely transported via the optic nerve to the cell soma of RGCs. Taking advantage of this effect we were able to measure the soma diameter of RGCs and to determine their fate over time after optic nerve crush. By observing the changes of cell soma swelling early after the lesion it was possible to predict whether a given RGC would later live or die (Rousseau and Sabel, 2001; Rousseau *et al.*, 1999).

With a similar approach, postlesional axonal transport has been investigated with ICON. The injection of two different fluorescent dyes (red and green latex beads) were spaced 8 days apart and the feasibility of the protocol was evaluated by analyzing the dye kinematics and fluorescence characteristics. This approach was then used to study axonal transport after optic nerve crush, revealing the dynamics of axon degeneration and regeneration.

As anticipated, functional imaging with calcium-sensitive dyes became a fruitful topic for ICON research. Calcium plays a central role in normal neuronal function and in postlesional dysfunction. By injecting Oregon Green BAPTA into the superior colliculus before optic nerve crush, we were able to visualize calcium activation of identified RGCs *in vivo* during the days after injury. The data obtained led to the conclusion that while fast and massive calcium activation leads to cell death, slower and more moderate calcium activation is a predictor of recovery and survival (Prilloff *et al.*, 2007).

As a new field of research the investigation of carrier-mediated transport across the blood–brain barrier has emerged. Nanoparticle-associated fluorescence in the retina can be visualized with nanoparticle formulations containing the fluorescent marker rhodamine. We could show that Lutrol-SDS particles with negative Zeta-potential could cross the blood–retina barrier (which is essentially identical to the blood–brain barrier) whereas DEAE–Lutrol particles with a positive Zeta-potential were detectable in the blood vessels only and no fluorescence reached the RGCs.

The biggest challenge regarding ICON remains the movement of the living animals—most importantly here the movement of the eye—due to breathing and heartbeat which makes it difficult to overcome a slightly blurred image. The rodent's eye optics also limit the maximum possible resolution, but new ideas and technical developments will hopefully lead to improvements regarding the current constrains.

However, even considering these restrictions, ICON still remains an outstanding method which enables scientists to repetitively visualize live neurons in their natural environment. Moreover, as ICON allows repetitive measurements in one and the same animal, less animals are needed to get results from time series experiments. This is a significant advantage regarding time spent and cost and ethical considerations.

Apart from technical improvements for imaging neuronal physiology and pathophysiology in animal models, we think that the future of ICON lies in translating this work into clinical applications.

II. Introduction

Anatomical investigations of animals usually require that the animal is sacrificed and the tissue of interest, such as the brain, is prepared with special procedures. However, these anatomical studies are only "snapshots" in time, with the one-time preparation of tissue precluding the possibility for repeated observations. For many reasons, it would be advantageous to study microscopic detail in the living organism, particularly in the brain. The use of confocal techniques for such *in vivo* imaging of neurons has been considered an advantageous approach since the early days of confocal microscopy. The optical sectioning ability and the rising availability of cell-specific markers especially provided opportunities for successful *in vivo* imaging in the nervous system. The first attempts for such *in vivo* imaging had to overcome challenges such as choosing the right animal models, assuring the stability of the markers, and overcoming the photodamaging effects following fluorescence excitation. Imaging with cranial window preparations or on the exposed spinal cord provided novel approaches for *in vivo* observation, but these kinds of experiments were highly invasive and often difficult to repeat. New two-photon excitation techniques might change this now, but tissue penetration ability and resolution are still limited because the observed tissue itself acts as an optical barrier.

An elegant way to overcome this problem is to find animal models in which it is possible to visualize nerve cells in a transparent environment. This is the case, for example, in amphibia during early development or in the adult zebrafish. Here, high-resolution imaging of nerve cells is possible in a noninvasive manner, allowing repeated observation of changes related to development or degeneration. While these models of *in vivo* imaging help to elucidate some fundamental neurobiological problems, they do not allow the observation of mammalian nerve tissue, as these animals are non-mammals. Therefore, there is still the need to find a mammalian tissue model for *in vivo* imaging of neurons. Two models have been proposed: one in the peripheral and one in the CNS. Imaging through the skin of mammals with fiber-optic confocal imaging (FOCI) allows subsurface observation of fluorescing nerve fibers around hair follicles, sometimes also related to nerve regeneration. The ICON method allows observation of RGCs in the eye. Because nerve cells of the retina derive directly from the brain during early development, the ICON procedure is the first method to observe mammalian CNS neurons in a noninvasive manner and without damaging effects.

Investigators interested in *in vivo* imaging have long been interested in the mammalian eye as a model, but early attempts using scanning laser ophthalmoscopes or modified tandem scanning microscopes had problems with the high refractive power that characterizes the eyes of a typical rodent laboratory animal. However, small rodents are the most frequently used animals in studies of development or degeneration in the nervous system. ICON now provides both the instrumental adaptation for cellular resolution imaging of embedded nerve cells and the use of a mammalian animal model for degeneration and plasticity studies. This article describes the instrumentation and experimental details of a typical ICON setup.

III. Equipment

In general, ICON requires versatile laser scanning microscopes and a mechanical and optical microscope basis (Fig. 1) that fits the physical requirements of the chosen animal model, in our case the rat. Similar sized animals may also be used. In our experiments, ICON was developed using a standard confocal laser scanning microscope (CLSM) with a complete motorized scan module. On this microscope, a special head holder for the rat was attached. The best instrumentation might vary depending on the commercial availability of the models, but as of 2009, the following components can be recommended.

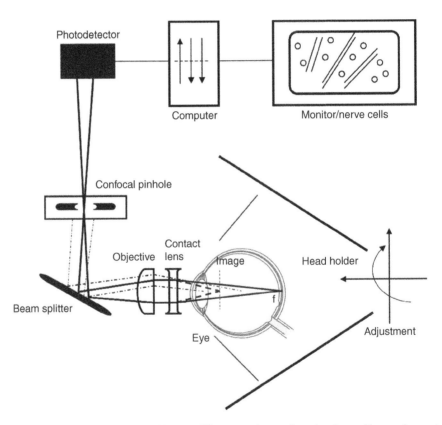

Fig. 1 Schematic drawing of an ICON setup. The contact lens projects the observed image plane onto the retina of the eye (f). Without such a lens, the image would be located in the vitreous, even with extremely long working distance objectives. The contact lens does not fully compensate the refractive power of the eye, which is useful in achieving further magnification. The confocal pinhole effectively eliminates out-of-focus information, even with the low-power optics used.

A. Equipment List

1. *Zeiss Axioskop 2 MOT microscope (LSM 5 PASCAL)*: The microscope was equipped with a mercury lamp (HBO 50) and three different lasers (argon laser—453–514 nm, green helium/neon laser—543 nm, red helium/neon laser—633 nm). The control of the microscope, the laser modules, the scanning module, and other accessories were operated by a high-performance digital signal processor (DSP).

2. *Animal size specimen holder plate*: The standard table and object guide are too small for a rodent, so the object guide has to be replaced by a simple plate with the approximate size of the animal.

3. *Narishige nontraumatic rat head holder (London, UK)*: For fixation and adjustment of the eye in the optical axis of the microscope, a head holder is needed. Narishige provides a nontraumatic version using the teeth and forehead for fixation. Some small ball heads and clamps will mount the head holder on the stage.

4. *Carl Zeiss basic software R2.8 (version 3.2 SP2) with slow scan mode*: Convenient, control, and configuration of all motorized microscope functions, the laser module and scanning module, storage and retrieval of application configuration. In general, slow scan modes are required, but because the animal will move slightly under the anesthesia due to heartbeat and breathing, it is recommended using moderate short scanning times and averaging a number of pictures.

5. *FUJITSU SIEMENS Workstation*: High-end PC ("CELSIUS") with ample RAM and hard drive, ergonomic high-resolution monitor, a lot of accessories operating system Windows XP (Intel Pentium 4), multiuser capability and the acquisition software AxioVision (software release 4.8), which offers all options needed for image analysis.

6. *Zeiss Epiplan 5X/10× ICS-lens*: The optical adaptation uses a low-power microscope lens (4–5×) with sufficient working space (10 mm or more), which will be combined with the animals eye optics using a contact lens by immersion with optical gel.

7. *Newport f = −12.5 (KPC-013) contact lens*: The best contact lens to be used is a planoconcave contact lens of −80 dpt. When combined with the objective described earlier, the resulting overall magnification power will be 20–25× in case of a rat eye. This may be adapted to study other species.

8. *Molecular Probes FluoSpheres 40 nm/5%*: The choice of the fluorescence label will depend on the topic to be studied, but as an example, FluoSpheres (Molecular Probes Eugene, OR) are particularly suitable for retrogradely labeling RGCs, to be able to analyze their morphology (such as cell number and size). FluoSpheres are nontoxic to cells and provide a strong retrograde label of about 90–95% of the RGCs. The label will not fade for several months, even with repeated illumination.

9. *Stoelting Lab Standard stereotactic instrument*: Any stereotaxic instrument is suitable for the standard intracranial injection of the retrograde label.

IV. Animal Preparation

In vivo imaging in general requires proper animal handling, otherwise tissue damage and bad image quality will result. However, in control experiments it was assured that typical laser illumination in order to excite fluorescent dyes will not cause any damage to the retinal cells. Anesthesia is the most critical step with ICON. For use with the hardware mentioned earlier, the following protocol can be followed (see Fig. 2).

A. Experiment Checklist

1. *Labeling*: Stereotactic injection of retrograde markers into the brain under standard anesthesia. The injections should be performed several days before the first imaging session to allow sufficient time for retrograde transport. Five to 7 days are sufficient for the fluorosphere transport.

2. *Anesthesia for imaging session*: Imaging requires a very stable anesthesia for about 20 min with a quick recovery. This anesthesia can be repeated every 5th day or so, preferably not less than 2–3 days to avoid anesthesia-related problems.

3. *Nontraumatic fixation*: The head of the animal has to be held in a fixed position such that the laser illumination of the retina is stable. Preferably, the head holder can be rotated such that larger retinal areas (typically 0–45° eccentricity) can be visualized. The head holder should not use ear bars; rather a head

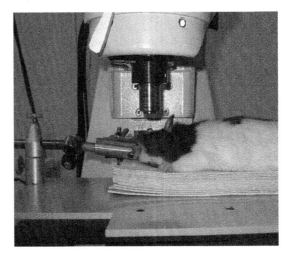

Fig. 2 Experimental setup with an animal. An anesthetized rat is lying on the modified microscope table and is adjusted in the optical path of the microscope with a nontraumatic head holder. A Hruby-style −80 dpt contact lens is attached to the surface of the cornea with the help of immersion gel, and a standard low-power microscope lens is focused on the retina.

holder such as the one from Narishige should be used, which uses the forehead and teeth for fixation.

4. *Protection of the eyes*: To increase the field of view and image brightness, iris relaxation drugs are very useful. We have tested different ones and found that standard Neosynephrin-POS eye drops (e.g., Ursapharm, Germany) are most useful. It is also necessary to protect the cornea during the imaging session.

5. *Contact lens*: The eye-sided surface should be immerged with optical gel, which will also protect the cornea. Any normal contact lens gel will work, for example, Vidisic optical gel (Dr. Mann Pharma, Germany).

6. *Illumination wavelength*: Use visible range excitation, with moderate laser power. Fluorescence will not be much stronger with high laser power.

7. *Scan speed*: A submillisecond scanning, with averaging of several pictures/second, provides the best image quality and sensitivity.

8. *Animals*: Wherever possible, dark-eyed rats (no albino rats) should be used to yield a better contrast between background and cellular fluorescence.

9. *Repeated observations with ICON*: The anatomy of the retinal blood vessels can be used as a landmark to identify individual cells over time for repeated measurements. In practice, it is helpful to employ transparent sheets to copy the vessels and the labeled cells from the computer screen and adjust the scanning position in the following experimental session in accordance with this drawing.

B. Typical Substances and Settings

- Green FluoSpheres 40 nm, 2.0 µl per brain injection site, 5–7 days transport time.
- Injection sites for the rat sup. colliculus: A/P: −6.9 mm, lat.: 1.2 mm, 4.0 mm below dura.
- Ketamine/Rompune anesthesia, 100 mg/10 mg per kg body weight intraperitoneally.
- Neosynephrin-POS 5% eye drops for iris relaxation, wait 5–10 min for maximum effect.
- Vidisic optical gel for protection of the cornea and contact lens immersion.
- Best images were obtained by using a scanning time of 24.5 s and an average of 16 scanning acquisitions.

V. Applications and Future of ICON

The use of confocal *in vivo* imaging techniques provides advantages especially for the observation of developmental and degenerative changes in the nervous system. Cells can be observed repeatedly in a noninvasive manner and in a totally

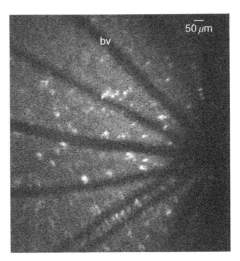

Fig. 3 Observation of living retinal ganglion cells in the rat. The image shows fluorescing ganglion cells (scale bar: 50 μm). The darker radial stripes are blood vessels (bv) of the retina, being about 20 μm in diameter. These blood vessels can be used as natural scale bars and will also help relocate the observed region in later imaging sessions. The ganglion cells visible are of different size and fluorescence intensity and belong to different types. This image was taken after an optic nerve damage (controlled crush), and here only the surviving or unaffected ganglion cells are labeled.

natural environment. Even with cell culture approaches and long-term two-photon imaging this cannot be achieved. As Fig. 3 shows, both the cell number and the cell size can be followed over time with ICON in the same animal. Because of this ability to image cells with microscopic resolution repeatedly over time, very small changes (e.g., in size and number) can be observed. Therefore, we hope that over time ICON will become an important tool for neurobiological research, as a number of questions can be studied *in vivo* that hitherto could only be investigated *in vitro*.

In the future, the use of ICON imaging will be influenced both by the available markers and by technical advances in imaging resolution. Some examples of new ICON applications might include the following. A new fluorescent marker may be found that, beyond retrograde labeling, may be useful to show other cell types or particular cell functions, including molecular alterations. It is conceivable that glial cells could be labeled via the blood system, and with GFP and viral vectors, even certain nerve cell types may be labeled specifically. Furthermore, while anatomical markers work well for describing entire cell populations, many markers exist that show specific cellular functions as well, for example, cell death markers or Ca^{2+} dyes for the monitoring of nerve cell activity.

The current drawback of ICON is that the light scattering properties of the surrounding tissue are not optimal. With deconvolution techniques and improved optical corrections, enhanced resolution may be possible. This is currently under development with the goal to visualize structural details such as nerve cell processes or even dendritic details.

ICON can be combined with other microscopic evaluation techniques as well. For example, after ICON has been applied in the living animal repeatedly, it is still possible to prepare the tissue and perform physiological or anatomical studies with this tissue. With the setup and procedures described, tissue slices or retinal whole mounts can be analyzed with conventional microscopic techniques, and the observed areas can be relocated without problems.

ICON thus provides the opportunity to visualize neurons of the mammalian CNS in a noninvasive manner repeatedly and without any damages. This opens up the opportunity to image neurons already in their natural environment and may be, at least for some questions, a very good alterative to *in vitro* experiments.

VI. Suppliers

The addresses of local distributors can be found at the following sites: confocal laser scanning microscope, www.zeiss.de; head holder, www.narishige.co.jp; contact lens, www.newport.com; and fluorescence markers, www.probes.com.

References

Prilloff, S., Noblejas, M. I., Chedhomme, V., and Sabel, B. A. (2007). Two faces of calcium activation after optic nerve trauma: Life or death of retinal ganglion cells in vivo depends on calcium dynamics. *Eur. J. Neurosi.* **25**(11), 3339–3346.

Prilloff, S., Fan, J., Henrich-Noack, P., and Sabel, B. A. (2010). In vivo confocal neuroimaging (ICON): non-invasive, functional imaging of the mammalian CNS with cellular resolution. *Eur. J. Neurosci.* **31**(3), 521–528.

Rousseau, V., and Sabel, B. A. (2001). Restoration of vision IV: Role of compensatory soma swelling of surviving retinal ganglion cells in recovery of vision after optic nerve crush. *Restor. Neurol. Neurosci.* **18**(4), 177–189.

Rousseau, V., Engelmann, R., and Sabel, B. A. (1999). Restoration of vision III: Soma swelling dynamics predicts neuronal death or survival after optic nerve crush in vivo. *Neuroreport* **10**(8), 3387–3391.

Sabel, B. A., Engelemann, R., and Humphrey, M. F. (1997). *In vivo confocal neuroimaging (ICON) of CNS neurons. *Nat. Med.* **3**(2), 244–247.

CHAPTER 18

Identification of Viral Infection by Confocal Microscopy

David N. Howell and Sara E. Miller

Department of Pathology
Duke University Medical Center
Durham, North Carolina
USA

I. Introduction

Confocal microscopes have found increasing use as diagnostic tools (Caruso *et al.*, 2010). In the clinic, confocal instruments have been used to perform *in vivo* biomicroscopy of tissues such as the cornea and skin. Surgical pathologists have employed confocal imaging to examine thick or difficult to section biopsy specimens. In the laboratory, confocal optics has been used to scan solid-phase arrays of oligonucleotide probes hybridized with DNA samples.

As an isolated technique, confocal laser scanning microscopy (CLSM) is of limited utility in the field of diagnostic virology. Confocal imaging has been employed in a few studies to detect *in situ* hybridization of virus-specific probes (Mahoney *et al.*, 1991; van den Brule *et al.*, 1991), but in most diagnostic settings offers little advantage over standard methods such as routine histology, enzyme immunohistochemistry, and conventional immunofluorescence microscopy. Used in concert with electron microscopy (EM), however, confocal microscopy provides a powerful survey tool, allowing ready identification of focal areas of pathologic alteration in large tissue samples for subsequent ultrastructural analysis.

Ultrastructural pathologists are handicapped by severe limitations on the size of tissue specimens that can be accommodated by electron microscopes. This is a particular problem for ultrastructural diagnosis of viral infections in tissues, which are often quite focal. Although a variety of survey methods have been developed to allow selection of tissue samples for EM (Miller *et al.*, 1997), many of these are cumbersome and time-consuming. The method employed most frequently, light microscopic examination of toluidine blue-stained survey sections produced from small epoxy-embedded blocks of randomly selected tissue, is labor-intensive and occasionally fruitless. Another popular alternative, light microscopic survey of sections of paraffin-embedded tissue with subsequent reprocessing of selected areas for EM, affords ultrastructural preservation insufficient for identification of any but the largest viral particles.

With their ability to produce optical sections of relatively thick tissues, confocal microscopes can be used to scan large tissue specimens for focal areas of interest, including foci of viral cytopathology. These foci can then be excised and examined by EM. Confocal microscopic examination can be performed either before (Chu *et al.*, 2010; Miller and Howell, 1997; Miller *et al.*, 1997) or after (Deitch *et al.*, 1991; Sun *et al.*, 1995) tissue embedment. Each approach has advantages and disadvantages. Preembedment scanning facilitates examination of a large amount of tissue for features of interest, but isolation and processing of the selected areas for EM are somewhat cumbersome. Postembedment scanning facilitates the subsequent production of sections for EM, but is best suited for relatively small specimens, as production of numerous large embedment blocks is tedious and time-consuming. Some fluorescent labels useful for localizing viral infections withstand EM embedment poorly, limiting their use in postembedment scanning techniques.

Because the search for foci of viral infection often requires examination of a considerable amount of tissue, we have concentrated our efforts on preembedment survey techniques. A standard method developed in our laboratories (Chu *et al.*, 2010; Miller and Howell, 1997; Miller *et al.*, 1997) is illustrated in Fig. 1. Fixed, unembedded tissue specimens are sliced using a vibrating microtome. The slices are stained by a variety of methods and examined with a confocal microscope for areas of pathologic alteration. These areas are dissected out, embedded, and examined by EM. An example of the application of this method, in which a focal adenovirus infection was identified in liver tissue, is provided in Fig. 2. With rapid processing techniques, the entire procedure can be accomplished in a day or less.

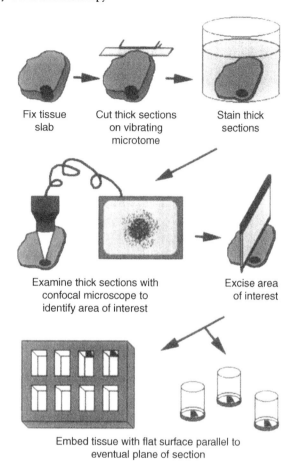

Fig. 1 Schematic diagram of the technique for confocal microscopic examination of unembedded tissue, with subsequent embedment of focal area of interest for EM. Reprinted, with permission and with minor modifications, from Miller *et al.* (1997).

This chapter provides a step-by-step discussion of the processing and examination of tissue specimens by confocal microscopy for subsequent ultrastructural study. Although the method was originally developed for examination of focal viral infections, it is equally applicable to the analysis of other focal pathologic features (e.g., foci of differentiation within poorly differentiated tumors).

II. Confocal Microscopy

Confocal microscopes are a relatively recent addition to the armamentarium of diagnostic pathologists and virologists. Many of the techniques for preparing specimens for CLSM, however, are similar to methods used in conventional

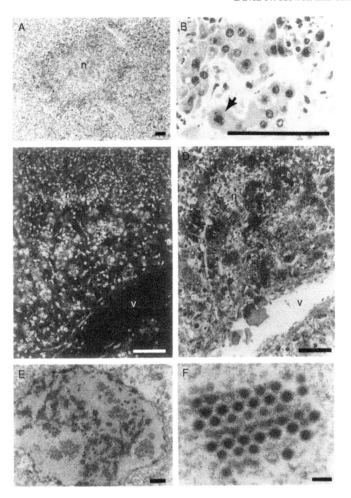

Fig. 2 Detection of adenovirus infection in liver tissue using CLSM and EM. (A) Light micrograph of an H&E-stained section of paraffin-embedded tissue showing a focus of necrosis (n). (B) Higher magnification view of the edge of necrotic focus shown in (A), demonstrating intranuclear inclusion in a hepatocyte (arrow). (C) Confocal micrograph of a glutaraldehyde-fixed, propidium iodide-stained tissue showing an area of necrosis (n) adjacent to a hepatic vein (v). (D) Toluidine blue-stained semithin section of focus shown in (C) following epoxy embedment, showing same area of necrosis (n), and hepatic vein (v). (E, F) Electron micrographs of infected cell from the edge of necrotic focus, showing paracrystalline arrays of 75-nm adenovirus particles within nucleus. Bars: (A–D) 100 μm, (E) 1 μm, (F) 100 nm. Reprinted, with permission, from Miller *et al.* (1997).

histology laboratories; a vast majority of the necessary reagents and supplies are readily available in such laboratories. The optical sections produced by confocal microscopes will look familiar to pathologists accustomed to performing conventional immunofluorescence microscopy on tissue sections produced with a microtome. Some confocal optical systems allow the microscopist to examine a specimen in real time, in much the same manner that a section would be viewed with a conventional microscope.

A. Specimen Selection and Fixation

Confocal microscopic survey is most useful for specimens too large to be embedded in a limited number of conventional EM blocks. (Using conventional methods, this limit is reached with fairly small specimens; complete embedment of a cube of tissue 3 mm on a side in 1-mm^3 tissue segments, a typical size for EM specimens, would require 27 blocks.) Although tissues with a variety of shapes can be accommodated, ideal specimens are rectangular solids measuring approximately 1 × 1 × 0.3 cm. Specimens of this shape and size can be fixed rapidly and are well suited for subsequent sectioning and examination (see later).

Rapid fixation of specimens is of paramount importance for examination by EM. Prompt fixation is necessary to achieve optimal ultrastructural preservation, particularly for diagnostic virology, where many of the potential pathogens are quite small and easily confused with endogenous cellular structures (e.g., ribosomes). Fixation has the added benefit of rendering tissues somewhat less pliable, facilitating the subsequent production of thick sections by vibrating microtomy.

Choice of fixative depends partially on the type of staining to be used for confocal microscopic examination. For many general histochemical stains, any good EM fixative can be used (Glauert, 1975; Hayat, 1981), A 2–4% (v/v) solution of glutaraldehyde in cacodylate buffer is an excellent choice, but other fixatives, such as 2–4% (w/v) paraformaldehyde, also work well. Ultrastructural preservation is usually quite acceptable in tissue fixed promptly with the 10% neutral-buffered formalin used by most general pathology services for routine specimen fixation. In cases where confocal examination is contemplated, a portion of a biopsy specimen can be reserved in buffered formalin and subsequently apportioned for CLSM/EM or processed for routine histologic sectioning as needed.

Glutaraldehyde-fixed tissues autofluorescence, a property that can actually be used as a basis for confocal imaging (see later), although it represents a severe impediment to subsequent immunofluorescence microscopy. Paraformaldehyde fixation is often a good choice for specimens being prepared for immunofluorescent staining. As with any form of immunohistochemical staining, many antigenic epitopes are affected adversely by fixation, and determination of optimal fixation times and fixative concentrations for preservation of antigenicity as well as ultrastructural detail may be required.

B. Preembedment Sectioning

Although confocal microscopes allow imaging beneath the surfaces of tissue specimens, the maximal depth for optical sectioning is dependent on a number of factors, including penetration of stain and inherent opacity of the tissue under study. With the exception of unusually clear tissues such as cornea, the practical limit is generally a few dozen micrometers at best. As a result, most of the interior of typical biopsy specimens is inaccessible to confocal imaging.

This problem can be circumvented by producing slices of the specimens with a vibrating microtome. Several versions of this device are available, including the Vibratome Tissue Sectioning System (Ted Pella, Redding, CA) and the Oscillating Tissue Slicer (Electron Microscopy Sciences, Ft. Washington, PA). All employ a rapidly oscillating blade (generally a conventional razor blade) to produce slices of wet, unembedded tissues. Slices produced in this manner are mounted easily on microscope slides and have flat surfaces that are ideally suited for rapid confocal scanning.

Specimens for vibrating microtomy are affixed to a glass slide or Teflon support; a fast-drying cyanoacrylate glue works well for this purpose. Handling and sectioning of small specimens can sometimes be facilitated by encasing them in molten gelatin and allowing it to solidify, forming a larger block. For this purpose, a 3–7% aqueous solution of agarose or gelatin with a hardness of 275–300 Bloom, melted and cooled to 42 °C, can be employed.

The support with attached tissue is immersed in a chamber in the vibrating microtome containing buffer such as phosphate-buffered saline. Slices float off into the surrounding buffer as they are produced and are picked up and stored (usually in their original fixative) pending staining and examination. The speed of the oscillating blade, rate of specimen advance, and thickness of the sections produced must be adjusted for the tissue type under study. With insufficient oscillating speed and/or excessive specimen advance speed, the tissue may be pushed aside or compressed rather than sliced by the advancing blade. In general, the thinner the slices produced, the greater the percentage of tissue accessible for confocal scanning. However, friability of the tissue generally imposes a lower limit of approximately 50 μm on slice thickness. Slices less than 100 μm thick, particularly tissues with a paucity of fibrous stroma such as brain and liver, are often quite fragile. Grasping such tissues with forceps frequently causes them to tear and crumble; they are most easily picked up and transferred by draping them over a plastic pipette tip, blunt probe, or camel-hair brush.

Small, irregular tissue fragments can also be examined by CLSM, although the results are less satisfactory than those achieved with vibrating microtome slices. Nonplanar specimens are difficult to mount with coverslips, and much of their volume is often inaccessible to confocal scanning. When examining such specimens with higher magnification objectives with short working distances, care must be taken to avoid racking the objective into the specimen while attempting to visualize features at deeper focal planes.

C. Staining

A wide variety of staining methods can be utilized to prepare tissues for examination by CLSM. These methods have variable applicability to the use of confocal microscopy as a survey tool for subsequent EM. Occasional tissue components (e.g., lipofuscin) are autofluorescent and can serve as landmarks for confocal imaging. Fixation with glutaraldehyde induces a more generalized autofluorescence, which in some cases can be employed to produce very detailed confocal images. This method has been used, for example, to study whole-mount preparations of human nerves (Reynolds *et al.*, 1994). An obvious additional benefit of this "staining technique" is the optimal ultrastructural preservation afforded by glutaraldehyde fixation.

Numerous tissue stains employed in conventional fluorescence microscopy can be applied to confocal imaging. The fluorescent nucleic acid-binding dye propidium iodide works well for visualizing many forms of viral cytopathology (Miller and Howell, 1997; Miller *et al.*, 1997); similar results should be achievable with bisbenzimide (Hoechst 33258) and 4,6-diamidino-2-phenylindole (DAPI). Tissues stained with propidium iodide and viewed with a green excitation filter exhibit strong nuclear staining as well as weaker cytoplasmic staining, the latter presumably due to binding of the dye to cytoplasmic RNA (Miller and Howell, 1997; Miller *et al.*, 1997). Erythrocytes, which contain low levels of residual cytoplasmic nucleic acids, frequently exhibit moderate fluorescence in propidium iodide-stained sections; care must be taken not to confuse them with nucleated cells. Several stains used in routine histochemistry (e.g., eosin, Congo red) also fluoresce when excited at appropriate wavelengths and can be used to stain tissues for CLSM.

Tissues can also be stained by immunohistochemical and *in situ* hybridization methods for confocal microscopy. Direct or indirect immunofluorescent staining with a variety of fluorochromes can be employed, for example, to pinpoint focal infections with known viruses for subsequent ultrastructural study (Miller *et al.*, 1997). Reflectance-mode confocal imaging has been used to study the localization of colloidal gold-labeled *in situ* hybridization probes specific for human papillomavirus (van den Brule *et al.*, 1991) and human immunodeficiency virus (Mahoney *et al.*, 1991). With proper embedment and sectioning, examination of probe localization in preparations stained in this manner by EM should be possible. Labeling can also be performed with a mixture of fluorescent and colloidal gold-tagged secondary reagents to allow sequential imaging of the same labeled feature by CLSM and EM (Sun *et al.*, 1995).

The development of techniques for the genomic incorporation of genes encoding green fluorescent protein (GFP) has allowed endogenous fluorescent tagging of a wide variety of tissue components. Experimental infections with viruses genetically engineered to express GFP can be monitored by confocal microscopy, and infected foci can be selected for subsequent examination by EM (Fig. 3) (Henry *et al.*, 2010).

Choice of stains for confocal examination of tissues depends on a variety of factors. If examination of portions of tissue beneath the most superficial layer is desired, penetration of the stain is an important consideration. This is obviously not a concern for endogenous fluorescent features (including tissue components expressing GFP). Small molecules such as glutaraldehyde and many histochemical stains also penetrate multiple cell layers with relative ease. Staining with large molecule probes such as antibodies, however, is generally confined to the most

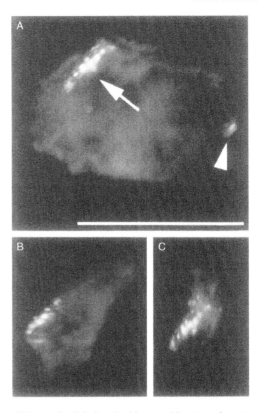

Fig. 3 Isolation of small focus of cells infected with recombinant murine cytomegalovirus expressing green fluorescent protein (GFP) from a small, irregularly shaped fragment of splenic tissue. (A) The position of the infected focus (arrow) in the tissue fragment is shown; a second, smaller focus of infection is also present (arrowhead). The larger GFP-expressing focus was excised in two successive steps, as illustrated in (B) and (C), and snap-frozen for ultracryomicrotomy. Tissue courtesy of Mr. Stanley C. Henry and Dr. John D. Hamilton, Veterans Affairs Medical Center, Durham, NC. Bar: 1 mm; magnification same in all panels.

superficial portion of the tissue. Techniques for enhancing the penetration of such reagents exist, including treatment of tissue with dilute detergent solutions or graded series of ethanol (Sun *et al.*, 1995) but their employment invariably sacrifices ultrastructural preservation to some extent.

In general, the identification of putative infectious foci within tissues for subsequent diagnostic EM is best performed using dyes, such as propidium iodide, which are capable of staining a broad range of cell types and tissue features (Miller and Howell, 1997; Miller *et al.*, 1997). Use of more specific labeling techniques such as immunofluorescent staining presupposes the knowledge of the infecting organism and is useful primarily in research settings (Miller *et al.*, 1997).

For staining, tissues are immersed in the stain of choice and incubated with gentle agitation. Care should be taken that both sides of tissue slices produced by vibrating microtomy are exposed to the stain, allowing staining and subsequent examination of both faces. The exact procedure followed depends on the stain employed. A 3-min incubation at 25°C with a 50-µg/ml aqueous solution of propidium iodide followed by a brief rinse in distilled water provides strong nuclear staining of superficial cell layers; deeper penetration of the stain can be achieved with longer incubations. Free propidium iodide fluoresces minimally, so exhaustive washing to remove unbound stain is unnecessary. Direct or indirect immunofluorescent staining can be achieved using staining protocols virtually identical to those used for routine histologic sections. In this setting, thorough washing to remove unbound reagents is advisable.

D. Mounting

Following staining, tissue samples are transferred to glass slides for examination. Specimens can be mounted in any suitable aqueous buffer; the solution used originally for fixation is a reasonable choice. If possible, a coverslip is applied; this provides an optically flat surface for viewing, protects the objective from inadvertent immersion in the buffer, and prevents desiccation of the tissue. Judicious blotting or aspiration of excess mounting buffer from around the edges of the coverslip minimizes slippage and facilitates the marking of areas of interest for subsequent excision (see later). Further stability can be achieved by sealing the edges of the coverslip with a melted mixture of equal volumes of paraffin wax, petroleum jelly, and lanolin. Vibrating microtome slices up to 200 µm in thickness can generally be coverslipped without difficulty. Thicker slices and irregular tissue fragments may be too large to allow the coverslip to rest flat on the slide. Such specimens can be scanned without a coverslip, but care must be taken to avoid drying of the tissue and contamination of the objective.

Thin tissue slices can also be mounted between two coverslips. The resulting "sandwich" is placed on a slide for confocal scanning and can be flipped, allowing easy examination of both faces of the slice.

E. Scanning

The goal of confocal scanning as a survey tool is to identify features of potential interest for subsequent ultrastructural study. For viral infections, the features in question are often quite focal, but can usually be identified at relatively modest magnification and resolution. As such, optimal results are obtained by scanning in a manner that allows rapid survey of large tissue areas.

Any available confocal laser scanning microscope can be used for tissue survey. Point-scanning microscopes provide images with optimal resolution, but generally can generate no more than one image per second. These images must be collected using a computer frame buffer and displayed on a monitor. Systematic scanning of a large piece of tissue is difficult using such an instrument; moving the stage to allow sequential viewing of overlapping fields covering the entire specimen surface requires considerable practice.

This problem can be circumvented by using a slit-scanning confocal microscope (e.g., Meridian Insight Plus, Genomic Solutions, Inc., Ann Arbor, MI). Such instruments, although not capable of the resolution achieved with point scanners, scan with sufficient rapidity to allow generation of real-time images, either on a computer monitor or directly through oculars. This capability greatly simplifies the systematic survey of an entire tissue surface and also facilitates focusing to identify features of interest in deeper tissue layers.

Confocal survey of large tissues is achieved most readily with a low-to-medium power objective (10–20×). The modest resolution afforded by such objectives is sufficient for the identification of most forms of viral cytopathology; their ample field of view and the relatively thick optical sections they produce are well suited to survey applications. The identity of features detected at scanning power can be confirmed by observation at higher magnification. A lower magnification objective (e.g., 4×) is also useful for capturing images of trimmed tissue blocks prior to embedment for EM (see later).

Given sufficient stain penetration and tissue translucency, serial images at progressive depths (z series) can be collected and used to produce three-dimensional reconstructions. Using this method, the exact location and depth of a feature within the tissue can be determined (Sun *et al.*, 1995). This process is fairly time-consuming, however, and is primarily useful for small specimens or selected areas within larger specimens.

While viruses themselves are generally beyond the resolution limits of optical microscopic methods, they produce a number of direct and indirect alterations in cell and tissue structure that can be detected. Large clusters of replicating virus particles can often be seen as inclusion bodies. With some exceptions, the inclusions of DNA viruses are generally intranuclear, whereas those of RNA viruses reside in the cytoplasm. Viral inclusions are typically rich in viral nucleic acids and, as a result, are stained easily with nucleic acid-binding dyes such as propidium iodide (Miller *et al.*, 1997).

Infection with viruses can induce a wide variety of other cell and tissue alterations, including the formation of multinucleate syncytial giant cells, cellular enlargement ("cytomegaly"), inflammation, hemorrhage, necrosis, and, in some cases, neoplastic transformation. While none of these features is specific for viral infection, they can all serve as keys in the search for areas harboring virus particles. Infected cells with optimal preservation for ultrastructural examination are found frequently at the peripheries of areas of necrosis, inflammation, or hemorrhage. Cytopathic alterations associated with specific virus infections have been catalogued (Caruso and Howell, 1999).

Although our initial applications of confocal microscopic survey have been in the area of diagnostic virology, the method has potential value as a survey tool for other forms of ultrastructural diagnosis. In particular, it can be used to identify

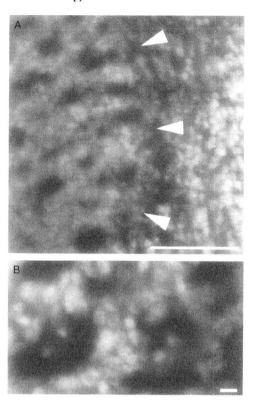

Fig. 4 Identification of a focal area of neuroblastoma-like differentiation within a vibrating microtome slice of an astrocytic neoplasm stained with propidium iodide. (A) A focus of tumor cell rosettes (arrowheads) is seen adjacent to tumor tissue without obvious differentiation; rosettes are shown at higher magnification in (B). Rosette-forming areas constituted less that 5% of the total tissue volume. Tissue courtesy of Dr. Christine M. Hulette, Duke University Medical Center. Bars: (A) 100 μm, (B) 10 μm.

focal areas of interest within tumors (e.g., foci of differentiation within poorly differentiated tumors, foci of viable tissue within largely necrotic masses) for subsequent examination by EM. An example is shown in Fig. 4 in which a small area of neuroblastoma-like rosette formation was isolated from an astrocytic brain tumor for ultrastructural study.

F. Trimming of Selected Tissue Specimens

Once a feature of interest has been identified within a tissue slice, the next goal is to trim away excess tissue, leaving the area of interest and a small amount of surrounding tissue. Because trimming of the specimen under direct confocal microscopic guidance is not practical, the selected area must be marked for subsequent trimming under a dissecting microscope. For specimens mounted with a coverslip, this can be accomplished using an inked object marker mounted on the microscope turret in place of an objective. This device marks the coverslip with an ink circle, approximately 2 mm in diameter, centered on the area of interest. Care must be taken to avoid slippage of the coverslip during this maneuver; this can be monitored by a second confocal imaging of the specimen after the circle has been marked.

The slide is next transferred to the stage of a dissecting microscope, and the portion of tissue inside the marked circle is examined for identifying landmarks (e.g., areas of visible hemorrhage, other discolorations, small blood vessels). The relationship of the circled area to the edges of the tissue slice is also noted. It is often helpful to sketch a diagram showing the approximate location of the area to be excised. The coverslip is removed carefully and the area of interest is reidentified under the dissecting microscope. A clean, fresh razor blade is then used to dissect away the surrounding tissue, leaving the area of interest in a small polygonal tissue fragment approximately 1–1.5 mm in maximal dimension. To allow distinction of the upper and lower faces of the tissue fragment (the former associated with the feature of interest) during subsequent steps, the fragment should be trimmed in an asymmetric shape. A trapezoid with two nonparallel sides and no axis of symmetry (e.g., the outline of the state of Nevada, USA) works well (Fig. 5).

After trimming, the specimen is returned to the confocal microscope and reexamined to ensure that the feature of interest has been retained. Although it is desirable to reapply a coverslip prior to scanning, small pieces of fragile tissue can be crushed by the weight of the glass. This problem can be circumvented by supporting the coverslip with the edges of additional coverslips or by mounting some of the larger, trimmed-off tissue fragments along with the selected fragment. If necessary, a second round of trimming can be performed to remove more excess tissue. Properly trimmed specimens will usually fit within the field of a 4× objective. An image captured with this objective is a useful guide for subsequent processing for EM; although the resolution of the image is insufficient for identifying most forms of viral injury, the shape of the block and the approximate location of the area of suspected infection can be documented (Fig. 5).

Trimming of small, irregular tissue fragments produced without the aid of a vibrating microtome poses a difficult challenge. Often, however, the irregular shape of the tissue provides landmarks to guide trimming. An example is shown in Fig. 3 in which a small patch of cells expressing GFP was dissected from a tissue fragment measuring approximately 1 mm in greatest dimension under confocal microscopic guidance.

When trimming is complete, the specimen is transferred to 2–4% glutaraldehyde in cacodylate buffer. At this stage, the tissue is generally too small and delicate to grasp with forceps, but can often be picked up and manipulated with the end of a wooden applicator stick sliced obliquely with a razor blade to produce a sharp, flat point. This implement is also useful during subsequent processing steps.

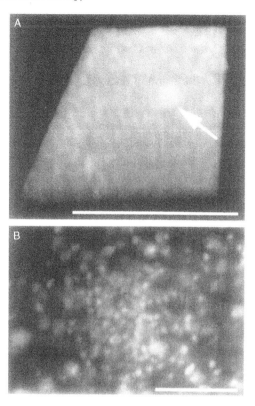

Fig. 5 Isolation of an inflamed focus in cerebral cortex from a patient with viral encephalitis. A trimmed block from a vibrating microtome section stained with propidium iodide is shown in (A); the inflamed area, visible as an indistinct white patch (arrow), is shown at higher magnification in (B). Bars: (A) 100 μm, (B) 10 μm.

III. Electron Microscopy

An exhaustive discussion of preparative techniques for EM and a review of viral ultrastructure are beyond the scope of this chapter. Thorough guides to preparative methods are available in several texts (Bozzola and Russell, 1992; Glauert, 1975; Hayat, 1981, 1989; Mackay, 1981) general principles of viral EM are provided in a variety of sources (Miller, 1986, 1991, 1995, 1999; Miller and Howell, 1988; Payne, 1997), and detailed descriptions of the ultrastructural features of individual viruses are available in several atlases (Connor *et al.*, 1997; Doane and Anderson, 1987; Murphy *et al.*, 1995; Palmer and Martin, 1988). A few guidelines will be offered, however, with specific reference to the processing and analysis of specimens selected by confocal microscopy.

A. Processing and Embedment

The method of processing and embedment selected depends on the type of EM examination to be performed. For routine ultrastructural study, specimens are postfixed with osmium tetroxide, stained *en bloc* with uranyl acetate, dehydrated, and infiltrated with epoxy resin by standard methods (Bozzola and Russell, 1992; Glauert, 1975; Hayat, 1989; Mackay, 1981). Spurr resin, available from most EM supply companies, is a good choice because it requires a relatively short baking time (8 h); other epoxy resins (e.g., EMbed-812 (Electron Microscopy Sciences), Polybed 812 (Polysciences, Warrington, PA)) are also acceptable, but require 16–24 h of baking.

During embedment, care must be taken to orient the tissue fragment so that the side containing the area of interest is parallel to and facing the eventual sectioning face of the embedment block. In initial experiments, we embedded specimens "on edge" in the ends of flat embedding molds. Specimens embedded in this manner frequently fell over and came to rest perpendicular to the block face before the resin hardened. To circumvent this problem, we now embed specimens in inverted cylindrical embedding capsules ("BEEM" capsules). To prepare the capsule for this form of embedment, the cap is sealed to the top of the capsule with paraffin film, and the conical tip is amputated with a razor blade. A small amount of liquid resin is placed in the inverted capsule, and the specimen is inserted with the surface containing the area of interest (the original "up" face) oriented face down. Positioning of the specimen in the capsule is best accomplished under a dissecting microscope with trans-illumination from a light box, using a trimmed applicator stick. When the specimen has been positioned adequately, additional resin is added, with care being taken not to disturb the specimen, and the capsule is baked.

Specimens destined for immunoelectron microscopy can be embedded in a hydrophilic resin or snap-frozen for ultracryomicrotomy as indicated. To ensure rapid and uniform freezing, tissue samples for ultracryomicrotomy are of necessity quite small (<0.5 mm in greatest dimension). Maintaining the orientation of such samples during freezing is generally not possible; in many cases, however, this problem is minimized by the fact that the area of pathologic alteration occupies most of the specimen (see Fig. 3).

B. Sectioning

In sectioning tissue specimens selected by confocal microscopy, extreme care must be taken to avoid cutting through the area of interest in the process of facing and orienting the block. Careful alignment of the block so that the plane of sectioning is parallel to the embedded tissue surface is crucial. Initial sectioning of epoxy blocks is performed using a glass knife. Serial 0.5-μm survey sections are produced, stained with toluidine blue, and examined by light microscopy. When the area of interest is identified, the glass knife is replaced with a diamond knife and thin sections are produced. A final toluidine blue-stained survey section can be cut after thin sectioning to ensure that the area of interest is still represented.

If the area of interest is not on the surface of the specimen, but at a known depth within the tissue, the microtome can be used to remove tissue to an appropriate depth before sections are collected. This should be done conservatively, however; one should err on the side of removing too little tissue rather than too much, for obvious reasons. This is particularly true as tissues can shrink during processing, diminishing the distance from the tissue surface to the desired area.

Sections should be mounted on fine-bar grids to minimize the chance of the area of interest lying on a grid bar. Alternatively, slotted grids with a Formvar support membrane can be used. However, they are technically more difficult to manipulate and are less stable in the electron beam. Stability can be considerably enhanced by carbon coating. It is preferable to place one section in the center of the grid rather than randomly positioning single sections or a ribbon, particularly if the block face is large (≥ 1 mm).

C. Electron Microscopy

Locating the area of interest in the electron microscope is facilitated greatly by a variable low-magnification setting for observing the whole section at once. A print of a low-magnification confocal micrograph or drawing of a toluidine blue-stained survey section can serve as a useful map of the thin section. Depending on how the section was picked up from the water (from underneath or above) and how the microscope specimen holder is inserted (flipping the grid 180 °C or not), the low-magnification image may be inverted from that observed by CLSM or light microscopy. Crossovers at various magnifications in the electron microscope will also invert the image. If this poses a problem, the confocal print or drawing can be viewed in strong light from the reverse side of the paper to provide an inverted image; computer inversion of confocal image files can also be accomplished easily with several software packages.

After the section has been oriented in the electron microscope, the structural landmarks identified by CLSM should be located. It is frequently possible to reidentify individual virus-infected cells initially detected by confocal imaging (Miller *et al.*, 1997). Once identified, areas suspected of harboring virus infection should be examined at a magnification of at least $40,000\times$. Keys for the ultrastructural detection and identification of virions are provided in the references cited previously.

IV. Summary

Confocal microscopy is a valuable adjunct to EM in the fields of diagnostic and investigative virology. Confocal imaging can be used to examine large amounts of tissue stained by a variety of methods for evidence of viral infection. Areas thus identified can then be processed for ultrastructural study, allowing a highly focused search for viral pathogens. With the possible exception of the vibrating microtome, all of the equipment and reagents necessary for the preparation of specimens for confocal scanning are available in any well-stocked histology laboratory. Although originally developed to facilitate viral diagnosis by EM, the methods described herein can be applied to the ultrastructural study of any focal pathologic process.

Acknowledgments

The authors are indebted to Dr. Emilie Morphew for critical review of the manuscript and to Dr. Charleen Chu for helpful discussions.

References

Bozzola, J. J., and Russell, L. D. (1992). "Electron Microscopy: Principles and Techniques for Biologists," Jones and Bartlett, Boston, MA.

Caruso, J. L., and Howell, D. N. (1999). In "Laboratory Diagnosis of Viral Infections," (E. H. Lennette, and T. F. Smith, eds.), 3rd edn., p. 21. Dekker, New York, NY.

Caruso, J. L., Levenson, R. M., and Howell, D. N. (2010). In "Biomedical Applications of Microprobe Analysis," (P. Ingram, J. D. Shelburne, V. L. Roggli, and A. LeFurgey, eds.). Academic Press, San Diego, CA (in press).

Chu, C. T., Howell, D. N., Morgenlander, J. C., Hulette, C. M., McLendon, R. E., and Miller, S. E. (2010). Am. J. Surg. Pathol. in press.

Connor, D. H., Chandler, F. W., Schwartz, D. A., Manz, H. J., and Lack, E. E. (eds.) (1997). "Pathology of Infectious Diseases," Appleton & Lange, Stamford, CT.

Deitch, J. S., Smith, K. L., Swann, J. W., and Turner, J. N. (1991). J. Electron Microsc. Technol. **18,** 82.

Doane, F. W., and Anderson, N. (1987). "Electron Microscopy in Diagnostic Virology," Cambridge Universiy Press, New York, NY.

Glauert, A. M. (1975). "Fixation, Dehydration, and Embedding of Biological Specimens," Elsevier, New York, NY.

Hayat, M. A. (1981). "Fixation for Electron Microscopy," Academic Press, New York, NY.

Hayat, M. A. (1989). "Principles and Techniques of Electron Microscopy: Biological Applications," CRC Press, Boca Raton, FL.

Henry, S. C., Schmader, K., Brown, T. T., Miller, S. E., Howell, D. N., Daley, G. G., and Hamilton, J. D. (2010). Submitted for publication.

Mackay, B. (ed.) (1981). "Introduction to Diagnostic Electron Microscopy," Appleton-Century-Crofts, New York, NY.

Mahoney, S. E., Duvic, M., Nickoloff, B. J., Minshall, M., Smith, L. C., Griffiths, C. E. M., Paddock, S. W., and Lewis, D. E. (1991). J. Clin. Invest. **88,** 174.

Miller, S. E. (1986). J. Electron Microsc. Technol. **4,** 265.

Miller, S. E. (1991). In "Medical Virology," (L. M. de la Maza, and E. M. Peterson, eds.), Vol. 10, p. 21. Plenum Press, New York, NY.

Miller, S. E. (1995). In "Diagnostic Procedures for Viral, Rickettsial and Chlamydial Infections," (E. H. Lennette, D. A. Lennette, and E. T. Lennette, eds.), p. 37. American Public Health Association, Washington, DC.

Miller, S. E. (1999). In "Laboratory Diagnosis of Viral Infections," (E. H. Lennette, and T. F. Smith, eds.), 3rd edn., p. 45. Dekker, New York, NY.

Miller, S. E., and Howell, D. N. (1988). J. Electron Microsc. Technol. **8,** 41.

Miller, S. E., and Howell, D. N. (1997). *Immunol. Invest.* **26**, 29.

Miller, S. E., Levenson, R. M., Aldridge, C., Hester, S., Kenan, D. J., and Howell, D. N. (1997). *Ultrastruct. Pathol.* **21**, 183.

Murphy, F. A., Fauquet, C. M., Bishop, D. H. L., Ghabrial, S. A., Jarvis, A. W., Martelli, G. P., Mayo, M. A., and Summers, M. D. (eds.) (1995). "Virus Taxonomy: Sixth Report of the International Committee on Taxonomy of Viruses," Springer, New York, NY.

Palmer, E. L., and Martin, M. L. (1988). "Electron Microscopy in Viral Diagnosis," CRC Press, Boca Raton, FL.

Payne, C. M. (1997). *In* "Pathology of Infectious Diseases," (D. H. Connor, F. W. Chandler, D. A. Schwartz, H. J. Manz, and E. E. Lack, eds.), Vol. 1, p. 9. Appleton & Lange, Stamford, CT.

Reynolds, R. J., Little, G. J., Lin, M., and Heath, J. W. (1994). *J. Neurocytol.* **23**, 555.

Sun, X. J., Tolbert, L. P., and Hildebrand, J. G. (1995). *J. Histochem. Cytochem.* **43**, 329.

van den Brule, A. J. C., Cromme, F. V., Snijders, P. J. F., Smit, L., Oudejans, C. B. M., Baak, J. P. A., Meijer, C. J. L. M., and Walboomers, J. M. M. (1991). *Am. J. Pathol.* **139**, 1037.

CHAPTER 19

Membrane Trafficking

Sabine Kupzig, San San Lee, and George Banting

Department of Biochemistry
School of Medical Sciences
University of Bristol, University Walk
Bristol, United Kingdom

I. Introduction

Immunofluorescence microscopy has played a central role in providing information concerning the localization of specific integral membrane proteins. However, immunofluorescence microscopy simply generates a "snapshot" of what is going on and fails to provide direct information concerning the movement of integral membrane proteins in living cells. Time course experiments, for example, designed to study the effects of specific hormones and drugs on intracellular morphology and membrane trafficking, are also particularly laborious to perform if based on the use of immunofluorescence microscopy. Such studies, and many others on the dynamic nature of membrane traffic within eukaryotic cells, would be much easier to perform and would not be subject to possible fixation artifacts if specific integral membrane proteins could be visualized in living cells. The expression of recombinant, green fluorescent protein (GFP)-tagged integral membrane proteins has provided us with precisely the right tools with which to perform such experiments. This chapter describes the use of a Leica (Leica Microsystems GmbH, Heidelberg, Germany;

http://www.llt.de/index1.html) DM IRBE inverted epifluorescence microscope attached to a Leica TCS-NT confocal laser scanning system to monitor the localization and movement of GFP-tagged integral membrane proteins in living cells; however, the general principles are applicable to a wide range of microscope systems.

II. Methods

A. Construct Design

Conventional molecular biological techniques should be used to generate constructs encoding GFP-tagged integral membrane proteins. We have found that GFP does not affect the intracellular distribution of the tagged membrane protein, and that GFP remains fluorescent when placed on either side of the membrane (i.e., cytosolic or within the lumen of the organelle to which the tagged membrane protein is localized) (Girotti and Banting, 1996). In the constructs we have made, we have introduced a short (four- to six-amino acid), flexible (glycine-rich) linker region between the GFP sequence and that of the tagged protein. This may or may not be necessary, but is done to increase the chances of the two components of the hybrid protein folding independently and correctly.

B. Construct Expression

The hybrid DNA construct should be placed in an appropriate eukaryotic expression vector. Our experience with GFP-tagged ratTGN38 (a type I integral membrane protein that is predominantly localized to the trans-Golgi network (TGN) at steady state; Luzio *et al.*, 1990) is that the protein is not correctly localized when expressed in transiently transfected cells (Girotti and Banting, 1996). However, when the same hybrid protein is expressed in stably transfected cells it is found to be correctly localized (Girotti and Banting, 1996), as has been described for another GFP-tagged, Golgi-localized integral membrane protein (*N*-acetylglucosamine transferase 1) expressed in stably transfected cells (Shima *et al.*, 1997). Others have expressed GFP-tagged Golgi membrane proteins in transiently transfected cells and observed appropriate localization (e.g., see Cole *et al.*, 1995, 1996). It is unclear why some constructs should be correctly localized only when expressed in stably transfected cells, but this may be due to the level of expression of the recombinant protein, because expression levels in stable transfectants are generally lower than those in transiently transfected cells. Thus, even if a GFP-tagged integral membrane protein is not appropriately localized in transiently transfected cells, it may be in stable transfectants.

C. Imaging

Stably transfected cells expressing the GFP-tagged membrane protein of interest are cultured on round (22-mm diameter) sterile glass coverslips until approximately 60–70% confluent. Immediately prior to image acquisition these coverslips are transferred to an appropriate holder for use on the laser scanning confocal microscope. We generally use a homemade aluminum holder as illustrated in Fig. 1. Approximately 0.2 ml of CO_2-independent medium (GIBCO-BRL, Gaithersburg, MD), prewarmed to 37 °C, is then added to bathe the cells on the upper surface of the coverslip. The slide holder is then placed into the center of a heated jacket (Medical Systems Corp., Greenvale, NY; http://www.medicalsystems.com/msc.htm) attached to a thermal regulator unit (Medical Systems Corp.) that has

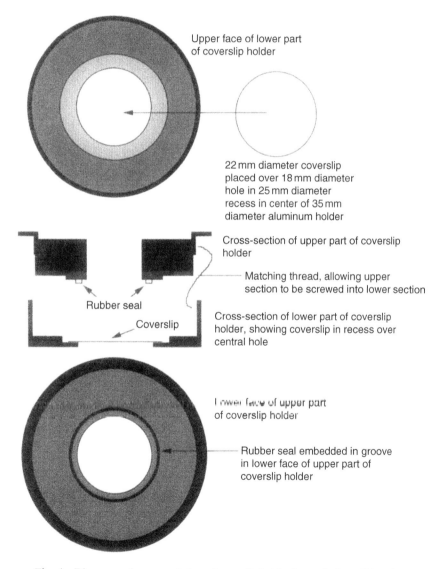

Upper face of lower part of coverslip holder

22 mm diameter coverslip placed over 18 mm diameter hole in 25 mm diameter recess in center of 35 mm diameter aluminum holder

Cross-section of upper part of coverslip holder

Matching thread, allowing upper section to be screwed into lower section

Rubber seal

Coverslip

Cross-section of lower part of coverslip holder, showing coverslip in recess over central hole

Lower face of upper part of coverslip holder

Rubber seal embedded in groove in lower face of upper part of coverslip holder

Fig. 1 Diagrammatic representation of coverslip holder for use in live cell imaging.

been precalibrated to maintain the temperature of the medium bathing the cells on the coverslip at 37 °C. This unit is then placed on the stage of a Leica DM IRBE inverted epifluorescence microscope attached to a Leica TCS-NT confocal laser scanning head equipped with a krypton–argon laser (488-, 568-, and 647-nm lines). The cells are initially viewed under illumination from a halogen lamp through a 63× (1.4 NA) oil immersion, objective lens using a fluorescein filter set. Once a suitable cell has been brought into focus and appropriately aligned on the stage, the microscope is switched to the confocal laser scanning format. The slide is then scanned at 488 nm and images acquired using the Leica TCS-NT software. Initial scanning should be performed at medium scan speed and with laser intensity set below 50% in order to minimize photobleaching of the sample prior to image acquisition. The relevant photomultiplier tube (PMT) can be used to adjust the intensity of the signal obtained; increasing the PMT value will give a brighter signal, but resolution will be lost. There is thus always a trade-off between low laser intensity plus high PMT value (minimal photobleaching, detectable signal, poor resolution) and high laser intensity plus low PMT value (increased photobleaching, detectable signal, increased resolution). Acceptable image resolution is generally obtained with the PMT value set below 600 (preferably closer to 300). The system can be converted to time-lapse image acquisition once a suitable image has been obtained. Note that the zoom facility in the Leica TCS-NT software allows the operator to increase the displayed image size up to 32 times that of the original scanned and displayed image. Before time-lapse image acquisition the scan speed should be changed to "slow" in order to obtain brighter images and the software should be configured (according to manufacturer instructions) to scan and record images at defined intervals (generally every 1–20 s). The Leica TCS-NT software allows the operator to set up a simple time-lapse procedure; for example, scanning a defined number of times (to generate an average image) at defined intervals for a specified number of frames. In reality, it is impractical to perform multiple scans at a particular point in time as there is movement within the sample during the time period in which the multiple scans, which are used to generate the average image, are performed; thus single scans are performed at defined intervals for a specified number of frames in order to generate a time-lapse movie. The Leica TCS-NT software also allows the operator to set up a more complex time-lapse procedure in which the frequency of frame acquisition may be varied. Once acquired, the movie can be played back using the TCS-NT software and can be saved in either scanner file format (for further manipulation using the TCS-NT software) or in export file format (for transfer as a folder of individual frames to other systems—Macintosh, PC, Silicon Graphics, etc.—for video compilation and/or image analysis).

This describes the basic procedures involved in imaging live cells expressing GFP-tagged proteins. Clearly, the effects of various factors (changes in temperature, different drugs, growth factors, hormones, etc.) on the morphology of the organelle in which a specific GFP-tagged integral membrane protein is primarily located can be assayed using this system (e.g., the effects of the fungal metabolite brefeldin A (BFA) on the morphology of the TGN can be demonstrated using

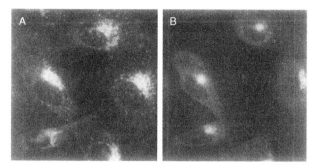

Fig. 2 Appearance of cells before and after BFA is administered. Live, stably transfected NRK cells expressing GFP-tagged ratTGN38 were imaged at 37 °C on a Leica TCS-NT confocal laser scanning microscope (as described in text) before (A) and after (B) incubation in brefeldin A (BFA, 5 μg/ml) for 30 min. These images show the "before and after" effects of BFA. In fact, the cells were imaged regularly during the incubation in BFA. This provided a series of images that have been used to generate a time-lapse movie showing what happens to the morphology of the TGN during incubation of cells in BFA.

GFP-tagged ratTGN38 as a reporter; Fig. 2), as also the movement of populations of GFP-tagged integral membrane proteins over time (e.g., GFP-tagged ratTGN38 in stably transfected normal rat kidney (NRK) cells; Fig. 3).

D. Fluorescence Recovery after Photobleaching

The basic procedure described in the preceding section can be modified to permit the study of the diffusional mobility in the plane of the lipid bilayer of GFP-tagged integral membrane proteins. Such studies involve fluorescence recovery after photobleaching (FRAP) (also known as fluorescence photobleaching recovery, FPR) analysis. The principle underlying FRAP is that transiently increasing the intensity of the light illuminating a particular region of a fluorescently labeled cell will lead to photobleaching of that region of the cell that has been subjected to increased illumination, that is, a "black hole" will be generated in the region of fluorescence. Once the photobleached area has been generated, further time-lapse imaging (under conditions of reduced light intensity) will allow analysis of how rapidly the photobleached area is repopulated with fluorescent molecules from the surrounding area, that is, how rapidly fluorescence recovers after photobleaching. This procedure has been used to monitor the lateral diffusional mobility of integral membrane proteins and lipids in the plane of the lipid bilayer at the cell surface (e.g., see Edidin and Stroynowski, 1990; Edidin *et al.*, 1994; Golan and Veatch, 1980; Sheetz *et al.*, 1980, 1987; Yeichel and Edidin, 1987), and more recently to study the lateral diffusional mobility of integral membrane proteins in the plane of the lipid bilayer in intracellular membranes (Cole *et al.*, 1996). The earlier work on lateral diffusional mobility of integral membrane proteins and lipids in the plane of the lipid bilayer at the cell surface used fluorescently labeled lipid, fluorescently labeled integral membrane proteins, fluorescently labeled lectins, or fluorescently

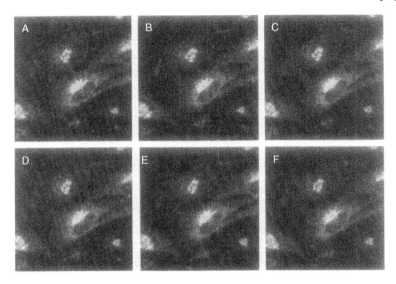

Fig. 3 Membrane movement in living cells. Live, stably transfected NRK cells expressing GFP-tagged ratTGN38 were imaged at 37 °C on a Leica TCS-NT confocal laser scanning microscope (as described in text). Images were recorded every 5 s over a 5-min period in order to visualize membrane movement within the cells. (A–F) The first six images in this series. Multiple vesicle-like profiles can be seen to have moved during this time period and the morphology of the TGN has also changed slightly. The directional "vesicle" movement is dependent on the presence of intact microtubules, because incubation of cells in the presence of the microtubule-depolymerizing drug nocodazole (20 μg/ml) leads to cessation of this movement.

labeled antibodies attached to target membrane proteins as reporter molecules; the more recent studies on the lateral diffusional mobility of integral membrane proteins in the plane of the lipid bilayer in intracellular membranes have used GFP-tagged integral membrane proteins as reporters. To perform FRAP experiments, cells are prepared and image acquisition proceeds initially as described in the previous section. However, once the first image has been obtained the zoom function should be used to focus on a specific, fluorescent region of the cell under study. This region should be positioned in the center of the screen prior to zooming in, since the zoom function will automatically zoom in on the center of the screen. The operator should zoom in either 16 or 32 times (depending on the area it is intended to photobleach). The scan should continue (at slow scan speed) and the laser power should be increased to maximum until the fluorescent signal from GFP is lost. It is advisable in these experiments to use a readily photobleachable version of GFP that is not expressed at particularly high levels. The reason for this is that the aim is to obtain maximal photobleaching while minimizing the potential for photodamage to the membranes; by using the S65T/Q80R/I167T triple mutant of GFP we have found that two or three slow scans at maximum laser power are generally sufficient to photobleach the desired region. The laser power should be returned to that previously used for image acquisition during scanning as soon as

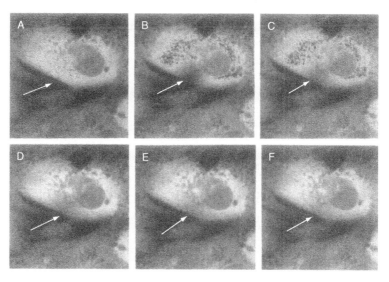

Fig. 4 Fluorescence recovery after photobleaching. Live, chick embryo fibroblast (CEF) cells expressing GFP-tagged sialyltransferase (ST) were incubated in the presence of BFA (5 µg/ml) for 10 min in order to redistribute the ST into the endoplasmic reticulum (ER), and were then subjected to FRAP analysis (as described in text). (A) A cell immediately prior to photobleaching of an area of the ER (the area to be photobleached is indicated by an arrow). (B) The same cell immediately after photobleaching of an area of the ER (the area that has been photobleached is indicated by an arrow). (C–F) Images (taken at 2-s intervals) of the same cell as it recovers from photobleaching.

photobleaching is complete. Time-lapse image acquisition should then proceed as already described in order to record fluorescence recovery of the photobleached area. An example of FRAP is given in Fig. 4.

E. Combining GFP Imaging with Immunofluorescence

We have observed that GFP fluorescence remains after both methanol and paraformaldehyde fixation of cells (Girotti and Banting, 1996). It is therefore possible to combine conventional immunofluorescence microscopy, using a tetramethylrhodamine B isothiocyanate (TRITC-) or rhodamine-conjugated secondary antibody, with detection of GFP-tagged membrane proteins for colocalization studies (Girotti and Banting, 1996).

III. Future Developments

The ever-expanding range of spectral variants of GFP will provide the opportunity to tag various integral membrane proteins with GFP molecules with various excitation and emission spectra. Thus, given a confocal microscope with a spectrophotometer scanner head that can be tuned to appropriate wavelengths and bandwidths, it will be possible to colocalize different integral membrane proteins in the same living cell. It should also be possible to perform fluorescence resonance energy transfer (FRET) experiments in living cells in order to address the proximity of different integral membrane proteins in living cells.

Acknowledgments

The authors thank David Stephens for critical reading of the manuscript, the Medical Research Council for providing an Infrastructure Award (G4500006) to establish the School of Medical Sciences Cell Imaging Facility, and Dr. Mark Jepson for assistance with confocal image processing. The authors also thank Prof. Colin Hopkins for providing the cells used for the experiment presented in Fig. 4.

References

Cole, N., Terasaki, M., Sciaky, N., and Lippincott-Schwartz, J. (1995). *Mol. Biol. Cell* **6**, 8.
Cole, N. B., Smith, C. L., Sciaky, N., Terasaki, M., Edidin, M., and Lippincott-Schwartz, J. (1996). *Science* **273**, 797.
Edidin, M., and Stroynowski, I. (1990). *J. Cell Biol.* **112**, 1143.
Edidin, M., Zuniga, M. C., and Sheetz, M. P. (1994). *Proc. Natl. Acad. Sci. USA* **91**, 3378.
Girotti, M., and Banting, G. (1996). *J. Cell Sci.* **109**, 2915.
Golan, D. E., and Veatch, W. (1980). *Proc. Natl. Acad. Sci. USA* **77**, 2537.
Luzio, J. P., Brake, B., Banting, G., Howell, K. E., Braghetta, P., and Stanley, K. K. (1990). *Biochem. J.* **270**, 97.
Sheetz, M. P., Schindler, M., and Koppel, D. E. (1980). *Nature (London)* **285**, 510.
Shields, M., La Celle, P., Waugh, R. E., Scholz, M., Peters, R., and Passow, H. (1987). *Biochim. Biophys. Acta* **905**, 181.
Shima, D. T., Haldar, K., Pepperkok, R., Watson, R., and Warren, G. (1997). *J. Cell. Biol.* **137**, 1211.
Yeichel, E., and Edidin, M. (1987). *J. Cell Biol.* **105**, 755.

PART III

Green Fluorescent Protein

CHAPTER 20

Monitoring of Protein Secretion with Green Fluorescent Protein

Christoph Kaether and Hans-Hermann Gerdes

Institute for Neurobiology
University of Heidelberg
Heidelberg
Germany

I. Introduction

In eukaryotic cells, the transport of secretory proteins from the endoplasmic reticulum (ER) to the plasma membrane (PM) is orchestrated by a number of vesicular transport steps. For each of these steps carrier vesicles bud from topologically separate compartments of the secretory pathway and specifically fuse with the membrane of the subsequent compartment (Palade, 1975). In addition, the steady state size of the various compartments is maintained by a membrane flux opposite to the exocytic route. Despite the complexity of this traffic, the delivery of proteins to the PM occurs with remarkable speed, emphasizing the dynamics as a prevalent feature of the secretory pathway.

Biochemical approaches largely based on *in vitro* reconstitution assays led to the discovery of molecules involved in the budding and fusion reactions along this pathway (Rothman, 1994). Furthermore, morphological studies on fixed cells provided snapshots of the underlying membranous structures. Despite this prog-ress, the spatial and temporal resolution of secretory traffic as it occurs in live cells remained largely elusive, because appropriate tools to address this issue were absent.

The availability of the green fluorescent protein (GFP) opened the door for real-time studies and made it possible to monitor distinct vesicular transport steps of the secretory pathway in real time (see also Chapter 19 in this volume; Kupzig et al., 1999). Transport from the ER to the Golgi was studied using the vesicular stomatitis virus (VSV) G protein tagged with GFP (Presley et al., 1997; Scales et al., 1997). Our group is focusing on the transport of proteins from the trans-Golgi network (TGN) to the PM (Wacker et al., 1997). This transport is either constitutive, if TGN-derived secretory vesicles fuse continuously with the PM, or regulated, if the vesicles fuse only in response to a stimulus. The regulated pathway is a special feature of certain cell types such as exocrine, endocrine, and neuronal cells, whereas the constitutive pathway is common to all eukaryotic cells. To monitor these two pathways we have tagged a secretory protein, human chromogranin B (hCgB), with GFP. hCgB is a soluble protein of neuroendocrine secretory granules (SGs) (Rosa and Gerdes, 1994), the storage organelles of the regulated pathway. Ectopic expression of this protein in cells with a regulated pathway results in its sorting to SGs (Krömer et al., 1998). Expression in cells lacking a regulated pathway of protein secretion, however, leads to its constitu-tive secretion (Kaether and Gerdes, 1995). Thus, depending on the cell type in which hCgB is expressed, it serves as a marker for SGs or constitutive secretory vesicles (CVs). In this chapter we describe methods for monitoring constitutive and regulated protein secretion in living cells, using hCgB tagged with GFP.

II. Constitutive Secretion

A. Principle of the Trans-Golgi Network-to-Plasma Membrane Secretion Assay

To study constitutive secretion from the TGN to the PM we use hCgB (Benedum et al., 1987) that is C-terminally tagged with GFP(S65T) (Heim et al., 1995). The fusion protein, hCgB–GFP(S65T) (Kaether and Gerdes, 1995), is expressed under the control of a cytomegalovirus (CMV) promoter in a CDM8 vector (Invitrogen, Leek, The Netherlands). We express hCgB–GFP(S65T) transiently in Vero or HeLa cells (Fig. 1), or take advantage of a stable Vero cell line, V7, expressing hCgB–GFP(S65T) (Wacker et al., 1997). Vero and HeLa cells have only a constitu-tive pathway of protein secretion.

Under normal culture conditions at 37 °C no fluorescent hCgB–GFP(S65T) is detected (Fig. 1, top row). However, when biosynthetic protein transport is arrested at the TGN by a 20 °C secretion block, fluorescent hCgB–GFP(S65T) is observed in the TGN (Fig. 1, middle row). Subsequent reversal of the secretion block results in the formation of fluorescent CVs (Fig. 1, bottom row), which can

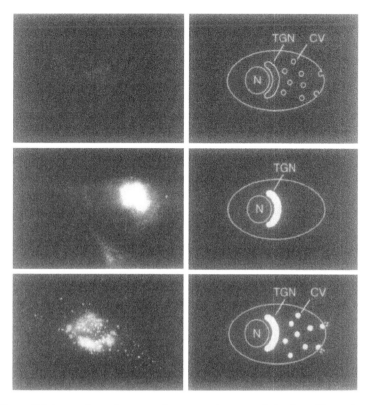

Fig. 1 Trans-Golgi network-to-plasma membrane secretion assay. *Left column*: Photomicrographs of HeLa cells transiently transfected with hCgB–GFP(S65T). *Right column*: Schematics symbolizing the principle of the assay; structures filled with white indicate the presence of GFP fluorescence. *Top row*: Cells incubated at 37 °C. *Middle row*: Cells incubated at 20 °C for 2 h. *Bottom row*: Cells incubated for 2 h at 20 °C and then for 30 min at 37 °C. N, Nucleus; TGN, trans-Golgi network: CV, constitutive secretory vesicle.

be analyzed by fluorescence microscopy. Thirty to 60 min after the release of the secretion block, no cellular GFP fluorescence is detected, reflecting the complete secretion of fluorescent hCgB–GFP(S65T).

1. Comment

By using hCgB–GFP(S65T) in combination with a secretion block and a subsequent release, we have detected green fluorescent CVs in all cell lines that have only a constitutive pathway of protein secretion. We have tested baby hamster kidney (BHK), HeLa, Madin–Darby canine kidney (MDCK), Vero, NIH 3T3, RK13, and human thymidine kinasedeficient (HuTK⁻) cells.

B. Experimental Procedure

1. Cell Culture and Transfection

Vero cells (African green monkey kidney cells, ATCC CCL81) are grown at 37 °C and 5% CO_2 in Eagle's minimal essential medium (EMEM; GIBCO-BRL, Eggenstein, Germany) supplemented with 10% (v/v) fetal calf serum and 2 mM glutamine. Transfection is performed either by standard calcium phosphate precipitation after plating the cells on coverslips in 3.5-mm dishes, or by electroporation using a Bio-Rad Gene Pulser (Bio-Rad Laboratories, Richmond, CA). For electroporation, confluent cells on a 15-cm dish are trypsinized, pelleted, and resuspended in 0.8 ml of phosphate-buffered saline (PBS). The cell suspension is transferred to a 0.8-ml cuvette and 50 µg of DNA purified on a Nucleobond AX500 column (Macherey-Nagel GmbH, Düren, Germany) is added. Thereafter the suspension is kept for 5 min at room temperature, with occasional mixing. Cells are then electroporated at 300 V and 960 µF, diluted in culture medium, and plated immediately.

2. Stable Cell Lines

For generation of a stable cell line expressing hCgB–GFP(S65T), Vero cells are cotransfected with hCgB–GFP(S65T)/pCDM8 linearized with *Sfi* I, and with pMC1-Neo (Stratagene, La Jolla, CA) linearized with *Sca*I, and then incubated in cell culture medium. Two days later the medium is replaced with fresh cell culture medium supplemented with G418 (0.5 mg/ml; GIBCO, Grand Island, NY). After 3–6 weeks of selection, single G418-resistant colonies are picked (by drawing them into a 200-µl pipette tip), triturated, and replated in 24-well dishes. Clones are analyzed for expression of GFP fusion proteins after treatment with 2 mM butyrate for 21 h (see Section II.B.3) by Western blotting, immunofluorescence staining, and detection of GFP fluorescence. For Western blotting a peptide antibody directed against the C terminus of GFP is used (Wacker *et al.*, 1997). Immunofluorescence microscopy is performed with a polyclonal anti-GFP antibody (Clontech Laboratories, Palo Alto, CA). For detection of GFP, fluorescence cells are analyzed after application of a 20 °C secretion block for 2 h (see Section II.B.5).

3. Sodium Butyrate Treatment

To enhance expression of the GFP fusion proteins we often use sodium buyrate, which induces expression of CMV promoter-driven genes. We have found that treatment with sodium butyrate leads to morphological changes of the cells (e.g., Vero and HeLa cells become spindle-shaped). For sufficient induction and minimal morphological changes in Vero cells the treatment has been optimized to a 21-h incubation period in 2 mM sodium butyrate. To prepare the induction medium we use cell culture medium and a stock solution of 500 mM sodium butyrate in H_2O stored at 4 °C.

4. Culture Chambers for Microscopy

To record living cells, disposable culture chambers that have a coverslip as bottom can be used (LabTek chambered coverglass; Nalge Nunc International, Naperville, IL). Alternatively, a culture chamber equipped with a 42-mm coverslip for holding cells (POC chamber; Bachofer, Reutlingen, Germany) may be assembled. The Bachofer chamber is fitted with an adaptor (available from Leica) for the stage of the microscope. For both setups the temperature is controlled by a custom-made, water-heated objective jacket.

5. Secretion Block and Release

For a 20 °C secretion block a heating bath is placed in a 4 °C cold room and adjusted to 20 °C. The block is started by replacing the culture medium with precooled medium supplemented with 10 mM HEPES, pH 7.4. The dishes are then placed for 2 h on a glass plate, which is located just below the water surface of the heating bath.

If cells are grown in LabTek chambers, the chambers are placed directly after the 20 °C incubation onto the microscope stage. The medium is subsequently replaced with prewarmed medium containing 10 mM HEPES, pH 7.4, or with prewarmed PBS. If cells are grown on 42-mm coverslips, the Bachofer chamber is assembled, perfused with prewarmed medium or PBS, and transferred to the microscope stage. Biogenesis of CVs is observed 10–15 min after release of the block and their movement is recorded.

a. Comment

In principle, the secretion block and its subsequent release can be accomplished on the microscope stage. However, during the warming from 20 to 37 °C thermal extensions of the objective and the microscope stage cause a continuous shift of the focal plane, disabling cell recordings during the warming-up period. Therefore, the secretion block is performed separately in a water bath, and thereafter the cells are transferred to the microscope, the objective of which has been prewarmed for at least 30 min.

6. Confocal Microscopy

The confocal microscope is used when optical sectioning or simultaneous detection of two different fluorescent probes is required. In this way the dynamics of the TGN can be monitored, and exo- and endocytic transport can be visualized simultaneously.

We use a Leica TCS-4D confocal system equipped with an argon–krypton laser (Leica Microsystems GmbH, Heidelberg, Germany) and a 63×/1.4 NA PlanApo objective. GFP fluorescence is detected with 488-nm excitation and a Leica filter set optimized for GFP detection (TK 500 and BP 525/50).

Images are recorded with 512 × 512 pixel resolution in the fast scan mode and four-times line averaging. These settings allow a time resolution of 1.7 s. Series of images are stored as TIFF files on a magnetooptical disk. We usually record series of 20–40 images per cell. During these recordings minor photobleaching and no effects of phototoxicity are observed.

7. Two-Color Imaging

To record exo- and endocytic traffic simultaneously, Vero cells expressing hCgB–GFP(S65T) are fed with Texas Red-labeled transferrin (Molecular Probes, Eugene, OR) during the 20 °C secretion block as described (Wacker *et al.*, 1997). The confocal settings for two-channel scanning are as follows: excitation 488/568 nm, double dichroic (DD) 488/568, BP 525/50 for GFP detection, LP 590 for Texas Red detection, fast scan mode, four-times line averaging, 512 × 512 pixel resolution. The images of the two channels can then be combined using the multicolor analysis option of the confocal software.

a. Comment

A two-color video can be viewed on the World Wide Web (http://www.nbio.uni-heidelberg.de/Gerdes.html). An example of two-color imaging of two different GFP mutants, using a Leica confocal system, is described by Zimmerman and Siegert (1998).

8. Fluorescence Video Microscopy

For better time resolution we employ a cooled charge-coupled device (CCD) camera. A Quantix-G2 camera (Photometrics GmbH, Munich, Germany) is mounted onto a Zeiss Axiophot (Carl Zeiss Jena GmbH, Jena, Germany) equipped with a 100-W HBO lamp and a Zeiss Plan Apochromat 63×/1.4 objective. GFP fluorescence is detected with a special filter set from Zeiss (BP 470/20, FT 493, BP 505–530). For the Axiophot a Bachofer chamber with a temperature-regulated holder frame (Bachofer) is used. The camera is controlled by IP Lab Spectrum software (Signal Analytics Corp., Vienna, VA) that runs on a MacIntosh 9500/150 Power PC. With this setup moving vesicles can be recorded with an exposure time down to 0.2 s, allowing sufficient resolution of rapid movements in the range of 2 μm/s.

a. Comment

When Vero cells are imaged with a CCD camera the secretion of hCgB–GFP (S65T) into the medium leads to strong background fluorescence. This can be reduced by frequent changes of the medium or by perfusion of the Bachofer chamber. Confocal imaging precludes background problems caused by secreted hCgB–GFP(S65T), owing to the confocal sectioning.

9. Image Processing

Images obtained by confocal or CCD video microscopy are already digitized and can be transferred to various image-processing programs. For quantitation of fluorescence intensities or animation of a series of images we use NIH-Image, a shareware program from the National Institutes of Health (available at http://rsb. info.nih.gov/nih-image/). Adobe Photoshop (Adobe Systems, Mountain View, CA) is used to lay out images, and Avid Videoshop (Avid Technology, Tewksbury, MA) is used to create video clips.

10. Tracking

To analyze series of images we track vesicles using a macro written for NIH-Image (Wacker *et al.*, 1997). From these tracks parameters such as velocities, travel distances, and direction of movement are calculated. In this way the effect of manipulations such as drug treatment or injection of antibodies on vesicle motility can be quantitated. By following this procedure we have analyzed the effect of nocodazole on vesicle motility and were able to demonstrate that CV movement (from the TGN to the PM) is microtubule dependent (Wacker *et al.*, 1997).

11. Fluorometry

Secretion of fluorescent hCgB–GFP(S65T) can be measured by fluorometry. For this the stable clone V7 and Vero wild-type cells (as a control) are grown for 1 day to subconfluency and then treated for 21 h with 2 mM sodium butyrate. To accumulate GFP fluorescence in the TGN, cells are incubated at 20 °C for 2 h in external solution (ES), consisting of 140 mM NaCl, 5 mM KCl, 2 mM $CaCl_2$, 1 mM $MgCl_2$, 1 mM glucose, and 10 mM HEPES (pH 7.4). Thereafter ES (20 °C) is replaced by ES (37 °C) and cells are incubated for 2 h at 37 °C. The ES is collected every 30 min during the 37 °C incubation. Emission spectra are recorded with excitation at 470 nm (Aminco Bowman fluorimeter) from the ES collected after various incubation times at 37 °C and from cells after 2 h of incubation at 37 °C. To obtain spectra of cellular hCgB–GFP(S65T), fluorescence cells are lysed by three cycles of freezing and thawing in hypotonic buffer (10 mM HEPES (pH 7.4), 1 mM EDTA, 1 mM magnesium acetate, 1.25 mM phenyl-methylsulfonyl fluoride (PMSF), leupeptin (10 μg/ml)) followed by preparation of high-speed (100,000 × *g*) supernatants. Emission spectra are normalized to total cellular protein. Difference spectra are calculated by subtracting the spectra of wild-type cells from those of V7 cells (Fig. 2). For calculation of secretion kinetics the peak values at 510 nm can be used (Wacker *et al.*, 1997).

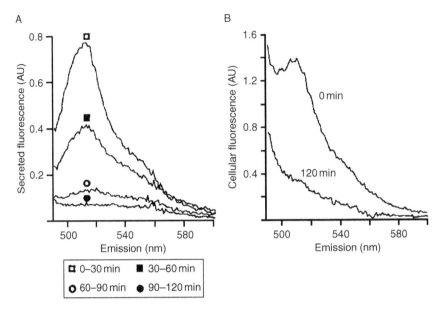

Fig. 2 Fluorometric analysis of hCgB–GFP(S65T) secretion. (A) Difference spectra of external solutions collected during release at 37 °C as indicated. (B) Difference spectra of cell lysates after a 2-h secretion block (0 min) or after 2 h of release at 37 °C (120 min). AU, arbitrary unit.

III. Regulated Secretion

A. Rationale

A crucial point for the *in vivo* analysis of SGs is the specificity of labeling. In the TGN of neuroendocrine cells regulated secretory proteins are sorted from constitutively secreted proteins and packaged into SGs (Kelly, 1985). Inefficient sorting of a regulated protein tagged with GFP would result in labeling of both SGs and CVs, thus disabling the unequivocal identification of SGs by fluorescence microscopy. Therefore, one must first determine the sorting efficiency of the GFP fusion protein at the level of the TGN. In our opinion, the most appropriate way to address this is by pulse–chase labeling in combination with gradient centrifugation. This method permits the separation of CVs from newly formed SGs on a gradient. A detailed protocol for this type of analysis has been described (Tooze and Huttner, 1992). By using this protocol we have been able to show that hCgB–EGFP (another GFP mutant; see Section III.B) exits from the TGN specifically into SGs, thus representing a highly specific fluorescent probe for SGs (Kaether *et al.*, 1997). With this probe movement of SGs, their docking and subsequent exocytosis can be addressed at the single-cell level. The fact that GFP elicits bright fluorescence in SGs demonstrates its applicability as an *in vivo* reporter under extreme conditions, that is, when it is aggregated and exposed to a pH 5.0–5.5.

B. Experimental Procedure

To fluorescently label SGs we use another GFP mutant, EGFP (Clontech Laboratories), fused to the C terminus of hCgB (hCgB–EGFP) (Kaether *et al.*, 1997). Because of the more efficient and temperature-insensitive folding of EGFP, a detectable amount of green fluorescence is generated without a 20 °C secretion block. We now also use hCgB–GFP(S65T) in cells with a regulated pathway of secretion in combination with a 20 °C secretion block (Salm *et al.*, 1998).

As a regulated cell line we use PC12 cells, a rat pheochromocytoma cell line (clone 251) (Heumann *et al.*, 1983); these cells are grown at 37 °C and 10% CO_2 in Dulbecco's modified Eagle's medium (DMEM) supplemented with 10% (v/v) horse serum and 5% (v/v) fetal calf serum. PC12 cells are prepared for electroporation in the same manner as Vero cells, except that electroporation is performed at a density of $2-3 \times 10^7$ cells/ml and at 250 V and 960 µF. Stable cell lines are generated as described for Vero cells. Expression of hCgB–EGFP can be enhanced by incubation in 10 mM sodium butyrate for 17 h.

Living cell recordings are done by confocal microscopy as described for constitutive secretion. To examine the motility of SGs near the PM we have taken confocal sections at the bottom of the cells, where the PM is attached to the coverslip (Kaether *et al.*, 1997).

1. Comment

An elegant way of visualizing GFP-labeled SGs in the immediate vicinity of the PM has been demonstrated by Lang *et al.* (1997), using evanescent-wave microscopy.

Acknowledgments

We thank D. Corbeil, J. Garwood, and R. Rudolf for critical comments on the manuscript. C. Kaether was supported by the Graduiertenkolleg Molekulare und Zelluläre Neurobiologie. H.-H. Gerdes is a recipient of a grant from the Deutsche Forschungsgemeinschaft (SFB 317/C7).

References

Benedum, U. M., Lamouroux, A., Konecki, D. S., Rosa, P., Hille, A., Baeuerle, P. A., Frank, R., Lottspeich, F., Mallet, J., and Huttner, W. B. (1987). *EMBO J* **6**, 1203.
Heim, R., Cubitt, A. B., and Tsien, R. Y. (1995). *Nature (London)* **373**, 663.
Heumann, R., Kachel, V., and Thoenen, H. (1983). *Exp. Cell Res.* **145**, 179.
Kaether, C., and Gerdes, H. H. (1995). *FEBS Lett* **369**, 267.
Kaether, C., Salm, T., Glombik, M., Almers, W., and Gerdes, H. H. (1997). *Eur. J. Cell Biol.* **74**, 133.
Kelly, R. B. (1985). *Science* **230**, 25.
Krömer, A., Glombik, M., Huttner, W. B., and Gerdes, H. H. (1998). *J. Cell Biol.* **140**, 1331.
Kupzig, S., Lee, S. S., and Banting, G. (1999). *Methods Enzymol.* **302**, Chapter 19, this volume.
Lang, T., Wacker, I., Steyer, J., Kaether, C., Wunderlich, I., Soldati, T., Gerdes, H. H., and Almers, W. (1997). *Neuron* **18**, 857.

Palade, G. E. (1975). *Science* **189,** 347.

Presley, J. F., Cole, N. B., Schroer, T. A., Hirschberg, C., Zaal, K. J. M., and Lippincott-Schwartz, J. (1997). *Nature (London)* **389,** 81.

Rosa, P., and Gerdes, H. H. (1994). *J. Endocrinol. Invest.* **17,** 207.

Rothman, J. E. (1994). *Nature (London)* **372,** 55.

Salm, T., Rudolf, R., and Gerdes, H. H. (1998). Manuscript in preparation.

Scales, S. J., Pepperkok, R., and Kreis, T. E. (1997). *Cell* **90,** 1137.

Tooze, S. A., and Huttner, W. B. (1992). *Methods Enzymol.* **219,** 81.

Wacker, I., Kaether, C., Krömer, A., Migala, A., Almers, W., and Gerdes, H. H. (1997). *J. Cell Sci.* **110,** 1453.

Zimmerman, T., and Siegert, F. (1998). *BioTechniques* **24,** 458.

CHAPTER 21

Comparison of Enhanced Green Fluorescent Protein and Its Destabilized Form as Transcription Reporters

Xiaoning Zhao, Tommy Duong, Chiao-Chian Huang, Steven R. Kain, and Xianqiang Li

Department of Cell Biology
CLONTECH Laboratories, Inc.
Palo Alto, California, USA

I. Introduction

Studies of signal transduction and gene expression, as well as of the regulation of these processes, often rely on an *in vivo* reporter system. The key element in such a system is the reporter gene that is used to measure the change in these biological processes. Commonly, this reporter gene is linked to the regulatory sequence of interest, such as an inducible promoter or enhancer. Assay of the reporter gene product provides a quantitative measure of the level of gene expression, and thus *cis*-acting sequences and *trans*-acting factors can be characterized.

Currently, several reporter genes are commonly used, such as β-galactosidase (An *et al.*, 1982), luciferase (Gould and Subramani, 1988), and secreted alkaline phosphatase (SEAP) (Cullen and Malim, 1992). The activities of such reporters are detected by measuring the light absorption or emission of the enzymatic products. These reporter enzymes have several advantages including high sensitivity, accuracy, and consistency. However, their assays require the preloading of substrate or the preparation of cell lysate. Therefore, they are not ideal for applications that require handling of a large number of samples, such as high-throughput screening. Also, for some reporters, the cost of substrates and other assay reagents can be quite high. Alternative reporter genes, ideally, would combine the sensitivity of the enzymatic analysis with a convenient assay involving minimal handling and, in particular, no preparation of cell lysate.

In this chapter, we describe the utility of green fluorescent protein (GFP) as a reporter gene in the study of gene expression. GFP emits fluorescent light on excitation, without the addition of substrate or cofactor, both *in vivo* and *in vitro* (Chalfie, 1995; Marshall *et al.*, 1995). The GFP fluorescence activity can be detected using a fluorescence microscope, fluorometer, fluorescence-activated cell sorting (FACS) machine, or imaging microplate reader. Enhanced GFP (EGFP) is a GFP mutant with brighter fluorescence that makes the detection much more sensitive (Yang *et al.*, 1996). Use of GFP as a reporter gene offers a number of advantages. These include real-time analysis, minimal sample handing, the possibility of large-quantity analysis, and high sensitivity. However, GFP is a stable protein, so that it is easily accumulated when expressed in cells. The accumulation makes the inducible expression of the reporter insensitive to any change in induction and thus it would be difficult to use in kinetics studies. In this chapter (Zhao *et al.*, 1999), we describe the generation of a destabilized EGFP (dEGFP) by fusing the degradation domain of mouse ornithine decarboxylase to EGFP. The fusion protein dEGFP, without any significant change in its fluorescence properties, has a short half-life of 2 h. In this chapter, we test the utility of EGFP and dEGFP as transcription reporters by fusing them with NF-κB-binding sequence and thymicline kinase (TK) promoter, and comparing the difference in expression between EGFP and dEGFP. We demonstrate that both EGFP and dEGFP can be used as reporters in transcription studies. We also show that dEGFP is more sensitive in response to changes in tumor necrosis factor (TNF) treatment owing to its faster turnover rate.

II. Principle of Methods

In this chapter, we describe the utility of EGFP and dEGFP as reporters to monitor TNF-mediated NF-κB activation. NF-κB is an inducible *trans*-activator of a large array of genes encoding cytokines, chemokines, other transcription factors, and receptors essential to the immune response (Baeuerle and Baltimore, 1996; Baldwin, 1996). NF-κB exists in an inactive form in the cytoplasm of most mammalian cells, and is activated in response to a variety of inducers, such as TNF treatment. On activation, NG-κB is rapidly released from its inhibitor, IκB, translocates to the nucleus, binds to specific recognition sequences in DNA, and induces transcription of the genes downstream of the NF-κB response element

(Peltz, 1997). We put the reporter gene EGFP or dEGFP under the regulatory control of four copies of the NF-κB response element and TK promoter. After transfection into cells, the reporter gene expression was induced by TNF-α treatment. The expression level of EGFP or dEGFP was quantitated by flow cytometry.

A. Materials

TNF-α (Clontech, Palo Alto, CA): Dissolve in phosphate-buffered saline (PBS) to make a stock of (concentration, 0.2 mg/ml); keep at 4 °C.
CalPhos Maximizer transfection kit (Clontech): For use in DNA transfection.
Luciferase assay system (Promega, Madison, WI).

III. Procedure

A. Time–Course of Induction

1. 293 cells are seeded sparsely in 6-well plates 1 day before transfection.
2. The culture medium is replaced with fresh medium 1 h before transfection.
3. The CalPhos transfection kit from Clontech is used for transfection. A master transfection mixture is made according to the manufacturer manual. One microgram of purified plasmid DNA is used for each well of the transfection. Two hundred microliters of the master transfection mixture is added to each well.
4. After 4 h of incubation at 37 °C, the transfection mixture is removed and replaced with fresh medium after rinsing once with PBS.
5. TNF-α is added to a final concentration of 0.1 μg/ml in 2-h intervals.
6. Before collection, cells are washed once with PBS. For the luciferase assay, cell lysates are prepared according to the manual of the provider (Promega). For flow cytometry, cells are treated with 1 ml of 2 mM EDTA in PBS, centrifuged at 10,000 rpm for 1 min, and resuspended in 0.5 ml of PBS.

Four copies of NF-κB-binding sequences are inserted in front of the simian virus 40 (SV40) early promoter of the plasmid pSEAP2-Promoter vector (Clontech) at the *Nhe*I and *Bgl*II sites. The TK minimal promoter is used to replace the SV40 early promoter. The SEAP gene of the plasmid is then substituted by cDNA encoding EGFP, dEGFP, or luciferase, resulting in pNF-κB–EGFP, pNF-κB–dEGFP, and pNF-κB–luciferase, respectively (Fig. 1). The reporter systems are used to monitor NF-κB activation induced by TNF-α.

On addition of TNF-α, the expression of the reporter genes is induced. EGFP- or dEGFP-transfected cells are analyzed by flow cytometry. As shown in Fig. 2, increased fluorescence intensity is observed in TNF-α-treated cells as a function of induction time. In both EGFP- and dEGFP-transfected cells, the total mean fluorescence of cells

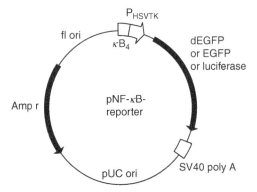

Fig. 1 Schematic map of luciferase, EGFP, and dEGPF reporter plasmid.

increases following the addition of TNF-α. However, they show different kinetic responses to TNF-α induction (Fig. 2A). In EGFP-transfected cells, a constant increase in EGFP expression is observed throughout the 8-h period. In contrast, dEGFP-transfected cells show a rapid increase in fluorescence activity for the first several hours and a plateau after 6 h of induction. After 8 h, a slight decrease in dEGFP expression occurs. This kinetic response is more comparable to that of the conventional reporter gene luciferase (Fig. 2B), although it reaches maximal induction 2 h later than luciferase. The delayed response could be due to the posttranslational folding of dEGFP, which requires more time and is necessary for its active fluorescence. These results demonstrate that dEGFP, with a faster degradation rate and less accumulation, responds to TNF-α more quickly than does the stable protein EGFP. Therefore, it could be a better reporter in kinetics studies of gene expression and regulation.

B. Dose–Response to TNF-α

1. 293 cells are seeded into a 6-well plate 1 day before transfection.
2. One microgram of each construct is transfected to each well of the 6-well plate.
3. Twenty-four hours after transfection, TNF-α is added to each well at a series of concentrations.
4. Cells are treated for an additional 6 h and collected for FACS analysis or luciferase assay.

As the concentration of TNF-α is elevated in the culture medium, the fluorescence activity of the dEGFP-transfected cells increases. However, no such dose-dependent response is observed when EGFP is used as the receptor (Fig. 3A). The stability of EGFP results in its accumulation even under the influence of the TK minimal promoter before TNF-α induction. The higher basal level of the

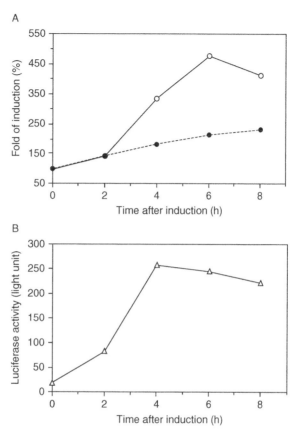

Fig. 2 Time-course study of TNF-α induction. Two hundred and ninety-three cells were transiently transfected with pNF-κB–luc, pNF-κB–EGFP, or pNF-κB–dEGFP. Twenty-four hours after transfection, cells were treated with TNF-α (0.1 μg/ml, final concentration) for various times, and then collected for FACS analysis or luciferase assay. (A) EGFP and dEGFP as reporters in time-course study. After TNF-α induction, cells were subjected to FACS analysis, and mean fluorescence intensity was measured. Data are represented by normalizing the uninduced cells to 100%. (○) dEGFP as reporter; (●) EGFP as reporter. (B) Luciferase as reporter in time-course study.

reporter, therefore, makes it insensitive to TNF-α-mediated induction. The dose-dependent response of dEGFP is more comparable to that of the conventional reporter gene luciferase (Fig. 3B), which shows maximum induction at a 0.1-μg/ml concentration of TNF-α.

IV. Conclusion and Remarks

In this chapter, we have compared the utility of EGFP and dEGFP as reporters in a kinetics study of gene expression. By coupling the reporters to TNF-α-mediated NF-κ B activation, we are able to monitor the activation of NF-κ B by

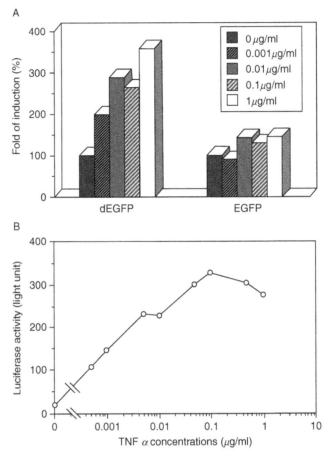

Fig. 3 Dose-dependent induction of reporters by TNF-α. Two hundred and ninety-three cells were transiently transfected with pNF-κB–luc, pNF-κB–EGFP, or pNF-κB–dEGFP. Twenty-four hours after transfection, various concentrations of TNF-α were added to the culture. Six hours after addition of TNF-α, cells were collected for FACS analysis or luciferase assay. (A) Dose-dependent induction of EGFP and dEGFP. For EGFP and dEGFP, data are represented by normalizing the uninduced cells to 100%. (B) Dose-dependent induction of luciferase.

measuring the changes in EGFP or dEGFP expression by flow cytometry. Furthermore, we have demonstrated that dEGFP is a better reporter than EGFP when used in a kinetics study. In any inducible expression system, expression of reporter genes includes both basal and regulated expression. If the reporter is a stable protein, such as EGFP, it will be easily accumulated even at the basal expression level. This accumulation leads to a narrowing of the range of induction. Hence, the induction becomes insensitive to the treatment of the inducer. In contrast, the accumulation of dEGFP is low, owing to its rapid turnover. Therefore, the induction range of dEGFP is greater than that of the stable protein and the kinetics of induction is faster. This rate determines the sensitivity of the reporter in the process of monitoring changes in treatment, as demonstrated in this study. In conclusion, this reporter system with dEGFP or EGFP can be used in both basic study and pharmaceutical screening. Because the detection is easy, of low cost, and reasonably sensitive, the system will be especially useful in high-throughput drug screening.

References

An, G., Hidaka, K., and Siminovitch, L. (1982). *Mol. Cell Biol.* **2,** 1628.

Baeuerle, P. A., and Baltimore, D. (1996). *Cell* **87,** 13.

Baldwin, A. S. (1996). *Annu. Rev. Immunol.* **14,** 649.

Chalfie, M. (1995). *Photochem. Photobiol.* **60,** 651.

Cullen, B. R., and Malim, M. H. (1992). *Methods Enzymol.* **216,** 362.

Gould, S. J., and Subramani, S. (1988). *Anal. Biochem.* **7,** 5.

Marshall, J., Molloy, R., Moss, G., Howe, J. R., and Hughes, T. E. (1995). *Neuron* **14,** 211.

Peltz, G. (1997). *Curr. Opin. Biotechnol.* **8,** 467.

Yang, T. T., Cheng, L., and Kain, S. R. (1996). *Nucleic Acids Res.* **24,** 4592.

Zhao, X., Jiang, X., Huang, C. C., Kain, S. R., and Li, X. (1999). *Methods Enzymol.* **302,** Chapter 21 (this volume).

CHAPTER 22

Measuring Protein Degradation with Green Fluorescent Protein

Stephen R. Cronin and Randolph Y. Hampton

Department of Biology
UCSD, La Jolla
California, USA

I. Introduction

Many cellular proteins are selectively degraded to control protein quality or quantity. Protein degradation can be highly selective, such that single point mutations or changes in physiological signals can drastically change the degradation rate of a given protein (Finger *et al.*, 1993; Hampton and Rine, 1994; Roitelman and Simoni, 1992; Tsuji *et al.*, 1992). It is now clear that the high specificity of protein degradation is brought about by a substantial and growing battery of proteins that are committed to targeting other proteins for degradation (Deshaies, 1995; Hampton *et al.*, 1996b; Hochstrasser, 1995; Huibregtse *et al.*, 1995). However, there remain many questions about the number and type of such factors, how they affect and control degradation, and how the numerous and disparate degradation pathways are coordinated and regulated in an intracellular milieu that is rich with proteins undergoing drastically different degradative fates, often in the same aqueous compartment.

The study of protein degradation has, until now, usually required some sort of invasive strategy for examining the half-life of a given degradation substrate, or the steady-state level of the substrate, which will vary with changes in degradation rate. Typical assays of protein stability capitalize on specific antibodies to immunoblot or immunoprecipitate the protein under study (see, e.g., Finger *et al.*, 1993; Hampton and Rine, 1994; Roitelman and Simoni, 1992; Tsuji *et al.*, 1992). In either case, the cells expressing the protein must be lysed to use these approaches. Immunological approaches can sometimes be used to assess the level of a protein *in situ* and so assess changes in the pool owing to changes in degradation (Stafford and Bonifacino, 1991), but even in such cases, fixation of the cells and immunostaining procedures could hardly be considered noninvasive.

A. Fusion Proteins as Degradation Substrates

Degradation determinants on proteins are often modular and can function autonomously when included in novel, engineered proteins (Glotzer *et al.*, 1991; Hochstrasser and Varshavsky, 1990). Numerous laboratories have capitalized on this feature by making a fusion gene that will express an easily measured reporter function, such as β-galactosidase, fused to a degradation-determining portion of a protein of interest. This approach allows the levels of the new degradation substrate, and the effects of altering degradation on these levels, to be conveniently measured by assaying the reporter activity that is now subject to the degradation process of interest. However, in most cases, the measurement of such reporter genes still requires lysis of the cells in which they are expressed.

The use of green fluorescent protein (GFP) allows a novel refinement of the reporter approach in the study of protein degradation. The idea is the same as with previous reporters. GFP is expressed as part of a protein that undergoes degradation, and so undergoes the same degradation as the authentic fusion partner. However, now the amount of the resulting protein can be directly measured in living cells by virtue of the fluorescence from the GFP moiety. In theory, this approach allows a totally noninvasive and integrated observation of protein degradation. Our interest is focused on the use of the GFP reporter in genetic studies of protein degradation (Hampton, 1998). However, the example that we describe in the following sections can in all likelihood be generalized to many other experimental circumstances.

II. HMG-CoA Reductase

HMG-CoA reductase (HMGR) is a key enzyme in the mevalonate pathway from which cholesterol and many other essential molecules are synthesized (Hampton *et al.*, 1996a). The degradation of HMGR is regulated by the mevalonate pathway as a mode of feedback control. When flux through the mevalonate pathway is high, HMGR degradation is fast. When flux is slowed, as for instance when cells are treated with the HMGR inhibitor lovastatin, HMGR degradation is slowed. HMGR is an integral membrane protein of the endoplasmic reticulum (ER), and HMGR degradation occurs without exit from the ER. Many molecular details about the degradation mechanism and the manner in which HMGR degradation is coupled to the mevalonate pathway remain to be discovered.

Regulated degradation of HMGR has been conserved between yeast and mammals (Chun and Simoni, 1992; Edwards *et al.*, 1983; Hampton and Rine, 1994). In yeast, the Hmg2p isozyme undergoes regulated ER degradation with many features similar to that observed in mammals (Gardner *et al.*, 1998; Hampton, 1998; Hampton and Bahkta, 1997; Hampton and Rine, 1994). This conservation has allowed us to launch a genetic analysis of yeast Hmg2p regulated degradation with the goal of understanding the molecular mechanisms operating in yeast, including those in common with mammals (Hampton *et al.*, 1996b). Our studies have revealed a set of genes required for the degradation of Hmg2p, called *HRD* genes (pronounced *herd*, for HMGR degradation). Loss-of-function *hrd* mutants such as *hrd1-1* or *hrd1*Δ fail to degrade Hmg2p. In the context of this chapter, the known regulatory behavior of Hmg2p degradation and the degradative phenotypes of *hrd* mutants provided strict criteria for the validity of using the GFP reporter as part of a novel fusion protein, Hmg2p–GFP, in the study of Hmg2p protein degradation. Our studies have indicated that the GFP faithfully reports the regulated degradation of Hmg2p, and so use of the Hmg2p–GFP fusion allows a variety of approaches that would otherwise not be possible. More generally, these studies indicate that the GFP reporter could have broad use in the study of protein degradation.

III. Hmg2p–GFP Reporter Protein

HMGR has a modular structure. The C-terminal portion of the protein provides the essential catalytic activity for which the protein is named. In contrast, the N-terminal, membrane-anchoring portion has no role in catalysis, but anchors the protein to the ER membrane, and is responsible for regulated degradation of the protein (Fig. 1). Accordingly, in both mammals and yeast, engineering an HMGR coding region with the catalytic region of the coding region replaced with a novel 3′ end encoding an enzyme (His4cp in yeast (Hampton and Rine, 1994) and β-galactosidase in mammals (Chun and Simoni, 1992)) results in a

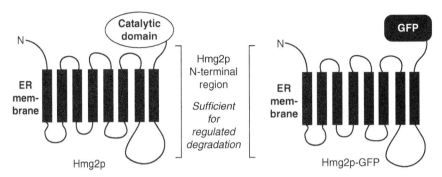

Fig. 1 Hmg2p and the fluorescent reporter Hmg2p–GFP. Hmg2p is anchored in the ER membrane of yeast by the hydrophobic N-terminus. The N-terminal anchor is responsible for regulated degradation. The C-terminal catalytic domain is attached to the membrane anchor by a poorly conserved linker region. Hmg2p–GFP, which still undergoes regulated degradation, has the authentic catalytic domain replaced by the GFP reporter. The *HMG2::GFP* fusion gene was produced by fusing the region encoding GFP to the *HMG2* coding region at codon 669 (Hampton *et al.*, 1996c). The S65T bright mutation was then introduced by subcloning that portion of the S65T-GFP coding region into the *HMG2::GFP* gene (see text) (Hampton *et al.*, 1996b).

fusion protein that still undergoes regulated degradation. GFP was tested as an *in vivo* reporter of Hmg2p degradation by similarly producing an *HMG2* coding region with the 3′, catalytic region codons replaced with the entire *GFP* coding region. The original *HMG2::GFP* coding region expressed the Hmg2p–GFP reporter protein, with the N-terminal 669 amino acids fused to the GFP protein, and a small number of amino acids between the two coding regions from the polylinker from which the GFP was cloned (Hampton *et al.*, 1996c) (Fig. 1).

The first Hmg2p–GFP optical reporter protein contained the wild-type GFP as cloned from jellyfish (Hampton *et al.*, 1996c). We have subsequently recloned the *HMG2::GFP* plasmids to include the brighter variant GFP with an S65T mutation (Gardner *et al.*, 1998; Hampton *et al.*, 1996b) (the S65T numbering refers to the GFP coding region considered alone, since that is the way that the mutation is referred to in most cases). Use of the S65T mutant results in an Hmg2p–GFP with approximately six or seven times the brightness of the original GFP (Heim *et al.*, 1994). The S65T mutation was introduced into our original fusion gene by subcloning the *MscI/SalI* fragment of the S65T-GFP coding region from the Clontech (Palo Alto, CA) pS65T-Cl plasmid into the corresponding sites of the original *HMG2::GFP* coding region. The *Msc*I site cuts the *GFP* coding region eight codons before the S65 position, making this approach generally useful for introducing the bright T65 mutation (or any other downstream mutations from other *GFP* genes) into a preexisting wild-type *GFP* coding region by nonmutagenic cloning techniques. Both the original and the brighter Hmg2p–GFP fusion proteins behave identically as degradation substrates. For the remainder of this work, *HMG2::GFP*, and Hmg2p–GFP, will refer to the gene and proteins, respectively, with the S65T mutation in the GFP.

In our studies, the promoter and plasmid type have been chosen to ensure that the level of Hmg2p–GFP protein is entirely controlled by protein degradation. The *HMG2::GFP* coding region is expressed from a strong, constitutive glyceralde-hyde-3-phosphate dehydrogenase (GAPDH) promoter. In this context, Hmg2p–GFP synthesis is constant, and unaffected by regulatory perturbations of the mevalonate pathway (such as addition of lovastatin; see below) that control the Hmg2p degradation rate.

Furthermore, expression plasmids with the *HMG2::GFP* coding region are always integrated into the yeast genome in single copy, so that the dose of the gene is not subjected to the variability that autonomously replicating yeast plas-mids can display (e.g., between one and four copies per cell for ARS/CEN plas-mids). These measures keep the expression of Hmg2p–GFP constant and stable among individual cells and across physiological or genetic conditions that affect degradation. Accordingly, the fluorescence intensity is directly controlled by the degradation rate of Hmg2p–GFP.

We have extensively compared the degradation of the Hmg2p–GFP reporter protein with that of authentic Hmg2p. In every way tested, the fluorescent reporter protein displays degradation behavior that mirrors the parent protein. The degra-dation of Hmg2p–GFP depends on *HRD* genes (Gardner *et al.*, 1998; Hampton *et al.*, 1996b). Also, Hmg2p–GFP degradation is regulated by signals from the mevalonate pathway (Gardner *et al.*, 1998; Hampton and Bahkta, 1997; Hampton *et al.*, 1996c). For example, inhibitors of early pathway enzymes such as the HMGR inhibitor lovastatin slow the degradation of the reporter protein. Finally, mutations in the N-terminal region of Hmg2p that alter the degradation cause identical alterations when present in Hmg2p–GFP (Gardner *et al.*, 1998). The focus of this chapter is to describe the techniques that are uniquely possible with the optical reporter. The features of Hmg2p–GFP degradation (*HRD* dependence, physiological regulation) are incorporated in the examples of data collected using the techniques uniquely permitted by the GFP reporter.

IV. Direct Visual Examination of Hmg2p–GFP

A strain of yeast expressing the Hmg2p–GFP reporter protein can be directly examined by fluorescence microscopy (Gardner *et al.*, 1998; Hampton *et al.*, 1996b,c) (Fig. 2). The Hmg2p–GFP is distributed in a manner identical to Hmg2p (Hampton *et al.*, 1996c; Koning *et al.*, 1996). At the levels attained with the strong GAPDH promoter, the ER is proliferated into stacks and whorls of membrane similar to the structures caused by high levels of Hmg2p (Hampton *et al.*, 1996c). The cells are typically grown and examined in yeast minimal medium, but the log-phase fluores-cence is quite independent of the exact medium or circumstances of growth. Hmg2p–GFP fluorescence can be observed with no preparation or fixation. The pictures shown in Fig. 2 were taken of samples removed from experimental cultures and placed directly on a microscope slide. In this particular case, a Nikon Optiphot II

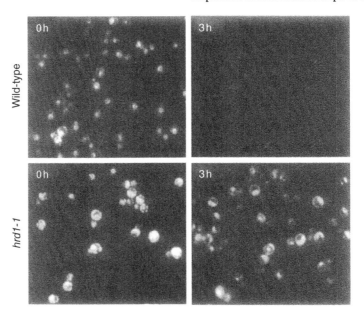

Fig. 2 Direct observation of Hmg2p– GFP degradation by fluorescence microscopy. A wild-type (top row) or degradation-deficient (bottom row) *hrd1-1* strain expressing Hmg2p– GFP were treated with cycloheximide and examined by fluorescence microscopy immediately or after a 3-h degradation period. The wild-type strain showed substantial loss of Hmg2p– GFP fluorescence. The dark, 3-h panel from the wild-type strain represents an identical exposure of a similar number of cells photographed in the 0-h panel. The loss of fluorescence is not observed in the degradation-deficient *hrd1-1* strain (bottom row). All panels were obtained by CCD and frame integrator, and printed on a thermal printer as described (Gardner *et al.*, 1998; Hampton *et al.*, 1996b). All panels were captured and printed using identical settings, and so can be compared for fluorescence intensity. Note that the initial level of fluorescence in the degradation-deficient *hrd1-1* strain (bottom left) is brighter than that of the otherwise identical wild-type strain (top left).

microscope with an Episcopic fluorescence attachment EFD-3 was used. The standard B-2A filter block, usually used for detection of the fluorescein moiety, allows excitation and filtration of the fluorescence signal (B-2A features: 450- to 490-nm excitation filter; 505-nm dichroic mirror; 520-nm barrier filter). Although we employ this standard filter block, various companies now offer filter blocks optimized for GFP excitation and emission. The light source is a mercury vapor lamp provided with the unit. The main point for direct microscopic examination is that the microscope and optics are all standard tools for fluorescence microscopy, and any of the numerous microscopes typically employed for this purpose will work as well. The more important issues for the success and/or utility of direct examination concern the actual features and expression levels of the fusion protein under study (see Section VII).

Figure 2 shows a typical experiment demonstrating direct examination of Hmg2p–GFP as a degradation substrate. When protein synthesis is blocked Hmg2p–GFP degradation continues and the cellular levels of the reporter protein drop as a function of the degradation rate (Hampton *et al.*, 1996c). Cells expressing the Hmg2p–GFP reporter were treated with cycloheximide (50 μg/ml) to block protein synthesis, and examined immediately or after a 3-h incubation period to allow protein degradation to proceed (Fig. 2). This experiment was run with otherwise identical strains that had a normal *HRD1* gene (wild-type; Fig. 2, top row) or the *hrd1-1* mutation (*hrd1-1*; Fig. 2, bottom row) that renders cells severely compromised for Hmg2p degradation. The top row in Fig. 2 shows a substantial reduction in the fluorescence intensity of the *HRD1* cells after a 3-h degradation period, caused by loss of the Hmg2p–GFP protein. The bottom row in Fig. 2 shows the results of the identical experiment when the *hrd1-1* mutation is present. The *hrd1-1* mutation strongly blocks the degradation of Hmg2p–GFP, preventing loss of fluorescence during the 3-h period. We have also used direct observation of Hmg2p–GFP degradation as one way of showing how zaragozic acid (ZA) hastens the degradation of the reporter by increasing the endogenous signal for degradation (Hampton and Bahkta, 1997).

To observe degradation by direct visual inspection, the production of the fluorescent reporter must be halted, in this case by global cessation of protein synthesis with cycloheximide. While Hmg2p degradation is not adversely affected by cycloheximide, it is possible that the degradation of a different fluorescent substrate could be affected by the drug. In that circumstance, other approaches can be employed to observe the degradation of the entire pool of protein, including removal of a required amino acid to halt protein synthesis (Hampton and Bahkta, 1997) or possibly use of a regulated promoter to stop expression of the fluorescent reporter.

Direct fluorescence microscopy can also be used to observe changes in the steady-state levels of Hmg2p–GFP caused by changes in degradation. As an example, the initial steady-state level of Hmg2p–GFP in the otherwise isogenic *hrd1-1* strain in Fig. 2 is clearly brighter (bottom left vs. top left), owing to stabilization of the protein. Similarly, treatment of strains expressing Hmg2p–GFP with ZA results in a new steady-state level of the protein, and thus cellular fluorescence brought about by the effect of ZA to hasten degradation (S. Cronin, H. Bhakta, and R. Hampton, unpublished observation, 1998). Importantly, using the steady-state level as a readout of changes in degradation requires that the rate of production of the protein from the fusion gene be unaffected by alterations in the degradation process brought about by genetic or pharmacological means.

V. Flow Microfluorimetry and Fluorescence-Activated Cell Sorting Analysis of Degradation

Flow microfluorimetry (FMF) is the technique whereby individual cells are measured for fluorescence while passing through a chamber in a laminar sheath of buffer solution. Fluorescence-activated cell sorting (FACS) devices perform FMF (analysis) on a population to query the fluorescence of individual cells, and then a subsequent collection step (sorting) of the individual cells according to designated optical properties ascertained by the FMF. Typically, one will sort the most or least fluorescent cells from the bulk population. Because GFP reports fluorescence in living cells, FACS sorting of cells expressing a degraded GFP fusion could be used to isolate mutants with altered degradation of fluorescent reporter protein, and hence atypical fluorescence from the bulk population. FACS sorters can typically process 1000–3000 cells/s, making it entirely reasonable to use this approach in the collection of degradation mutants. Most of our work has been in the use of FMF as an analytical tool, and that is the emphasis in the remainder of this chapter. However, the FMF data indicate that the effects of perturbing degradation on the fluorescence histograms are significant enough to allow enrichment of appropriate mutants by FACS.

FMF results are usually plotted as a histogram, showing individual cell fluorescence on the horizontal axis and number of cells with a given level on the vertical axis (Fig. 3). The histogram represents the range and distribution of fluorescences present in the population of cells under analysis. Typical histograms are plotted using the logarithm of fluorescence (in arbitrary units) and represent the collected data from 5000 to 20,000 cells. Yeast cells expressing Hmg2p–GFP can be directly analyzed by FMF from living cultures with no fixation or preparation. Furthermore, since tens of thousands of cells are typically analyzed, the amount and density of living culture required for analysis can be very small, usually less than 50 μl (a typical early log-phase culture has a cell density on the order of 2×10^6 cells/ml), and often far less than would be amenable to more standard modes of protein analysis. Use of FMF to assess cellular fluorescence is more quantitative and faster than direct microscopic examination, although less informative about cellular localization.

We have used FMF to study the effect of altering degradation on the steady-state levels of Hmg2p–GFP (Gardner *et al.*, 1998) caused by either pharmacological or genetic means. In the top row ("steady state" in Fig. 3) are fluorescence histograms of yeast expressing Hmg2p–GFP under various circumstances that affect degradation of the reporter. The left panel in the top row demonstrates the effect of physiological regulation of Hmg2p–GFP degradation on the steady-state fluorescence ("lovastatin treatment" in Fig. 3). When the mevalonate pathway is slowed with a small dose (25 μg/ml) of the HMGR inhibitor lovastatin, the degradation of Hmg2p–GFP is slowed. (The HMGR activity is being provided in this strain by a separately expressed HMGR.) The superimposed, steady-state fluorescence histograms of this strain grown for 3 h in the absence or presence of lovastatin ("no drug" and "lova," respectively, in Fig. 3) show the expected rightward shift of the drug-treated cells to brighter fluorescence. The right panel in the top row shows the effect of a genetic deficiency in degradation on Hmg2p–GFP fluorescence. Otherwise identical strains with a normal or null allele of *HRD1* ("wild-type" and "*hrd1*Δ," respectively, in Fig. 3) were compared by FMF. The degradation-deficient *hrd1*Δ strain has significantly higher steady-state Hmg2p–GFP fluorescence, as indicated by the shift of the histogram to the right when overlaid with that from the otherwise identical strain with a normal *HRD1* gene.

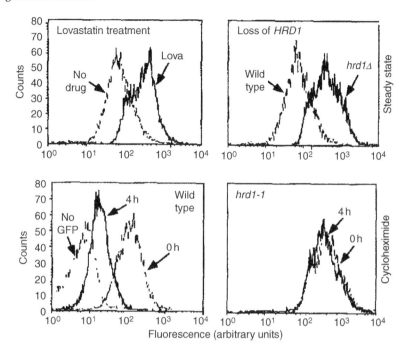

Fig. 3 Use of flow microfluorimetry (FMF) to evaluate steady-state (top row) and time-dependent (bottom row) fluorescence of strains expressing the Hmg2p–GFP reporter. *Top row*: Effect of physiological or genetic alterations of Hmg2p–GFP on steady-state levels of cellular fluorescence. *Left*: The effect of lovastatin treatment on steady-state fluorescence. Cells expressing the Hmg2p–GFP protein were grown for 3 h in the presence or absence of lovastatin (25 µg/ml) and then subjected to FMF to examine individual cellular fluorescence. The lovastatin slows degradation of Hmg2p–GFP and so causes an increase in the steady-state levels of cellular fluorescence, as indicated by a rightward shift of the fluorescence histogram. *Right*: Growing cultures of two strains—one with the normal (wild-type) allele and the other with a null allele (*hrd1Δ*) of the *HRD1* gene—were subjected to FMF. The degradation-deficient *hrd1Δ* strain is significantly more fluorescent as indicated by the histogram shift to the right. *Bottom row*: Direct observation of time-dependent loss of cellular fluorescence due to degradation of the Hmg2p–GFP reporter. *Left*: A strain with a normal *HRD1* gene (wild-type) was treated with cycloheximide and examined immediately after treatment, or following a 4-h degradation period. Degradation of the Hmg2p–GFP protein results in a lowering of cellular fluorescence and a leftward shift of the FMF histogram. The histogram of an identical strain, except that it has no *HMG2::GFP* expression plasmid, is also included to show cellular autofluorescence (peak marked *no GFP*). *Right*: The same experiment run with an identical strain, except that it has an *hrd1-1* allele rendering it degradation deficient.

FMF can also be used to observe directly the degradation of the reporter protein after cessation of protein synthesis. When protein synthesis is halted with cycloheximide in strains expressing Hmg2p–GFP, the fluorescent reporter is degraded, and cellular fluorescence drops accordingly. The decline in fluorescence can be monitored by FMF after various degradation periods have been allowed to pass, also shown in Fig. 3. The bottom row ("cycloheximide" in Fig. 3) shows the effect of a 4-h incubation with cycloheximide on Hmg2p–GFP in a wild-type strain (Fig. 3, left panel) or in a degradation-deficient *hrd1-1* strain (Fig. 3, right panel). In each panel, the initial and 4-h histograms are superimposed to show the effects of inhibition of protein synthesis. Cycloheximide causes a leftward shift in the wild-type strain fluorescence histogram to lower mean cell fluorescence (Fig. 3, bottom row, "wild-type"). This shift is totally dependent on having a normal *HRD1* gene, and so is completely abrogated in the otherwise identical *hrd1-1* strain that cannot degrade Hmg2p–GFP (Fig. 3, bottom row, "*hrd1-1*").

In testing a new GFP fusion for utility in FMF or FACS, it is important to evaluate cellular autofluorescence. This is best done by assaying otherwise identical strains (or cells of whatever sort being tested) that do not express the GFP fusion being measured. As an example, the histogram of the parent strain to those in Fig. 3, but not expressing Hmg2p–GFP, has been included with the histograms of the wild-type strain in the bottom panel (Fig. 3, bottom row, "no GFP").

An FMF (or FACS) apparatus represents a fairly large investment, so the best strategy to explore using this powerful tool is to find a local laboratory that employs one regularly. Fortunately this is usually quite easy, since most immunology laboratories, and many cell cycle laboratories, own such a device. The histograms shown in Fig. 3 were acquired on a Becton Dickinson (Mountain View, CA) FACScalibur flow microfluorimeter, using fluorescence settings appropriate for GFP (488-nm excitation, 512-nm emission). Each histogram represents individual fluorescence from 10,000 cells. The samples were run directly into the flow system from the experimental cultures in each case. Data were next graphically organized into overlaid histograms with the graphical software used to run the device. Although not the focus of our work, the data collected by FMF are amenable to detailed statistical analysis, and could be used to evaluate protein degradation in quantitative detail.

VI. Direct Examination of Hmg2p–GFP in Living Colonies

We are particularly interested in using yeast to perform a complete genetic analysis of Hmg2p regulated degradation. A significant part of our effort is devoted to obtaining yeast mutants that are deficient in various aspects of Hmg2p regulated degradation (Hampton, 1998). The data presented in Figs. 2 and 3 show that the fluorescence of Hmg2p–GFP in a strain of yeast is affected in a predictable manner by alterations in the degradation of the protein, either by genetic or pharmacological means. Because the fluorescence can be observed in living yeast cells, we developed a method to evaluate the GFP fluorescence in living yeast colonies on agar plates in order to facilitate screening for mutants with abnormal levels of Hmg2p–GFP due to alterations in regulated degradation.

To examine fluorescence of colonies on agar plates, they must be illuminated with intense light at 488 nm (the absorbance maximum for S65T-GFP). This is accomplished with a narrow band-pass filter custom-made from Omega Optical

(Brattelboro, VT; www.omegafilters.com) with maximal transmittance at 488 nm, and an added blocking filter to remove unwanted red and higher wavelength light from the excitation source. The filter is square in shape with 5-cm sides, the size of a standard photographic slide, allowing it to be placed in the slide holder of a Kodak (Rochester, NY) Carousel 4400 projector. With an FMS bulb, the light from the projector has ample intensity in the 488-nm range, which when passed through the in-place filter results in a field of intense blue light sufficient in size to illuminage multiple agar plates simultaneously. The plates are examined on a bench top in a dark room. The green fluorescence of the illuminated colonies can be directly seen by examination with a long band-pass filter (Kodak Wratten gelatin filter, 75 × 75 mm, No. 12) that removes the substantial contribution of the reflected blue excitation light. The long band-pass filter was purchased at a local camera store and can be taped to a clear face mask or over the lenses of safety goggles.

Figure 4 shows patches of cells grown on agar medium, and expressing the Hmg2p–GFP reporter, that are normal ("w.t.") or degradation-deficient ("hrd1-1"). The top row in Fig. 4 ("GFP") shows the green fluorescence of each patch, photographed by illuminating the agar plate with the blue-filtered slide projector, and placing the long band-pass filter in front of a Polaroid camera lens. The bottom row in Fig. 4 ("cells") represents the same pair of patches photographed with normal white light and no filtration, demonstrating that the density of cells in the two patches is the same. The degradation-deficient strain ("hrd1-1" in Fig. 4) is far brighter than the wild-type strain.

The extreme difference in the fluorescence signal compared with what one would expect from the intensity of the wild-type versus hrd1-1 strains in Fig. 2 is due to a convenient feature of Hmg2p degradation that we have exploited in numerous ways. The agar on which these patches are grown is a typical minimal medium made with yeast nitrogen base, glucose, and the supplements that are required by the auxotrophies of the strain, including the amino acid methionine. On this medium, the cells deplete the amino acid before the glucose, and so cease making protein. Degradation, however, still proceeds, and so the normal strain that can degrade Hmg2p–GFP grows very dark. In contrast, the degradation-deficient

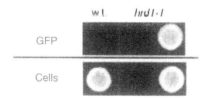

Fig. 4 Colony fluorescence of Hmg2p–GFP. Patches of normal (w.t.) or degradation-deficient (hrd1-1) strains expressing Hmg2p–GFP were grown on supplemented minimal agar medium, Hampton et al. (1996b) and subjected to illumination and filtration as described in text to observe green fluorescence (top row). The two identical patches were also photographed with room light and no filtration to show the cell density of each patch (bottom row). Each patch was prepared by spotting a 50-μl suspension of ~1 × 10^5 cells onto the agar medium, and allowing growth for 2 days at 30 °C.

(*hrd1-1*) strain cannot degrade the Hmg2p–GFP, and so the patch remains bright despite a similar cessation of Hmg2p–GFP synthesis. Thus, by simply allowing patches to grow for a day or two on supplemented minimal medium, we can score the propensity of that strain to degrade Hmg2p–GFP.

The ease of whole colony analysis offered by the optical colony assay has allowed us to devise numerous, previously infeasible genetic and molecular biological screens relevant to Hmg2p degradation. The protocols are far more straightforward than previous methods to assay protein levels as a gauge of degradation, such as colony immunoblotting. Thus, when an optical reporter works in colony assays such as this, it should certainly be considered as a pheno-type for genetic analysis. We are currently looking for mutants deficient in Hmg2p–GFP degradation, mutants deficient in regulation of Hmg2p–GFP stabil-ity, variations of the Hmg2p–GFP sequence itself that affect regulated degradation of the protein, high-copy plasmids that alter normal regulated degradation, and mammalian genes that alter degradation- or complement-recessive yeast mutants. Although we use yeast in our current studies, it is likely the same ideas can be applied to the genetics of degradation in other organisms as well.

VII. Caveats and Considerations

The preceding data demonstrate a single example of using GFP as a reporter function in the study of degradation. The majority of people interested in this technique will want to apply it to proteins other than yeast Hmg2p. With this in mind, there are some considerations and caveats that should be entertained by those who want to harness this powerful tool. These are not to dissuade future fluorophiles from attempting to use GFP, but rather to help allow the greatest chance of success in the exploration of this approach.

A. GFP Fluorogenesis Rate Versus Fusion Degradation Rate

GFP in its final folded form is fluorescent. The acquisition of this structure, or fluorogenesis, involves both folding and an auto-oxidative covalent reaction with molecular oxygen. The rate of this process can vary with the type of GFP employed (Heim *et al.*, 1994), the availability of O_2, and most likely cellular factors that influence protein processing. Many of the mutants that are brighter than the original (and now infrequently used) jellyfish GFP acquire the correct autofluor-escent structure fairly rapidly. The steady-state fluorescence of a degraded protein with a GFP fusion will be influenced by the rate at which the protein becomes fluorescent. If the degradation rate of the fusion protein is much faster than the rate of GFP fluorogenesis, then only a small fraction of the fusion molecules will survive long enough to become fluorescent. Of course, fluorogenesis of GFP also has its own half-life, so this level of consideration is by no means rigorous. Never-theless, a degradation rate that is significantly faster than the rate at which the GFP reporter becomes fluorescent will cause diminution of the steady-state signal below that predicted from considering the simple effect of degradation on final steady-state level of the reporter alone.

On the other hand, the requirement that GFP be fully folded to be optical can be useful in degradation studies. For example, in the case of Hmg2p–GFP, we have shown that the drug ZA (an inhibitor of squalene synthase) can stimulate the degradation of Hmg2p and Hmg2p–GFP (Hampton and Bahkta, 1997; Hampton et al., 1996c). Even in the absence of protein synthesis, the addition of ZA to fluorescent cells causes rapid loss of signal that is dependent on the *HRD* genes. This effect indicates that in all likelihood the *HRD*-encoded degradation machinery can seize and destroy previously made, fully folded, and previously stable Hmg2p–GFP.

B. Is Loss of Fusion GFP Fluorescence Due to Degradation?

In the examples herein, we show that various physiological or genetic manipulations have effects on Hmg2p–GFP signal levels and loss rates consistent with *bona fide*, relevant degradation of the optical reporter. However, these optical effects do not alone confirm that processive degradation is occurring. It is important to verify that the fusion protein is actually undergoing degradation, as opposed to unfolding or cleavage or other unknown processes that at the optical level would be indistinguishable from degradation. This is best done by performing traditional biochemical degradation assays on the fusion itself (see, e.g., Chun and Simoni, 1992; Edwards et al., 1983; Finger et al., 1993; Gardner et al., 1998; Glotzer et al., 1991; Hampton and Bahkta, 1997; Hampton and Rine, 1994; Hampton et al., 1996b,c; Hochstrasser and Varshavsky, 1990; Huibregtse et al., 1995; Roitelman and Simoni, 1992; Stafford and Bonifacino, 1991; Tsuji et al., 1992). The most common assay of degradation is the pulse–chase protocol, in which cells are labeled with radioactive amino acids (the pulse step, usually with radioactive methionine), after which incorporation is blocked by addition of excess unlabeled amino acid (the chase), and the protein under study is immunoprecipitated, resolved on sodium dodecyl sulfate–polyacrylamide gels, and autoradiographed to ascertain the rate at which the pulsed sample disappears. Another common assay, similar to that used in Fig. 2, is a "cycloheximide chase," in which protein synthesis is stopped with the drug cycloheximide, and the degradation of the protein of interest is determined by immunoblotting. In either case, antibodies against the fusion protein are needed. Suitable antibodies may be already available against a portion of the authentic protein of interest that is included in the optical reporter. If not, the GFP moiety can also be used for immunodetection or precipitation, either with anti-GFP antibodies made by individual investigators, or with those obtained from a growing number of commercial sources. The original description of Hmg2p–GFP included biochemical analysis of regulated degradation of the new fusion with techniques previously employed in the studies of authentic Hmg2p (Hampton et al., 1996c).

C. Does the GFP Fusion Protein Go Down the Same Degradation Pathway?

It is not given that addition of a novel, folded, 20-kDa domain to a protein will preserve the recognition processes that allow the recognition and degradation of the original protein. If possible, ascertain enough features of the mechanism of degradation to verify that a novel optical reporter behaves in a way that is similar to the authentic protein of interest. Such tests could include examining the effects of physiological perturbations that alter degradation rate, measuring stability of the fusion in degradation-deficient mutants (such as *hrd1-1*), and examining the effects of degradation-altering *in-cis* mutations when introduced into the reporter. Of course, if one knew everything possible about the degradation mechanism of the original protein, the utility of the optical degradation reporter might be small. Thus, in developing the GFP reporter as a discovery tool, one should verify that the optical protein behaves as expected from information obtained through more invasive and cumbersome methods. Then new discoveries made with the more facile optical approach can be confirmed in traditional studies on the actual protein of interest.

D. Where to Put the GFP Moiety?

GFP is a remarkably flexible fusion partner. It can be put at the N-terminus of proteins, the C-terminus (as in Hmg2p–GFP), or even in the middle of some protein sequences. What is most important is to use as much information about the degradation of the protein as possible to inform the placement of the *GFP* coding region, and if possible, to try several different fusions at the early stages of developing the optical approach to identify the most useful version.

E. Which GFP to Use?

There are a growing number of novel sequence variants of the original GFP protein that can be applied to a given research problem. Variations include altered spectral properties, differing folding rates in particular organisms, and coding regions optimized to express a given version optimally in a desired organism, and even a version that is itself degraded rapidly to allow more dynamic analysis of changes in promoter and translational activity. It is safe to say that the generation of novel variations of the original GFP is far from complete, perhaps including the development of completely novel proteins that function in the same manner. Numerous companies now offer a variety of such variations of the DNA and protein sequence of GFP. Some of these companies include Clontech (www.clontech.com) and Packard (Downers Grove, IL; www.packardinst.com), to name two. It is advisable to perform an Internet search to mesh specific needs with the constantly growing availability of custom GFPs. It can also be useful to take advantage of the GFP newsgroup (for information, e-mail biosci-server@net.bio.net) to investigate the newest developments and to interact with many experienced researchers.

Acknowledgments

The authors thank Dr. Robert Rickert (UCSD Department of Biology) for the use of the FACScalibur flow microfluorimeter, software, and color printer, Dr. Michael Coady (University of Montreal) for valuable suggestions about the colony fluorescence assay, and the members of the Hampton laboratory for continuous effort, interest, and interaction. R. Y. H. also thanks L. Rampulla for consultations without charge, and supportive field work. This work is supported by AHA Grant 96013020, NIH Grant DK5199601, and a Searle Scholarship.

References

Chun, K. T., and Simoni, R. D. (1992). *J. Biol. Chem.* **267**, 4236.

Deshaies, R. J. (1995). *Curr. Opin. Cell Biol.* **7**, 781.

Edwards, P. A., Lan, S. F., and Fogelman, A. M. (1983). *J. Biol. Chem.* **258**, 10219.

Finger, A., Knop, M., and Wolf, D. H. (1993). *Eur. J. Biochem.* **218**, 565.

Gardner, R., Cronin, S., Leader, B., Rine, J., and Hampton, R. Y. (1998). *Mol. Biol. Cell* **9**, 2611.

Glotzer, M., Murray, A. W., and Kirschner, M. W. (1991). *Nature (London)* **349**, 132.

Hampton, R. Y. (1998). *Curr. Opin. Lipidol.* **9**, 93.

Hampton, R. Y., and Bahkta, H. (1997). *Proc. Natl. Acad. Sci. USA* **94**, 12944.

Hampton, R. Y., and Rine, J. (1994). *J. Cell Biol.* **125**, 299.

Hampton, R., Dimster-Denk, D., and Rine, J. (1996a). *Trends Biochem. Sci.* **21**, 140.

Hampton, R. Y., Gardner, R., and Rine, J. (1996b). *Mol. Biol. Cell* **7**, 2029.

Hampton, R. Y., Koning, A., Wright, R., and Rine, J. (1996c). *Proc. Natl. Acad. Sci. USA* **93**, 828.

Heim, R., Prasher, D. C., and Tsien, R. Y. (1994). *Proc. Natl. Acad. Sci. USA* **91**, 12501.

Hochstrasser, M. (1995). *Curr. Opin. Cell Biol.* **7**, 215.

Hochstrasser, M., and Varshavsky, A. (1990). *Cell* **61**, 697.

Huibregtse, J. M., Scheffner, M., Beaudenon, S., and Howley, P. M. (1995). *Proc. Natl. Acad. Sci. USA* **92**, 2563.

Koning, A. J., Roberts, C. J., and Wright, R. L. (1996). *Mol. Biol. Cell* **7**, 769.

Roitelman, J., and Simoni, R. D. (1992). *J. Biol. Chem.* **267**, 25264.

Stafford, F. J., and Bonifacino, J. S. (1991). *J. Cell Biol.* **115**, 1225.

Tsuji, E., Misumi, Y., Fujiwara, T., Takami, N., Ogata, S., and Ikehara, Y. (1992). *Biochemistry* **31**, 11921.

CHAPTER 23

Studying Nuclear Receptors with Green Fluorescent Protein Fusions

Gordon L. Hager

Head, Hormone Action and Oncogenesis Section
Laboratory Chief
Bethesda, Maryland, USA

I. Introduction

Subcellular trafficking of nuclear receptors is an important component in the biological control of these regulatory molecules. The steroid receptors form one major group of the nuclear receptor family, and are the best understood in terms of subcellular distribution. Three distribution patterns have been described classically. The glucocorticoid receptor (GR) is located in the cytoplasm in the absence of ligand and translocates to the nucleus when occupied by ligand (Mendel *et al.*, 1990; Pratt, 1990). The GR seems to be unique in its exclusive cytoplasmic compartmentalization. A fraction of the progesterone receptor (PR) is also present in the cytoplasm, but most of this receptor is found in the nucleus under ligand-free conditions (Guiochon-Mantel *et al.*, 1992, 1994; Savouret *et al.*, 1993). The thyroid receptor (TR) and retinoic acid receptor (RAR) subfamily (Song *et al.*, 1995), as well as the estrogen receptor (ER) (Gorski *et al.*, 1993; Greene and Press, 1986;

379
DOI: 10.1016/B978-0-12-384658-7.00023-0

Yamashita, 1998), are usually described as bound to the nucleus both in the presence and absence of hormone.

These views of receptor localization are derived from two experimental approaches, subcellular fractionation and immunolocalization. Both of these approaches require major disruption of cellular architecture: either fixation of cells to render epitopes accessible to large antibody complexes or complete separation of cellular compartments. While data gathered using these techniques are frequently in agreement on the subcellular distribution for a given receptor, there are frequent discrepancies. For example, the mineralocorticoid receptor (MR) has been variously reported as cytoplasmic and nuclear (Lombes *et al.*, 1990; Sasano *et al.*, 1992), or exclusively in the nucleus (Pearce *et al.*, 1986). Likewise, using fixed cell preparations, the GR has been reported to be exclusively in the nucleus (Brink *et al.*, 1992).

Some confusion regarding intracellular distribution of macromolecules undoubtedly arises from artifacts inherent in these two approaches. Breakage of cells for preparation of subcellular compartments inevitably disrupts the integrity of these compartments, and may lead to inappropriate localization of complexes. Similarly, fixation of cells can either destroy native macromolecular interactions or mask the epitopes for antibodies directed against purified proteins; both problems would lead to incorrect identification of subcellular location. With the cloning and characterization of the green fluorescent protein (GFP), direct labeling of proteins in living cells is now feasible. Intracellular studies with these chimeras would avoid the major artifacts associated with cell fractionation and immunolocalization in fixed cells.

Successful tagging of the GR was first reported by Umesono and colleagues for a truncated version of the receptor (Ogawa *et al.*, 1995) and by Hager and coworkers (Htun *et al.*, 1996) for the full-length GR. An extensive analysis of the full-length GFP–GR fusion chimera (Htun *et al.*, 1996) indicated that its transactivation potential and ligand specificity were similar to that of the normal receptor, indicating that the GFP fusions would be highly useful reagents for the study of intracellular receptor trafficking.

II. Tagging Nuclear Receptors with GFP

GFP-fusion chimeras for several of the major nuclear receptor groups have now been reported (see Fig. 1). An example of subcellular localization studied with a GFP–GR fusion is given in Fig. 2. This receptor is localized uniquely in the cytoplasm in the absence of ligand, then undergoes complete nuclear translocation after addition of dexamethasone (dex) (Htun *et al.*, 1996). Furthermore, as noted in Htun *et al.* (1996), there appears to be an association of the receptor with a fibrillar network in the cytoplasm. Once the receptor has moved into the nucleus, there is also a discrete intranuclear organization of the receptor. A strong focal distribution of the receptor is observed when the GR is activated by an agonist

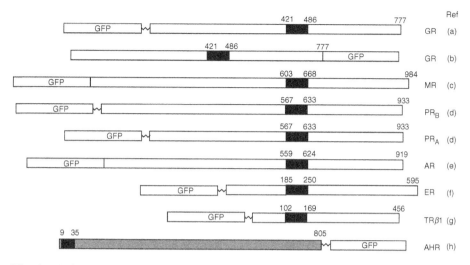

Fig. 1 Nuclear receptor–GFP fusions that have been characterized as of this writing. Schematic presentations of the receptors show the DNA-binding domains (solid fill) and position of the GFP tag. Wavy lines between the GFP domain and the receptor indicate the presence of a peptide linker (see text). Citations for these studies are as follows: Htun *et al.* (1996, 1999), Carey *et al.* (1996), Fejes-Toth *et al.* (1998), Lim *et al.* (1999), Georget *et al.* (1997), Zhu *et al.* (1998), and Chang and Puga (1998).

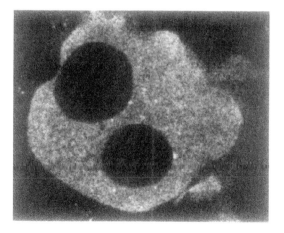

Fig. 2 Intracellular distribution of the glucocorticoid receptor, utilizing GFP–GR. Expression of GFP–GR is shown in a murine C127 cell line. This form of GFP–GR (Htun *et al.*, 1996) contains the green fluorescent protein fused to the C656G mutant glucocorticoid receptor, which is approximately 20-fold more sensitive to ligand (Chakraborti *et al.*, 1991). Cells shown here were untreated with ligand. Images from living cells expressing GFP–GR were taken with a Leica confocal laser scanning microscope, with fluorescent excitation produced by the 488-nm line of a krypton–argon laser, and using a 100× oil-phase objective. The image from uninduced cells shows complete cytoplasmic localization of the receptor; the receptor in hormone-stimulated cells is found in the nucleus (Htun *et al.*, 1996).

such as dexamethasone (Htun *et al.*, 1996). When induced with an antagonist (Hager *et al.*, 1998; Htun *et al.*, 1996), uniform nuclear localization is observed. There are several important implications from these results. First, subnuclear distribution of steroid receptors suggests the existence of specific intranuclear targets. Second, correct recognition of these targets is dependent on the nature of the activating ligand. Finally, the interrelationship between these focal structures and receptor regulatory elements in chromatin is poorly understood, but now becomes an important area of investigation.

Although C-terminal GFP chimeras for several other cellular proteins have been described, most of the nuclear receptors have been prepared as N-terminal additions (Fig. 1). One exception is a C-terminal GR–GFP fusion described by Macara and colleagues (Carey *et al.*, 1996). Using this reagent, these authors studied the involvement of RAN/TC4 GTPase in receptor translocation. A second C-terminal example is provided by the work of Chang and Puga (1998) who utilized a C-terminal fusion to characterize intracellular distribution of the aromatic hydrocarbon receptor (AHR), and identify constitutive nuclear localization signals. Although ligand binding and transactivation features of these receptors were not completely characterized, these examples suggest that C-terminal labeling also yields a functional receptor.

Preliminary findings on the intracellular distribution of several other members of the major receptor classes are now available. Fejes-Toth *et al.* (1998) labeled the MR at the N-terminus and studied cytoplasmic/nuclear partitioning of this receptor. Results from this living cell investigation indicated a complex distribution pattern, with the unliganded receptor observed both in the nucleus and the cytoplasm. A similar pattern of distribution was found for the B form of the PR by Lim *et al.* (1999). These investigators found PR-B in both compartments in the unliganded state. These two receptors (PR and MR) seem to fall in a separate class of receptors that cycle between nucleus and cytoplasm in the absence of hormone.

Georget *et al.* (1997) also using an N-terminal fusion, examined trafficking for the androgen receptor (AR) and found a distribution similar to that for GR. That is, nonstimulated cells contained the AR uniquely in the cytoplasm, and hormone activation led to rapid nuclear translocation. For the TR, use of an N-terminal GFP fusion led to the surprising observation that a major subpopulation of cells contained significant amounts of the GFP–TR in the cytoplasm Zhu *et al.* (1998), and this receptor moved to the nucleus in a hormone-dependent manner. This last result is controversial, given that the TR has classically been described as constitutively present in the nucleus. The findings with GFP–TR suggest that accepted models for intracellular distribution of the nuclear receptors may be incomplete, and in some cases in error. Furthermore, the intranuclear focal localization observed with many of the GFP-tagged receptors indicates a level of intranuclear organization not addressed in current paradigms of receptor function.

One final point should be stated concerning the fusion between GFP and a target receptor. The crystal structure of GFP (Ormo *et al.*, 1996) indicates the protein has a rather rigid globular domain. The potential exists that fusion of GFP directly into the open reading frame of a given protein may disrupt an essential feature of the target structure. M. Moser (personal communication, 1996) first suggested that isolation of the GFP moiety from the candidate protein with a peptide linker could minimize the impact of the GFP globular domain. This feature was first incorporated in GFP–GR (Htun *et al.*, 1998), and subsequently included in the design of several tagged receptors (Chang and Puga, 1998; Htun *et al.*, 1999; Lim *et al.*, 1999; Zhu *et al.*, 1998) (linker indicated in Fig. 1 by a wavy line between the receptor and GFP). A comprehensive study to determine whether inclusion of this linker serves to preserve native structure of the target protein has not been performed. However, inclusion of such a linker is a relatively simple modification, and can easily be included in the chimeric design.

III. Characterization of Fusion Chimera

Prior to use of a given receptor–GFP fusion for studies on intracellular distribution, it is imperative that the basic parameters of the chimeric receptor be verified. The size of the fusion should conform to the predicted molecular weight, and should be confirmed by Western blot analysis. Many of the chimeric proteins described to date include an epitope tag, such as the hemagglutinin (HA) antigen, that facilitates discrimination between the fusion chimera and the endogenous, native receptor (Htun *et al.*, 1998, 1999; Lim *et al.*, 1999; Zhu *et al.*, 1998). To be considered a useful reagent, the fusion chimera should obviously be expressed as a relatively stable protein of the predicted size, with little evidence of breakdown products. As an example of a well-behaved chimera, a Western analysis of GFP–GR is shown in Fig. 3.

The GFP chimera should also be characterized for its transactivation potential with an appropriate hormone-responsive promoter and reporter. This analysis will demonstrate that the basic transactivation domains of the fusion protein are intact and functional. It is useful in this analysis to evaluate the chimeric receptor with

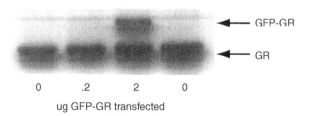

Fig. 3 Expression of the GFP–GR fusion protein. Western blot analysis shows detection of endogenous GR and transfected GFP–GR. Extracts from untreated cells, or from cells transfected with the indicated amounts of GFP–GR DNA, were treated with the BuGR-2 monoclonal GR antibody (Gametchu and Harrison, 1984) and subjected to ECL (Amersham) analysis. Reproduced with permission from Htun *et al.* (1996).

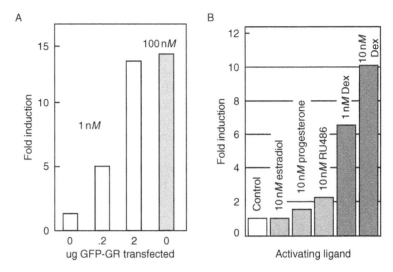

Fig. 4 Transactivation properties of GFP–GR. (A) The response of a transient MMTV reporter plasmid to GFP–GR in 1471.1 cells is shown for 1 nM dex (which activates only the GFP–GR fusion; open bars) and 100 nM dex (which activates both GFP–GR and the endogenous GR; closed bar). Cell line 1471.1 is described in Archer *et al.* (1989). (B) Ligand specificity of GFP–GR is presented for activation of endogenous MMTV-LTR-CAT sequences present in cells. Ligand identity and the respective concentrations are displayed above the bars. Reproduced with permission from Htun *et al.* (1996).

both agonists and antagonists, to ensure insofar as possible that the tagged receptor retains proper ligand dependency. An evaluation of GFP–GR transactivation potential and ligand specificity is presented in Fig. 4. It is clear that the tagged receptor retains full transactivation potential, and also maintains a ligand specificity similar to that of the unsubstituted receptor.

The fusion chimera should be tested for ligand affinity by carrying out a dose–response analysis in an *in vivo* response system. The chimeric receptor should manifest a K_d for ligand affinity similar to that measured for the wild-type receptor. Finally, electrophoretic mobility shift analysis (EMSA) with the cognate DNA response element will demonstrate that the fusion maintains native DNA recognition specificity.

IV. Choice of Fluorescent Protein to be Utilized for Receptor Labeling

Essentially all of the first-generation nuclear receptor–GFP fusions were prepared with the high-efficiency S65T variant of GFP described by Heim *et al.* (1995). This mutation renders the GFP protein stable at mammalian cell growth temperatures, resulting in a much more efficient chromophore. This form of the protein enables the preparation of receptor chimeras that are easily visualized in single-cell analysis.

There are now many color variants of GFP available, permitting for the first time the design of multiple expression experiments in real time in living cells. The first variants available had an excitation spectrum shifted into the ultraviolet (UV) range, and emitted in the visible blue range. These early blue variants were strongly susceptible to photobleaching, and marginally useful for real-time visualization. These forms of GFP were also susceptible to improper folding when fused to target proteins, resulting both in low fluorescence efficiency and inappropriate localization in subcellular compartments. This last point is, of course, a serious problem.

Fortunately, several stable color-shifted variants with high quantum yield are now available (see Table I for an overview of emission/excitation spectra for variants in current use). Pavlakis and colleagues have described a blue GFP that is relatively resistant to photobleaching and bright enough for abundant proteins (Stauber *et al.*, 1998). The W7 variant (Heim and Tsien, 1996) is one of the most stable blue-shifted forms, with an excitation maximum at 433 nm. Enhanced green forms of GFP are commercially available from Clontech (Palo Alto, CA). Also, a "yellow" version (YFP) of the molecular has appeared. Although the excitation/ emission spectra for this variant are not widely separated from those of the GFP spectra, it is in fact possible to discriminate between YFP and GFP expressed in the same cell by the use of appropriate instrumentation. Baumann *et al.* (1998) describe the use of a new-generation confocal microscope (Leica, Exton, PA) to visualize YFP and GFP separately. This instrument features a continuously adjustable monochromator on the emission side that permits the selection of any desired emission window.

These reagents now provide a powerful new approach in real-time imaging. Multiple protein species can be labeled in different colors, and visualized separately when expressed in the same cell. For example, a given nuclear receptor and a suspected cofactor can be examined for subcellular distribution when coexpressed. In principle, a great deal of experimentation is now possible regarding the distribution, colocalization, and interaction of all the nuclear receptors, cofactors, and other interacting species.

Table I
Excitation/Emission Maxima for Green Fluorescent Protein Variants

GFP variant	Excitation maximum (nm)	Emission maximum (nm)	References
GFP-S65T	488	511	Cubitt *et al.* (1995)
EGFP	490	507	Living Colors User Manual (1997)
YGFP	513	527	Living Colors User Manual (1997)
BGFP	387	450	Stauber *et al.* (1998)
GFP–W7	433	473	Heim and Tsien (1996)

V. Mode of Expression

Almost all experimentation to date with nuclear receptor–GFP chimeras has been performed by transient introduction of the GFP receptor. While this is the mode most readily available, there are potentially serious drawbacks to this method of expression. First, it is now clear that transiently expressed steroid receptors often function aberrantly in gene activation. Smith, Hager, and colleagues have described pronounced differences between the activation potentials of the transient and endogenous PR (Hager *et al.*, 1993; Smith and Hager, 1997; Smith *et al.*, 1993). Simons et al. also describe major shifts in the dose–response curve for transient versus endogenous receptors (Szapary *et al.*, 1996). Second, transient transfection often produces high levels of expression per cell, and these levels may induce inappropriate localization of the receptor chimera, or improper interaction with other molecules.

For these reasons, stable introduction of a receptor chimera into a cell line of choice is a much preferred method for study of receptor behavior in that cell. This will likely have three positive effects: the levels of expression per cell are more likely to be in the normal physiological range, all cells in the population will have a similar complement of chimeric receptor, and the stably expressed receptor has a better chance of forming appropriate associations with other components of multiprotein complexes in which receptors usually reside.

Well-characterized systems for the controlled expression of introduced chimeras are also now available. The tetracycline-inducible expression system (Gosen and Bujard, 1992) has been widely used, and has been found to work well with GFP–GR (Walker *et al.*, 1998). This approach provides several additional advantages. Levels of receptor expression can be regulated to whatever level is desired and subcellular trafficking can be observed after expression of the receptor is induced.

VI. Instrumentation

Much of the preliminary characterization for subcellular distribution of a new receptor chimera can be carried out with a standard epifluorescence microscope. These instruments are now widely available, and when equipped with a high-quality oil emersion lens (60–100×) provide excellent images with sufficient resolution for initial experimentation. For higher resolution studies, however, particularly in attempts to define intranuclear receptor trafficking patterns, a state-of-the-art laser scanning confocal microscope is indispensable. Furthermore, when attempting to discriminate between multiple chromophores, the excitation wavelength selections available on these modern instruments are invaluable.

Newer technology promises to bring more sophistication to the analysis of receptor function with GFP chimeras. Collection of three-dimensional images through "optical deconvolution" or "extensive reassignment of light" is rapidly emerging as an alternative to confocal imaging. Deconvolution offers the significant advantage of utilizing essentially all of the fluorescence emitted from a sample, whereas confocal imaging is only about 1% efficient. This is a major advantage when using chromophores excited in the UV band of the spectrum, because these wavelengths can be quite damaging to cells. The even more recent development of "two-photon" imaging also promises to expand the boundaries of possible experimentation.

VII. Future Potential

Studies on the distribution and trafficking of nuclear receptors using chimeras with the GFP and its color variants are in their infancy. Two general points have emerged. First, addition of the large GFP peptide to most nuclear receptors does not seriously impact activities of the receptors, insofar as these properties have been examined. Indeed, in this laboratory we have now tagged eight different receptors (GR, ER, PR, TR, AHR, PPAR [peroxisome proliferator receptor] α, β, and γ); all of these molecules retain relatively normal transcriptional activation efficiencies and ligand-binding profiles. The fusion chimeras are also efficiently expressed as appropriately sized polypeptides with little evidence of degradation. Second, these reagents clearly provide a powerful new tool with which to examine receptor trafficking and subcellular distribution in living cells. In particular, new insights into the intranuclear distribution of the nuclear receptors have already emerged with the introduction of this technology.

Several applications are obvious for the immediate future. Real-time observation of intracellular receptor movement is now possible. This opens a new approach for investigations into several difficult issues such as receptor cycling and cytoplasmic/nuclear shuttling. The use of fluorescence recovery after photobleaching (FRAP) will permit a direct approach to issues such as receptor cycling off and on chromatin targets. One major avenue of investigation will involve studies of the colocalization of receptors, coactivators, and transcription factors. These reagents will provide a major new tool in exploring architecture of the interphase nucleus, which is poorly understood. The potential exists to detect direct molecular interaction between receptors and interacting factors by the use of fluorescence resonance energy transfer (FRET), or fluorescence correlation spectroscopy. FRET analysis for energy transfer between GFP variants has been reported (Heim and Tsien, 1996). It is not yet clear whether these sophisticated approaches for the detection of direct molecular interactions will emerge as a major tool in the study of nuclear receptors. Finally, the ability to monitor receptor function in living cells in response to ligand stimulation provides new opportunities to characterize ligand effects on compartmentalization, and to conduct searches for new classes of agonists and antagonists. In summary, the application of GFP methodology has enormous implications in the study of nuclear receptor biology.

References

Archer, T. K., Cordingley, M. G., Marsaud, V., Richard-Foy, H., and Hager, G. L. (1989). *In* "Proceedings: Second International CBT Symposium on the Steroid/Thyroid Receptor Family and Gene Regulation," (J. A. Gustafsson, H. Eriksson, and Carlstedt-Duke, eds.), p. 221. Birkhauser-Verlag, Berlin.

Baumann, C. T., Lim, C. S., and Hager, G. L. (1998). *J. Histochem. Cytochem.* **46**(9), 1073.

Brink, M., Humbel, B. M., de Kloet, E. R., and van Driel, R. (1992). *Endocrinology* **130**, 3575.

Carey, K. L., Richards, S. A., Lounsbury, K. M., and Macara, I. G. (1996). *J. Cell Biol.* **133**(5), 985.

Chakraborti, P. K., Garabedian, M. J., Yamamoto, K. R., and Simons, S. S. J. (1991). *J. Biol. Chem.* **266**, 22075.

Chang, C. Y., and Puga, A. (1998). *Mol. Cell. Biol.* **18**, 525.

Cubitt, A. B., Heim, R., Adams, S. R., Boyd, A. E., Gross, L. A., and Tsien, R. Y. (1995). *Trends Biochem. Sci.* **20**, 448.

Fejes-Toth, G., Pearce, D., and Naray-Fejes-Toth, A. (1998). *Proc. Natl. Acad. Sci. USA* **95**, 2973.

Gametchu, B., and Harrison, R. W. (1984). *Endocrinology* **114**, 274.

Georget, V., Lobaccaro, J. M., Terouanne, B., Mangeat, P., Nicolas, J. C., and Sultan, C. (1997). *Mol. Cell. Endocrinol.* **129**, 17.

Gorski, J., Furlow, J. D., Murdoch, F. E., Fritsch, M., Kaneko, K., Ying, C., and Malayer, J. R. (1993). *Biol. Reprod.* **48**, 8.

Gosen, M., and Bujard, H. (1992). *Proc. Natl. Acad. Sci. USA* **89**, 5547.

Greene, G. L., and Press, M. F. (1986). *J. Steroid Biochem.* **24**, 1.

Guiochon-Mantel, A., Loosfelt, H., Lescop, P., Christin-Maitre, S., Perrot-Applanat, M., and Milgrom, E. (1992). *J. Steroid Biochem. Mol. Biol.* **41**, 209.

Guiochon-Mantel, A., Delabre, K., Lescop, P., Perrot-Applanat, M., and Milgrom, E. (1994). *Biochem. Pharmacol.* **47**, 21.

Hager, G. L., Archer, T. K., Fragoso, G., Bresnick, E. H., Tsukagoshi, Y., John, S., and Smith, C. L. (1993). *Cold Spring Harbor Symp. Quant. Biol.* **58**, 63.

Hager, G. L., Smith, C. L., Fragoso, G., Wolford, R. G., Walker, D., Barsony, J., and Htun, H. (1998). *J. Steroid Biochem. Mol. Biol.* **65**, 125.

Heim, R., and Tsien, R. Y. (1996). *Curr. Biol.* **6**, 178.

Heim, R., Cubitt, A. B., and Tsien, R. Y. (1995). *Nature (London)* **373**, 663.

Htun, H., Barsony, J., Renyi, I., Gould, D. J., and Hager, G. L. (1996). *Proc. Natl. Acad. Sci. USA* **93**, 4845.

Htun, H., Walker, D., and Hager, G. L. (1998). *In* "Structure, Motion, Interaction and Expression of Biological Macromolecules," (R. H. Sarma, and M. H. Sarma, eds.), p. 157. Adenine Press, Schenectady, NY.

Htun, H., Holth, L. T., Walker, D., Davie, J. R., and Hager, G. L. (1999). *Mol. Biol. Cell.* (in press).

Lim, C. S., Baumann, C. T., Htun, H., Xian, W., Irie, M., Smith, C. l., and Hager, G. L. (1999). *Mol. Endocrinol.* (in press).

Living Colors User Manual (1997). PT2040-1 Clontech Laboratories, Palo Alto, CA.

Lombes, M., Farman, N., Oblin, M. E., Baulieu, E. E., Bonvalet, J. P., Erlanger, B. F., and Gasc, J. M. (1990). *Proc. Natl. Acad. Sci. USA* **87**, 1086.

Mendel, D. B., Orti, E., Smith, L. I., Bodwell, J., and Munck, A. (1990). *Prog. Clin. Biol. Res.* **322**, 97.

Ogawa, H., Inouye, S., Tsuji, F. I., Yasuda, K., and Umesono, K. (1995). *Proc. Natl. Acad. Sci. USA* **92**(25), 11899.

Ormo, M., Cubitt, A. B., Kallio, K., Gross, L. A., Tsien, R. Y., and Remington, S. J. (1996). *Science* **273**, 1392.

Pearce, P. T., McNally, M., and Funder, J. W. (1986). *Clin. Exp. Pharmacol. Physiol.* **13**, 647.

Pratt, W. B. (1990). *Prog. Clin. Biol. Res.* **322**, 119.

Sasano, H., Fukushima, K., Sasaki, I., Matsuno, S., Nagura, H., and Krozowski, Z. S. (1992). *J. Endocrinol.* **132**, 305.

Savouret, J. F., Perrot-Applanat, M., Lescop, P., Guiochon-Mantel, A., Chauchereau, A., and Milgrom, E. (1993). *Ann. NY Acad. Sci.* **684,** 11.

Smith, C. L., and Hager, G. L. (1997). *J. Biol. Chem.* **272**(44), 27493.

Smith, C. L., Archer, T. K., Hamlin-Green, G., and Hager, G. L. (1993). *Proc. Natl. Acad. Sci. USA* **90,** 11202.

Song, C., Hiipakka, R. A., Kokontis, J. M., and Liao, S. (1995). *Ann. NY Acad. Sci.* **761,** 38.

Stauber, R. H., Horie, K., Carney, P., Hudson, E. A., Tarasova, N. I., Gaitanaris, G. A., and Pavlakis, G. N. (1998). *Biotechniques* **24**(3), 462.

Szapary, D., Xu, M., and Simons, S. S., Jr. (1996). *J. Biol. Chem.* **271,** 30576.

Walker, D., Htun, H., and Hager, G. L. (1998). Unpublished observations.

Yamashita, S. (1998). *Histol. Histopathol.* **13,** 255.

Zhu, X., Hanover, J. A., Hager, G. L., and Cheng, S. Y. (1998). *J. Biol. Chem.* **273**(42), 27058.

CHAPTER 24

Signaling, Desensitization, and Trafficking of G Protein–Coupled Receptors Revealed by Green Fluorescent Protein Conjugates

Larry S. Barak, Jie Zhang, Stephen S. G. Ferguson, Stephane A. Laporte, and Marc G. Caron

Howard Hughes Medical Institute
Duke University Medical Center
Durham, North Carolina
USA

I. Introduction

G protein-coupled receptors (GPCRs) are components of signal transduction cascades that regulate the physiological responses to a wide variety of hormones and neurotransmitters (Ross, 1990; Watson and Arkinstall, 1994). Included among

DOI: 10.1016/B978-0-12-384658-7.00024-2

the many processes GPCRs regulate are cardiac and vascular tone, odorant perception, digestion, and central nervous system transmission (Alberts *et al.*, 1998;
Nef, 1993; Presti and Gardner, 1993). Approximately 200 GPCRs have been
identified and it is estimated that hundreds to thousands more may exist (Larhammar *et al.*, 1993; Presti and Gardner, 1993). Because these receptors respond to a
diverse array of ligands and are coupled to multiple types of G proteins and second-
messenger pathways, the characterization of newly identified GPCRs represents a
formidable challenge (Blitzer *et al.*, 1993; Cotecchia *et al.*, 1990). However,
biochemical studies of rhodopsin and the β_2-adrenergic receptor (β2AR) suggest
that in addition to the many common properties that exist in these signaling
mechanisms, GPCRs also follow a common paradigm to dampen or terminate
their signaling activity. This phenomenon, which is referred to as *desensitization*,
involves at least two distinct types of proteins, G protein-coupled receptor kinases
(GRKs) and arrestins (Lohse *et al.*, 1990; Sterne-Marr and Benovic, 1995).
In particular, β-arrestins not only bind to GRK-phosphorylated β2ARs and
uncouple them from G proteins, but they also initiate agonist-mediated β2AR
sequestration (internalization) and resensitization of its signaling ability. Owing to
the difficulties in obtaining sufficiently large quantities of purified GPCRs, GRKs,
and β-arrestins for *in vitro* studies, and the difficulties in observing these proteins in
live cells, data concerning the cell biology of their interactions are lacking. A means
to follow the real-time behavior and to assess the activity of GPCRs, GRKs, and
β-arrestins in living cells should have a profound impact for understanding the
regulation of GPCR signal transduction, and has become a major goal of our
laboratory. Therefore, we have constructed and characterized green fluorescent
protein (GFP) conjugates of GPCRs, GRKs, and β-arrestins and have demonstrated
their utility in real-time studies of GPCR signal transduction in living cells.

II. Properties of G Protein-Coupled Receptors, G Protein-Coupled Receptor Kinases, and β-Arrestins

All GPCRs are structurally similar. They contain seven membrane-spanning
domains connected by alternating intra- and extracellular loops, extracellular
amino termini, cytoplasmic carboxy termini, and ligand-binding sites that can be
transmembrane or extracellular. Through interactions with one of several classes
of heterotrimeric G proteins at their cytoplasmic domains, GPCRs may be coupled
to any one of several second-messenger systems (Sterne-Marr and Benovic, 1995).

GPCRs are homologously desensitized by only a limited number of GRKs and
arrestin proteins. GRKs compose a family of serine–threonine kinases with significant sequence homology (Ferguson *et al.*, 1996a; Palczewski and Benovic, 1991;
Sterne-Marr and Benovic, 1995). The phosphorylation sites for these kinases are
contained in the C-terminal tails and intracellular cytoplasmic loops of the receptors. Whereas several sites on a particular receptor may be phosphorylated by a
single GRK, agonist-mediated phosphorylation of a single site may be sufficient to

initiate homologous receptor desensitization (Fredericks *et al.*, 1996; Ohguro *et al.*, 1993). GRKs have a variable tissue distribution. GRK1 (rhodopsin kinase) is restricted to the visual system and pineal gland. GRK2 (β-adrenergic receptor kinase 1) is most abundant in brain, heart, lung, and white blood cells, and has a critical role in normal cardiac physiology (Arriza *et al.*, 1992; Kong *et al.*, 1994; Sterne-Marr and Benovic, 1995). Transgenic mice overexpressing GRK2 demonstrate decreased cardiac contractility, whereas mice expressing a single copy of the GRK2 gene or a GRK2 inhibitor protein show enhanced contractility (Jaber *et al.*, 1996; Koch *et al.*, 1994, 1995). Of the remaining GRKs, GRK3 (β-adrenergic receptor kinase 2) is found in the olfactory system, brain, spleen, heart, and lungs; GRK4 in the testes, GRK5 in the heart and lungs, and GRK6 in a broadly distributed tissue pattern (Sterne-Marr and Benovic, 1995). Four different arrestins have been identified and characterized: they are visual arrestin, cone arrestin, β-arrestin 1, and β-arrestin 2. Visual arrestins are predominantly found in the eye. β-Arrestin 1 and 2 are widely distributed and especially elevated in brain, heart, lungs, and blood cells (Sterne-Marr and Benovic, 1995).

Little is known about the dynamic regulation of β-arrestins in tissue. The binding of β-arrestin 1 and 2 to the β2AR has been well studied *in vitro* (Gurevich *et al.*, 1995). However, their abilities to interact with many other GPCRs in cells have been only recently demonstrated by our laboratory, using a β-arrestin 2–GFP conjugate (Barak *et al.*, 1997a). Splice variants of β-arrestin 1 and 2 also exist, but it is unknown how this affects their biochemistry. The regulation of β-arrestin activity does seem to be dependent on GPCR type, GRK activity, their own phosphorylation by protein kinase C (PKC), and by their intracellular compartmentalization. For example, even though β-arrestins are generally cytosolic, they have been found concentrated at neuronal synapses with GRKs (Attramadal *et al.*, 1992; Ferguson *et al.*, 1996b).

GRK- and β-arrestin-mediated desensitization of GPCRs has been shown to require agonist occupancy of the receptor, in contrast to second messenger-mediated phosphorylation of receptors by protein kinase A (PKA) or PKC. As a result of GRK phosphorylation, receptor affinity for β-arrestins can increase more than 10-fold, as was demonstrated *in vitro* for the equimolar binding of the β2AR to β-arrestin 2 or in whole cells for the ability of β-arrestin to enhance receptor internalization (Ferguson *et al.*, 1996b; Gurevich *et al.*, 1995). For a functional interaction to occur between predominantly cytoplasmic β-arrestins and membrane-localized GPCRs, one of them must change its cellular compartmentalization. Assessment of the kinetics and magnitude of this redistribution would provide measures of both receptor activation and desensitization.

GRK phosphorylation followed by β-arrestin binding contributes to both β2AR desensitization and initiation of sequestration, a process leading to receptor resensitization (Barak *et al.*, 1995; Pippig *et al.*, 1993; Yu *et al.*, 1993). Agonist-dependent interaction of β-arrestin with the receptor serves as the switch that initiates receptor migration to a clathrin-coated pit, and once receptors are assembled at coated pits they can be internalized. This paradigm may be common to different types of

GPCRs, because many of them rapidly sequester in response to agonist. Therefore, the direct regulation of sequestration by β-arrestin or the ability of a particular receptor to be phosphorylated by a GRK or to interact with a β-arrestin may contribute substantially to its rate of resensitization.

Moreover, receptor resensitization should require β-arrestin dissociation from the receptor. It has been suggested that receptor dephosphorylation occurs in internalized acidic endosomes prior to receptor recycling to the plasma membrane, so that it would be expected that β-arrestin dissociation from the receptor probably occurs prior to these events (Krueger *et al.*, 1997). Where the process occurs and at which step it takes place should be readily discernable, if the cellular dynamic behaviors of receptors and β-arrestins can be simultaneously followed. While all GPCRs may not internalize through clathrin-coated pits, activation of all GPCRs so far tested promote the association of β-arrestin (Barak *et al.*, 1997a; Zhang *et al.*, 1996).

The use of N-terminal, epitope-tagged GPCRs has greatly facilitated the study of sequestration. Tagged GPCRs can be specifically labeled and visualized by immunofluorescence using monoclonal antibodies directed against these accessible surface epitopes. Flow cytometry measurements of the time-dependent agonist-mediated redistributions of various GPCRs suggest that they do not all internalize by the same pathway. β2ARs apparently sequester via a clathrin-dependent mechanism whereas the angiotensin II type 1A receptor internalizes despite blockade of clathrin-mediated endocytosis (Zhang *et al.*, 1996). While experiments indicate that β_2-adrenergic sequestration absolutely requires β-arrestin binding and that GRK-mediated phosphorylation is only a facilitative step, the degree to which many other GPCRs are regulated by these mechanisms remains unclear (Zhang *et al.*, 1996). For the β2AR the interdependent contributions of GRK phosphorylation and β-arrestin binding to sequestration are evident from the behavior of the phosphorylation-impaired Y326A–β2AR mutant. In HEK-293 cells under conditions of endogenous GRK expression this mutant receptor sequesters poorly, but its sequestration can be enhanced to levels of wild-type receptor by the overexpression of GRK2 (Barak *et al.*, 1994, 1995; Ferguson *et al.*, 1995). Thus, determination of the time-dependent redistributions of receptors, GRKs, and β-arrestins would provide significant insight into the biology and kinetics of signal transduction regulation.

Attempts to study GPCRs, GRKs, and β-arrestins in live cells have been limited by the comparatively low endogenous expression levels of most GPCRs and the permeability barrier provided by the cell plasma membrane. Transfection of cells with GPCR cDNA may increase GPCR membrane expression many fold and render receptors visible using immunofluorescence. However, these techniques cannot guarantee the permanent labeling of receptors or provide easy access to intracellular antigens. Thus, the behaviors of intracellular signal transduction proteins have generally remained out of reach to real-time fluorescence microscopy. The use of GFP technology may provide insight into how GPCR signal transduction regulatory proteins function in vivo. Our experience in this area is described in the following sections.

III. Design of GFP-Conjugated G Protein-Coupled Receptors, G Protein-Coupled Receptor Kinases, and β-Arrestins

Ideally, the properties of GFP-conjugated proteins should recapitulate the biochemical and biophysical characteristics of the unconjugated forms. Our experiences in creating and characterizing adrenergic receptor mutants indicate that their N and C termini are the two most promising positions for GFP conjugation, because the transmembrane domains and loops generally govern receptor conformation and provide binding sites for ligands and associated signal transduction proteins. Our decision to place GFP at the C terminus of the β_2-adrenergic receptor was motivated by two further considerations. First, the β2AR has a relatively long cytoplasmic tail, and minor truncation of this tail has relatively little effect on receptor biochemistry. Second, additions of certain epitopes at the receptor N terminus have been demonstrated to reduce receptor expression drastically. However, this problem was circumvented, at least for the Flag epitope, by placing before it a cleavable signal peptide, and perhaps a similar strategy could be considered for N-terminal GFP conjugation (Guan *et al.*, 1992). Using a C-terminal conjugation strategy and site-directed mutagenesis, we have constructed GFP variants of the β2AR, the Y326A–β2AR, and the angiotensin II type 1A receptor.

GFP variants have been engineered that possess spectral properties significantly different from the original GFP isolated from the jellyfish *Aequorea victoria* (Prasher, 1995; Prasher *et al.*, 1992). These modified GFPs have been engineered for enhanced brightness, visible light excitation maxima, and red-shifted emission spectra (Heim and Tsien, 1996). In this chapter, protein-conjugated GFP refers to the S65T variant, which is at least 10–20 times brighter than the original form, has a 480-nm excitation maximum, and can be excited and observed with filter sets optimized to view fluorescein fluorescence.

GFP was conjugated to the C terminus of an N-terminal epitope-tagged (hemagglutinin, HA) β2AR. The terminal stop codon of the receptor was replaced with a *Sal*I restriction site by directed mutagenesis. The resulting receptor cDNA was ligated in frame between the *Sad*I/*Sal*I polylinker sites of a eukaryotic GFP expression vector obtained from Clontech (Palo Alto, CA), which we had modified to express S65T-GFP (at the time an S65T version was unavailable) (Barak *et al.*, 1997a). However, numerous GFP expression vectors containing mutated and enhanced GFP variants have since become commercially available.

The proliferation of GFP expression vectors has decreased the need for complicated cloning strategies. Therefore, the design of GFP fusion proteins should depend predominantly on two factors: (1) which end of the protein is a more important determinant of the behavior or activity that needs to be preserved, and (2) which biological functions of the protein can be readily assessed. We followed this strategy for creating GFP-conjugated GRK2 (βARK1). GRK2 can be divided into three functional regions. The N-terminal third binds cellular components, the middle third constitutes the catalytic domain, and the C-terminal third segment interacts with G protein $\beta\gamma$ subunits. Of the six known GRKs, only GRK2 and GRK3 contain similar C-terminal $\beta\gamma$-binding domains, and the interaction of GRK2 with G protein $\beta\gamma$ subunits constitutes an important regulatory step in GPCR phosphorylation and desensitization. The majority of GRK2 is cytosolic, so that phosphorylation of the receptor by GRK2 (or GRK3) requires a functional translocation of kinase to the plasma membrane directed by G protein $\beta\gamma$ subunits and phospholipids (Ferguson *et al.*, 1996a). Because this function of GRK2 can be readily assessed, GFP was added to the C terminus of GRK2 in order to avoid modifying the presumed receptor-binding domain, and in a manner similar to that described above using the same S65T-GFP expression vector. Interestingly, $G_{\beta\gamma}$ binding is not critical for the kinase activity of other GRKs, because GRKs 4–6 are either palmitoylated or contain a polybasic domain at their C termini that allows them to associate with the plasma membrane. In such a case, GFP conjugation might be more judiciously performed at the N-terminal portion of the protein.

Arrestin protein binding is the terminal step in the homologous desensitization of GPCRs. The crystal structure of visual arrestin predicts that the receptor-binding domain is located in the first quarter of the molecule (Granzin *et al.*, 1998). These data agree with a conclusion from mutagenesis studies that the receptor-binding site of β-arrestin 2 is in its N-terminal domain (Sterne-Marr and Benovic, 1995). Moreover, our findings that C-terminal GFP constructs of β-arrestin 1 and β-arrestin 2, conjugated as described above with S65T-GFP, are biologically active also support this notion.

IV. Characterization of GFP-Conjugated G Protein-Coupled Receptors, G Protein-Coupled Receptor Kinases, and β-Arrestins

GFP conjugates of signal transduction proteins should be characterized using known biochemical or functional assays prior to their use in addressing new biological problems. Our experiences with the three types of signal transduction proteins suggest that their conjugation to GFP probably does not significantly change their biochemical properties. Although only minor differences have been observed between the conjugated and native proteins, this conclusion applies only to those properties that have been examined. This is surprising in light of the relatively large, although smaller, size of the GFP in relation to the native proteins (Fig. 1). However, the crystal structure of GFP suggests a compact tertiary structure (Ormo *et al.*, 1996). Thus, a sufficiently long linker between GFP and its conjugate may be appropriate to minimize potential changes in the biochemical properties of the chimeric protein.

The biochemistry of the β2AR has served as a paradigm for other GPCRs, and its characterization also provides insight into how the GFP conjugates should be evaluated. At a minimum the following assays should be performed to assess receptor pharmacology, antagonist binding, agonist binding, second-messenger coupling (in this case to adenylyl cyclase); and if possible receptor phosphorylation.

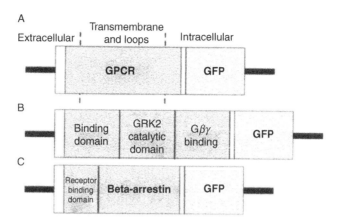

Fig. 1 Linear models indicating the relative sizes of the GFP conjugates of the (A) β2AR, (B) G protein-coupled receptor kinase 2, and (C) β-arrestin.

Little difference was seen between the wild-type receptor and the β2AR–GFP conjugate (β2AR–GFP) evaluated for these properties. Saturation binding measurements using ^{125}I-labeled pindolol showed that the dissociation constants for both were equal to 70 ± 10 pM. Isoproterenol, a pure βAR agonist with nanomolar affinity, displayed the same ability to compete the radiolabeled antagonist (pindolol) binding from the wild-type receptor and β2AR–GFP. Moreover, the ability of β2AR–GFP to couple to G protein, and activate the adenylyl cyclase, was equal to or slightly greater than that of wild-type β2AR. These data indicate that the GFP group does not substantially change the receptor conformation responsible for ligand binding or significantly interfere with receptor coupling to G protein. The ability of β2AR–GFP to be phosphorylated was also directly evaluated by measuring the agonist-mediated incorporation of ^{32}P into the protein. β2AR–GFP was maximally phosphorylated in response to agonist to approximately 80% of that observed with wild-type β2AR. When tested in the paradigm of agonist-promoted receptor internalization, which is dependent on both receptor phosphorylation and β-arrestin binding, β2AR–GFP sequestered as well as the wild-type receptor. Flow cytometry measurement showed that the decrease in surface receptor after 30 min of agonist exposure in HEK-293 cells was $57 \pm 5\%$ for wild-type receptor and $62 \pm 11\%$ for β2AR–GFP (Barak *et al.*, 1997b).

Fewer tools or assays are available to characterize GRKs and arrestins as opposed to GPCRs because comparatively less is known about their cell biology and biochemistry. This has somewhat simplified our evaluations of GRK2–GFP and β-arrestin–GFP (βarr–GFP). GRK2–GFP was characterized by its ability to enhance the internalization of the sequestration-impaired Y326A–β2AR (i.e., phosphorylate the mutated receptor) in HEK-293 cells and to promote $G_{\beta\gamma}$-mediated phosphorylation of rhodopsin *in vitro* (Pitcher *et al.*, 1998). Each of these functions might be expected to be impaired more readily than other activities of GRK2, because these functions of GRK2 depend on C-terminal motifs in close proximity to the GFP moiety. Nevertheless, we demonstrated that in HEK-293 cells, overexpression of GRK2–GFP promotes sequestration of the Y326A–β2AR nearly as well as wild-type GRK2. In addition, cell homogenates containing GRK2–GFP, when combined with purified rhodopsin and varying concentrations of $G_{\beta\gamma}$ subunits, were able to phosphorylate rhodopsin 60% as well as wild-type GRK2. These results indicate that while the functional kinase activity of the GRK2–GFP–$G_{\beta\gamma}$ complex is mildly impaired, the ability of GRK2–GFP to interact with GPCRs remains.

βarr2–GFP function was characterized using a sequestration assay based on the observation that β2ARs sequester poorly in COS-7 cells as a result of the relatively low levels of endogenous β-arrestins (Menard *et al.*, 1997). Overexpression of wild-type β-arrestin 1 or 2 generally increases β2AR sequestration three- to fourfold, and overexpression of β2AR2–GFP was observed to increase it to a similar degree. These data suggest that addition of GFP to the C terminus of β-arrestin does not interfere with its receptor binding and trafficking functions (Barak *et al.*, 1997b). In contrast, addition of the Flag epitope slightly proximal to the C terminus of β-arrestin 2 resulted in a β-arrestin construct unable to enhance β2AR sequestration normally (our unpublished data). Both the amino and carboxy termini of β-arrestin contain regulatory domains, so that it is unexpected, considering the relatively small size of the Flag epitope compared with GFP, that βarr2–GFP is even functional. This retention of function supports the observation that the GFP moiety is noninteracting and also suggests a general approach for the C-terminal modification of arrestins. Additional studies have since demonstrated that an N-terminal βarr2–GFP conjugate is also functional.

V. Instrumentation and Methodology

Once the GFP conjugates have been characterized their cell biology can be evaluated using conventional optical methods. We have observed β2AR–GFP fluorescence by flow cytometry, fluorescence spectrophotometry, conventional fluorescence microscopy, and laser scanning confocal microscopy. Many of our current experiments now measure the time-dependent distribution of GFP-conjugated receptors or β-arrestins in live cells in response to agonists, and our best results have been acquired using the confocal microscope.

While we have developed permanent clones expressing GFP-conjugated receptors, GRKs, and β-arrestins, most experiments are performed using live HEK-293 or COS-7 cells that are transiently transfected with the appropriate cDNA. The advantage of a transient expression system rests in its versatility and the variety of conditions that can be explored within a given experiment. For these experiments cells are plated at 50% confluence in 100-mm tissue culture dishes on day 0. On day 1, the cDNA of interest in a GFP expression plasmid is transfected into the cells by coprecipitation with calcium phosphate. In this method, 5–10 μg of plasmid cDNA is resuspended in 450 μl of 18-MΩ water to which 50 μl of 2.5 M calcium chloride is added. Five hundred microliters of 2× Hanks' balanced salt solution is then added dropwise to this mixture. The resulting suspension is then added dropwise over the cells, which are now usually between 60% and 80% confluent. On day 2 the medium is replaced with fresh medium, and the cells are returned to the incubator for 24–48 h for use on days 3 or 4. For confocal microscopy, transfected cells are plated into 35-mm dishes (MatTek, Ashland, MA) containing centered, 1-cm wells formed from glass coverslip-sealed holes. For live cell experiments, cells are placed in minimal essential medium supplemented with 20 mM N-2-hydroxyethylpiperazine-N'-2-ethanesulfonic acid (HEPES). The experiments are performed with a Zeiss (Thornwood, NY) LSM-410 confocal microscope using the time series function. A baseline scan of GFP fluorescence is performed at 488 nm. Ligand is then added in volumes of 50–100 μl (5–10% of the total volume), and scans of the same field are typically made every 30 s to 2 min. Fluorescence images acquired on the Zeiss LSM-410 confocal microscope are analyzed using the LSM-image software or transferred to Adobe Photoshop (Adobe, Mountain View, CA).

VI. Cell Distribution of β2AR–GFP and Y326A–β2AR–GFP

GFP technology offers significant advantages over conventional fluorescence techniques for the study of the cellular regulation of GPCRs. These advantages include the following:

1. Receptors or proteins are stoichiometrically labeled by GFP with a ratio of 1:1.
2. Every GFP-tagged molecule is potentially visible despite its intracellular location.
3. The labeling and visualization of intracellular proteins does not require permeabilization or extraction of the cell membrane, thus preventing the introduction of fixation artifacts.
4. The same set of live cells can be observed over extended intervals (days) since GFP-labeled protein is a product of cellular synthesis.

Figure 2 shows video images acquired on a Leitz (Deerfield, IL) model DM fluorescence microscope connected to an Optronics (Goleta, CA) VI-470 CCD video camera system and a Sony (New York, NY) model UP-5600 color video printer with a UPK-5502SC digital interface board. The panels display images of 4% paraformaldehyde-fixed HEK-293 cells that permanently express either β2AR–GFP (Fig. 2A–D, H) or the sequestration- and phosphorylation-impaired Y326A–β2AR–GFP conjugate (Fig. 2E–G). These cells were exposed to saturating concentrations of the agonist isoproterenol at 37 °C for various periods of time in order to examine receptor sequestration and downregulation. β2AR–GFP should

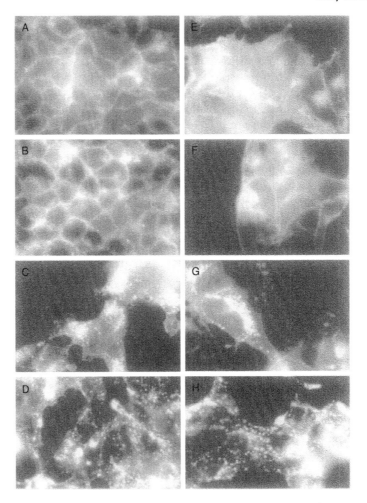

Fig. 2 Visualization of the time course of agonist-mediated internalization of the β2AR and the Y326A–β2AR. Permanent clones of HEK-293 cells containing each receptor were treated as described in text and visualized using standard fluorescence microscopy. Receptors can be observed in at least three different cellular compartments (plasma membrane, endosomal, and perinuclear) in this series of experiments (see text for details).

respond to the agonist in a manner similar to unconjugated β2AR, as discussed previously. Figure 2A shows the plasma membrane distribution of β2AR–GFP in cells never exposed to agonist. In these permanent clones receptor fluorescence is homogeneous and uniform from cell to cell. In addition, there is a paucity of intracellular fluorescence. In contrast, cells exposed only to isoproterenol for 15 min (Fig. 2C) have large amounts of internalized receptors. These cells demonstrate a loss of plasma membrane staining at their periphery, and internalized receptors appear as bright patches and white dots in the cell interior. The cells in

Fig. 2B appear roughly similar to those in Fig. 2A. However, they were first exposed to 20 μM isoproterenol for 15 min, washed, and then incubated with saturating concentrations of the antagonist propranolol for 1 h. Cells in Fig. 2D and H were treated for 3 h with either agonist followed by antagonist (Fig. 2D) or agonist alone (Fig. 2H). In contrast to results shown in Fig. 2B, in which sequestered β2AR–GFP returns to and remains on the cell surface in the presence of antagonist, a large intracellular receptor population persists in Fig. 2D. Moreover, these receptors have the same perinuclear appearance as those in Fig. 2H, which are treated with agonist only. The failure of the GFP-conjugated receptors to return to the cell surface in the long-term presence of agonist and their aggregated perinuclear distributions are most consistent with receptor downregulation. In contrast to β2AR–GFP, Y326A–β2AR–GFP does not sequester. Figure 2E–G shows cells permanently expressing Y326A–β2AR–GFP after 0 min, 15 min, and 3 h of exposure to the agonist isoproterenol. No internalization of receptor has occurred after 15 min, in agreement with previous sequestration measurements of the Y326A–β2AR (Barak *et al.*, 1994). However, after 3 h of exposure to isoproterenol the distribution of the Y326A–β2AR–GFP as seen in Fig. 2G is similar to that of β2AR–GFP (Fig. 2D and H). The cells in Fig. 2G show a considerable loss of staining along the plasma membrane. Furthermore, perinuclear aggregates of receptors are now apparent. These experiments demonstrate the utility of GFP-conjugated GPCRs for studying receptor behavior. Moreover, the results suggest that sequestration is not a prerequisite for downregulation.

VII. Determination of Receptor Activation by GFP-Conjugated G Protein–Coupled Receptor Kinase and β-Arrestin

Phosphorylation by GRKs followed by β-arrestin binding is perhaps a common step in the homologous desensitization of most GPCRs. These processes should occur within seconds of receptor activation, and should coincide with the translocation of GRK and β-arrestin from the cytosol to the plasma membrane. We have attempted to verify by confocal microscopy in live HEK-293 cells that translocations of GRK2–GFP and βarr2–GFP do indeed occur. Translocation should be reflected as an increase in plasma membrane fluorescence that coincides with a decrease in fluorescence from the cell interior. An optimal signal-to-noise ratio will result following agonist treatment if the cytoplasmic fluorescence signal is reduced to levels of cellular autofluorescence and the plasma membrane now contains all of the GFP-conjugated protein. These limits can be practically approached by increasing receptor expression (usually >1 pmol/mg of cell protein) with respect to the amount of expressed GFP-conjugated GRK or β-arrestin protein, and by using cells with relatively small diameters. For example, HEK-293 cells produce better fluorescence signals than the larger COS-7 cells for the same amount of GFP expression. Our attempts to monitor GRK2–GFP translocation under these conditions have failed to demonstrate any persistent GRK2–GFP translocation to membrane-bound β2ARs or angiotensin II type 1A receptors within 10–15 s of agonist exposure, although preliminary experiments have indicated that a sustained translocation to the substance P receptor can be observed within that time frame. Most unexpected, however, was the novel observation made with GRK2–GFP that GRK2 colocalizes with tubulin in mitotic spindles (Pitcher *et al.*, 1998).

The translocation and binding of βarr2–GFP to an agonist-activated receptor is a more persistent process, and is consistent with the slower time course observed for receptor sequestration. The agonist-mediated translocation of βarr2–GFP is shown in Fig. 3. Figure 3A is a confocal scan of HEK-293 cells that have not

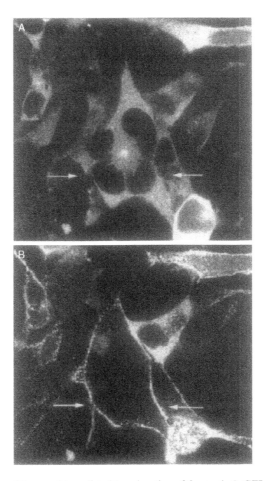

Fig. 3 Visualization of the agonist-mediated translocation of β-arrestin 2–GFP to membrane-bound receptors: (A) image before agonist is administered; (B) 90 s after agonist is added. In these confocal images the disappearance of GFP fluorescence from the cytosol and the increase in GFP fluorescence at the plasma membrane (arrows) is a measure of receptor activation, receptor desensitization, and the recompartmentalization of β-arrestin 2.

been agonist treated and that express receptors at the plasma membrane and βarr2–GFP in the interior of the cell. The nucleus excludes βarr2–GFP and so appears dark, whereas the cytosol, containing the vast majority of βarr2–GFP, appears white (fluorescent). Figure 2B shows the redistribution of βarr2–GFP 90 s following the addition of agonist. The majority of GFP fluorescence now appears at the edge of the cell as a result of βarr2–GFP binding to plasma membrane-localized receptor. Depending on the experimental conditions, βarr2–GFP redistribution can easily occur within 10–15 s. However, because images result from the averaging of between 4 and 16 scans that require approximately 1 s each to obtain, it is unlikely that this technique can resolve cellular processes that occur in time intervals of less than 1–3 s. Controlling variables such as the temperature or coexpression of GRKs can decrease or accelerate the rate of βarr–GFP translocation.

VIII. Analysis of β-Arrestin 2–GFP Translocation that Occurs in Single Cells

In this section, we describe how numerical and kinetic data can be obtained from confocal images using the scans presented in Fig. 3. Many computer programs such as NIH Image are available over the Internet or commercially to analyze these types of images. Figure 4 demonstrates a simple type of analysis. Histograms of the average fluorescence at a point, along a line or in a defined area, can be made using the imaging software available with the Zeiss LSM-410 microscope. Histogram trace A in Fig. 4 corresponds to a line drawn through the lower portion of the cell in Fig. 3A and trace B corresponds to a similar line drawn through the image of the agonist-treated cell in Fig. 3B. It is apparent that the decrease in fluorescence in the cell interior (Fig. 3B) following agonist treatment is reflected by the concave appearance of the corresponding histogram (Fig. 4B). Similarly, the two peaks in the lower histogram reflect the increase in fluorescence at the cell membrane (Fig. 3B).

As a result of the stoichiometric labeling of βarr2–GFP, the measured fluorescence signal should be proportional to the amount of activated receptor. Background noise is now mainly a result of cell autofluorescence rather than of nonspecific or nonuniform probe labeling as is often the case with conventional fluorescence. Therefore, we can use translocation of βarr2–GFP as a measure of any process that changes or affects the population of activated receptors at the cell membrane. For example, the pharmacology of an unknown agonist can be determined by integration of the fluorescence originating from the plasma membrane of a single cell. Known increases in the extracellular agonist concentration over time would produce measurable changes in the magnitude of membrane-translocated βarr2–GFP. A plot of agonist concentration versus plasma membrane fluorescence intensity would thus yield a dose–response curve. We have performed this

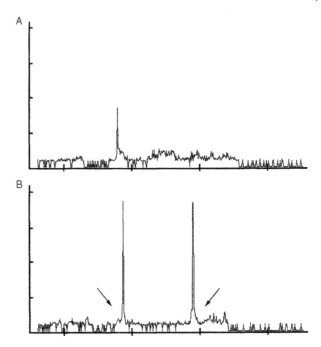

Fig. 4 Measurement of the distribution of β-arrestin 2–GFP fluorescence. The relative fluorescence intensity of β-arrestin 2–GFP along a line drawn through the before and after agonist exposure is proportional to the height of the curves at each point. Note that the concave shape of the curve in (B) corresponds to the movement of β-arrestin 2–GFP to the plasma membrane.

experiment for the β2AR and the agonist isoproterenol, and the results are in agreement with direct radioligand binding data (our unpublished data).

IX. Limitations

Study of the biology of β-arrestins and GRKs in live cells had been all but impossible prior to the generation of the corresponding GFP conjugates. Thus, there are only limited data concerning the effects of β-arrestins on GPCRs other than the β2AR and the m₂-muscarinic receptor. Although in its infancy, this new GFP technology will contribute immensely to the study of the biochemistry and cell biology of GPCRs. While the technology will circumvent and solve many practical problems, it will still suffer some of the same limitations that currently confine fluorescence measurements in general: the resolving power of the optics, cell autofluorescence, and fluorescence collected from out-of-the-plane of focus.

Cellular autofluorescence, which diminishes with longer wavelength excitation, perhaps remains the greatest impediment for obtaining larger signal-to-noise ratios. For most experiments we obtain signal-to-noise ratios of 5–10 to 1, and can easily discriminate probe distribution from other cell fluorescence. Under the best of circumstances signal-to-noise ratios of 50:1 have been measured. In either case, these signals are more than large enough to measure probe kinetics. Moreover, the dynamic range of the signal is well within the 8-bit range (256-gray scale) of most imaging systems. Further improvements will perhaps require GFP variants that are excited by green light and fluoresce in the red portion of the spectrum.

The relative expression levels and biological fidelity of the transfected proteins are other important factors limiting signal quality. For example, the magnitude of the βarr–GFP fluorescence signal at the plasma membrane increases with increasing amounts of plasma membrane-localized receptor. Therefore, for the purpose of measuring processes such as receptor activation, more receptor means larger signals until some other component of the signal transduction system becomes limiting, such as the amount of G protein available to produce high agonist-affinity receptor complexes. At the least, we have been able to measure translocation in cells containing 50,000–100,000 receptors (i.e., with 0.5–1 pmol/mg cell protein). The sensitivity is somewhat improved since the receptors are measured at the edge of the cell, reducing the autofluorescence. Sensitivity can also be improved by using smaller cells with the same receptor expression. However, the ultimate goal should be to make measurements at endogenous receptor expression levels, which are at least 5- to 10-fold lower.

Each GFP-conjugated protein utilized in our laboratory undergoes either a biochemical or functional characterization. However, an unappreciated biochemical or biological property of the GFP conjugate can still be affected. For example, the G protein-related signal transduction properties of the β2AR are essentially unaffected by conjugation of GFP at its C terminus. However, properties that require a free C-terminal motif might well be affected. Therefore, for proteins such as β-arrestin, construction and characterization of both N- and C-terminal GFP conjugates should perhaps be undertaken in spite of theoretical considerations. Similar reasoning applies when only segments rather than full-length constructs of these proteins are expressed conjugated to GFP. While it is unlikely that new binding specificity will be created by conjugation of a protein to GFP, it is not unlikely that existing ones will be affected, and the degree to which that occurs must be determined in order for the construct to be useful for a particular assay.

X. Conclusion

GFP technology will significantly change the way GPCR signal transduction proteins are studied in cells. In this chapter we have presented the methodology to construct and characterize GPCRs, GRKs, and β-arrestins conjugated to GFP. In addition, we have demonstrated that their behaviors can be studied in living cells in real time. Even though this technology is in its infancy, at the least it will contribute to the study of GPCR-based diseases, the search for new classes of drugs, and the identification of ligands that bind to orphan receptors.

Acknowledgments

This work was supported by National Institutes of Health Grants HL 03422 (to L. S. B.) and NS 19576 (to M. G. C.), a Heart and Stroke Foundation of Ontario Grant NA-3349 (to S. S. G. F.), a Heart and Stroke Foundation of Canada Grant (to S. A. L.), and an unrestricted Neuroscience Award from Bristol-Myers Squibb and an unrestricted grant from Zeneca Pharmaceutical Co. (to M. G. C.). M. G. C. is an investigator of the Howard Hughes Medical Institute.

References

Alberts, B., Bray, D., Johnson, A., Lewis, J., Raff, M., Roberts, K., and Walter, P. (1998). *In* "Essential Cell Biology," (E. Lawrence, and V. Neal, eds.), p. 481. Garland Publishing, New York.

Arriza, J. L., Dawson, T. M., Simerly, R. B., Martin, L. J., Caron, M. G., Snyder, S. H., and Lefkowitz, R. J. (1992). *J. Neurosci.* **12,** 4045.

Attramadal, H., Arriza, J. L., Aoki, C., Dawson, T. M., Codina, J., Kwatra, M. M., Snyder, S. H., Caron, M. G., and Lefkowitz, R. J. (1992). *J. Biol. Chem.* **267,** 17882.

Barak, L. S., Tiberi, M., Freedman, N. J., Kwatra, M. M., Lefkowitz, R. J., and Caron, M. G. (1994). *J. Biol. Chem.* **269,** 2790.

Barak, L. S., Menard, L., Ferguson, S. S., Colapietro, A. M., and Caron, M. G. (1995). *Biochemistry* **34,** 15407.

Barak, L. S., Ferguson, S. S. G., Zhang, J., and Caron, M. G. (1997a). *J. Biol. Chem.* **272,** 27497.

Barak, L. S., Ferguson, S. S. G., Zhang, J., Martenson, C., Meyer, T., and Caron, M. G. (1997b). *Mol. Pharmacol.* **51,** 177.

Blitzer, R. D., Omri, G., De Vivo, M., Carty, D. J., Premont, R. T., Codina, J., Birnbaumer, L., Cotecchia, S., Caron, M. G., and Lefkowitz, R. J. (1993). *J. Biol. Chem.* **268,** 7532.

Cotecchia, S., Kobilka, B. K., Daniel, K. W., Nolan, R. D., Lapetina, E. Y., Caron, M. G., Lefkowitz, R. J., and Regan, J. W. (1990). *J. Biol. Chem.* **265,** 63.

Ferguson, S. S. G., Menard, L., Barak, L. S., Koch, W. J., Colapietro, A. M., and Caron, M. G. (1995). *J. Biol. Chem.* **270,** 24782.

Ferguson, S. S. G., Barak, L. S., Zhang, J., and Caron, M. G. (1996a). *Can. J. Physiol. Pharmacol.* **74,** 1095.

Ferguson, S. S. G., Downey, W. E., Colapietro, A. M., Barak, L. S., Menard, L., and Caron, M. G. (1996b). *Science* **271,** 363.

Fredericks, Z. L., Pitcher, J. A., and Lefkowitz, R. J. (1996). *J. Biol. Chem.* **271,** 13796.

Granzin, J., Wilden, U., Choe, H. W., Labahn, J., Krafft, B., and Buldt, G. (1998). *Nature (London)* **391,** 918.

Guan, X. M., Kobilka, T. S., and Kobilka, B. K. (1992). *J. Biol. Chem.* **267,** 21995.

Gurevich, V. V., Dion, S. B., Onorato, J. J., Ptasienski, J., Kim, C. M., Sterne-Marr, R., Hosey, M. M., and Benovic, J. L. (1995). *J. Biol. Chem.* **270,** 720.

Heim, R., and Tsien, R. Y. (1996). *Curr. Biol.* **6,** 178.

Jaber, M., Koch, W. J., Rockman, H., Smith, B., Bond, R. A., Sulik, K. K., Ross, J., Jr., Lefkowitz, R. J., Caron, M. G., and Giros, B. (1996). *Proc. Natl. Acad. Sci. USA* **93,** 12974.

Koch, W. J., Hawes, B. E., Inglese, J., Luttrell, L. M., and Lefkowitz, R. J. (1994). *J. Biol. Chem.* **269,** 6193.

Koch, W. J., Rockman, H. A., Samama, P., Hamilton, R. A., Bond, R. A., Milano, C. A., and Lefkowitz, R. J. (1995). *Science* **268,** 1350.

Kong, G., Penn, R., and Benovic, J. L. (1994). *J. Biol. Chem.* **269,** 13084.

Krueger, K. M., Daaka, Y., Pitcher, J. A., and Lefkowitz, R. J. (1997). *J. Biol. Chem.* **272,** 5.

Larhammar, D., Blomqvist, A. G., and Wahlestedt, C. (1993). *Drug Design Discov.* **9,** 179.

Lohse, M. J., Benovic, J. L., Codina, J., Caron, M. G., and Lefkowitz, R. J. (1990). *Science* **248,** 1547.

Menard, L., Ferguson, S. S. G., Zhang, J., Fang-Tsyr, L., Lefkowitz, R. J., Caron, M. G., and Barak, L. S. (1997). *Mol. Pharmacol.* **51,** 800.

Nef, P. (1993). *Receptors Channels* **1,** 259.

Ohguro, H., Palczewski, K., Ericsson, L. H., Walsh, K. A., and Johnson, R. S. (1993). *Biochemistry* **32,** 5718.

Ormo, M., Cubitt, A. B., Kallio, K., Gross, L. A., Tsien, R. Y., and Remington, S. J. (1996). *Science* **273,** 1392.

Palczewski, K., and Benovic, J. L. (1991). *Trends Biochem. Sci.* **16,** 387.

Pippig, S., Andexinger, S., Daniel, K., Puzicha, M., Caron, M. G., Lefkowitz, R. J., and Lohse, M. J. (1993). *J. Biol. Chem.* **268,** 3201.

Pitcher, J. A., Hall, R. A., Daaka, Y., Zhang, J., Ferguson, S. G., Hester, S., Miller, S., Caron, M. G., Lefkowitz, R. J., and Barak, L. S. (1998). *J. Biol. Chem.* **273,** 12316.

Prasher, D. C. (1995). *Trends Genet.* **11,** 320.

Prasher, D. C., Eckenrode, V. K., Ward, W. W., Prendergast, F. G., and Cormier, M. J. (1992). *Gene* **111,** 229.

Presti, M. E., and Gardner, J. D. (1993). *Am. J. Physiol.* **264**(1), G399.

Ross, E. M. (1990). *In* "Goodman and Gilman's The Pharmacological Basis of Therapeutics," p. 33. Pergamon Press, New York.

Sterne-Marr, R., and Benovic, J. L. (1995). *Vitam. Horm.* **51,** 193.

Watson, S., and Arkinstall, S. (1994). The G-Protein Linked Receptor Facts Book. Academic Press, San Diego, CA.

Yu, S. S., Lefkowitz, R. J., and Hausdorff, W. P. (1993). *J. Biol. Chem.* **268,** 337.

Zhang, J., Ferguson, S. S. G., Barak, L. S., Menard, L., and Caron, M. G. (1996). *J. Biol. Chem.* **271,** 18302.

CHAPTER 25

Fluorescent Proteins in Single- and Multicolor Flow Cytometry

Lonnie Lybarger and Robert Chervenak

Department of Microbiology and Immunology
LSU Health Sciences Center-Shreveport
Shreveport, Louisiana, USA

I. Update

Since the publication of our original paper, it is hardly an exaggeration to state that the field of fluorescent proteins has exploded. A simple PubMed search for reviews on the subject turns up over 4300 papers as of January 2010, and original articles utilizing fluorescent protein technology are simply impossible to enumerate. Besides an ever-expanding creativity in the application of fluorescent proteins for biomedical exploration, two new and ongoing developments promise to keep these molecules at the forefront of biological research for the foreseeable future (Muller-Taubenberger and Anderson, 2007; Nienhous and Wiedenmann, 2009; Shaner et al., 2007). The first is the continued development of new fluorescent proteins with unique spectral properties that can be used singly, or in tandem with other fluorescent proteins (as well as conventional fluorescently tagged antibodies) to give an ever-clearer picture of complex biologic processes. The singular significance of this effort to the scientific community was brought home in 2008 by the awarding of the Nobel Prize in Chemistry to Drs. Osamu Shimormura, Martin Chalfie, and Roger Tsien for their discovery, cloning, and development of green fluorescent protein and its numerous spectral variants. The ongoing discovery and development of fluorescent proteins from new species represents an intense effort within the scientific community that promises to further expand how these molecules can be used to answer critical questions in biomedical research. The second development has to do with the instrumentation, specifically flow cytometers, that are used for the detection of fluorescent proteins. Improvements in these instruments, especially in optics and lasers have been critical in the practical application of fluorescent protein technologies. The increased sensitivity of flow cytometers makes the detection of even weakly fluorescent molecules practical and the evolution of even the most basic flow cytometers into multilaser instruments makes the detection of multiple fluorescent proteins in a single sample feasible for nearly every modern flow laboratory.

Yet, as much as the field has changed, the basic principles discussed in this chapter still apply. We are still met with the challenge of distinguishing fluorescent molecules whose spectral properties are quite similar and, to a large extent, overlapping. This obstacle can sometimes be overcome by the use of two fluorescent proteins whose emission spectra are similar, but whose excitation properties are such that each is excitable by the unique wavelength of two different lasers. At other times, such as the circumstance described in this chapter with eGFP and eYFP, a careful examination of emission spectra can lead to the design of optical filters that can distinguish between two fluorescent proteins whose emission overlaps to the point that the human eye cannot tell them apart under normal circumstances. Even with the further development of novel fluorescent proteins for biologic research, these issues are unlikely to abate in the near future and the principles outlined in this chapter remain a solid roadmap to their resolution.

II. Introduction

Advances in green fluorescent protein (GFP) expression technology have fueled the drive to develop efficient and sensitive methods for GFP detection. Flow cytometry, at least in principle, should be applicable to the detection of GFP-derived fluorescence in mixed populations of cells to obtain quantitative and qualitative data regarding those cells. This could include simply monitoring the activity of a single genetic regulatory element that is engineered to control GFP expression, or combining this information with cell surface phenotypic data and/or additional fluorescent protein reporter construct data in multiparameter analyses. Experimental data from several laboratories have now demonstrated that all of these possibilities can be realized with the proper reagents and instrumentation. In large part, this success can be ascribed to the generation of GFP variants that possess spectral properties that are superior to those of wild-type GFP, in terms of flow cytometric detection.

III. Choice of Fluorescent Proteins

Flow cytometry can be used to detect wild-type GFP expression in mammalian cells subsequent to transfection, using the commonly employed 488-nm wavelength from an argon laser for excitation (Lybarger et al., 1996; Ropp et al., 1995). However, the signal obtained is modest, even when the protein is expressed from relatively powerful promoter elements. Several mutants of GFP, differing from wild-type GFP by one or a few amino acids, are better suited for flow cytometry owing to their increased efficiency of fluorescence and their altered excitation and emission spectra. For single fluorescent protein (FP) detection, one of the green-emitting variants that excites efficiently in the 488-nm (blue light) range is probably a better choice as compared with the ultraviolet (ff)-excitable mutants. Such variants are generally brighter (although effective UV-excitable FPs do exist; Anderson et al., 1996; Stauber et al., 1998) and can be used with the simplest analyzers. These include GFP-S65T (Heim et al., 1995), EGFP (enhanced GFP; Cormack et al., 1996; Yang et al., 1996), GFPsg25 (Stauber et al., 1998), and EYFP [enhanced yellow fluorescent protein (Ormö et al., 1996; 1997) (clone 10C variant; Ormö et al., 1996)]. In particular, EYFP (527-nm emission maximum) is attractive because this variant emits bright fluorescence with 488-nm excitation and optimal detection of this fluorescence does not require modification of the cytometer optics (Lybarger et al., 1998), which may be required for optimal detection of the other variants.

Multicolor FP detection has been most frequently accomplished by combining one of the green-emitting variants with a UV-excitable mutant (Ropp et al., 1996; Stauber et al., 1998; Yang et al., 1998). Although this approach can be useful, many of the UV mutants are relatively dim in comparison with the blue-excitable forms or are simply not excited well by the UV light emitted from an argon laser. If the cytometer is equipped with both a krypton laser and an argon laser, FPs are available that permit sensitive detection of two distinct regulatory elements (Anderson et al., 1996). In addition, our laboratory utilizes two blue-excitable variants, EGFP and EYFP, for two-color analyses (Lybarger et al., 1998). Certainly, the choice of FP variants will depend on the experimental system and the available flow cytometry hardware. This chapter focuses on single- and two-color FP analyses using EGFP and EYFP in conjunction with 488-nm stimulation, because fluorescence detection of these two variants can be achieved on virtually any cytometer.

IV. Transfection of Mammalian Cells

Published reports have described successful expression of FPs following transfection of cells via all of the standard transfection methods. Thus, essentially any method should provide cells suitable for flow cytometric analysis. With each method, especially transfection with some liposome reagents, it is imperative to carry negative (i.e., mock-transfected) control cells through the experimental procedure, as transfection may alter the light-scattering and autofluorescent properties of the cells and, therefore, affect the subsequent analyses. A procedure is detailed below for electroporation of NIH 3T3 murine fibroblasts using a BTX (San Diego, CA) model T820 Electrosquareporator.

1. *Harvest and count cells.* It is important that the cells be in the exponential growth phase at the time they are transfected. Resuspend the cell pellet in complete growth medium [Dulbecco's modified Eagle's medium supplemented with glutamine (0.2 mM), gentamicin (50 mg/l), 2-mercaptoethanol (0.05 mM), and 10% fetal calf serum] at a concentration of 10^7 cells/ml. Keep the cells on ice.

2. *Prepare all reagents.* For each transfection, label a 35-mm tissue culture plate and fill each plate with 6 ml of growth medium. If several transfections are to be performed, a six-well plate may be more convenient. Pipette 200 µl (2 million cells) of the cell suspension into a sterile, 4-mm electroporation cuvette for each reaction.

3. *Add DNA to each cuvette just prior to electroporation.* DNA should be of the highest purity, obtained by CsCl-gradient centrifugation or ion-exchange chromatography. Five to 50 µg of DNA (or of each DNA construct in the case of cotransfection) may be used per reaction, depending on the strength of the promoter in the cells to be transfected. The optimal quantity of DNA will have to be determined for each cell type. DNA should be at a concentration of ≥ 1 mg/ml. Pipette DNA into the cuvette and mix thoroughly. Remove any bubbles that form and place the cuvette in the electrode chamber.

4. Using the low-voltage (LV) mode, set the appropriate pulse amplitude (voltage), pulse length in milliseconds, and pulse number. For NIH 3T3 cells, we use 240 V, 20 ms, and one pulse, respectively. After the pulse is delivered, remove the cuvette from the chamber.

5. With a Pasteur pipette and bulb, remove the cells from the cuvette and transfer to a culture plate that contains growth medium. Rinse the cuvette (two or three times) with medium to remove all of the cells. There will be cell clumps.

6. Place the cells in an incubator and culture overnight. Analyze the cells for FP expression after 20–24 h. With most cell lines we have analyzed, maximal expression is reached by 24 h, and a 48-h incubation is unnecessary. Harvest the cells and wash in a $3\times$ volume of phosphate-buffered saline (PBS). Use of trypsin to detach adherent NIH 3T3 cells will not affect FP detection, but trypsin can damage cell surface proteins. Therefore, it may be necessary to detach the cells by more gentle means [incubation in an EDTA solution (Versene; Life Technologies, Gaithersburg, MD)] if the expression of surface molecules is to be monitored in combination with GFP.

7. Resuspend the cell pellets in fluorescence-activated cell sorting (FACS) buffer (PBS–2% newborn calf serum–0.1% sodium azide). Sodium azide is a metabolic inhibitor that is added to prevent the alteration of cell surface phenotype during analysis. If cells are to be placed in culture following analysis, omit sodium azide

from the buffer. In any case, keep the cells on ice after they are harvested. If possible, cells should be at a concentration of 5×10^5–2×10^6 ml^{-1}, with a minimum sample volume of 300 μl.

V. Stable Fluorescent Protein Expression

For some experimental situations, stable FP expression is required. For example, we have generated stable FP-expressing cells that are useful for exploring critical parameters in FP detection by flow cytometry. This provides an abundant source of FP-positive cells without the need to perform transfections prior to each analysis. Using retroviral vectors, stable FP expression can be readily achieved through drug selection or on the basis of FP-derived fluorescence and cell sorting. Typically, stably expressing cell populations exhibit more uniform levels of expression, albeit lower levels, compared with transiently transfected cells. One problem with transiently transfected cells is that a broad spectrum of fluorescent intensities can be seen, and this range can exceed the capabilities of a 4-log amplifier that is standard on many cytometers.

A brief protocol to obtain stable FP-expressing cells is listed below. In this example, a retrovirus-based vector is utilized that encodes a hygromycin resistance (hygR)/GFP fusion gene. This vector can be used to generate stable cell lines following transient transfection of cells, or following infection of target cells with retrovirus particles derived from the vector (Lybarger et al., 1996). For a detailed discussion of FP expression from retroviral vectors, readers are encouraged to refer to Chapters 28 and 30 in this volume.

1. Harvest NIH 3T3 cells 48 h after infection with retrovirus or transfection with the retroviral construct. Fluorescence microscopic analysis should reveal GFP-positive cells.

2a. *Drug selection.* Split cells 1:4 into normal growth medium that is supplemented with hygromycin B (500 μg/ml; Boehringer Mannheim, Indianapolis, IN). Cell death should become evident 2/3 days after the addition of hygromycin. Dead cells will become rounded and eventually detach from the culture surface. After 1 week, selection will be nearly complete, with virtually all of the viable cells expressing GFP. Infection should yield a stable cell line more quickly than transfection, because every infected cell will possess a stable copy of the provirus. In contrast, only a minor fraction of the transfected cells will harbor integrated copies of the plasmid vector. Therefore, it may take longer with transfection (2–3 weeks) to obtain sufficient numbers of stable clones to perform analyses.

2b. *Cell sorting.* GFP fluorescence can also be used as a means of selection for stable expression. After infection, GFP-positive cells can be isolated via cell sorting and placed back into culture. These cells maintain expression of GFP over time, in a manner nearly identical to that of stable GFP-expressing cells obtained by hygromycin selection (Lybarger et al., 1996). In some cases, additional rounds of cell sorting may be required to select pure populations of GFP-expressing cells.

A. Single–Color Analysis

Reporter studies that require a single FP are most easily performed using mutants that excite well with 488-nm light from the argon laser, such as EGFP or EYFP. EGFP is widely used in reporter studies and this variant can be detected flow cytometrically using the standard filter configuration for fluorescein detection. However, better signals may be achieved with an emission filter that captures slightly shorter wavelengths. For example, we have found that replacement of the standard 530/30-nm band-pass filter on a FACS Vantage (Becton Dickinson, San Jose, CA) with a 510/20-nm band-pass filter significantly improves the signal-to-noise ratio for GFP-S65T and EGFP detection (Lybarger *et al.*, 1996). Although this is a simple and inexpensive modification, the optics is not easily accessed on some analyzers, especially those certified for clinical analysis. For this reason, EYFP represents a sound alternative to EGFP. EYFP is sufficiently stimulated at 488 nm to produce a signal that is comparable in intensity to that obtained with EGFP (Lybarger *et al.*, 1998). Furthermore, because the emission maximum of EYFP is nearly identical to that of fluorescein, no optical alterations are necessary to achieve efficient fluorescence detection. Regardless of which FP is used, the subsequent analyses are performed in a similar manner.

The FP signal should be treated no differently than signals from typical fluorochromes. A negative control is required, preferably one that has been mock transfected or mock infected in order to set the light scatter and background fluorescence gating. Most of our analyses are performed on a FACS Vantage equipped with an Enterprise (Coherent Inc., Santa Clara, CA) argon ion laser producing 125 mW of 488-nm light. We have not observed a significant fluctuation in the signal-to-noise ratio at different laser powers. In fact, we have achieved similar results with an EPICS Profile II (Coulter Corp., Miami, FL), which uses a 15-mW argon ion laser.

One important consideration is the intensity of the FP signal. Transient transfection will generate a population of cells that express widely variable levels of FP. In fact, some vectors that employ strong promoters (such as the cytomegalovirus (CMV) immediate-early promoter/enhancer) will generate FP levels that are detected as off-scale even when the photomultiplier tube (PMT) voltages are adjusted so that the negative control cells are largely off-scale in the negative direction. This is particularly problematic when two-color FP analyses are performed (see below). It must be determined which cells are most important for the experiment, the brightest, the dimmest, or the best average of the population, and the PMT voltages should be adjusted accordingly. Alternatively, less DNA can be used in the transfections to help diminish the signal. The result will be that fewer of the transfected cells will cross the detection threshold (detected as positive), but this should be acceptable for most experiments.

Cells cultured *in vitro* tend to be highly autofluorescent and this fact can obscure relatively weak FP signals. To visualize more clearly low-level FP expression, two-parameter analysis of cells that are singly FP positive can be helpful. An example is

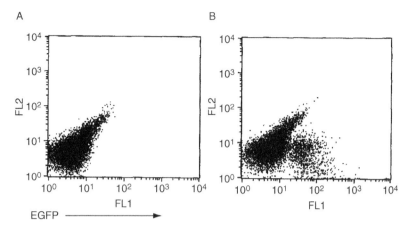

Fig. 1 Two-parameter analysis to distinguish GFP-derived fluorescence from background fluorescence. Murine splenic B cells were stimulated for 48 h with LPS (25 μg/ml) and then mock infected (A) or
infected with EGFP retrovirus (B) (Lybarger *et al.*, 1998). Cells were analyzed 48 h after infection.
EGFP fluorescence was measured in FL1 (510/20 band-pass filter) and cellular autofluorescence was
monitored in FL2 (585/42 band-pass filter normally used to detect PE). Electronic compensation (FL2–
18.0% FL1) was applied during analysis.

shown in Fig. 1. Splenic B cells were stimulated to divide with lipopolysaccharide
(LPS) and then infected with a retroviral vector encoding a hygromycin resistance/
EGFP fusion protein. By analyzing the EGFP signal (FL1) versus an irrelevant
signal in another PMT (FL2, normally used for phycoerythrin (PE) detection) in a
two-parameter fashion, it is easier to visualize low-level EGFP fluorescence above
the autofluorescence background. As noted in the mock-infected sample (Fig. 1A),
LPS treatment causes the cells to become autofluorescent. Because many of these
cells are as bright in FL1 as the true EGFP-positive cells (Fig.1 B), two-parameter
analysis reveals EGFP-expressing cells that would otherwise go undetected. Here,
electronic compensation is needed to subtract the EGFP signal from FL2, because
EGFP fluorescence does cross over into the PE filter.

B. Two–Color Fluorescent Protein Detection

As mentioned previously, various strategies have been developed to permit the
detection of multiple FPs. In Fig. 2, a scheme is presented to analyze simultaneously the expression of two distinct FPs, EGFP and EYFP, but many of the general
considerations introduced here should be applicable to other systems. Our initial
experiments with EYFP revealed that this mutant yields a signal that is comparable in intensity to that of EGFP. For this reason, we sought to determine whether
the fluorescence emission from both proteins could be separated, in spite of the
similar emission spectra of these proteins. Figure 2 depicts the optical

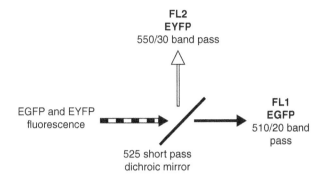

Fig. 2 Optics for dual detection of EGFP and EYFP. In this scheme, fluorescence from both proteins is elicited with 488-nm light and the signals are separated using the indicated filters.

configuration we found to be useful in the simultaneous analysis of EGFP and EYFP fluorescence (Lybarger *et al.*, 1998). This scheme requires 488-nm light for excitation and the indicated filters to segregate fluorescence from the two proteins. This type of analysis should be possible using any analyzer that has optics that is easily modified.

The minimal experimental controls required for each analysis include mock-transfected/infected cells, cells that express only EGFP, and cells that express only EYFP. Mock-transfected cells are used to determine the background fluorescence levels that are detected in each PMT (FL1 and FL2 in Fig. 2). The single-color controls are critical for setting electronic compensation, because crossover of each fluorescent signal into the opposite PMT will occur. For most analyses, compensation levels of <50% in each direction are required. These levels may vary from cytometer to cytometer, depending on such factors as the sensitivity inherent in each PMT, the exact optical filters that are employed, and the PMT voltages that are applied during analysis.

Figure 3 depicts the analysis of a population of cells that express EGFP or EYFP, both before and after compensation is applied. NIH 3T3 cells were infected with hygR/EGFP or hygR/EYFP retroviruses and completely selected with hygromycin. Each FP-expressing population can be resolved once the compensation is set. In this experiment, each FP was expressed at moderate levels. When much higher levels of FPs are expressed, such as the levels observed following transient transfection, simultaneous detection is still feasible (Lybarger *et al.*, 1998) (also see Chapter 17 in this volume; Lybarger and Chervenak, 1999). However, it becomes extremely difficult to determine the appropriate compensation levels when many of the events are off-scale. Therefore, as mentioned for single-color analyses, the DNA quantities used for transfection should be adjusted to obtain reasonable signals.

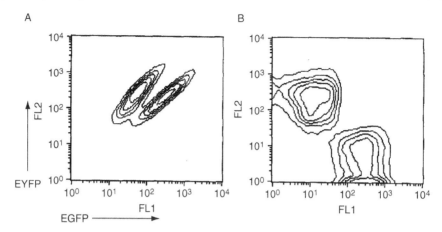

Fig. 3 Resolution of the EGFP and EYFP signals. NIH 3T3 cells were infected with retrovirus vectors encoding either hygR/EGFP or hygR/EYFP fusion proteins, and then completely selected with hygromycin. A 1:1 mixture of cells from each population was then analyzed using the optical configuration shown in Fig. 2, both before (A) and after (B) compensation was applied (adapted from Lybarger *et al.*, 1998).

C. Three–Color Fluorescent Protein Detection

With the increasing number of available GFP variants, it seems likely that three-color (or more) FP analysis will be possible. Indeed, we have performed three-color detection using the EGFP/EYFP system outlined here in combination with a UV-excitable mutant, enhanced blue fluorescent protein (EBFP) (Yang *et al.*, 1998). Figure 4 represents analysis of NIH 3T3 cells that were infected with a mixture of hygR/EGFP and hygR/EYFP retroviruses. A pool of stable cell lines was generated, some of which express both proteins (see Fig. 4A and B). Subsequent to transfection of these cells with an EBFP expression construct, signals from each FP can be detected in a fraction of the cells. However, fluorescence is not strongly stimulated from EBFP with UV light from the argon laser. For this reason, EBFP may be most useful as a second FP to use with EGFP in dual-color fluorescence microscopy. By combining EGFP and EYFP fluorescence with other FPs that possess spectra better suited to flow cytometric analysis, effective three-color detection may be possible.

VI. Combining Fluorescent Proteins with Common Fluorochromes

The versatility and power of FP reporter studies can be greatly extended by combining cell surface phenotypic analysis with FP reporter protocols. In theory, FP-derived fluorescence should be compatible with many of the common fluorochromes used in flow cytometry. Fluorescence from the green-emitting variants could be analyzed using the PMT that is normally used to detect fluorescein. However, this

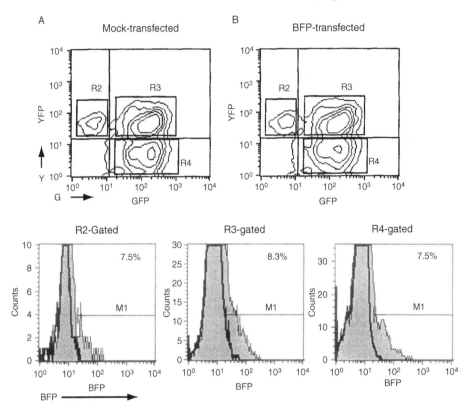

Fig. 4 Two- and three-color analysis. An EGFP/EYFP-expressing cell line was generated by coinfection of NIH 3T3 cells with a mixture of EGFP- and EYFP-encoding retroviruses. This cell line was then either mock transfected (A) or transfected (B) with 30 μg of an EBFP expression construct (pEBFP-C1; Clontech). EBFP fluorescence (380-nm maximum excitation; 440-nm maximum emission) was stimulated with UV light from the argon laser and collected with a 450/65 band-pass filter. EBFP fluorescence is shown in the bottom row of histograms for cells that fall within each indicated region (R1–R3) for the mock-transfected (heavy line) and EBFP-transfected (shaded peak) cells. The percentage of EBFP-transfected cells minus the percentage of mock-transfected cells in region M1 is given.

precludes the use of fluorescein-labeled probes for analyzing cell surface proteins. If UV excitation is available, especially if the UV and 488-nm lines are not colinear, more possibilities exist through the use of UV-excitable proteins. We have analyzed EGFP fluorescence in conjunction with single-color cell surface staining using phycoerythrin, allophycocyanin, or red-670 (Lybarger *et al.*, 1996). In addition, one report describes FP signal detection along with eight-color cell surface staining (Roederer *et al.*, 1997). Thus, cell surface staining and FP reporter signals can be analyzed simultaneously to produce an additional level of information about the target cells.

Undoubtedly, this methodology will become prominent in many future applications that involve FPs.

VII. Conclusions

Flow cytometry represents a high-throughput method to rapidly obtain quantitative data regarding FP-expressing cell populations, and cell sorting may be employed to obtain pure preparations of cells with the desired phenotype. Furthermore, multiple transcriptional elements that are coupled to distinct FPs may be independently analyzed and this information can be interpreted in the context of cell surface phenotypic data. The choice of which FPs, and which expression technologies, to use will be dictated by the demands of each experimental system. With careful forethought, the available (and yet to be discovered) spectral variants of GFP should meet the requirements of a wide range of experimental designs.

Acknowledgments

The authors are grateful to Deborah Dempsey for flow cytometry assistance and to Dr. Steve Kain (Clontech) and Jim Grobholz (Omega Optical) for providing critical reagents. This work was supported by National Institutes of Health Grant R01 AI31567.

References

Anderson, M. T., Tjioe, I. M., Lorincz, M. C., Parks, D. R., Herzenberg, L. A., Nolan, G. P., and Herzenberg, L. A. (1996). *Proc. Natl. Acad. Sci. USA* **93,** 8508.

CLONTECHniques, L. A. (1997). **XII,** p. 10.

Cormack, B. P., Valdivia, R. H., and Falkow, S. (1996). *Gene* **173,** 33.

Heim, R., Cubitt, A. B., and Tsien, R. Y. (1995). *Nature (London)* **373,** 663.

Lybarger, L., and Chervenak, R. (1999). *Methods Enzymol.* **302,** Chapter 17 (this volume).

Lybarger, L., Dempsey, D., Franek, K. J., and Chervenak, R. (1996). *Cytometry* **25,** 211.

Lybarger, L., Dempsey, D., Patterson, G. H., Piston, D. W., Kain, S. R., and Chervenak, R. (1998). *Cytometry* **31,** 1.

Muller-Taubenberger, A., and Anderson, K. I. (2007). *Appl. Microbiol. Biotechnol.* **1,** 1.

Nienhous, G. U., and Wiedenmann, J. (2009). *Chemphyschem* **10,** 1369.

Ormö, M., Cubitt, A., Kallio, K., Gross, L., Tsien, R., and Remington, S. (1996). *Science* **273,** 1392.

Roederer, M., De Rosa, S., Gerstein, R., Anderson, M., Bigos, M., Stovel, R., Nozaki, T., Parks, D., Herzenberg, L., and Herzenberg, L. (1997). *Cytometry* **29,** 328.

Ropp, J., Donahue, C., Wolfgang-Kimball, D., Hooley, J., Chin, J., Hoffman, R., Cuthbertson, R., and Bauer, K. (1995). *Cytometry* **21,** 309.

Ropp, J., Donahue, C., Wolfgang-Kimball, D., Hooley, J., Chin, J., Cuthbertson, R., and Bauer, K. (1996). *Cytometry* **24,** 284.

Shaner, N. C., Patterson, G. H., and Davidson, M. W. (2007). *J. Cell Sci.* **120,** 4247.

Stauber, R. H., Horie, K., Carney, P., Hudson, E. A., Tarasova, N. I., Gaitanaris, G. A., and Pavlakis, G. N. (1998). *BioTechniques* **24,** 462.

Yang, T. T., Cheng, L., and Kain, S. (1996). *Nucleic Acids Res.* **24,** 4592.

Yang, T. T., Sinai, P., Green, G., Kitts, P. A., Chen, Y. T., Lybarger, L., Chervenak, R., Patterson, G. H., Piston, D. W., and Kain, S. R. (1998). *J. Biol. Chem.* **273,** 8212.

CHAPTER 26

Jellyfish Green Fluorescent Protein: A Tool for Studying Ion Channels and Second-Messenger Signaling in Neurons

L. A. C. Blair, K. K. Bence, and J. Marshall

Department of Molecular Pharmacology
Physiology, and Biotechnology, Brown University
Providence, Rhode Island, USA

I. Introduction

A major aim of modern biology is to understand the regulation of cellular function. One of the most powerful tools with which to dissect intracellular interactions is the expression of specific genes encoding proteins of altered function. These altered proteins may be inactive for a variety of reasons. For example, they may be enzymatically inactive, unable to bind a normal target protein, or unable to achieve the proper subcellular localization through loss of myristoylation sites (e.g., "dominant-negatives"). Conversely, the altered proteins may be constitutively active, retaining the normal type of activity, but performing normal function(s) at a much higher level. However, whether the activity of a protein increases or decreases, its role in specific intracellular pathways and its importance to overall cellular function can be assessed by an array of techniques if it is first possible to transfect the cells of interest and identify unambiguously the transfectants.

DOI: 10.1016/B978-0-12-384658-7.00026-6

Several problems and restrictions have until recently effectively limited the ability to express exogenous proteins in mammalian cell lines. Moreover, the absence of reliable methods to differentiate between transiently transfected and nontransfected cells led many researchers to establish stable cell lines carrying the gene of interest. Unfortunately, many of these genes encode proteins that are involved in survival and/or the regulation of long-term properties of the cells. The long-term expression of altered proteins or even overexpression of normal, wild-type proteins can dramatically alter basal cellular properties, making it difficult to establish a frame of reference, or can even confer lethality. Furthermore, stable transfection requires the maintenance of dividing cells in long-term culture, eliminating most primary cells and all nondividing cells.

Researchers have addressed these issues in several ways. The development of viral transfection systems has greatly expanded the types of cells that can be transfected and the introduction of inducible promoters is ameliorating the problems associated with extended expression of potentially harmful genes. In addition, to identify successfully living transfectants, several groups have established techniques for transiently cotransfecting cells with a reporter gene that encodes a nontoxic marker, the jellyfish green fluorescent protein (GFP) (Cubitt *et al.*, 1995; Marshall *et al.*, 1995; Prasher *et al.*, 1992). Importantly, this approach can be applied to cell lines, dividing primary cells, and postmitotic primary cells (Blair and Marshall, 1997).

GFP-based transfection techniques have proved particularly useful for the study of second-messenger pathways in neurons. Here, we focus on how to transfect neurons in primary culture, providing examples of several kinds of constructs (cotransfection with separate gene of interest and GFP cDNA constructs, transfection with a single construct containing the gene of interest fused to a GFP gene, and transfection with a single construct containing a separate gene of interest and GFP cassettes), and of several types of biological questions that can be resolved using this approach. We also address how to transfect neuronal ion channels into cell lines and some of their uses.

II. Primary Cultures of Cerebellar Granule Neurons

For most of our experiments on primary neurons we utilize cerebellar granule cell cultures. These cells were chosen for several reasons. First, it is possible to obtain large numbers of cerebellar granule neurons and to maintain them in relatively homogeneous cell culture; this enables the employment of biochemical as well as electrophysiological and immunocytochemical assays. In addition, because the cerebellum develops relatively late, cells are taken out for culture postnatally, obviating the necessity of operating to remove embryos. Specific to our work, our previous research using neuronal cell lines strongly suggested that receptor tyrosine kinases rapidly regulate specific ion channel activities (Selinfreund and Blair, 1994); it was known that granule neurons express both the receptor tyrosine kinases and the specific ion channel subtypes. Described here is a method largely based on Messer (1972) with the helpful additions from L. Nowak (personal communication, Cornell University, 1998). We include sufficient detail to enable researchers familiar with mammalian gross neuroanatomy to obtain a reasonable yield the first time. (On request, we will provide the complete four-page protocol.)

Prior to culturing, the culture surfaces, either tissue culture plastic or glass coverslips, are coated with a solution of high molecular weight poly-D-lysine (MW > 150,000, dissolved in sterile double-distilled water (ddH$_2$O)). Omission of pretreatment drastically reduces the yield; few granule neurons will attach to untreated tissue culture plastic and essentially none to untreated glass. Therefore, before starting the dissection, we coat culture dishes or glass coverslips with sterile-filtered, 0.1–1 mg/ml poly-D-lysine (~1 ml/35-mm tissue culture plate for at least 3–4 h at room temperature, or overnight at 4 °C); the higher concentrations are used to coat glass surfaces. We then aspirate the polylysine solution, rinse with sterile ddH$_2$O, aspirate the ddH$_2$O, and dry the open plates under ultraviolet light for approximately 1 h to ensure sterilization. Using this procedure, granule cells attach well (i.e., survive extensive washing), put out neurites within 6 h of plating (often sooner), and survive and grow for at least 3 weeks.

For the dissection, cerebella are typically removed from 5-day-old rat pups (p5; tested range, p4–p10). The rats are decapitated, the skin over the top and sides of the skull is removed, and the back section of the skull is removed to reveal the cerebellum. Because the skin is a major source of bacterial infection, the head is repeatedly rinsed with 70% ethanol during all stages of the gross dissection. Note that the cerebellum is an outfolding of the myelencephalon; it floats on a stalk behind the colliculi and above the brainstem. Cerebella (usually 10–20) are collected and stored on ice in ice-cold Hanks' buffered saline solution (HBSS; pH 7.4, with penicillin–streptomycin) until all are obtained.

Cerebella are then washed in ice-cold HBSS and diced into small chunks (~1 mm^3). Tissue chunks are incubated in 10 ml of trypsin (0.5 mg/ml) in HBSS at 37 °C for 10 min (agitating every 2–3 min). Protease digestion is stopped by competition with serum. The chunks are gently pelleted (30 s, lowest speed in a clinical centrifuge, ~600 rpm) and, after aspirating the trypsin–HBSS, resuspended in growth medium (Dulbecco's modified Eagle's medium (DMEM), 10% fetal bovine serum (FBS), 25 mM KCl, glucose (6 g/l), 2 mM glutamine plus 10 U of penicillin and streptomycin (10 μg/ml)). The tissue pellet is then triturated using sterile, flame-polished, silanized pasteur pipettes until the solution is slightly cloudy; overtrituration destroys cells and reduces the overall yield. This solution is then briefly recentrifuged (lowest speed, ~30 s) to pellet the larger chunks while keeping the single cells in solution in the supernatant. The supernatant containing the dissociated cells is plated into growth medium on the poly-D-lysine-coated culture surfaces. As shown in Fig. 1, fully dispersed and easily identifiable neurons are detected by 1 day in culture. Interestingly, though, granule neurons migrate in the dish, forming increasingly large clusters (compare Figs. 1 and 2A, left panel

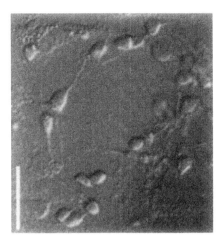

Fig. 1   Cerebellar granule neurons, 1 day *in vitro*, cultured from p5 rats. Scale bar: 10 μm.

with Fig. 2A, right panel, where after 4 days in culture the neurons have already formed a massive group). For many applications, such as kinase assays and enzymatic survival assays, isolated versus "lumped" cells makes no difference. However, for other types of assays, in particular, electrophysiological recordings, the success rate of the experiments depends on being able to utilize isolated cells; for these experiments, cells are plated at the lowest possible density and typically used before 7 days *in vitro*.

Granule neurons can be maintained for up to 3–4 weeks in a humidified 37 °C incubator (5% CO_2). Because nonneuronal support cells appear to condition the medium and aid in granule neuron survival, we often add, but only at the time of plating, basic fibroblast growth factor (bFGF, 20 ng/ml) to stimulate initially the support cells to help the neurons to get a strong start. Subsequently, however, cytosine arabinoside can be added to the culture medium (final concentration, 10 μM; starting on days 1–2 *in vitro* and maintained continuously thereafter) to eliminate the dividing cells and maintain a pure culture of granule neurons.

To maximize the yield, the trypsinization/trituration procedures are repeated several times on the remaining cell chunks, gradually reducing the volume of and exposure time to the trypsin solution and gradually switching to using smaller tip-bore, flame-polished, silanized pasteur pipettes. Each time the trituration is stopped as soon as the solution becomes cloudy, and those cells plated. Typically, approximately five rounds of trypsinization/trituration are performed. Importantly, the time spent triturating and the number of trypsinization/trituration rounds depend on more than the initial size of the tissue chunks; the age of the rat pup is also significant. Cerebella from older pups, especially >p7, have considerably more extracellular matrix and require more trituration.

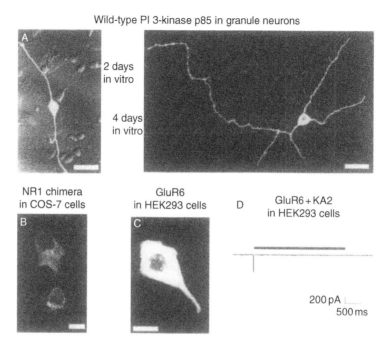

Fig. 2 Transient transfection of primary neurons and cell lines. (A) Cotransfection of cultured cerebellar granule neurons with cDNAs encoding a wild-type p85 subunit of PI 3-kinase and GFP. Overlays of fluorescence micrographs superimposed on the Nomarski images of the fields: *left*, 2 days in culture; *right*, 4 days in culture. Confocal microscopy: *z* sectioning was used to ensure full representation of the neurites, which, especially within clumps of neurons, traversed many optical sections. Cells were transfected by calcium phosphate precipitation. (B) Transfection of COS-7 cells with a single construct containing a single chimeric GFP-NR1 NMDA receptor subunit gene. Note that because the receptor subunit is directly tagged by the marker, fluorescence is detected only where NR1 is expressed. Confocal microscopy: a single optical section. The construct is shown in Fig. 4A. Transfection was via Lipofect-AMINE. (C) Transfection of HEK293 human embryonic kidney cells with a single construct containing, in separate cassettes, both the kainate receptor GluR6(Q) subunit and GFP genes. Although cells are again transfected with a single construct, the GFP is translated separately from the GluR6, resulting, as in (A), in diffuse, cytoplasmic expression of the marker. Confocal microscopy: one optical plane. The construct is shown in Fig. 4B. Transfection was by LipofectAMINE. (D) Coexpression of two kainate receptor subunits, GluR6(Q) and KA2, with GFP in HEK293 cells generates glutamate-induced currents that show rapid activation and rapid and complete desensitization. Glutamate, 100 μM (indicated by the bar), was applied by rapid superfusion. Cells were transfected by calcium phosphate precipitation. Scale bars: (A–C) 10 μm. (Data from Garcia *et al.*, 1999.)

III. Green Fluorescent Protein and Gene-of-Interest Constructs

GFP, a 28-kDa autofluorescent protein that is naturally produced by the jellyfish *Aequorea victoria*, serves as the marker of successfully transfected cells. It is particularly useful because it requires no enzymatic reaction or cofactors to function and no special processing of the cells, meaning that it effectively serves as a vital dye to identify living cells. Moreover, because it is introduced into cells not as a protein, but as DNA, special injection procedures are not required. Cells are merely transfected with the GFP gene along with the gene of interest; the normal cellular machinery then transcribes and translates both messages. Detection is via standard fluorescein fluorescence optics.

A variety of GFP genes is now available. In addition to the wild-type gene, genetically engineered forms have been constructed to increase brightness and alter the emission spectrum of the chromophore to produce an autofluorescent protein emitting blue light (BFP; Quantum Biotechnologies, Montreal, Quebec, Canada) (Heim and Tsien, 1996; Wachter *et al.*, 1997). Potentially, utilizing two proteins directly tagged with different chromophores could allow simultaneous tracing of the movements of two proteins of interest within cells. Currently, we employ Super-Glo (sg25) GFP (Quantum Biotechnologies) (Palm *et al.*, 1997; Stauber *et al.*, 1998), which produces one of the brightest GFP proteins, detectable as early as 12 h post-transfection in mammalian neurons and cell lines (our unpublished data, 1998). Early detection is especially important when testing overexpression of molecules expected to be harmful to the cells when produced over extended periods, such as dominant-negatives of proteins known to be involved in survival (see Section VI and Fig. 3). Our earlier constructs utilized the pGreenLantern GFP (GIBCO-BRL, Gaithersburg, MD). Both of these GFP genes contain point mutations to increase fluorescence intensity and codon usage modifications (silent mutations) to stabilize the RNA and increase translation efficiency in mammalian cells.

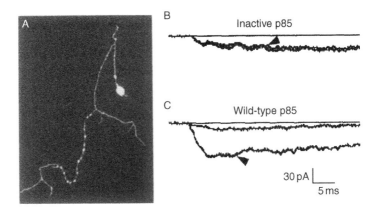

Fig. 3 Transfection of granule neurons with inactive PI 3-kinase subunits blocks IGF-I potentiation of calcium channel currents. (A) A cerebellar granule neuron cotransfected in culture by cDNAs encoding an inactive PI 3-kinase subunit (Δp85, gift of M. Kasuga; Δp85 is described in detail in Dhand *et al.*, 1994) and GFP to identify the transfected cell. (B) IGF-I failed to potentiate calcium channel currents in a granule neuron transfected with Δp85. Currents recorded before and 10 and 60 s after the addition of IGF-I (100 ng/ml) superimpose. (B, C) Voltage-activated barium currents were evoked by depolarizing voltage pulses from −80 to 0 mV. Arrowheads indicate currents recorded after IGF-I addition. (Data from Blair and Marshall (1997), with permission.) (C) Calcium channel currents recorded 10 s after IGF-I (20 ng/ml) addition increased manyfold in a wild-type p85- and GFP-transfected cell.

GFP genes have been placed primarily into three kinds of constructs: (1) Alone in expression vectors such as pcDNA3 (Invitrogen, San Diego, CA) under control of the cytomegalovirus (CMV) or simian virus 40 (SV40) promoter (Blair and Marshall, 1997; Marshall *et al.*, 1995). In this case, the GFP construct is cotransfected with a second construct carrying the gene of interest, and the GFP protein is expressed throughout the cytoplasm (Figs. 2A, C and 3A). (2) A single construct that contains the gene of interest directly ligated to the GFP gene to create a fusion protein (Figs. 4A and 2B; Marshall *et al.*, 1995) can also be made. Although the addition of the 28-kDa GFP to the protein of interest could potentially modify its function, this approach has the advantage that GFP is detected only where the protein of interest is expressed. It has proved particularly useful for examining the subcellular localization of proteins and the movements of proteins within cells (Burke *et al.*, 1997; Ellenberg *et al.*, 1997; Wolter *et al.*, 1997). (3) A number of groups have begun utilizing a third method. Here, a single construct that contains separate cassettes for the gene of interest and the GFP gene (Figs. 4B and 2C) is made. It contains a promoter (in our case, CMV), the gene of interest, an internal ribosome entry site (IRES), and the GFP gene. In this situation, a single transcript

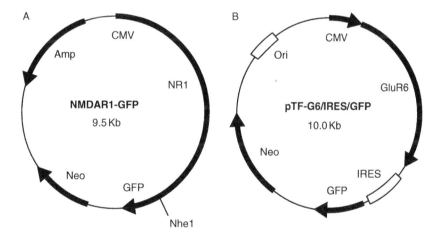

Fig. 4 Constructs used to express GFP tagged proteins of interest and bicistronic transcripts encoding separately GFP and the protein of interest. (A) A construct encoding the NR1 NMDA receptor subunit directly tagged with GFP was made as described by Marshall *et al.* (1995). Briefly, to create the chimera, sequence encoding the last 13 amino acids of NR1 was replaced by sequence encoding alanine, followed by the GFP sequence encoding the full protein minus the first amino acid. In addition, to obtain brighter fluorescence, the wild-type GFP gene was replaced with that from pGreenLantern (GIBCO-BRL). Constructed by J. Edgerton. (B) A construct, pTF-G6/IRES/GFP, containing the coding sequences for the GluR6 kainate receptor (gift of J. Boulter and S. Heinemann) and the sg25 SuperGlo GFP (Quantum Biotechnologies), and driven by the CMV promotor-enhancer to ensure a high level of expression in mammalian cells, was made. It also contains the internal ribosome entry site (IRES) of the encephalomyocarditis virus, located between the GluR6- and GFP-coding sequences; this allows both the GFP and receptor proteins to be translated individually from a single bicistronic mRNA. Constructed by T. Fukushima.

is formed, but the protein of interest and GFP are translated separately, leading to a diffuse distribution of the GFP marker in the cytosol. The advantage of this approach is that it allows correlating the level of expression of the protein of interest with the level of GFP fluorescence without affecting the function of the protein of interest. Although we are currently using a pBK-CMV (Stratagene, La Jolla, CA)-based expression vector (Fig. 4B; constructed by T. Fukushima, Brown University), this type of construct can also be put into viral vectors to increase transfection efficiency, as demonstrated by the use of an adenoviral expression system to transfect olfactory neurons *in situ* with odorant receptors (Zhao *et al.*, 1998).

IV. Transient Transfection of Neurons by Calcium Phosphate Precipitation

To transfect cerebellar neurons, we use calcium phosphate precipitation procedures slightly modified from Chen and Okayama (1987). The yield of successful transfectants, as indicated by the presence of GFP fluorescence, is typically 5–10%. We have also tried lipid-mediated transfection techniques, which in our cases give a high efficiency in cell lines (see below), but a lower efficiency in primary neurons than calcium phosphate techniques.

Granule neurons are usually transfected 24–48 h postplating. The efficiency of transfection sharply declines after 48 h in culture owing, most likely, to secretion of extracellular matrix. Efficiency is also low when cells cultured less than 24 h are transfected, possibly because the cells, still recovering from the dissociation procedures, are not yet strong enough to support the burden of overexpression of exogenous proteins. Typically, we transfect cells plated onto glass coverslips or directly into 35×10 mm dishes. To make the transfection cocktail, the following proportions are used per 35-mm dish: Up to 10 μg of DNA is diluted into 100 μl of $0.25~M$ $CaCl_2$ in a sterile tube; depending on the experiment, 1–10 μg of the gene of interest and 1–5 μg of GFP cDNA are used. Then, 100 μl of N,N-bis-(2-hydroxyethyl)-2-aminoethanesulfonic acid (BES)-buffered saline (50 mM BES, 280 mM NaCl, 1.5 mM Na_2HPO_4, pH 6.97; the pH is critical for maximizing the transfection efficiency) is added dropwise to the DNA solution while gently bubbling the mixture with a sterile Pasteur pipette. Starting solutions should be at room temperature and the resultant mixture should appear slightly cloudy. It is left at room temperature for 30 min and then added dropwise to the culture dish. Cells are exposed to the DNA solution for 6 h in a 37 °C/3% CO_2 incubator. At the end of the 6 h, the DNA-containing precipitate should appear very fine. If larger crystals appear, the efficiency of transfection will be significantly lower. Using solutions that have been warmed to room temperature will help the formation of a fine precipitate. After 6 h, we replace the transfection medium with standard growth medium, and return the cells to a 5% CO_2 incubator. For most of the genes of interest that we have tested, expression is optimal 12–48 h posttransfection.

V. LipofectAMINE Transfection in Cell Lines

We routinely transiently transfect cell lines using LipofectAMINE reagent (GIBCO-BRL). The efficiency varies with the line, but is high when using the kidney-derived COS-7 or HEK293 cells (typically 50–80%). It is, however, much lower (10–40%) when we transfect neuroblastoma lines, such as 401L or B104.

Cells are first plated onto poly-D-lysine-coated glass coverslips or tissue culture dishes and grown to 75% confluence in standard growth medium (DMEM, 5% FBS, 10 U of penicillin, and streptomycin (10 μg/ml)). For each 35-mm dish, the following transfection cocktail is prepared in sterile 12×75 mm polystyrene tubes (do not use polypropylene tubes because lipids adhere to their surfaces): 1–2 μg of DNA is diluted into 100 μl of serum-free MEM and mixed with a solution of 5 μl of LipofectAMINE reagent in 100 μl of serum-free MEM. This mixture is left at room temperature for 15 min to allow DNA–liposome complexes to form; while complexes are forming, cells are rinsed once with serum-free MEM. For each transfection, 0.8 ml of Opti-MEM reduced-serum medium (GIBCO-BRL) is added to the DNA–liposome mixture, mixed gently to avoid breaking the relatively fragile DNA–liposome complexes, and gently pipetted over cells. Cells are then incubated in the presence of the DNA–liposome mixture at 37 °C/15% CO_2 for 4 h; this mixture is then aspirated and replaced with posttransfection medium (DMEM, 10% FBS, with antibiotics). Peak expression of most proteins of interest is obtained 24–48 h posttransfection.

VI. Applications to Biological Problems

Cellular processes within the central nervous system have proved, for a variety of reasons, particularly difficult to study. In the past, nearly all work that attempted to examine second-messenger pathways and the regulation of ligand-gated and voltage-sensitive ion channels has utilized cell lines transfected either with regulatory receptors (G protein-coupled receptors or receptor tyrosine kinases, called RTKs) or ion channels or both. Studies of the movements of intracellular proteins have typically required "freeze-frame" immunocytochemistry using fixed tissue. The transfection of primary neurons with a nontoxic marker such as GFP provides a powerful approach to addressing these questions in the mammalian nervous system. Here, to provide specific examples of raw data, we focus on two of the applications that we have tested, the expression of ligand-gated ion channels in cell lines and the expression of signaling components that potentially could mediate the RTK regulation of voltage-sensitive ion channels.

Briefly, we had found that stimulation of the insulin-like growth factor I (IGF-I) RTK in cerebellar granule neurons rapidly induced a large increase in N and L calcium channel currents. However, the pathway mediating this response was unknown. To identify essential signaling components, we first used traditional pharmacological approaches, testing whether inhibitors of conceivable signaling components would inhibit the IGF-I/RTK regulation of the channels (Blair and Marshall, 1997). These experiments strongly implicated phosphatidyl-inositol 3-kinase (PI 3-kinase, EC 2.7.1.127) as a second messenger. Unfortunately, though, few pharmacological agents are as specific as researchers would usually prefer. We, therefore, adopted a molecular approach, transfecting the granule neurons in culture with cDNAs encoding an inactive, dominant-negative PI 3-kinase subunit (Δp85; Dhand *et al.*, 1994) or the wild-type p85 subunit (Blair and Marshall, 1997) (Fig. 3). Coexpression with GFP (pGreenLantern; GIBCO-BRL) was used to detect transfected neurons. As described above, neurons were cultured from p4–p5 rats. All cDNAs were subcloned into pcDNA3 expression vectors (Invitrogen) and were under the control of CMV promoters. Cells were transfected by the calcium phosphate precipitation techniques also described above. The ability of transfectants to respond to IGF-I was assessed by the permeabilized-patch variation of standard whole-cell recording techniques (Hamill *et al.*, 1981; Rae *et al.*, 1991) and, to avoid the harmful long-term effects of overexpression of the inactive enzyme (PI 3-kinase is involved in granule neuron survival; Dudek *et al.*, 1997), responsiveness was assayed as soon as possible after transfection.

We found that transient expression of the inactive, dominant-negative PI 3-kinase subunit fully blocked the rapid IGF-I potentiation of N and L channels (Fig. 3B). In addition, we found that overexpression of wild-type PI 3-kinase subunits not only permitted the IGF-I modulation to occur (Fig. 3C), but significantly increased IGF-I responsiveness over the level observed in untransfected and GFP-vector-alone-transfected cells (Blair and Marshall, 1997). The results obtained using this approach provide strong evidence that PI 3-kinase is an essential and rate-limiting intracellular messenger in the signaling pathway used by the neuronal IGF-I/RTKs. The general biological interpretation of these experiments would be that the PI 3-kinase-dependent modulation, by controlling calcium channels, might also control an enormous variety of cellular calcium-dependent processes, including neurotransmitter release and IGF-I-dependent differentiation and/or survival.

The transfection approach differs from conventional pharmacological tests in important ways. First, by circumventing problems of pharmacological specificity, transfection with genes encoding inactive proteins of interest provides a fully independent means of assessing the importance of the normal protein. Moreover, the ability to transfect and detect wild-type and/or constitutively active proteins enormously expands the possible conclusions that can be drawn. For the experiments described above, the conclusion that PI 3-kinase is rate limiting would have been impossible if only pharmacological inhibitors had been used; it depends on being able to overexpress the wild-type protein in addition to the dominant-negative form (the equivalent of a pharmacological inhibitor). The expression of constitutively active proteins permits a separate set of conclusions. Taken in comparison with wild-type and vector-alone transfections, analysis of cells transfected with a constitutively active enzyme allows a reasonable estimate of how a maximally driven signaling pathway will control the activity of downstream effectors (Blair *et al.*, 1998). In short, the expression of altered proteins of interest is a new and uniquely powerful tool for understanding second messenger-mediated systems.

We have, in addition, utilized GFP fluorescence to detect transfectants expressing specific glutamate receptor subunits (Figs. 2B–D and 4) and neuronal N and L calcium channel subunits in a variety of cell lines (K. Bence, unpublished data, Brown University, 1998). The purpose of the glutamate receptor experiments was to determine how coexpression of receptors with synapse-associated proteins (SAPs) regulates receptor clustering and ligand-induced responses (Garcia *et al.*, 1999). Although GFP-tagged subunits can be used to detect receptors directly (Fig. 2B), the GFP marker was primarily used here to detect the expression of the gene of interest in cell lines (COS-7 and HEK293) for electrophysiological analysis (Fig. 2C and D). Glutamate responses, as regulated by expression of the different receptor subunit and SAP combinations, were determined. An example of the responses, assessed by standard techniques (Hamill *et al.*, 1981; Rae *et al.*, 1991) that are obtained when glutamate receptor subunits are expressed in the absence of SAP proteins, is shown in Fig. 2D.

These few examples represent only a fraction of a multitude of possible applications. GFP, because it is a nontoxic and easily detectable marker of living transfected cells, is being applied to the study of dynamic processes within cells, including membrane and vesicular trafficking (Burke *et al.*, 1997) and the movements/targeting of intracellular proteins (Ellenberg *et al.*, 1997; Wolter *et al.*, 1997). Importantly, improvements in autofluorescent protein technologies are rapidly expanding their applications. In particular, the development of autofluorescent proteins with different chromophore emission spectra (e.g., BFP; Quantum Biotechnologies) (Heim and Tsien, 1996; Wachter *et al.*, 1997) should provide a new and unique tool for studying dynamic protein interactions, the development of the brighter GFPs will enhance their usefulness for dissecting signal transduction pathways via the expression of dominant-negative components, and the development of GFP- and gene-of-interest-containing viral vectors should dramatically enhance their usefulness for the study of that most recalcitrant of systems, central nervous function.

Acknowledgments

We thank Dr. John Wallenberg of Quantum Biotechnologies for information on the new GFP genes. We also thank Dr. Irwin Levitan for suggesting the IRES-containing construct, Teruyuki Fukushima for permission to discuss his pTF-G6/IRES/GFP construct, Jeremy Edgerton for making the construct for the NR1 GFP chimera and for assistance with the confocal microscopy, and Sunil Mehta for helpful advice. This work was supported by R29 NS 33914-02 (NIH) and SA047 (Council for Tobacco Research Scholar Award) to J. M.

References

Blair, L. A. C., and Marshall, J. (1997). *Neuron* **19,** 421.

Blair, L. A. C., Bence, K. K., and Marshall, J. (1998). *Soc. Neurosci. Abstr.* **24.**

Burke, N. V., Han, W., Li, D., Takimoto, K., Watkins, S. C., and Levitan, E. S. (1997). *Neuron* **19,** 1095.

Chen, C., and Okayama, H. (1987). *Mol. Cell. Biol.* **7,** 2745.

Cubitt, A. B., Heim, R., Adams, S. R., Boyd, A. E., Gross, L. A., and Tsien, R. Y. (1995). *Trends Biochem. Sci.* **20,** 448.

Dhand, R., Hara, K., Hiles, I., Bax, B., Panayotou, G., Fry, M. J., Yonezawa, K., Kasuga, M., and Waterfield, M. D. (1994). *EMBO J.* **13,** 511.

Dudek, H., Datta, S. R., Franke, T. F., Birnbaum, M. J., Yao, R., Cooper, G. M., Segal, R. A., Kaplan, D. R., and Greenberg, M. E. (1997). *Science* **275,** 661.

Ellenberg, J., Siggia, E. D., Moreira, J. E., Smith, C. L., Presley, J. F., Worman, H. J., and Lippincott-Schwartz, J. (1997). *J. Cell Biol.* **138,** 1193.

Garcia, E. P., Mehta, S., Blair, L. A. C., Wells, D. G., Shang, J., Fukushima, T., Fallon, J. R., Garner, C. C., and Marshall, J. (1999). *Neuron* **21,** 727–739.

Hamill, O. P., Marty, A., Neher, E., Sakmann, B., and Sigworth, F. J. (1981). *Pfluger's Arch.* **391,** 85.

Heim, R., and Tsien, R. (1996). *Curr. Biol.* **6,** 178.

Marshall, J., Molloy, R., Moss, G. W. J., Howe, J. R., and Hughes, T. E. (1995). *Neuron* **14,** 211.

Messer, A. (1972). *Brain Res* **130,** 1.

Palm, G. J., Zdanov, A., Gaitanaris, G. A., Stauber, R., Pavlakis, G. N., and Wlodawer, A. (1997). *Nature Struct. Biol.* **4,** 361.

Prasher, D. C., Eckenrode, V. K., Ward, W. W., Prendergast, F. G., and Cormier, M. J. (1992). *Gene* **111,** 229.

Rae, J., Cooper, K., Gates, P., and Watsky, M. (1991). *J. Neurosci. Methods* **37,** 15.

Selinfreund, R. H., and Blair, L. A. C. (1994). *Mol. Pharmacol.* **45,** 1215.

Stauber, R. H., Horie, K., Carney, P., Hudson, E. A., Tarasova, N. I., Gaitanaris, G. A., and Pavlakis, G. N. (1998). *BioTechniques* **24,** 462.

Wachter, R. M., King, B. A., Heim, R., Kallio, K., Tsien, R. Y., Boxer, S. G., and Remington, S. J. (1997). *Biochemistry* **36,** 9759.

Wolter, K. G., Hsu, Y. T., Smith, C. L., Nechushtan, A., Xi, X. G., and Youle, R. J. (1997). *J. Cell Biol.* **139,** 1281.

Zhao, H., Ivic, L., Otaki, J. M., Hashimoto, M., Mikoshiba, K., and Firestein, S. (1998). *Science* **279,** 237.

CHAPTER 27

Expression of Green Fluorescent Protein and Inositol 1,4,5-Triphosphate Receptor in *Xenopus laevis* Oocytes

Atsushi Miyawaki, Julie M. Matheson, Lee G. Sayers, Akira Muto, Takayuki Michikawa, Teiichi Furuichi, and Katsuhiko Mikoshiba

Department of Molecular Neurobiology
Institute of Medical Science
University of Tokyo, Minato-ku, Tokyo, Japan

I. Introduction

Green fluorescent protein (GFP) is a sensitive and reliable fluorescent marker for examining levels and subcellular sites of expression of functional proteins in living cells. We report here efficient expression of GFP in *Xenopus laevis* oocytes by direct microinjection of its plasmid DNA into the nucleus. Application of this method to studies on optical imaging of inositol 1,4,5-triphosphate (IP$_3$)-sensitive intracellular Ca^{2+} store sites and expression of IP$_3$ receptor/Ca^{2+} release ion channels are discussed.

IP$_3$ is a ubiquitous second messenger responsible for the release of Ca^{2+} from intracellular stores (Berridge, 1993), IP$_3$, following its release into the cytosol through phosphoinositide turnover stimulated with various extracellular signals such as growth factors, hormones, and neurotransmitters, binds to an IP$_3$-gated intracellular Ca^{2+} channel, the IP$_3$ receptor (IP$_3$R; type 1, type 2, and type 3) (Furuichi and Mikoshiba, 1995), which in turn mediates the release of Ca^{2+} from the intracellular stores into the cytosol. IP$_3$-induced Ca^{2+} release is involved in spatiotemporal Ca^{2+} dynamics within a cell. The organization of intracellular Ca^{2+} stores is complex (Meldolesi *et al.*, 1990). It is now generally accepted that the IP$_3$R resides on specialized regions, mostly the smooth endoplasmic reticulum (sER) (Otsu *et al.*, 1990; Ross *et al.*, 1992; Satoh *et al.*, 1990), which constitutes IP$_3$-sensitive Ca^{2+} stores (Pozzan *et al.*, 1994). Intracellular Ca^{2+} dynamics is closely associated with organization of the ER network which can be altered depending on cell dynamics. Differences in subcellular distribution among IP$_3$R types as well as in organella as Ca^{2+} store present IP$_3$Rs need to be considered. Therefore, specific store sites and IP$_3$R types in certain cell-types and/or throughout cell stages have to be identified. In addition, little is known of mechanisms underlying intracellular targeting of the IP$_3$R to Ca^{2+} stores. We make use of GFP, originally isolated from the jellyfish *Aequorea victoria* (Prasher *et al.*, 1992), to monitor the expression level (Matheson *et al.*, 1996) and intracellular localization (Sayers *et al.*, 1997) of mouse type 1 IP$_3$R (mIP$_3$R1) (Furuichi *et al.*, 1989). We expressed GFP and chimeric GFP–IP$_3$R fusion protein in *X. laevis* oocytes. GFP fluorescence was observed, using a confocal laser scanning microscope, and relative expression levels of exogenous mIP$_3$R1 could be elucidated based on the GFP fluorescence intensity. Subcellular localization of GFP–mIP$_3$R1 can be monitored according to the GFP fluorescence pattern. It proved to be like that of the endogenous *Xenopus* IP$_3$R (xIP$_3$R) (Kume *et al.*, 1993).

II. Procedures

A. Green Fluorescent Protein Expression Plasmids

pMT3 (Genetics Institute, Cambridge, MA) (Kaufman *et al.*, 1989; Swick *et al.*, 1992) carries the adenovirus major late promoter and some sequence elements that increase mRNA stability and translatability (Kaufman *et al.*, 1989; Swick *et al.*, 1992) and provides a system for expressing high levels of exogenous proteins in *X. laevis* oocytes by direct nuclear injection (Swick *et al.*, 1992). We used this system to examine expression of exogenous GFP and IP$_3$R proteins in *Xenopus* oocytes.

pMT3-S65T: The coding region of GFP-S65T, a brighter form of the fluorophore by replacing chromophoretic Ser65 with Thr (S65T) (Heim *et al.*, 1995), was amplified by PCR using as a template pRSET$_B$-S65T (GFP-S65T cloned in the *Bam*HI site of pRSET$_B$ (Invitrogen); provided by Drs. R. Heim, A. B. Cubbit, and R. Y. Tsien, UCSD), and cloned into the *Not*I and *Sal*I sites of pMT3 as described (Matheson *et al.*, 1996).

pMT3-S65T-IP$_3$R(ES): The coding region of GFP-S65T amplified by PCR using pRSET$_B$-S65T as a template and the IP$_3$R1(ES), the *Eco*RI fragment (named ES fragment; amino acids 2216–2749) of mouse type 1 IP$_3$R (mIP$_3$R1) (Furuichi *et al.*, 1989) were fused in frame and cloned into the *Not*I and *Kpn*I site of pMT3 (Sayers *et al.*, 1997). The resultant pMT3-S65T-IP$_3$R(ES) contained six additional amino acids (Gly-Ser-Pro-Gly-Leu-Gln) which linked S65T to IP$_3$R(ES).

pMT3-IP$_3$R1: The *Sal*I fragment of the full length mIP$_3$R1 cDNA was cloned into the *Sal*I site of pMT3 (Matheson *et al.*, 1996).

B. Procurement and Microinjection of *Xenopus laevis* Oocytes

Adult female *X. laevis* (purchased from Hamamatsu Seibutsu Company, Shizuoka, Japan) were subjected to hypothermia (by placing them on ice for approximately 20 min). Ovarian fragments were surgically removed, separated into small clumps, and washed in modified Barth's solution (MBS) (Colman, 1984) (88 mM NaCl, 1 mM KCl, 2.4 mM NaHCO$_3$, 10 mM HEPES (pH 7.5), 0.82 mM MgSO$_4$, 0.33 mM Ca(NO$_3$)$_2$, 0.41 mM CaCl$_2$; passed through a 0.22-μm Millex-GV filter (Millipore)). Fully grown stage VI oocytes (1.2–1.3 mm diameter) were obtained by manual defolliculation using watchmaker's forceps and a stereoscopic microscope, then were maintained in MBS at 18 °C until microinjection.

To prepare microinjection glass needles with gradual shanks, a borosilicate glass rod (G-1, Narishige, Japan) was pulled out with a micropipette puller (PN-3, Narishige) and the resultant tip was broken obliquely into a 6 μm outside diameter by gently bumping against a tip of microforge (MF-79, Narishige) or something else, for example, tip of a forceps. Microinjection of individual oocytes was done using a stereoscopic microscope equipped with a double arm fiber optic system and a microinjection glass needle connected to a nitrogen air pressure (10 kgf/cm^2) microinjector (IM-300, Narishige) and a micromanipulator (M-152, Narishige). Before use of the glass needles, calibration of microinjection volumes was empirically estimated by the measurement of water drop sizes made on the tip of a needle, using a stereoscopic microscope with a micrometer while adjusting the pressure and time levels of the microinjector; drop volumes 1, 2, 5, and 10 nl, respectively, corresponded to diameters 124, 156, 212, and 267 μm. One to two microliters of plasmid DNA dissolved in injection buffer (88 mM NaCl, 1 mM KCl, 15 mM Tris–HCl, pH 7.5; filtrated through a 0.22-μm filter) placed onto Parafilm M (American National Can) was drawn into a needle [or backfilled with a Microloader (Eppendorf)]. Oocytes were immobilized by placing their animal hemisphere diagonally upward on a stainless-wire gauze (with approximately 1 mm mesh size, a little smaller than oocyte diameter) in a plastic petri dish containing MBS. Nuclei are generally present underneath the oocyte membrane of the animal pole. The glass needle set diagonal to the ground, that is, perpendicular to the oocyte pole, was microinjected into the nucleus of the oocyte by impaling the animal hemisphere (approximately 1–2 ng; e.g., 1–2 nl of 1 mg/ml DNA solution). All oocytes were maintained at 18 °C in MBS during all microinjection procedures and in MBS containing antibiotics (50 units/ml penicillin and 50 μg/ml streptomycin; filtrated through a 0.22-μm filter) after microinjection.

C. Fluorescence Measurements Using Confocal Microscopy

GFP fluorescent images after microinjection were monitored on either a Carl Zeiss laser scanning confocal microscope system (LSM 410) adapted to a Carl Zeiss inverted microscope Axiovert 135 TV (objective lens: 10×, 20×, 40×) using the Carl Zeiss LSM software package or a Bio-Rad laser scanning confocal microscope system (MRC-500) adapted to an Olympus inverted microscope IMT-2 (objective lens: 4×, 10×) using CoMOS and TCSM software package (Bio-Rad). The focus of light was adjusted to the surface of the oocyte placed on a glass bottom plastic dish (35 mm diameter; Mat Tek Corp., Ashland, MA) and the confocal aperture was fully opened to maximize the collection of light. Conditions for FITC observation (light of excitation wavelength 488 nm and emission 510 nm, through a band path or long path filter) were utilized to visualize GFP fluorescence from microinjected oocytes. Oocytes positive in GFP fluorescence were selected for further experiments. Oocytes microinjected with vector pMT3 alone (vector control oocytes) were also monitored for fluorescence. All fluorescence measurements were carried out at room temperature.

GFP fluorescence from individual oocytes was monitored for 1 week after microinjection. Ten images were averaged by Kalman algorithm for each measurement, and relative fluorescence levels were determined by measuring the average value over the entire area of each oocyte, using the program NIH Image (NTIS, Springfield, VA).

D. Fixation of Oocytes for Confocal Imaging

Oocytes positive in fluorescence were fixed in 4% (w/v) paraformaldehyde in phosphate-buffered saline (PBS) for 2 h at 4 °C and then washed in PBS. The oocytes were then manually sliced in half under a stereoscopic light microscope using a razor blade (thickness 0.1 mm) and along the plane perpendicular to the boundary between the animal and vegetal hemispheres of the oocyte. This enabled both hemispheres to be visualized using confocal microscopy. The sliced oocytes were then mounted on a *glass coverslip* and visualized for GFP fluorescence by confocal microscopy using a 10×, 20×, and 40× objective lens.

E. Preparation of *Xenopus* Oocyte Microsomes

Microsomes of oocytes were prepared by a method modified from Parys *et al.* (1992). Approximately 20 oocytes were homogenized in a glass-Teflon Potter homogenizer containing 4 volumes of ice-cold buffer A (50 mM Tris–HCl (pH 7.25), 250 mM sucrose, 1 mM EGTA, 1 mM DTT, 0.1 mM PMSF, 10 μM leupeptin, and 1 μM pepstatin A). The homogenate was centrifuged at 4500 × g for 15 min at 2 °C and the resulting supernatant was recovered. The pellet contained yolk and melanosomes (Parys *et al.*, 1992). The supernatant was centrifuged at 160,000 × g for 1 h at 2 °C to sediment the microsomal fraction. The microsomal pellet was resuspended in ice-cold buffer B (20 mM Tris–HCl (pH 7.25), 300 mM sucrose, 1 mM EGTA, 1 mM DTT, 0.1 mM PMSF, 10 μM leupeptin, 1 μM pepstatin A), placed in liquid nitrogen and stored at −80 °C until use. Protein concentrations were measured using the Bio-Rad Protein Assay Kit according to the manufacturer's instructions.

F. Electrophoresis and Immunoblot Analysis

Microsomal and soluble fractions were analyzed by SDS–polyacrylamide gel electrophoresis (SDS–PAGE) (7.5%) according to the method of Laemmli (1970) (~10 μg protein/well). Following electrophoresis, proteins were transferred to a nitrocellulose membrane (Hybond ECL, Amersham) overnight at 2 °C, then the membranes were blocked for 1 h at room temperature in 5% skimmed milk in 0.1% Tween 20/PBS. The nitrocellulose sheet was then washed in 0.1% Tween 20/PBS and the primary antibody, monoclonal antibody (mAb) 18A10 (raised against mouse cerebellar IP$_3$R1; Maeda *et al.*, 1990), was applied to the sheet for 1 h at room temperature. After washing the nitrocellulose membrane by gentle shaking in 0.1% Tween 20/PBS three times for 10 min each, the membrane was incubated with a secondary antibody (horseradish peroxidase conjugated with antirat IgG) for 1 h at room temperature. The membrane was then washed three times (10 min each) in 0.1% Tween 20/PBS and bound the mAb 18A10, which is specific for exogenous mIP$_3$R1, and at an even higher concentration very weakly cross-reacted with endogenous xIP$_3$R, and was visualized using the Enhanced Chemi-Luminescence system (ECL, Amersham).

III. Use of GFP for Evaluation of Relative Expression Levels of Exogenous Protein in *Xenopus* Oocytes

There was heterogeneity in the fluorescence intensity of oocytes microinjected with pMT3-S65T (Fig. 1). Fluorescence from GFP-S65T expressed was visible in some oocytes within 24 h. By days 2 and 3 after microinjection, oocytes expressing GFP-S65T were easily identified as bright and dimly fluorescent ones.

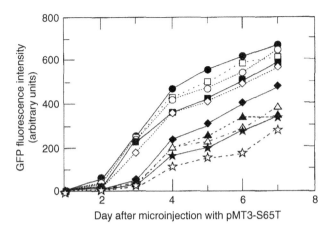

Fig. 1 Temporal changes in GFP-S65T fluorescence from 10 individual oocytes following pMT3-S65T microinjection. Each plot represents data from 10 individual oocytes.

The fluorescence signal continued to increase over a 7-day period as GFP-S65T accumulated. Because GFP-S65T fluorescence is resistant to photobleaching, bright field and epifluorescence illumination can be used to identify positively expressing oocytes. Using conventional optics for FITC, GFP-S65T expression is observed as bright green fluorescence and nonfluorescence indicates a failure of expression or failure of injection into the nucleus. State-of-the-art recombinant GFPs, such as enhanced or humanized GFPs, provided a stronger fluorescence in this amphibian expression system (data not shown).

When pMT3-S65T is coinjected with pMT3-IP$_3$R1, GFP-S65T fluorescence can be used not only to identify oocytes capable of transcription and translation of the IP$_3$R1 protein but also to estimate the relative expression level (Matheson *et al.*, 1996). The correlation coefficient of fluorescence intensity from GFP-S65T and expression level of IP$_3$R1 protein estimated through Western blotting was 0.93 ($n = 5$) (Fig. 2). This high correlation makes it possible to evaluate relative levels of IP$_3$R expression in fluorescent oocytes. These results indicate that one can select oocytes highly expressing ion channels such as IP$_3$R1 or other molecules by coinjecting pMT3-S65T.

IV. GFP Localization of Intracellular IP$_3$R–Ca^{2+} Stores in *Xenopus* Oocytes

A chimeric fusion protein between GFP-S65T protein and truncated mouse IP$_3$R1 (mIP$_3$R1) mutant protein, the ES region (amino acids 2216–2749) containing the putative transmembrane channel domain composed of six

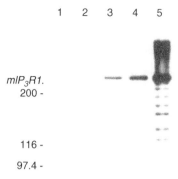

Fig. 2 Western blot analysis of expression levels of IP$_3$R1 in differently fluorescing oocytes, coinjected with pMT3-S65T and pMT3-IP$_3$R1. IP$_3$R1-specific monoclonal antibody 18A10, which barely cross-reacted with xIP$_3$R, was used. Lane 1: 5 μg of control microsome from uninjected oocytes; lanes 2–4: 5 μg of microsomes from oocytes prepared 3 days after coinjection of pMT3-S65T and pMT3-IP$_3$R1 exhibiting no fluorescence, weak fluorescence, and bright fluorescence, respectively; lane 5: 0.5 μg of mouse cerebellar microsomes as positive control. Molecular size markers (in kDa) are indicated on the left.

membrane-spanning segments (M1–6) and the following C-terminal tail region, was expressed in *Xenopus* oocytes by nuclear microinjection with pMT3-S65T-IP$_3$R1(ES) (Fig. 3) (Sayers *et al.*, 1997). This system facilitated monitoring of the intracellular distribution of IP$_3$R1-Ca^{2+} pools, as GFP-S65T fluorescence within the cell. Following microinjection, the oocytes were incubated for 3 days in MBS and observed under a confocal microscope. The expression of GFP–IP$_3$R1(ES) protein (calculated molecular weight 77,000) in oocytes microinjected with pMT3-S65T-IP$_3$R1(ES) positive in GFP-S65T fluorescence was confirmed by the presence of an intense band of approximately 80 kDa in Western blot analysis (data not shown). Using a confocal microscope, we detected the localization of GFP-S65T fluorescence on reticular structures under the animal pole (AP) of the oocyte microinjected with pMT3-S65T-IP$_3$R1(ES), with increased intensity in the perinuclear region (PN) and just beneath the plasma membrane (Fig. 4, see color insert).

The finding that expressed exogenous GFP–IP$_3$R1(ES) is selectively localized on the ER network, like endogenous *Xenopus* IP$_3$R1 (xIP$_3$R) (Kume *et al.*, 1993; Parys *et al.*, 1992) provides circumstantial evidence that GFP–IP$_3$R1(ES) and xIP$_3$R share similar targeting/retention mechanisms and are localized on the same Ca^{2+} store. Moreover, this study using the GFP-tagged IP$_3$R1(ES) paves the way for subsequent studies on not only the structure and function of mIP$_3$R1 necessary for ER targeting, but will facilitate studies on *in vivo* dynamics of IP$_3$R1-Ca^{2+} pools which apparently differ before and after fertilization (Kume *et al.*, 1993).

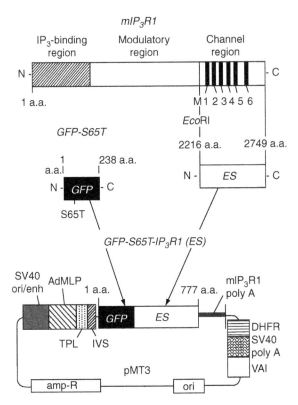

Fig. 3 Chimeric GFP–IP₃R1(ES) fusion protein. GFP-S65T was fused to the N-terminal end of a truncated mouse IP₃R1 protein (amino acids 2216–2749), IP₃R1(ES), which contained all six of the putative transmembrane-spanning domains (M1–6) and the following C-terminal tail, and cloned into the *Xenopus* oocyte expression vector pMT3, which is composed of the SV40 origin and enhancer element (SV40 ori/enh), the adenovirus major late promoter (AdMLP) containing the majority of the tripartite leader (TPL) present on adenovirus late mRNAs, a hybrid intervening sequence (IVS), a multicloning site, a DHFR-coding region, the SV40 early polyadenylation signal (SV40 poly A), and the adenovirus VAI gene.

Acknowledgments

This study was supported by grants from the Ministry of Education, Science, Sports and Culture of Japan, the Ministry of Health and Welfare of Japan, and RIKEN (the Institute of Physical and Chemical Research).

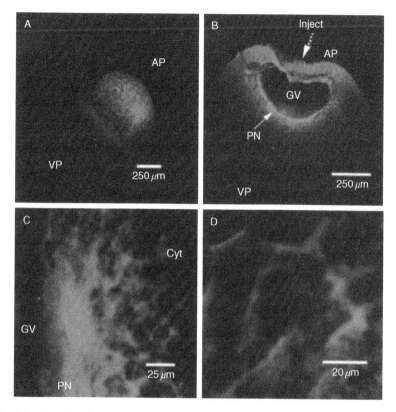

Fig. 4 Subcellular localization of GFP–IP$_3$R1(ES) in *Xenopus* oocytes as determined by fluorescence confocal microscopy. (A) Observation of GFP-S65T fluorescence in an intact, fully grown stage VI *Xenopus* oocyte, as viewed through a 10× objective lens 3 days after microinjection with pMT3-S65T-IP$_3$R1(ES); enriched fluorescence under the animal pole (AP) of the oocyte is demonstrated, with little detectable fluorescence under the vegetal pole (VP). (B) Intracellular observation of GFP-S65T fluorescence from an oocyte microinjected with pMT3-S65T-IP$_3$R1(ES), viewed using a 10× objective lens. Fluorescence is apparent from the AP (dashed arrow) into the nucleus, sliced in half after fixation in 4% paraformaldehyde. Intense areas of fluorescence can be observed in the perinuclear region (PN) around the germinal vesicle (GV) and in regions of the AP under the plasma membrane of the oocyte. Little fluorescence can be observed inside the VP. (C) Enlargement of a portion of the PN (40× objective lens). Note the intense fluorescence observed in the PN surrounding the GV, which appears to exhibit continuity with a reticular network of fluorescence in the cytoplasm (Cyt) distant from the GV. (D) Enlargement of the reticular fluorescence pattern observed under the AP (40× objective lens). The white arrow depicts regions of intense GFP-S65T fluorescence, possibly representing intracellular membranes that are the site of GFP–IP$_3$R1(ES) targeting (probably IP$_3$-sensitive Ca^{2+} pools).

References

Berridge, M. J. (1993). *Nature (London)* **361,** 315.

Colman, A. (1984). *In* "Transcription and Translation: A Practical Approach," pp. 271–302. IRL Press, Oxford.

Furuichi, T., and Mikoshiba, K. (1995). *J. Neurochem.* **64,** 953.

Furuichi, T., Yoshikawa, S., Miyawaki, A., Wada, K., Maeda, N., and Mikoshiba, K. (1989). *Nature (London)* **342,** 32.

Heim, R., Cubitt, A. B., and Tsien, R. Y. (1995). *Nature (London)* **373,** 663.

Kaufman, R. J., Davies, M. V., Pathak, V. K., and Hershey, J. W. B. (1989). *Mol. Cell. Biol.* **9,** 946.

Kume, S., Muto, A., Aruga, J., Nakagawa, T., Michikawa, T., Furuichi, T., Nakade, S., Okano, H., and Mikoshiba, K. (1993). *Cell* **73,** 555.

Laemmli, U. K. (1970). *Nature (London)* **227,** 680.

Maeda, N., Niinobe, M., and Mikoshiba, K. (1990). *EMBO J.* **9,** 61.

Matheson, J. M., Miyawaki, A., Muto, A., Inoue, T., and Mikoshiba, K. (1996). *Biomed. Res.* **17,** 221.

Meldolesi, J., Madeddu, L., and Pozzan, T. (1990). *Biochim. Biophys. Acta* **1055,** 130.

Otsu, H., Yamamoto, A., Maeda, N., Mikoshiba, K., and Tashiro, Y. (1990). *Cell Struct. Funct.* **15,** 163.

Parys, J. B., Sernett, S. W., DeLisle, S., Snyder, P. M., Welsh, M. J., and Campbell, K. P. (1992). *J. Biol. Chem.* **267,** 18776.

Pozzan, T., Rizzuto, R., Volpe, P., and Meldolesi, J. (1994). *Physiol. Rev.* **74,** 595.

Prasher, D. C., Eckenrode, V. K., Ward, W. W., Prendergast, F. G., and Cormier, M. J. (1992). *Gene* **111,** 229.

Ross, C. A., Danoff, S. K., Schell, M. J., Snyder, S. H., and Ullrich, A. (1992). *Proc. Natl. Acad. Sci. USA* **89,** 4265.

Satoh, T., Ross, C. A., Villa, A., Supattopone, S., Pozzan, T., Snyder, S. H., and Meldolesi, J. (1990). *J. Cell Biol.* **111,** 615.

Sayers, L. G., Miyawaki, A., Muto, A., Takeshita, H., Yamamoto, A., Michikawa, T., Furuichi, T., and Mikoshiba, K. (1997). *Biochem. J.* **323,** 273.

Swick, A. G., Janicot, M., Cheneval-Kastelic, T., McLenithan, J. C., and Lane, M. D. (1992). *Proc. Natl. Acad. Sci. USA* **89,** 1812.

CHAPTER 28

Confocal Imaging of Ca^{2+}, pH, Electrical Potential, and Membrane Permeability in Single Living Cells

John J. Lemasters, Donna R. Trollinger, Ting Qian, Wayne E. Cascio, and Hisayuki Ohata

Department of Cell Biology and Anatomy
University of North Carolina at Chapel Hill
Chapel Hill, North Carolina
USA

TECHNIQUES IN CONFOCAL MICROSCOPY
Copyright © 1999 by Elsevier Inc. All rights reserved.

DOI: 10.1016/B978-0-12-384658-7.00028-X

I. Introduction

Responses of single cells to stimuli are often heterogeneous. Bulk measurements by conventional biochemical and physiological techniques may fail to represent accurately the magnitude and time course of individual cell changes. Spatial heterogeneity of responses within single cells also occurs. For this reason, microscopic techniques with good three-dimensional resolution are needed to study individual cells as they respond to imposed stimuli and stresses. Increasingly, confocal microscopy of parameter-specific fluorophores is permitting direct observation of single-cell physiology with high spatial and temporal resolution.

II. Formation of Optical Sections by Confocal Microscopy

A. Pinhole Principle

The lateral resolving power of conventional optical microscopy approaches 0.2 μm, but axial resolution is much less because the effective depth of field for all but the smallest objects is 2–3 μm even at highest magnification. Thus, superimposition within this thick plane of focus obscures structures that might otherwise be resolved. In addition, in specimens more than a few microns thick, light arising from out-of-focus planes is projected on the in-focus image. Out-of-focus light decreases image contrast and limits applications that require quantitative analysis. This effect of out-of-focus light is especially severe in fluorescence microscopy, where imaging of thick, densely stained specimens is almost impossible by wide-field microscopy.

The technique of laser scanning confocal microscopy acts to eliminate out-of-focus light and produces remarkably detailed images whose depth of field is less than 1 μm. The principle of confocal microscopy, originally described by Minsky (1961) is quite simple (Fig. 1). Excitation light, most often from a laser, is focused by the microscope objective to a small spot within the specimen. At the crossover point, spot diameter is diffraction limited, or about 0.2 μm for a high numerical aperture (NA) objective lens. Light reflected or fluoresced by the specimen is collected by the objective lenses, separated from the excitation light using a partially reflecting prism or dichroic mirror, and focused onto a small pinhole. This light, which originates from the in-focus crossover point of the illuminating light beam, then passes through the pinhole to a light detector, typically a photomultiplier. However, light from above and below the focal plane strikes the edges of the pinhole and is not transmitted (Fig. 1). This selective transmission by the pinhole of in-focus light and rejection of out-of-focus light forms the basis for creating thin optical sections through thick specimens. Pinhole diameter determines the z-axis resolution of the microscope. As the pinhole is made smaller, the thickness of the confocal slice decreases until the system becomes diffraction limited, after which further decreases of pinhole diameter produce no additional improvement of z-axis resolution. Theoretical z-axis resolution is (Inoué, 1995):

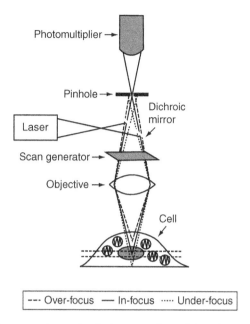

--- Over-focus — In-focus ····· Under-focus

Fig. 1 Principles of laser scanning confocal microscopy. Laser light reflects off a dichroic mirror and is focused to a small spot by the microscope objective, which is scanned across the specimen by the scan generator. Light fluoresced by the specimen passes through the dichroic mirror and is focused by the microscope objective lens on a pinhole aperture. The pinhole aperture transmits in-focus light but blocks out-of-focus light. Thus, the photomultiplier detects only light arising from the in-focus specimen plane.

$$z_{min} = \frac{2\lambda\eta}{(NA)^2} \tag{1}$$

where z_{min} is diffraction-limited z-axis resolution, λ is the wavelength of emitted light, η is the index of refraction of the object medium, and NA is the numerical aperture of the objective lens. Assuming that λ is 0.54 μm (fluorescein emission), η is 1.33 (water), and NA is 1.4, then z_{min} is about 0.7 μm.

B. Scanning

To create two-dimensional images, the beam must be translated across the specimen, either by moving the specimen underneath a stationary beam or, more commonly, by scanning the beam in x and y directions using vibrating mirrors in the light path (White *et al.*, 1987). The scan generator, which moves the beam across the specimen, also "descans" the returning light so that it can be focused on the stationary pinhole. The output of the photomultiplier is then stored in computer memory and displayed on a video monitor as the confocal image. In theory, any light source can be used to generate the scanning beam, but in practice only lasers provide powerful enough point illumination. The laser, scan generator, microscope objective, and pinhole together form the essential elements of the laser scanning confocal microscope (Fig. 1).

III. Precautions Against Photodamage and Photobleaching

Confocal microscopy requires the use of brighter excitation light than conventional wide-field fluorescence microscopy for the simple reason that confocal microscopy images light from a thin optical slice rather than from the entire specimen itself. In addition, the more complicated optics of confocal scan generators invariably involve some additional light loss compared with conventional fluorescence microscopy. Finally, the quantum efficiency for photon detection of photomultipliers is less than that of the best cooled charge-coupled devices (CCDs) used in conventional fluorescence microscopy, although the temporal response of photomultipliers remains far superior. As a consequence, specimens viewed by confocal microscopy are more vulnerable to photobleaching and photodamage, especially during serial imaging of living cells. The following strategies are helpful to minimize photobleaching and photodamage.

1. Light collection should be optimized by using a high-NA objective lens. For oil immersion, an NA of 1.3–1.4 should be used. For water immersion lenses, which are needed to image deeper than about 10 μm into aqueous samples, an NA greater than 1.0 is desirable. Empirical measurement of light transmission by individual lenses may be necessary, because light throughput also depends on the design of the lens and the glasses used in its construction. In particular, objectives made from ordinary glass transmit ultraviolet light poorly.

2. Laser power should be attenuated to the maximum extent possible using neutral density filters. Photomultiplier voltage (sensitivity) must be increased to compensate for the weaker fluorescent signal. Because lasers are so powerful, excellent image quality can often be obtained using an attenuation of 300- to 3000-fold. However, too much attenuation will increase spatial signal to noise and produce a snowy (noisy) image, much like reception of a distant television signal. This is because too few photons are measured from each picture element (pixel).

3. Because a larger pinhole transmits more light, the pinhole may be opened to increase image brightness and decrease spatial signal to noise without increasing laser illumination. Most laser scanning confocal microscopes incorporate adjustable pinhole apertures for this purpose. For a particular application, the resulting small loss of axial resolution may be acceptable. However, for maximum axial resolution, the pinhole diameter should be set to slightly less than the size of the diffraction-limited Airy disk projected on the pinhole. Airy disk diameter at the pinhole increases as the magnification factor of the objective lens increases but decreases as NA increases. Thus, the pinhole diameter needs to be readjusted when the objective lens is changed. Many commercial systems use software to calculate the Airy disk and adjust the pinhole diameter automatically.

4. Casual viewing, which might cause unnecessary photobleaching, should be avoided. A typical beginner's mistake is to view a sample at high laser excitation. Initially, excellent images are obtained, but photobleaching soon makes collection of any image almost impossible. Instead, survey images should be collected at the lowest possible illumination. When an area of interest is found, laser intensity may then be increased to achieve the desired image brightness and spatial signal to noise.

5. The zoom feature of laser scanning confocal microscopes permits magnification to be varied continuously over a 5- to 10-fold range. Zooming simply limits the x and y excursion of the beam across the specimen. As zoom magnification increases, the same laser energy concentrates into a smaller volume of the sample, causing accelerated photobleaching and photodamage. Thus, zoom magnification should be minimized to the maximum extent that is consistent with the resolution requirements of the experiment. However, pixel size must remain less than half the diameter of the smallest object to be discerned.

6. The choice of fluorophores is also critically important. Stable fluorophores can be serially imaged with little phototoxicity. These include many of the rhodamine derivatives. Other fluorophores, such as fluorescein and acridine orange, rapidly bleach or generate toxic photoproducts. A simple strategy to reduce photobleaching is to use higher fluorophore concentration, which permits greater attenuation of the excitation light for a given intensity of fluorescent signal.

IV. Preparation of Cells for Microscopy

Cells must be viewed through glass coverslips for virtually all types of high-resolution light microscopy. For most objective lenses, cells should be attached to $1\frac{1}{2}$ coverslips, a designation that means that the glass is 170 μm thick. Generally, cultured cells grow poorly on glass. To improve both adherence and growth on glass, coverslips must be treated with materials such as type I collagen or laminin. Coverslips are first rinsed in ethanol for sterilization and dried. Subsequently, 100 μl (1–2 drops) of collagen (1 mg/ml in 0.1% acetic acid) or laminin (0.1 mg/ml in Tris-buffered saline) is spread across the glass surface. After air drying overnight, the coverslips are placed inside petri dishes for plating with cells in the usual fashion. For viewing in the microscope, a chamber is needed into which the coverslip is placed. These chambers can be built in a machine shop or purchased commercially (FCS2 chamber; Bioptechs, Butler, PA). An open chamber requires use of an inverted microscope, a configuration that permits easy access to the cell medium. Closed chambers usually permit continuous perfusion with buffers of interest and can be adapted to both upright and inverted microscopes. For studying living cells, adequate temperature regulation is important, which heaters built into the coverslip chamber or a stream of tempered air can provide.

V. Multicolor Fluorescence

Commercially available lasers for confocal microscopy emit narrow lines of light at fixed wavelengths, a fact that dictates the types of fluorophores that can be used. The most common lasers and their wavelengths of emission are as follows: argon, 488 and 514 nm; argon–krypton, 488, 568, and 647 nm; helium–cadmium, 442 nm; helium–neon, 543 nm; and UV–argon, 351–364 nm. The argon–krypton laser with well-separated blue, yellow, and red emissions is perhaps the most versatile single laser for biological applications. Using a multiline laser, multiple spectrally distinct fluorophores can be excited simultaneously. Fluorescence emitted from multiple fluorophores can be separated by wavelength and directed to different photomultipliers for simultaneous detection.

VI. Imaging Electrical Potential with Cationic Fluorophores

In living cells under normal conditions, the plasma membrane and mitochondrial inner membrane maintain a negative-inside electrical potential difference ($\Delta\Psi$). As a consequence lipophilic cations, such as rhodamine 123, tetramethylrhodamine methylester (TMRM), and tetramethylrhodamine ethylester (TMRE), accumulate in the cytosol and the mitochondrial matrix in response to these membrane potentials (Ehrenberg *et al.*, 1988; Emaus *et al.*, 1986; Johnson *et al.*, 1981). At equilibrium, accumulation of permeant monovalent cations is related to $\Delta\Psi$ by the Nernst equation:

$$\Delta\Psi = -60 \log\left(\frac{F_{in}}{F_{out}}\right) \tag{2}$$

where F_{in} and F_{out} represent the concentration of monovalent cation inside and outside the membrane, respectively. By quantifying the intracellular distribution of cationic fluorophores using confocal microscopy, a map of intracellular electrical potential can be generated using Eq. (2).

For most cells, plasma membrane $\Delta\Psi$ is -30 to -90 mV and mitochondrial $\Delta\Psi$ is -120 to -180 mV. These $\Delta\Psi$ values are additive. Thus, mitochondria are as much as 270 mV more negative than the extracellular space. From Eq. (2), this potential difference corresponds to a cation concentration ratio of more than 10,000 to 1 inside mitochondria relative to the cell exterior. With a conventional memory of 256 gray levels per pixel (8 bits), the large gradients of fluorescence corresponding to this concentration ratio cannot be stored using a linear scale. Either 16-bit memory or a nonlinear logarithmic (gamma) scale must be used instead (Chacon *et al.*, 1994). Gamma scales, long used in scanning electron microscopy, condense the input signal into the available 256 gray levels of pixel memory.

To measure electrical potential, cells are first loaded with a potential-indicating fluorophore (50–500 nM), such as TMRM. A loading time of 15 min is usually adequate. The loading buffer is then replaced with experimental medium containing a small amount of fluorophore (20–150 nM) to maintain equilibrium distribution of the fluorophore. Equation (2) assumes ideal electrophoretic distribution of the fluorophore, but some probes do not behave ideally. Rhodamine 123, for example, binds nonspecifically to the mitochondrial matrix (Emaus *et al.*, 1986). Moreover, the fluorescence of rhodamine 123 and other fluorophores becomes quenched as accumulation into mitochondria occurs. This quenching is concentration dependent and may be decreased by using smaller loading concentrations. Fluorophores can reach millimolar concentrations inside mitochondria, which may cause metabolic inhibition. For example, rhodamine 123 at such high concentrations strongly inhibits the oligomycin-sensitive mitochondrial F_1F_0-ATPase that catalyzes ATP synthesis during oxidative phosphorylation (Emaus *et al.*, 1986). DiOC(6), a cationic fluorophore frequently used in flow cytometry, is an even more potent inhibitor of mitochondrial respiration (Rottenberg and Wu, 1997). TMRM and TMRE seem to lack many of these undesirable characteristics (Ehrenberg *et al.*, 1988; Farkas *et al.*, 1989), but low fluorophore concentrations (<500 nM) should always be employed to minimize quenching and metabolic effects.

In cells loaded with a permeable cationic fluorophore, confocal images are collected. The stored images are then processed to produce a map of intracellular electrical potential. Average fluorescence intensity in the extracellular space is divided into intracellular fluorescence on a pixel-by-pixel basis to determine F_{in}/F_{out} of Eq. (2). Equation (2) is then applied to each pixel to calculate the electrical potential difference between each point inside the cell and the extracellular space. Finally, the calculations are displayed as a pseudocolor map that shows the intracellular distribution of electrical potential (Chacon *et al.*, 1994).

Figure 2 illustrates intracellular electrical potential measured by this procedure for a cultured hepatocyte. In the cytosol and nucleus, pseudocoloring reveals an electrical potential of between −20 and −40 mV. Because the electrical potential of the extracellular medium is zero, plasma membrane $\Delta\Psi$ is about −30 mV, as expected for hepatocytes. Throughout the cytoplasm are also distributed electronegative mitochondria. These mitochondria show an apparent heterogeneity in their electrical potential. Heterogeneity is due, at least in part, to the fact that not all mitochondria extend completely through the confocal section. Consequently, fluorophore uptake for many mitochondria is underestimated. In Fig. 2, maximum mitochondrial potential calculated by Eq. (2) was −160 mV. The difference between this value and the cytosolic potential, −130 mV, represents a minimum estimate of mitochondrial $\Delta\Psi$.

VII. Principle of Ratio Imaging

A wide variety of fluorophores is available to measure ions and other parameters within individual living cells. Typically these fluorophores are loaded as their membrane-permeant ester derivatives. Inside cells, esterases release and trap the

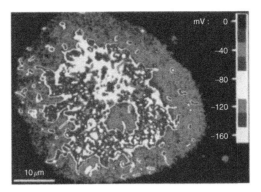

Fig. 2 Distribution of electrical potential in a cultured rat hepatocyte. A 1-day cultured rat hepato-cyte was loaded with 500 nM TMRM for 15 min, and distribution of electrical potential was determined by laser scanning confocal microscopy using 568-nm excitation from an argon–krypton laser, as described in text. After J. J. Lemasters, E. Chacon. J. M. Reece, A.-L. Nieminen, and G. Zahrebelski, unpublished.

parameter-sensitive free acid form of the fluorophores. The fluorescence of these fluorophores depends not only on the parameter measured but also on the amount of fluorophore present in the light path. With some fluorophores, a rationing proce-dure may be used to correct for variations in fluorophore concentration provided that parameter binding causes a shift of the fluorescence excitation or emission spectrum. Images are acquired at two different excitation or emission wavelengths, one that increases as the parameter of interest increases and one that decreases or stays the same. Background images of cell-free regions must also be collected at both wave-lengths at the same microscope settings. These backgrounds are then substracted on a pixel-by-pixel basis from the experimental images. After background subtraction, the image at one wavelength is divided by the image at the second wavelength to form a ratio image. Such rationing removes the contribution of fluorophore concentration to signal strength. The ratios are then compared with a standard curve and converted to units signifying parameter concentration. These values are often displayed in the image using a pseudocolor scale. The advantage of ratio imaging is the elimination of variations in signal due to differences in regional fluorophore concentration, dye leakage over time, photobleaching, and accessible volume.

For laser scanning confocal microscopy, rationing of emission wavelengths must usually be employed. Lasers provide only one or a few narrow lines of monochro-matic light, which severely limits the selection of excitation wavelengths. A good ratiometric fluorophore for confocal microscopy is carboxyseminaphthorhoda-fluor 1 (SNARF 1), a pH indicator. When SNARF 1 is excited with the 568-nm line of an argon–krypton laser, fluorescence emission at 640 nm increases as pH increases but emission at 585 nm remains unchanged (Chacon *et al.*, 1994). Thus, the ratio of the two wavelengths is proportional to pH. For *in situ* calibration, SNARF-1-loaded cells are incubated with 10 µM nigericin and 5 µM valinomycin

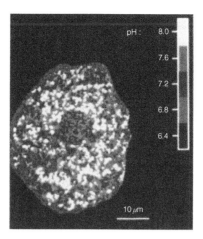

Fig. 3 Intracellular pH in a cultured rat hepatocyte. A 1-day cultured rat hepatocyte was loaded with 5 μM SNARF 1 AM for 45 min at 37 °C. Using 568-nm excitation light, a confocal fluorescence image at 584 nm (10-nm bandpass) was divided by the image at >620 nm after background subtraction. Using an *in situ* calibration, distribution of pH within the cell is represented in pseudocolor. After J. J. Lemasters, E. Chacon, J. M. Reece, A.-L. Nieminen, and G. Zahrebelski, unpublished.

in modified Krebs–Ringer buffer, in which NaCl and KCl are replaced by their corresponding gluconate salts to minimize swelling (Chacon *et al.*, 1994). SNARF 1 readily enters both the cytosolic and mitochondrial compartments after ester loading, and confocal ratio imaging of SNARF-1 demonstrates marked heterogeneity of pH within cells (Fig. 3). Cytosolic and nuclear regions have a pH near 7.2, whereas mitochondrial pH is close to 8. Thus, a ΔpH of 0.8 pH units exists across the mitochondrial membrane, as predicted by P. Mitchell's chemiosmotic hypothesis (Mitchell, 1966). Because the protonmotive force (Δp) equals $\Delta\Psi - 59\,\Delta$pH, Δp in intact living hepatocytes can be estimated to be approximately -180 mV.

VIII. Compartmentation of Ester-Loaded Fluorophores

A. Warm Versus Cold Loading

Many fluorophores useful for confocal microscopy are multivalent organic anions. These multiple charged molecules are impermeant to membranes. To load these molecules into cells, the carboxyl groups of these acids must be neutralized by forming acetate or acetoxymethyl esters, which are membrane permeable. When cells are incubated with these derivatives, endogenous esterases regenerate the free acid form of the fluorophores, which becomes trapped inside the cell. Esterases are found within several organelles, including the cytosol, mitochondria, and lysosomes. Thus, after ester loading, fluorophores can be heterogeneously distributed in several different compartments.

For many fluorophores, the temperature of loading strongly affects intracellular distribution (Nieminen *et al.*, 1995). This is illustrated for calcein, a pentacarboxylic acid fluorophore whose fluorescence is independent of physiological changes of intracellular ions. When primary cultured hepatocytes or cardiac myocytes are incubated with the acetoxymethyl ester (AM) of calcein at 37 °C, calcein fluorescence enters the cytosol and nucleus (Fig. 4A). Under these loading conditions, calcein is excluded from mitochondria, leaving small round voids in the green fluorescence of calcein. These voids correspond exactly to the punctate red fluorescence of TMRM-labeled mitochondria (Fig. 5). By contrast, after ester loading at 4 °C, calcein enters both the cytosol and the mitochondria (Fig. 4B). Presumably during warm loading, cytosolic esterases are so active that calcein AM is hydrolyzed before it has a chance to enter mitochondria. At 4 °C, however, enzymatic activity is inhibited, allowing movement of unhydrolyzed ester into the mitochondria where mitochondrial esterases can cleave the ester and trap the free acid. The temperature dependence of loading is both probe and cell type specific. For example, the Ca^{2+}-indicating fluorophore Fluo 3 loads similarly to calcein, whereas Indo 1, another Ca^{2+} indicator, and SNARF 1 load into mitochondria even at 37 °C (Chacon *et al.*, 1994; Ohata *et al.*, 1998).

B. Cold Loading Followed by Warm Incubation

Fluorophores loaded into the cytosol gradually leak across the plasma membrane during warm incubation through an organic anion carrier (Wieder *et al.*, 1993). However, fluorophores trapped in mitochondria and other organelles are lost much more slowly. Thus, when cold ester loading is followed by several hours of warm incubation, fluorophore loading is almost exclusively mitochondrial

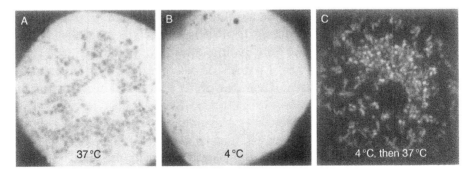

Fig. 4 Temperature dependence of calcein loading in hepatocytes. One-day cultured rat hepatocytes were loaded with 1 μM calcein AM for 15 min at 37 °C (A), with 1 μM calcein AM for 2 h at 4 °C (B), or with 5 μM calcein AM on ice for 2 h followed by incubation at 37 °C for 12 h (C). Green fluorescence was imaged by laser scanning confocal microscopy. Note dark round voids of fluorescence in (A), representing mitochondria. These voids are absent in (B). In (C), mitochondria themselves are labeled with calcein with little fluorescence in the cytosol. After Nieminen *et al.* (1995) and T. Qian and J. J. Lemasters, unpublished, 1997.

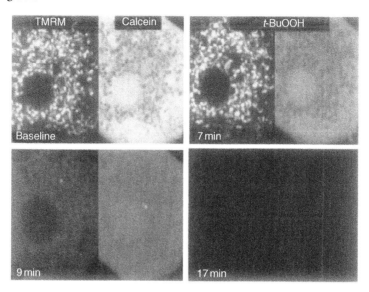

Fig. 5   Onset of the MPT in hepatocytes induced by *tert*-butylhydroperoxide (*t*-BuOOH). A cultured rat hepatocyte was coloaded with TMRM and calcein. Pairs of red TMRM and green calcein fluorescence images were collected by laser scanning confocal microscopy. In the baseline images, note that TMRM-labeled mitochondria correspond to dark voids in the calcein fluorescence. Within 9 min of adding 100 μM *t*-BuOOH to produce oxidative stress, calcein redistributed from the cytosol into the mitochondria and TMRM fluorescence was lost from the mitochondria, events signifying the onset of the MPT. These changes were followed by loss of cell viability, documented by loss of cytosolic calcein fluorescence after 17 min. After Nieminen *et al.*, 1995.

(Fig. 4C) (Trollinger *et al.*, 1997). Thus, by manipulating the temperature of loading and the duration of the subsequent incubation, ester-loaded fluorophores can be directed to the cytosol, the cytosol plus mitochondria, or just the mitochondria (Nieminen *et al.*, 1995; Ohata *et al.*, 1998; Trollinger *et al.*, 1997).

IX. Membrane Permeability

Directed compartmental loading of calcein allows direct observation in cells of changes of permeability of membranes. For example, at the onset of cell death after toxic and hypoxic injury, trapped cytosolic calcein is lost almost instantaneously (see Fig. 5) (Zahrebelski *et al.*, 1995). Similarly, when calcein is in the extracellular space, the fluorophore enters the cell interior at onset of cell death. Other extracellular fluorophores enter also, such as propidium iodide, which binds to DNA with an enhancement of fluorescence.

After warm ester loading of calcein, calcein redistribution from the cytosol into the mitochondria indicates onset of the mitochondrial permeability transition (MPT).

The MPT is caused by opening of a high-conductance pore in the mitochondrial inner membrane with a molecular weight cutoff of about 1500 (Zoratti and Szabo, 1995). Onset of the MPT leads to mitochondrial depolarization and uncoupling of oxidative phosphorylation. At onset of the MPT in hepatocytes coloaded with TMRM and calcein, mitochondria release TMRM, which indicates depolarization, and fill with calcein, which indicates permeability of the mitochondria to small solutes (Fig. 5) (Nieminen *et al.*, 1995). The MPT is an important event leading to both necrotic and apoptotic cell death (Lemasters *et al.*, 1998).

X. Confocal Imaging of Ca^{2+}

A. Nonratiometric Imaging

Green-fluorescing Fluo 3 and red-fluorescing Rhod 2 are useful visible-wavelength fluorophores for imaging Ca^{2+} by laser scanning confocal microscopy. Ca^{2+} binding produces a greater than 50-fold increase in fluorescence with a K_d of 300–600 nM (Haugland, 1996; Minta *et al.*, 1989). No shift of the fluorescence spectrum of Fluo 3 and Rhod 2 occurs after Ca^{2+} binding, and calibration relies on measurements *in situ* of maximal and minimal fluorescence after respective additions of calcium ionophore and EGTA (Harper *et al.*, 1993; Minta *et al.*, 1989). Even without calibration, Fluo 3 and Rhod 2 are useful to determine relative changes of free Ca^{2+}.

B. Measurement of Ca^{2+} Transients Using Line Scanning

To increase temporal resolution of rapid Ca^{2+} transients, confocal images can be collected in the line-scanning mode whereby the *x*-axis is scanned every 25 ms at the same *y*-axis position to create *x*-versus-time images (Cheng *et al.*, 1993; Ohata *et al.*, 1998). Figure 6 illustrates simultaneous red and green line-scan images for a cardiac myocyte coloaded with TMRM and Fluo 3 at 4 °C (Ohata *et al.*, 1998). An *x*-axis position is selected that crosses the cytosol, interfibrillar mitochondria, perinuclear mitochondria, and the nucleus, as shown in Fig. 6A. In the line-scan images, red and green fluorescence is collected simultaneously (Fig. 6B). Red TMRM fluorescence appears as wavy vertical stripes. Each stripe is a mitochondrion, and each wave in the stripes is cell movement from an electrically stimulated contraction. By contrast, green Fluo 3 fluorescence does not show vertical striations but only horizontal banding corresponding to Ca^{2+} transients after each field stimulation. Importantly, the transients of Fluo 3 fluorescence occur both in pixels corresponding to TMRM-labeled mitochondria and in those of the TMRM-unlabeled cytosol. These results demonstrate that both cytosolic and mitochondrial Ca^{2+} transients occur after field stimulation of cardiac myocytes.

Using image analysis software, regions corresponding to the cytosol, the nucleus, and interfibrillar and perinuclear mitochondria can be selected on the basis of the

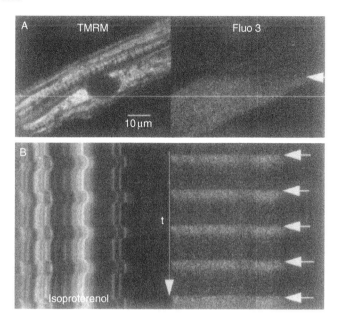

Fig. 6   Line-scanning confocal microscopy of mitochondrial and cytosolic Ca^{2+} transients during electrical stimulation. An adult rabbit cardiac myocyte was loaded with 10 µM Fluo 3-AM for 2 h at 4 °C followed by 600 nM TMRM, and simultaneous red (TMRM) and green (Fluo 3) fluorescence confocal images were collected. (A) x–y images of TMRM (*left*) and Fluo 3 (*right*), and the effect of a single-field stimulation (arrow) as the 1.5-s scan was collected. In (B), 1 µM isoproterenol was added, and the myocyte was stimulated at 0.5 Hz (arrows) as a line-scan image at 25 ms per line was collected. The region scanned is indicated by the white line in (A). After Ohata *et al.*, 1998.

brightness of TMRM fluorescence and the position in the cell. Bright regions in the red TMRM image are mitochondria, whereas areas of low fluorescence intensity are cytosol and nucleus. Pixels of intermediate intensity are excluded from analysis because they likely represent the overlap of mitochondrial and cytosolic domains. Using morphological criteria, high-TMRM regions are subdivided into interfibrillar and perinuclear mitochondria, and low-fluorescence regions are assigned to the nucleus and cytosol. The corresponding regions in the green Fluo 3 fluorescence image are then averaged in each horizontal line. Relatively inexpensive software packages, such as NIH Image (National Institutes of Health, Bethesda, MD) for Macintosh computers and Image PC (Scion Corp., Frederick, MD) for Windows-based personal computers, can perform this analysis. In this way, average Fluo 3 fluorescence for the cytosol, nucleus, and interfibrillar and perinuclear mitochondria is calculated every 25 ms. When the repeating Ca^{2+} transients are averaged and normalized as a percentage of baseline fluorescence, the peak change in fluorescence after field stimulation is greatest in the cytosol and least in perinuclear mitochondria (Fig. 7). Isoproterenol increases the peak intensity and rate of decay of Fluo 3 fluorescence in the cytosol and nucleus with a lesser effect on interfibrillar

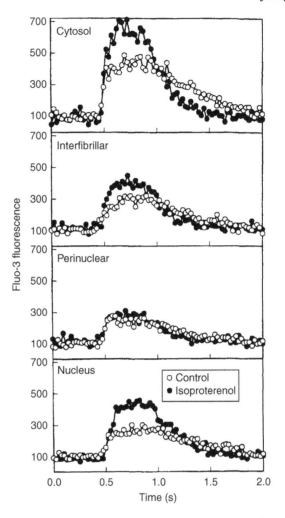

Fig. 7 Averaged and normalized plots of Fluo 3 fluorescence in different subcellular regions during electrical stimulation of a cardiac myocyte. From line-scan images in the presence and absence of 1 μM isoproterenol (see Fig. 6B), TMRM fluorescence was used to identify cytosol, interfibrillar mitochondria, perinuclear mitochondria, and the nucleus. Pixels corresponding to each region were superimposed on the corresponding Fluo 3 image, and Fluo 3 pixel intensity was calculated using NIH Image. Average pixel intensity minus background for each region in the line-scan image was determined. Fluorescence transients after electrical stimulation in each region were averaged, normalized as a percentage of diastolic intensity, and plotted versus time. After Ohata *et al.*, 1998.

mitochondria and virtually no effect on perinuclear mitochondria. Because TMRM and Fluo 3 fluorescence are collected simultaneously, these calculations automatically correct for movement of mitochondria laterally or into and out of the optical section during cell contraction (Ohata *et al.*, 1998).

C. Quantification of Free Ca^{2+} by Ratio Imaging of Indo 1-Loaded Myocytes

Ca^{2+} binding causes an increase of Fluo 3 fluorescence but no wavelength shift, preventing calibration by ratio imaging. To estimate absolute Ca^{2+} concentrations, myocytes can be loaded with Indo 1 AM (Grynkiewicz et al., 1985), which distributes into both the cytosol and the mitochondria even after ester loading at 37 °C. Using 351-nm excitation from a UV argon laser, Indo 1 fluorescence increases at 405 nm and decreases at 480 nm as Ca^{2+} increases. Ratio images of Indo 1 fluorescence show a relatively uniform intracellular free Ca^{2+} concentration of about 200 nM in both the cytosol and mitochondria of resting myocytes (Fig. 8, see color insert). This value is uncorrected for possible increases in the binding constant of Indo 1 for Ca^{2+} that occurs when Indo 1 is in the intracellular environment (Baker et al., 1994; Bassani et al., 1995; Roe et al., 1990). Ca^{2+} sensitivity of Indo 1 and other BAPTA-derived probes is also affected by pH, but significant changes in K_d for Ca^{2+} occur only when the pH falls below 6.8 (Kawanishi et al., 1991; Lattanzio and Barschat, 1991). Because mitochondrial pH is alkaline relative to the cytosol (Fig. 3), pH should not affect the K_d for Ca^{2+} binding inside mitochondria. In any case, rationing of Indo 1 fluorescence corrects for the effect of variable loading of fluorophore into various cellular compartments and confirms that Ca^{2+} transients occur both in the cytosol and the mitochondria after electrical stimulation (Fig. 8).

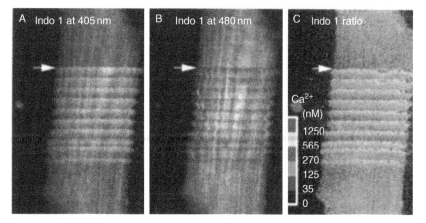

Fig. 8 Indo 1 ratio imaging of cytosolic free Ca^{2+} in a cardiac myocyte during electrical stimulation. Adult rabbit cardiac myocytes were loaded with 5 µM Indo 1 AM for 1 h at 37 °C. Using 351-nm excitation from a UV-argon laser, fluorescence images of Indo 1 were collected at emission wavelengths of 395–415 nm (A) and 470–490 nm (B) as the cell was electrically stimulated at 0.5 Hz for 10 s during a 40-s scan. (C) The ratio image after background subtraction scaled to Ca^{2+} concentration. Electrically stimulated Ca^{2+} transients occurred in both mitochondrial and cytosolic regions, as identified by rhodamine 123 labeling (not shown). (After Ohata et al., 1998).

D. Cold Loading of Rhod 2 Followed by Warm Incubation

Another approach to measure mitochondrial Ca^{2+} uses cold loading of Rhod 2 AM followed by warm incubation. After loading in this way, Rhod 2 fluorescence in unstimulated myocytes shows a distinctly mitochondrial pattern, comparable to that observed when myocytes are loaded with TMRM or rhodamine 123 (Fig. 9, compare with Fig. 6) (Trollinger *et al.*, 1997). When the myocytes are electrically stimulated as 16-s confocal scans are collected, mitochondrial Rhod 2 fluorescence increases after each electrical stimulation, producing horizontal bands in the image (Fig. 9). When the pacing frequency is doubled, banding frequency also doubles. The largest fluorescence transients occur in the bright regions corresponding to mitochondria. Fluorescence in areas between mitochondria either remains dark or increases moderately after electrical stimulation. Addition of the Ca^{2+} ionophore Br-A23187 verifies Rhod 2 localization (Fig. 9). After raising Ca^{2+} in all compartments to levels that saturate Rhod 2, the distribution of red fluorescence represents the intracellular distribution of the fluorophore. In this experiment, CCCP, a powerful protonophoric uncoupler, is also added to depolarize mitochondria and block cycling of free Ca^{2+} after Br-A23187 treatment. Glucose and oligomycin

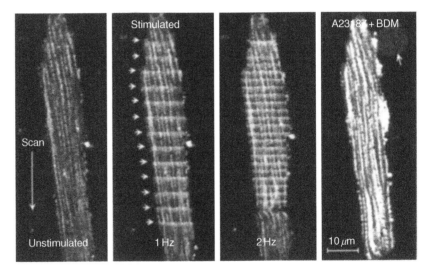

Fig. 9 Mitochondrial Ca^{2+} transients after cold Rhod 2 loading followed by warm incubation. An adult rabbit cardiac myocyte was cold loaded with Rhod 2 AM and then incubated at 37 °C for 3 h prior to confocal fluorescence imaging. In the unstimulated myocyte, Rhod 2 fluorescence has a mitochondrial pattern. After stimulation at 1 Hz (arrows) and then 2 Hz, Rhod 2 fluorescence transiently increased in the bright mitochondrial regions and then returned to baseline. Oligomycin (10 μM), CCCP (10 μM), butanedione monoxime (20 mM), and then Br-A23187 (20 μM) were added to increase free Ca^{2+} in all compartments. Under these conditions, saturated Rhod 2 fluorescence was confined to mitochondria. Cytosolic areas, such as in blebs (arrow), were only faintly fluorescent. Each 16-s confocal scan proceeds from top to bottom. The images are pseudocolored using the black body lookup table of Photoshop (Adobe Systems, San Jose, CA). (After Trollinger *et al.*, 1997.)

are also added. In the presence of a glycolytic substrate, oligomycin prevents ATP depletion by inhibiting the uncoupler-stimulated mitochondrial ATPase (Bers *et al.*, 1993; Nieminen *et al.*, 1990, 1994). In addition, butanedione monoxime (BDM) is added prior to Br-A23187 to prevent myocyte contracture (Armstrong and Ganote, 1991). After Br-A23187 treatment in this way, Rhod 2 fluorescence shows a characteristic mitochondrial pattern of staining (Fig. 9). By contrast, cytosolic areas between mitochondria remain dark. Occasional toxic blebs form after Br-A23187 treatment (Fig. 9, arrow). The contents of these blebs are an extension of the cytosol (Lemasters *et al.*, 1981). These blebs in Br-A23187-treated myocytes remain dark, confirming the virtual absence of cytosolic Rhod 2 loading.

During electrical stimulation, Rhod 2 fluorescence increases sharply in the bright regions corresponding to mitochondria. However, areas between mitochondria sometimes appear to increase moderately in fluorescence, an apparent contradiction of the conclusion that Rhod 2 is confined virtually exclusively to mitochondria. This inconsistency is explained by the fact that the confocal slice thickness is finite, in the range of 0.8–0.9 μm. Cardiac mitochondria are approximately 1 μm in diameter and many mitochondria are only partially sectioned within the confocal slice. Parts of confocal images containing partially sectioned mitochondria will be darker than areas with fully sectioned mitochondria. However, fluorescence will nonetheless increase in darker regions containing partially sectioned mitochondria as Ca^{2+} increases. The observation that ruthenium red, an inhibitor of mitochondrial Ca^{2+} uptake, blocks mitochondrial but not cytosolic Ca^{2+} transients after electrical stimulation further confirms the mitochondrial localization of the fluorophore (Trollinger *et al.*, 1998).

Cold loading of Rhod 2 followed by warm incubation to measure mitochondrial Ca^{2+} has advantages over coloading with Fluo 3 and TMRM. With the latter approach, the contribution of partially sectioned mitochondria to the analysis must be excluded. Only Fluo 3 pixels corresponding to the brightest and darkest TMRM pixels can be relied on to represent, respectively, mitochondrial and cytosolic Ca^{2+}. Pixels corresponding to intermediate TMRM fluorescence must be discarded. With cold loading of Rhod 2 followed by warm incubation, all pixels can be used for analysis, which increases signal strength and signal to noise.

XI. Conclusion

Confocal microscopy has become an essential tool to study the physiology of single living cells. As more parameter-indicating fluorophores are discovered, the range of applications of confocal microscopy can only increase. Uniquely, confocal microscopy permits observation of the physiology and metabolism of single organelles inside cells, as illustrated here for measurements of Ca^{2+}, pH, $\Delta\Psi$, and membrane permeability of mitochondria. Overall, the impact of confocal microscopy on experimental cell physiology may someday compete with that of single-cell electrical recording.

Acknowledgments

This work was supported, in part, by Grants DK37034, HL27430, AG07218, and AA11605 from the National Institutes of Health and by Grant N00014-96-1-0283 from the Office of Naval Research.

References

Armstrong, S. C., and Ganote, C. E. (1991). *J. Mol. Cell Cardiol.* **23,** 1001.

Baker, A. J., Brandes, R., Schreur, J. H., Camacho, S. A., and Weiner, M. W. (1994). *Biophys. J.* **67,** 1646.

Bassani, J. W., Bassani, R. A., and Bers, D. M. (1995). *Biophys. J.* **68,** 1453.

Bers, D. M., Bassani, J. W., and Bassani, R. A. (1993). *Cardiovasc. Res.* **27,** 1772.

Chacon, E., Reece, J. M., Nieminen, A. L., Zahrebelski, G., Herman, B., and Lemasters, J. J. (1994). *Biophys. J.* **66,** 942.

Cheng, H., Lederer, W. J., and Cannell, M. B. (1993). *Science* **262,** 740.

Ehrenberg, B., Montana, V., Wei, M. D., Wuskell, J. P., and Loew, L. M. (1988). *Biophys. J.* **53,** 785.

Emaus, R. K., Grunwald, R., and Lemasters, J. J. (1986). *Biochim. Biophys. Acta* **850,** 436.

Farkas, D. L., Wei, M. D., Febbroriello, P., Carson, J. H., and Loew, L. M. (1989). *Biophys. J.* **56,** 1053.

Grynkiewicz, G., Poenie, M., and Tsien, R. Y. (1985). *J. Biol. Chem.* **260,** 3440.

Harper, I. S., Bond, J. M., Chacon, E., Reece, J. M., Herman, B., and Lemasters, J. J. (1993). *Basic Res. Cardiol.* **88,** 430.

Haugland, R. P. (1996). "Handbook of Fluorescent Probes and Research Chemicals," 6th edn. Molecular Probes, Eugene, OR.

Inoué, S. (1995). *In* "Handbook of Biological Confocal Microscopy," 2nd edn. (J. B. Pawley, ed.), p. 1. Plenum Press, New York.

Johnson, L. V., Walsh, M. L., Bockus, B. J., and Chen, L. B. (1981). *J. Cell Biol.* **88,** 526.

Kawanishi, T., Nieminen, A. L., Herman, B., and Lemasters, J. J. (1991). *J. Biol. Chem.* **266,** 20062.

Lattanzio, F. A., and Barschat, D. K. (1991). *Biochem. Biophys. Res. Commun.* **171,** 102.

Lemasters, J. J., Ji, S., and Thurman, R. G. (1981). *Science* **213,** 661.

Lemasters, J. J., Nieminen, A. L., Qian, T., Trost, L. C., Elmore, S. P., Nishimura, Y., Crowe, R. A., Cascio, W. E., Bradham, C. A., Brenner, D. A., and Herman, B. (1998). *Biochim. Biophys. Acta* **1366,** 177.

Minsky, M. (1961). U.S. Patent 3,013,467.

Minta, A., Kao, J. P. Y., and Tsien, R. Y. (1989). *J. Biol. Chem.* **264,** 8171.

Mitchell, P. (1966). *Biol. Rev.* **41,** 445.

Nieminen, A. L., Dawson, T. L., Gores, G. J., Kawanishi, T., Herman, B., and Lemasters, J. J. (1990). *Biochem. Biophys. Res. Commun.* **167,** 600.

Nieminen, A. L., Saylor, A. K., Herman, B., and Lemasters, J. J. (1994). *Am. J. Physiol.* **267,** C67.

Nieminen, A. L., Saylor, A. K., Tesfai, S. A., Herman, B., and Lemasters, J. J. (1995). *Biochem. J.* **307,** 99.

Ohata, H., Chacon, E., Tesfai, S. A., Harper, I. S., Herman, B., and Lemasters, J. J. (1998). *J. Bioenerget. Biomembr.* **30,** 207.

Roe, M. W., Lemasters, J. J., and Herman, B. (1990). *Cell Calcium* **11,** 63.

Rottenberg, H., and Wu, S. (1997). *Biochem. Biophys. Res. Commun.* **240,** 68.

Trollinger, D. R., Cascio, W. E., and Lemasters, J. J. (1997). *Biochem. Biophys. Res. Commun.* **236,** 738.

Trollinger, D. R., Cascio, W. E., and Lemasters, J. J. (1998). *Biophys. J.* **74,** A356.

White, J. G., Amos, W. B., and Fordham, M. (1987). *J. Cell Biol.* **105,** 41.

Wieder, E. D., Hang, H., and Fox, M. H. (1993). *Cytometry* **14,** 916.

Zahrebelski, G., Nieminen, A. L., Al-Ghoul, K., Qian, T., Herman, B., and Lemasters, J. J. (1995). *Hepatology* **21,** 1361.

Zoratti, M., and Szabo, I. (1995). *Biochim. Biophys. Acta* **1241,** 139.

PART IV

Laser Capture Microdissection

CHAPTER 29

Laser Capture Microdissection and Its Applications in Genomics and Proteomics

James L. Wittliff

Hormone Receptor Laboratory
Department of Biochemistry and Molecular Biology
Brown Cancer Center
University of Louisville, Louisville
Kentucky, USA

I. Background

Human tissue collection, handling, and analyses present specific problems for developing clinically reliable genomic and proteomic tests unlike studies with animal tissues or homogeneous cell lines grown in culture. For example, determinations of levels of clinically relevant analytes in tissue biopsies, used as markers for detection, diagnosis, prognosis, or therapeutic response of a cancer patient are performed either biochemically or by immunohistochemistry currently (e.g., Fleisher *et al.*, 2002; Wittliff *et al.*, 1998). If the analyte (e.g., estrogen receptor, HER-2/neu oncoprotein) is measured biochemically, a tissue specimen consisting of a heterogeneous cell population is homogenized and the final concentration of the analyte extracted from the cancer cells is "diluted" by the contribution of other

DOI: 10.1016/B978-0-12-384658-7.00029-1

proteins released from noncancerous cells (e.g., epithelium, histiocytes, macrophages, and connective tissue cells). Therefore, an underestimate of the analyte concentration is likely to be determined compromising the appropriate cut-off value between disease and normal states.

While certain tumor markers in tissue biopsies have well served the clinical management of cancer patients (e.g., estrogen receptors in the selection of tamoxifen-responsive breast cancer; Fleisher *et al.*, 2002; Wittliff *et al.*, 1998), many questions of analyte expression in normal and neoplastic cells remain. Likewise, immunohistochemistry is used to measure certain proteins in cancer tissue sections for clinical application in spite of reports indicating the results are often highly operator and antibody dependent and, at best, semiquantitative (e.g., Hammond *et al.*, 2010; Igarashi *et al.*, 1994). As the senior author (Wittliff *et al.*, 1998) and Cole *et al.* (1999) have noted, collection and processing of human tissue biopsies have focused on their clinical purpose (e.g., diagnosis, staging, prognosis, therapy selection) with little emphasis on sampling and cryopreservation for sophisticated genomic (e.g., microarrays) and proteomic analyses (e.g., protein chips). The obvious problem of cellular heterogeneity in the tissue section, which may result in misleading or confusing molecular findings (Cole *et al.*, 1999) complicates these issues. Therefore, a reproducible method for obtaining homogeneous cell populations from normal tissue or from cancer biopsies was required to obtain accurate information from molecular analyses.

Laser capture microdissection (LCM) was initially conceived by a team of investigators at the National Institutes of Health, lead by Lance Liotta, Robert Bonner, and Michael Emmert-Buck, to address this need (Bonner *et al.*, 1997; Emmert-Buck *et al.*, 1996). LCM provides a rapid and direct method for procuring homogeneous subpopulations of cells or complex structures for biochemical and molecular biological analyses. Arcturus Engineering, Inc. (Mountain View, CA) developed the first commercial LCM instrument, made available in 1997, in collaboration with the NIH group as part of a Cooperative Research and Development Agreement (CRADA).

A. Laser Capture Microdissection Instrumentation

LCM represents a major advancement in nondestructive cell sampling technology that can be applied to genomic and proteomic studies. Studies conducted in our laboratories (Andres and Wittliff, submitted; Kerr *et al.*, 2008; Wittliff *et al.*, 2000, 2008) utilize the PixCell IIe™ LCM System (Arcturus Engineering, Inc., now MDS Analytical Technologies) composed of the LCM instrument with fluorescence microscopy, the CapSure™ Transfer Film Carrier, and the PixCell IIe™ Image Archiving Workstation (Fig. 1). Briefly, the LCM station integrates a research grade-inverted microscope, a low-power infrared laser, a joystick-controlled stage, and custom CapSure™ LCM Cap handling mechanism with a video monitor and controller.

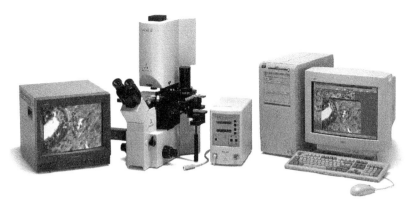

Fig. 1 Components of a laser capture microdissection instrument. The LCM PicCell II station integrates a research grade inverted microscope, a low-power infrared laser with trigger control, a joystick-controlled stage, and custom CapSure™ LCM Cap handling mechanism (cassette module and placement arm). The LCM also employs a video camera connected to an image archiving unit for annotation, storage, and review of the microdissection process.

II. Protocol for Processing Human Tissue Specimens

To evaluate differences between normal and diseased cells, one must first isolate the cells or structures by LCM (Fig. 2) and extract them independently for either DNA, RNA, or protein analyses (Fig. 3) (Andres and Wittliff, submitted; Espina *et al.*, 2006; Kerr *et al.*, 2008; Simone *et al.*, 1998; Wittliff *et al.*, 2000, 2008). Refined protocols and reagents (Fig. 3) have been developed to perform micro-genomic analyses on small quantities of RNA extracted from LCM-procured cells (Andres and Wittliff, submitted; Kerr *et al.*, 2008; Wittliff *et al.*, 2000, 2008). A comprehensive review of additional downstream applications was report by Espina et al. (2006). Proper tissue procurement, specimen handling, and cryopreservation are essential for the collection of quality information from these analyses (Fleisher *et al.*, 2002; Wittliff *et al.*, 1998). Briefly, biopsy specimens should be excised expeditiously and without trauma during the surgical procedure. Specimens must be chilled on ice, and then well trimmed of necrotic tissue, leaving normal tissue present with the lesion in question. The tissue specimen should either be frozen on dry ice in the pathology suite within 20–30 min of collection or rapidly transported chilled in a Petri dish or plastic bag immersed in ice prior to cryopreservation and frozen section preparation in the LCM laboratory, to retain the biological integrity of macromolecules. Any procedure avoiding RNase contact is desirable. It is preferable to freeze the specimen immediately on dry ice after collection at the time of frozen section diagnosis if studies requiring RNA are to be conducted. With the advent of LCM and sensitive technologies of genomics and proteomics requiring nondestructive isolation of pure cell populations, new surgical pathology approaches and methods must be developed as recommended by Cole *et al.* (1999) and Wittliff *et al.* (Andres and Wittliff, submitted; Kerr *et al.*, 2008; Wittliff *et al.*, 1998, 2000, 2008).

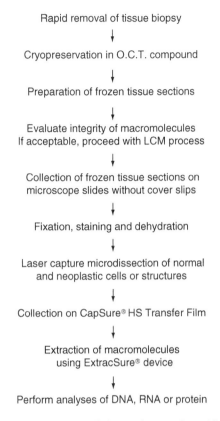

Rapid removal of tissue biopsy
↓
Cryopreservation in O.C.T. compound
↓
Preparation of frozen tissue sections
↓
Evaluate integrity of macromolecules
If acceptable, proceed with LCM process
↓
Collection of frozen tissue sections on
microscope slides without cover slips
↓
Fixation, staining and dehydration
↓
Laser capture microdissection of normal
and neoplastic cells or structures
↓
Collection on CapSure® HS Transfer Film
↓
Extraction of macromolecules
using ExtracSure® device
↓
Perform analyses of DNA, RNA or protein

Fig. 2 Sequence for LCM procurement of cells from a tissue section exhibiting cellular heterogeneity. After tissue collection, macromolecular integrity is assessed before fixation, staining, and dehydration as described in the text. Cells of interest are located in the stained tissue section and the CapSure® optically transparent Cap is placed on the tissue. A laser pulse releases the cell from surrounding structures transferring it to the thermoplastic film. The intact cell bound to the CapSure® device is lifted and placed into a 500 μl microcentrifuge tube for subsequent extraction and analysis if a Macro LCM cap is employed. If a CapSure® HS Cap is used, the ExtracSure® Device is required.

Specimens are processed according to accepted biohazard policies in clean rooms prepared to reduce RNase contamination. Specimens are frozen in Optimum Cutting Temperature compound (TissueTek® OCT medium, VWR Scientific Products Corp.) and stored at −86 °C until LCM is performed. Prior to LCM, we prepare a slide for H & E staining to assess the integrity and cellular composition of the tissue section. If the tissue quality and composition meet the experimental criteria, frozen sections are collected on sterile microscope slides without a coverslip and the slides are retained frozen by placing on a flat surface of dry ice to preserve labile macromolecules. We recommend that glass slides without coatings (uncharged) be used to enhance LCM of selected cells. Frozen tissue biopsies or tissue sections collected on slides and stored in sterile plastic slide holders may be shipped to a distant laboratory for LCM analyses, if the specimens are retained on dry ice during transfer. Preservation of the biological integrity

Protocol for frozen tissue analyses

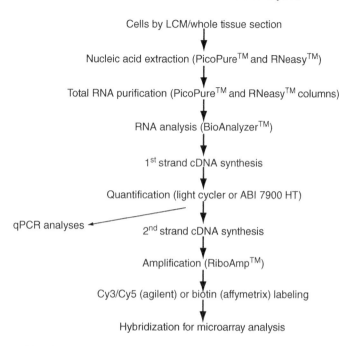

Cells by LCM/whole tissue section

↓

Nucleic acid extraction (PicoPure™ and RNeasy™)

↓

Total RNA purification (PicoPure™ and RNeasy™ columns)

↓

RNA analysis (BioAnalyzer™)

↓

1st strand cDNA synthesis

↓

Quantification (light cycler or ABI 7900 HT)

qPCR analyses ←

↓

2nd strand cDNA synthesis

↓

Amplification (RiboAmp™)

↓

Cy3/Cy5 (agilent) or biotin (affymetrix) labeling

↓

Hybridization for microarray analysis

Fig. 3 Protocol for microgenomic analyses of frozen tissue sections or LCM-procured cells. The sequence outlined represents a useful protocol for RNA extraction, purification, and analysis in the downstream process for assessment of individual gene expression levels by qPCR or for global gene profiling using microarrays. Reagents and instruments listed have been evaluated and shown to be reproducible and reliable for genomic analyses if RNA from either intact tissue sections or LCM-procured cells.

of the biopsy tissue prior to arrival in the LCM laboratory is the shared responsibility of the pathologist and the surgeon, if proteomic and genomic analyses are to become routine clinical tests (Wittliff *et al.*, 2008). In addition, sections of the tissue procured must be representative of the lesion.

A. Fixation, Staining, and Dehydration

Frozen sections mounted on uncoated glass slides are handled according to the following procedures depending upon the type of staining reagent used. The intercalating dye, ToPro3 (Molecular Probes, Inc., Eugene, OR), which binds tightly to double stranded nuclei acids and exhibits a peak fluorescence at 661 nm, has been used to assess the integrity of DNA in LCM-procured cells and structures (Wittliff *et al.*, 2000).

1. ToPro3 Staining Protocol

1. Place frozen section in 70% ethanol for 1 min.
2. Transfer to phosphate-buffered saline (PBS) for 30 s.

3. Place slide in tray and stain with 10 µM ToPro3 for 2 min.

4. Transfer to PBS for 2 min.

5. Transfer to deionized water for 30 s.

6. Transfer to 70% ethanol for 30 s.

7. Transfer to 95% ethanol for 30 s.

8. Transfer to 100% ethanol for 30 s.

9. Transfer to xylene for 5 min.

10. Air dry for 20 min, store desiccated.

PBS and deionized water are used in all wash steps. All of the steps utilizingethanol employed ethyl alcohol UPS (Aaper Alcohol and Chemical Co., Shelbyville, KY).

2. H & E Staining Protocol

Due to the lability of mRNA, H & E staining is accomplished in a fume hood, in RNase-free 50 ml plastic conical tubes using a manual protocol that expedites fixation, dehydration and staining.

1. Place frozen tissue section in 70% ethanol for 1 min.

2. Transfer to RNase-free water for 6 dips.

3. Drip hematoxylin I (Richard-Allen Scientific) from a 0.45 µm filter (Fisher) syringe to cover tissue for 3-5 s.

4. Transfer to RNase-free water for 6 dips.

5. Transfer to 70% ethanol for 6 dips.

6. Tranfer to Eosin Y alcoholic (ThermoShandon) for 10 dips.

7. Transfer to 95% ethanol for 10 dips.

8. Transfer to 100% ethanol for 10 dips.

9. Transfer to 100% ethanol for 10 dips.

10. Transfer to 100% ethanol for 30 s.

11. Transfer to 100% ethanol for 1 min.

12. Transfer to Xylene (Fisher) for 30 s.

13. Transfer to Xylene (Fisher) for 1 min.

14. Air dry for 2–5 min.

Desiccate only if slide will not be used for RNA extraction. Tissue slides to be used for total RNA extraction and gene expression profiling must be used within 1–2 h for LCM procurement of cells. We have refined the staining, processing, and extraction procedures for microgenomic studies of human cancer (Andres and Wittliff, submitted; Kerr *et al.*, 2008; Wittliff *et al.*, 2008).

Before performing LCM, we evaluate the structural status of the frozen tissue biopsy after sectioning and H & E staining (Fig. 4). As illustrated in Fig. 4A, the

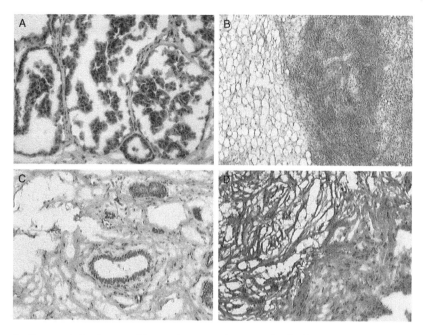

Fig. 4 Representative images of human breast carcinomas stained with H & E. (A) Typical human breast carcinoma showing infiltration of breast carcinoma cells into stromal elements; (B) breast carcinoma adjacent to a large population of adipose and inflammatory cells; (C) tissue specimen containing normal breast epithelia; and (D) breast carcinoma biopsy exhibiting freezing artifact.

section indicates the biopsy, which contains carcinoma cells infiltrating the stroma, is acceptable to proceed with LCM and gene expression profiles. The section shown in Fig. 4B also indicates an acceptable specimen but considerable caution must be exercised to avoid removing unwanted adipose and inflammatory cells with the carcinoma cells of interest. The tissue section shown in Fig. 4C illustrates a specimen received in the laboratory that contained considerable areas of normal ductal epithelia although the diagnosis indicated invasive ductal carcinoma. Therefore, the specimen would not be used for genomic analyses if LCM-procured carcinoma cells were the principal focus. However, LCM collection of the normal epithelial cells and gene expression analyses would serve as a reasonable control specimen compared to analyses of cancer tissues. This illustrates the value of pathology confirmation on the portion of the tissue biopsy received in the LCM laboratory. Finally, the tissue section shown in Fig. 4D depicts significant freezing artifact indicating the biopsy was unsatisfactory for LCM and gene expression profiling.

   Prior to LCM, other types of tissue preparations have been utilized (Kerr *et al.*, 2008). These include either formalin-fixed or alcohol-fixed sections that are paraffin-embedded, as well as cytospin preparations of cells from blood or ascites fluids. Fixation conditions are dictated by the nature of the antigen of interest. Precipitive reagents such as acetone and methanol are used for intracellular antigens while cross-linking fixatives such as glutaraldehyde or paraformaldehyde are used for cell-surface antigens (Shi *et al.*, 2000). Immunohistochemistry of protein analytes has been performed to guide cell selections by LCM (e.g., Fend *et al.*, 1999). We routinely perform immunohistochemistry of clinically relevant analytes such as estrogen and progestin receptors, EGF receptors, and HER-2/neu oncoprotein to direct the procurement of cells expressing particular tumor markers (Wittliff *et al.*, 1998).

B. Steps in Laser Capture Microdissection

The sequence of tissue collection, cell procurement by LCM, and macromolecular extraction is depicted in Fig. 5. Avoid the presence of moisture (e.g., frost, fingertips, breath, room humidity) during all steps prior to RNA extraction. Since our LCM laboratories are used for proteomic and gene expression studies, all procedures are conducted under RNase-free and protease-free conditions, including cleaning of the stage and related areas of the LCM instrument and surrounding bench with RNase AWAY® (Molecular BioProducts, San Diego, CA). Gloves and lab coats are worn at all times.

Prior to performing LCM, the joystick should be positioned perpendicular to the bench and the CapSure™ LCM Cap should be placed over the tissue under examination. The operator locates the cell or structure to be microdissected from the tissue

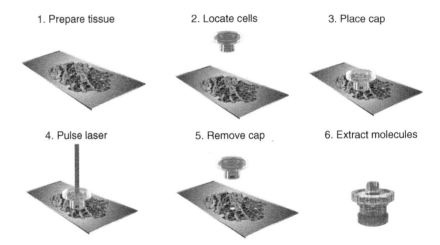

Fig. 5 Simple four-step process to capture cells and recover macromolecules. After locating the cells of interest, a CapSure™ Macro or CapSure™ HS LCM Cap is placed over the target area. Using the PixCell IIe instrument, pulsing the laser through the cap activates the thermoplastic film to form a thin protrusion that bridges the gap between the cap and tissue, and adheres to the target cell. Lifting of the cap removes the target cell(s) from the fixed tissue section since the target cells are now attached to the LCM-cap. Macromolecules, such as total RNA, may be extracted from as few as 200–1000 cells with appropriate reagents using the ExtracSure™ Sample Extraction Device which accommodates small volumes.

section by viewing the histology on the monitor of the PixCell IIe™ LCM System (Wittliff *et al.*, 2000). After test firing the IR Laser in an area devoid of cells and observing the features of the melted plastic ring, the settings for Power and Duration are adjusted to obtain the desired spot size. This laser "spot" may also serve as a point of orientation for the large tissue section. Also a "Map" image is also taken for later reference. Typically, if one is using CapSure™ HS LCM Caps, the following adjustments are suggested: spot size of 7.5 μm (Power setting = 65–75 mW; Duration setting = 650–750 μs), spot size of 15 μm (Power setting = 35–45 mW; Duration setting = 2.5–3.0 ms), and spot size of 30 μm (Power setting = 45–55 mW; Duration setting = 6.0–7.0 ms). If CapSure™ Macro LCM Caps are employed, the following adjustments are suggested: spot size of 7.5 μm (Power setting = 40–50 mW; Duration setting = 550–650 μs), spot size of 15 μm (Power setting = 30–40 mW; Duration setting = 1.5–2.0 ms), and spot size of 30 μm (Power setting = 25–35 mW; Duration setting = 5.0–6.0 ms). To correctly return to the initial area of cell capture ("Map" or "Before" images) for making photo records of "After" and "Cap" images (Fig. 6),

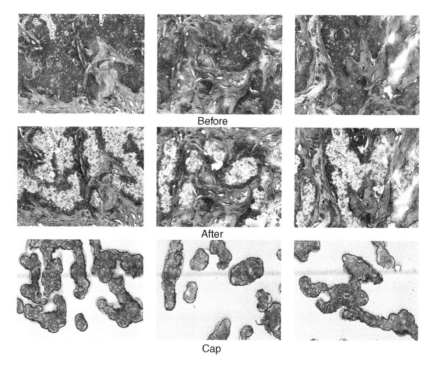

Fig. 6 Representative collection of human breast cancer cells by LCM. Three different regions of the same breast carcinoma biopsy are shown in the top three panels marked (Before). The regions of the tissue section where carcinoma cells were removed by LCM are shown in the images marked (After), while the isolated cells adhering to the CapSure™ LCM Caps are shown in the images marked (Capture). Each Cap, containing 200–300 carcinoma cells, is extracted for RNA that is quantified and amplified before microarray analyses.

the authors "mark" the region by placing several spots on areas devoid of cells adjacent to the cells of interest. This is particularly helpful when collecting multiple caps of cells from distant regions in the same tissue section for later comparison of proteomic and genomic analyses. Using the image archiving unit, characteristics of the tissue section are recorded before and after LCM, and of the cells procured on each Cap (Fig. 6).

The cells or structures are microdissected after firing the IR laser and lifting the CapSure™ LCM Cap with the intact cells collected on the transfer film. The CapSure™ Macro and CapSure HS™ consist of a proprietary thermoplastic polymer film hermetically sealed to the bottom of a precision optical grade plastic cap. In certain experiments requiring extraction of small numbers of cells (200–1000) in low microliter volumes, we utilize the ExtracSure™ Sample Extraction Device (Arcturus/MDS) and the CapSure™ HS LCM Caps for efficient removal of total RNA. The CapSure™ Macro LCM Caps containing the cells fit directly onto standard reagent tubes (500 μl Eppendorf) in preparation for cell extraction. Typically, 1–6 ng of total RNA may be extracted in this manner using either PicoPureTM Extraction Buffer (XB) (Arcturus/MDS) or Buffer RLT™ (Qiagen, Valencia, CA). The transfer process does not damage the captured cells nor the surrounding cells remaining on the slide containing the original tissue preparation (Fig. 6). Usually there is no undesirable cellular contamination since the IR laser beam may be focused between 7.5 and 30 μm providing accurate selection. If necessary, we employ the CapSure™ Pads to remove debris (e.g., stromal elements) from the CapSure™ LCM Caps prior to extraction (Andres and Wittliff, submitted). However, cellular decontamination may be accomplished with a Post-It® note (3M).

Forces involved in an efficient LCM manipulation include (a) those between tissue and slide, (b) those between tissue and activated film, (c) tissue–tissue interactive forces, and (d) the force between tissue and inactivated film. The dynamics of the IR focusing and the melting properties of the thermoplastic transfer film on the CapSure™ LCM Caps are optimized with those of cells in 5–10 μm tissue sections.

After collection of cells on the CapSure™ LCM Cap, macromolecules are extracted using a variety of procedures depending upon whether the analyses are focused on DNA, RNA (Fig. 3), or protein, as described in other chapters of this volume. Gene expression as measured by analyses of mRNA provides an understanding of the manner in which normal cells respond to endocrine changes, malignant transformation, and environmental insults (Andres and Wittliff, submitted; Glasow et al., 1998; Goldsworthy et al., 1999; Kerr et al., 2008; Luo et al., 1999; Sgroi et al., 1999; Simone et al., 1998; Wittliff et al., 2000, 2008). Determination of the level of gene expression as well as the size and structure of RNA molecules requires retention of biological integrity. Because of the lability of mRNA, several workers have studied the effects of tissue fixation on RNA extraction and amplification after LCM (Fend et al., 1999; Goldsworthy et al., 1999), providing some insight into the stability of these labile molecules using current procedures.

C. Advantages of LCM

Manual microdissection techniques, which require tedious manipulation, significant manual dexterity, and a lengthy training program, are slow and the variability in tissue collection is significant. However, LCM, which uses standardized technology, allows rapid sample procurement of tissue structures with awkward geometry and efficient isolation of different cell types in close proximity or adjacent to each other (Bonner *et al.*, 1997; Cole *et al.*, 1999; Emmert-Buck *et al.*, 1996; Espina *et al.*, 2006; Igarashi *et al.*,1994; Wittliff *et al.*, 2000). Furthermore, the transfer process is nondestructive and cell morphology is retained (Wittliff *et al.*, 2000). Of particular importance in molecular diagnostics and gene discovery, there is a record of the original location of cells in the tissue and visual verification of cell capture. My laboratory retains both electronic tissue section and cap images of the LCM process as well as the H & E stained slides after LCM.

Some investigators have reported successful DNA analyses using 300–500 cells (e.g., Simone *et al.*, 1998; Sirivatanaukorn *et al.*, 1999), while 500–1000 cells have been used to isolate RNA (e.g., Glasow *et al.*, 1998; Goldsworthy *et al.*, 1999; Luo *et al.*, 1999). Examinations of proteins using a single technology have employed 1000–5000 cells isolated by LCM (e.g., Banks *et al.*, 1999; Emmert-Buck *et al.*, 2000). Extraction and 2D-PAGE of proteins from representative samples requires capturing 20,000–30,000 cells although new nanotechnology approaches are being developed (Banks *et al.*, 1999; Emmert-Buck *et al.*, 2000; Espina *et al.*, 2006).

D. RNA Isolation, Characterization, and Amplification for Microarray

In our laboratories, total RNA is isolated using the PicoPure® (Arcturus/MDS) kits, which are optimized for extracting RNA from cells procured by LCM (Fig. 3). Routinely 1–6 ng of total RNA may be isolated from 200–300 human breast cancer cells procured by LCM, using these reagents. The intactness of RNA in tissue sections is evaluated prior to proceeding with LCM by a variety of procedures including electrophoresis incorporating a series of markers of different base pair lengths. More recently we utilize the Agilent Bioanalyzer™ and the Nanodrop™ Instrument to assess the RNA quality and concentration. In general, a RIN value of greater than 7 correlates with high-quality RNA acceptable for genomic analyses. For investigations of gene expression profiles of human tissues, we procure cells of interest (e.g., normal vs. neoplastic) from at least three different regions of a single tissue section (Fig. 6). Note that the carcinoma cells were removed from each of the three regions of interest and procured cleanly and retained on the CapSure™ LCM Caps (Fig. 6, lower images). Each cell capture (usually containing 200–1000 cells) is treated as an independent evaluation in that the RNA is extracted, purified, and amplified, then subjected to microarray (Fig. 7) or to individual gene expression analysis by qPCR (Andres and Wittliff, submitted; Kerr *et al.*, 2008; Wittliff *et al.*, 2008).

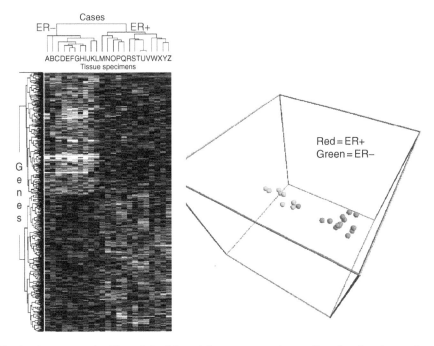

Fig. 7 A representative Eisen Color Map of the gene expression profiles of various human breast carcinomas. Although the microarray performed contained more than 12,000 genes, only a portion of the gene expression profile of each breast cancer is shown using the GeneMaths program (Applied Maths, Austin, TX). Note that without preconceived selection of criteria, gene clustering was observed. Through preliminary bioinformatic analyses, molecular signatures are being identified for several types of human breast cancers, such as those expressing estrogen receptors (ER+) compared with carcinomas lacking the receptor (ER−), which is a marker of antiestrogen responsiveness. Principal component analysis (diagram on right) was performed using the data matrix shown on the left, and the collective results of the breast specimens are projected onto the three-dimensional space diagram using the first three components.

RNA isolated from cells procured by LCM is amplified efficiently with the RiboAMP® kits (Arcturus/MDS) enabling the production of microgram amounts of RNA from nanogram quantities isolated from breast carcinoma and normal cells. Amplification requires preparation of double stranded cDNA from the mRNA fraction of total RNA followed by transcription *in vitro*. The use of exogenous primers maximizes reliability in the synthesis of cDNA template while reducing reaction times. The aRNA prepared by this protocol is ready for labeling and hybridization necessary for microarray analyses. Preliminary studies of micro-array analyses of independent amplifications from the same RNA preparation indicate an excellent correlation. RT-PCR was used to detect low-, medium-, and high-abundance genes within the amplified RNA population. Our laboratory has demonstrated amplification of mRNA in all abundance classes insures that differential gene expression patterns will be identified. We have used a microarray containing approximately 12,000 genes of which 10% were included through "knowledge-based selection" based on reported alterations in cancer. From studies of more hundreds of human breast cancers, we have demonstrated that the RNA isolated from LCM-procured cells is intact for use in amplification of mRNA and subsequent microarray (Fig. 7). We are employing this approach to derive molecular signatures (gene expression profiles) to advance the classification of various cancers and assessment of patient prognosis and therapeutic response using both qPCR and microarrays (Affymetrix GeneChip U133 platform) (Andres and Wittliff, submitted; Kerr *et al.*, 2008; Wittliff *et al.*, 2008).

III. Additional Applications of Laser Capture Microdissection

LCM is rapidly becoming the method of choice for selecting diseased cells from normal cells of the same tissue specimen for genomic and proteomic analyses (Andres and Wittliff, submitted; Banks *et al.*, 1999; Emmert-Buck *et al.*, 2000; Espina *et al.*, 2006; Glasow *et al.*, 1998; Goldsworthy *et al.*, 1999; Kerr *et al.*, 2008; Luo *et al.*, 1999; Sgroi *et al.*, 1999; Sirivatanaukorn *et al.*, 1999; Wittliff *et al.*, 2000, 2008). Some of the applications of LCM in these areas related to molecular diagnostics and prognostics of human cancer are shown below.

Genomics:

> Differential gene profiling
> Loss of heterozygosity
> Microsatellite instability
> Gene quantification
> clonal analysis/clonal analysis

Proteomics:

> Two-dimensional PAGE
> Western blots
> Immunoquantitation of proteins
> MALDI-TOF mass spectrometry

The ability to procure homogeneous populations of specific cell types, for example, normal, premalignant, and malignant cells, and to accumulate data from each cell type advances our understanding of the underlying causes of tumor formation. Furthermore, LCM approaches permit the exploration and tracking, at the molecular level, of cell progression from the normal phenotype into a carcinoma with a metastatic phenotype. Efforts are well underway to produce cDNA libraries that catalog genes that are differentially expressed during tumor progression (e.g., Peterson *et al.*, 1998). The Cancer Genome Anatomy Project (CGAP) has utilized LCM to obtain normal and premalignant samples from human prostate, breast, ovary, colon, kidney, and lung tissue to name a few. Information from CGAP is publicly available through the CGAP-NIH web site (http://www.ncbi.nlm.nih.gov/CGAP/).

LCM has proven to be a powerful tool for research into the cellular basis of disease, and is increasingly being employed in drug discovery and clinical diagnostics, such that CPT codes have been issued in 2008 by the College of American Pathologists (www.cap.org). Physiological changes occurring during normal cell and organ development as well in the progression of normal cells to diseased-based lesions may be explored easily with LCM and proteomics and gene expression profiling. For clinical diagnosis, the ability to sample specific types of cells creates a new analytical paradigm, which will allow patients to be diagnosed based on qualitative and quantitative gene expression as well as on levels of cell-specific proteins.

As the author suggested previously (Wittliff et al., 1998), a new generation of laboratory tests is rapidly evolving in which analyses will be performed directly on human tissue biopsies. It is envisioned that IRB-approved, de-identified tissue banks, such as the Biorepository at the Hormone Receptor Laboratory, will be developed for long-term preservation of human tumor samples. Recently, the National Cancer Institute/NIH developed guidelines now entitled "Best Practices for Biospecimen Resources" describing protocols for tissue collection, processing, storage, retrival and dissemination concepts of good laboratory practices (http://biospecimens.cancer.gov/practices/). These approaches will allow assessment of genetic and biochemical properties of the stored tumor tissues as new clinical, chemical, and molecular biological probes are developed for cancer management, and as technologies such as LCM are utilized to simultaneously separate and explore the molecular phenotypes of normal and diseased cells. It is likely that the use of LCM will allow the individual patient to contribute both clinical and control tissues for comparative analyses enhancing diagnosis, assessment of prognosis, and therapy selection resulting in improved management.

References

Andres, S. A., and Wittliff, J. L. (submitted).

Banks, R. E., Dunn, M. J., Forbes, M. A., Stanley, A., Pappin, D., Naven, T., Gough, M., Harnden, P., and Selby, P. J. (1999). *Electrophoresis* **20,** 689.

Bonner, R. F., Emmert-Buck, M. R., Cole, K., Pohida, T., Chuaqui, R., Goldstein, S., and Liotta, L. A. (1997). *Science* **278,** 1481.

Cole, K. A., Krizman, D. B., and Emmert-Buck, M. R. (1999). *Nat. Genet.* **21,** 38.

Emmert-Buck, M. R., Bonner, R. F., Smith, P. D., Chuaqui, R. F., Zhuang, Z., Goldstein, S. R., Weiss, R. A., and Liotta, L. A. (1996). *Science* **274,** 998.

Emmert-Buck, M. R., Gillespie, J. W., Paweletz, C. P., Ornstein, D. K., Basrur, V., Appella, E., Wang, Q. H., Huang, J., Hu, N., Taylor, P., and Petricoin, E. F., III (2000). *Mol. Carcinog.* **27,** 158.

Espina, V., Wulfkuhle, J. D., Calvert, V. S., VanMeter, A., Zhou, W., Coukos, G., Geho, D. H., Petricoin, E. F., III., and Liotta, L. A. (2006). *Nat. Protoc.* **1,** 586.

Fend, F., Emmert-Buck, M. R., Chuaqui, R., Cole, K., Lee, J., Liotta, L. A., and Raffeld, M. (1999). *Am. J. Pathol.* **154,** 61.

Fleisher, M., Dnistrian, A. M., Sturgeon, C. M., and Wittliff, J. L. (2002). *In* "Tumor Markers: Physiology, Pathobiology, Technology and Clinical Applications," (D. P. Diamandis, H. A. Fritsche, H. Lilja, D. W. Chan, and M. K. Schwartz, eds.), pp. 33–63. AACC Press, Washington, DC.

Glasow, A., Haidan, A., Hilbers, U., Breidert, M., Gillespie, J., Scherbaum, W. A., Chrousos, G. P., and Bornstein, S. R. (1998). *J. Clin. Endocrin. Metab.* **83,** 4459.

Goldsworthy, S. M., Stockton, P. S., Trempus, C. S., Foley, J. F., and Maronpot, R. R. (1999). *Mol. Carcinog.* **25,** 86.

Hammond, E. H., Hayes, D. F., Dowsett, M., *et al.* (2010). *J. Clin. Oncol.* epub April 19, 2010.

Igarashi, H., Sugimura, H., Maruyama, K., *et al.* (1994). *APMIS* **102,** 295.

Kerr, D. A., II., Eliason, J. F., and Wittliff, J. L. (2008). *Adv. Exp. Med. Biol.* **617,** 377.

Luo, L., Salunga, R. C., Guo, H., Bhittner, A., Joy, K. C., Galindo, J. E., Xiao, H., Rogers, K. E., Wan, J. S., Jackson, M. R., and Erlander, M. G. (1999). *Nat. Med.* **5,** 117.

Peterson, L. A., Brown, M. R., Carlisle, A. J., Kohn, E. C., Liotta, L. A., Emmert-Buck, M. R., and Krizman, D. B. (1998). *Cancer Res.* **58,** 5326.

Sgroi, D. C., Teng, S., Robinson, G., LeVangie, R., Hudson, J. R., Jr., and Elkahloun, A. G. (1999). *Cancer Res.* **59,** 5656.

Shi, S. R., Gu, J., and Taylor, C. R. (2000). Antigen Retrieval Techniques: Immunohistochemistry and Molecular Morphology Eaton Publishing, Natick, MA.

Simone, N. L., Bonner, R. F., Gillespie, J. W., Emmert-Buck, M. R., and Liotta, L. A. (1998). *Trends Genet.* **14,** 272.

Sirivatanaukorn, Y., Sirivatanauksorn, V., Bhattacharya, S., Davidson, B. R., Dhillon, A. P., Kakkar, A. K., Williamson, R. C. N., and Lemoine, N. R. (1999). *J. Pathol.* **189,** 344.

Wittliff, J. L., Pasic, R., and Bland, K. I. (1998). *In* "The Breast: Comprehensive Management of Benign and Malignant Diseases," (K. I. Bland, and E. M. Copeland, III, eds.), p. 458. W. B. Saunders Co., Philadelphia, PA.

Wittliff, J. L., Kunitake, S. T., Chu, S. S., and Travis, J. C. (2000). *J. Clin. Ligand Assay* **23,** 66.

Wittliff, J. L., Kruer, T. L., Andres, S. A., and Smolenkova, I. A. (2008). *In* "Proceedings: Fifth Annual Symposium on Hormonal Carcinogenesis," (J. J. Li, S. A. Li, S. Mohla, H. Rochefort, and T. Maudelone, eds.), pp. 349–357. Springer Science.

CHAPTER 30

Going *In Vivo* with Laser Microdissection

Anette Mayer, Monika Stich, Dieter Brocksch, Karin Schütze, and Georgia Lahr

Humangenetik für Biologen
Universität Frankfurt
Frankfurt/Main, Germany

I. Introduction

Tissue microdissection and single-cell isolation is one of the most advanced techniques in modern gene analysis and is especially useful for studying expression of genes in isolated tumor cells. Till now, microdissection methods have been limited to cells from fixed or frozen tissues (Bonner *et al.*, 1997; Emmert-Buck *et al.*, 1996; Kubo *et al.*, 1995; Lahr, 2000; Lahr *et al.*, 2000; Meier-Ruge *et al.*, 1976; Schindler *et al.*, 1985; Schütze and Lahr, 1998; Schütze *et al.*, 1998). An old dream of cell biologists is to isolate living cells from tissue culture or unfixed and unfrozen sections of living tissues (Schindler, 1998). But to date, laser-based microdissection of living cells resulted in the destruction of the isolated cells (Schindler *et al.*, 1985). Now, a modification of the laser microbeam microdissection (LMM) method in combination with the laser pressure catapulting (LPC) technique (Schütze and Lahr, 1998) and a newly developed cell

DOI: 10.1016/B978-0-12-384658-7.00030-8

culture protocol allows microdissection and "ejection" of living single cells or cell clusters with ongoing cultivation for potential treatment and analysis. We established a unique technique in which cultured cells were microdissected and afterward catapulted by LPC into the cap of a microfuge tube. The viability of the catapulted cells is not affected as they enter the cell cycle and proliferate. Applying this protocol—select, microdissect, eject, and clone living cells—to biopsy slices will come true in the near future. As this, "going *in vivo*" opens up a broad spectrum of applications.

II. Step I: Cell Culture Preparation

A prerequisite for isolation of single living cells from cell cultures by laser-assisted cell picking is the growth of the cells on a supporting membrane. The membrane is mounted in a specific cell culture chamber, the ROC chamber. For microdissection the membrane around the cell or cell clusters of interest is cut by the focused laser beam in a sufficient distance from the cell. Then the cell-membrane stack is catapulted by the laser beam into a conventional cap of a microfuge tube centered directly above the selected area (Fig. 1A and B).

A. Buffers, Reagents, and Equipment

ROC chamber, Round Open Closed (PeCon and LaCon, Erbach-Bach, Germany)

Polyethylene–naphthalene membrane, 1.35 μm (PEN membrane; P.A.L.M. Microlaser Technologies AG, Bernried, Germany)

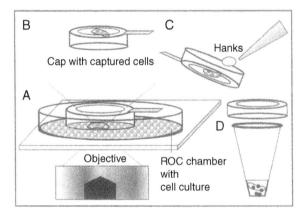

Fig. 1 Schematic drawings of a cell culture grown on a PEN membrane in an ROC chamber. The chamber is attached to the microscope stage and the microfuge cap is centered above the line of laser fire directly in the ROC chamber (A). The selected cell-membrane stacks are microdissected by the laser beam (LMM). The cell-membrane stacks are catapulted by LPC directly into the cap of the sample tube supplied with a droplet of Hanks' solution (B). The captured cells are covered with 25 μl of Hanks' solution (C). The cap is topped with the remaining tube and the assembled tube is centrifuged to collect captured cells at the bottom of the microfuge tube (D).

EJ28 cells, a bladder carcinoma cell line

TPC-1 cells, a thyroid carcinoma cell line

Dulbecco's modified Eagle's medium (DMEM; Invitrogen GmbH (GIBCO-BRL), Karlsruhe, Germany)

DMEM Nutrient Mixture F12-Ham (DME/F12 Hams) (Sigma-Aldrich GmbH, Deisenhofen, Germany)

200 mM L-glutamine (Sigma-Aldrich GmbH)

10% fetal calf serum (FCS; Sigma-Aldrich GmbH)

100× antibiotic–antimycotic solution (Sigma-Aldrich GmbH)

Hanks' solution (Sigma-Aldrich GmbH)

Trypsin–EDTA solution (Sigma-Aldrich GmbH)

Conventional culture dish plates for cell culture

Gassed incubator for cell culture

Laser microscope, Robot-MicroBeam (P.A.L.M. Microlaser Technologies AG)

Inverted microscope Axiovert 135 (Carl Zeiss, Göttingen, Germany)

Microfuge tubes (P.A.L.M. Microlaser Technologies AG)

B. Procedure

1. Assembly of the ROC Chamber

   1. Cover the glass bottom of the ROC chamber with the polyethylene–naphthalene membrane (PEN membrane) by using a droplet of 100% ethanol for mounting it onto the glass.
   2. Expose the opened chamber with the membrane to UV light for 20 min to change the hydrophobic nature of the membrane into a more hydrophilic one.
   3. Assemble the ROC chamber totally and autoclave it at 121 °C for 20 min.

2. Cell Culture

   The EJ28 bladder carcinoma cell line and the papillary thyroid tumor cell line TPC-1 (a generous gift of Dr. B. Mayr, Med. Hochschule Hannover, Germany) were used for the experiments. EJ28 cells were grown in DMEM and TPC-1 cells were grown in DMEM Nutrient Mixture F12-Ham (DME/F12 Hams), both supplemented with 5 mM L-glutamine, 10% FCS, and 1× antibiotic–antimycotic solution. Seed the cell culture cells at the desired density onto the membrane-covered ROC chamber in their appropriate medium.

   4. Incubate the cells in the ROC chamber at 37 °C in a gassed incubator. After 1–2 days in culture the cells are ready for microdissection.

3. Laser Microdissection and Catapulting

   5. Remove the medium completely from the ROC chamber before laser microdissection (Figs. 1–4A).

   6. Microdissect the desired cell-membrane sample. The parameters concerning laser energy and laser focus during microdissection (LMM) are dependent on the laser microscope system used and have to be optimized before use (Figs. 1A, 2B, 3B, 4C, and 5B).

   7. Apply a 10-μl droplet of Hanks' solution on top of the selected cells to facilitate LPC. Be careful not to wash away the microdissected specimen.

   8. Pipette a 5-μl droplet of Hanks' solution into the center of the cap of a microfuge tube and place the cap directly above the selected cells into the ROC chamber (Fig. 1A).

   9. Catapult the cell-membrane stack with one single laser shot positioned at the border of the circumscribed membrane. For the catapulting the laser is focused below the microdissected target specimen.

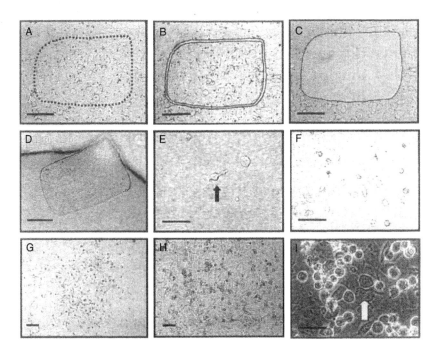

Fig. 2 Images using LMM and LPC to capture 40 EJ28 cells. Cells before microdissection (A), after microdissection (B), cells remaining after LPC (C), catapulted membrane with the cells (D), 11 h after plating (E), after 1 day (F), after 5 days (G), after 8 days (H), and after 12 days (I). Black arrow: cell filopodium; white arrow: mitotic cell; dotted line: area to be microdissected. Bar = 100 μm in (A)–(D), (F), and (H); 50 μm in (E), (G), and (I). Objective lenses in (A)–(D) and (F): 20×; (E) and (I): 40×; (G): 5×; and (H): 10×.

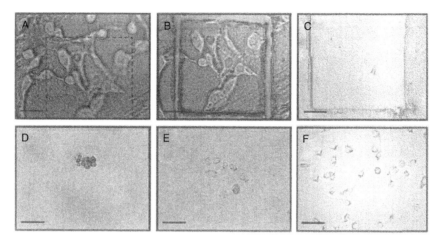

Fig. 3 LMM and LPC images of a small TPC-1 cell cluster (10 cells). The sequence shows cells before LMM (A), after LMM (B), and remaining cell culture after LPC (C). (D) Aggregated cells within the "hanging droplet" 9 h after collection. One day in culture the catapulted and aggregated cells begin to adhere to the bottom of the culture dish (E). Proliferating cells shown after 12 days in culture (F). Dotted line: area to be microdissected. Bar = 50 μm in (A)–(C); 100 μm in (D)–(F). Objective lenses in (A)–(C): 40×; (D)–(F): 20×.

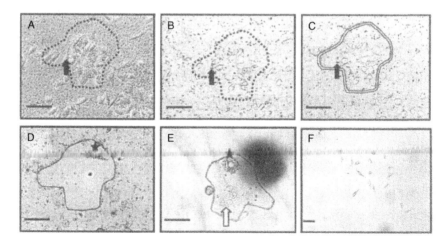

Fig. 4 Images of an experiment to destroy an "undesired cell" before LMM and LPC of about 18 living EJ28 cells. The sequence shows the cells before microdissection (A), after destruction of one specific cell (B), and after microdissection (C), the remaining cells after LPC (D), and the catapulted membrane with the cells (E). (F) Cells after 12 days in culture. Black arrow: cell destroyed by a precise laser shot; white arrow: cells on the membrane after catapulting; dotted line: area to be microdissected. Bar = 100 μm in (A)–(E); 50 μm in (F). Objective lenses in (A)–(E): 20×; (F): 10×.

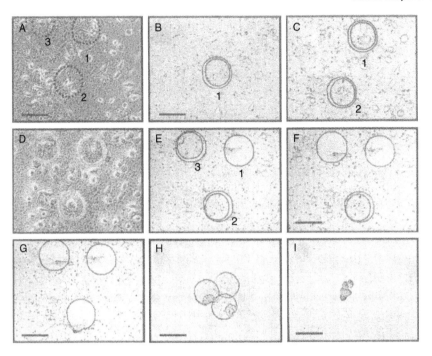

Fig. 5 Images of five pooled single EJ28 cells. The sequence shows the cells before LMM (A), after LMM of cell group 1 (B), after LMM of cell groups 1 and 2 (C), after LMM of cell groups 1–3 (D). The remaining cells after LPC of cell group 1 (E), 1 and 2 (F), and after LPC of cell group 3 (G). (H) Three catapulted membranes with cells. Aggregated cells within the "hanging droplet" (I). Dotted lines: areas to be microdissected. Bar = 100 μm in (A)–(H); 50 μm in (I). Objective lenses in (A)–(H): 20×; (I): 40×.

10. Energy settings should be sufficiently high to catapult the microdissected cells with the membrane into a cap (Fig. 1A and B). Even large cell-membrane stacks (e.g., 385 × 248 μm) can be catapulted (Fig. 2C).

11. After LPC, remove the ROC chamber from the microscope stage, take the cap, and inspect the catapulted cells in the cap now fixed within the manipulator (Figs. 2D, 4E, and 5H).

C. Notes to Step I

(a) To reduce the chance of contamination wear gloves during the whole cell culture procedure. Do not keep the cell cultures outside the incubator longer than necessary. In case of several experiments allow cells in the ROC chamber to recover from dryness by adding medium back to the cells. This medium has to be removed totally, otherwise during the cutting process the laser energy will be absorbed by the aqueous solution. This results in local heating of the medium and visible steam bubbles, which destroy the viable cells. In addition, be aware that after several microdissection events the medium is entering the microspace between the membrane and the glass bottom of the ROC chamber. This makes further micro-dissection and catapulting more and more difficult and finally impossible.

(b) With the focused laser beam single cells or cell clusters are precisely separated together with the membrane from their surrounding (Figs. 2B, 3B, 4C, and 5B–D) using 20× (Figs. 2, 4, and 5) or 40× (Fig. 3) objective lenses. Laser circumscription of the cell-membrane stack results in a gap, free of any other biological material, separating the target from its surroundings (Schütze and Lahr, 1998). The width of the gap is about 3–5 μm, depending on the objective lens and the absorption behavior of the specimen. Where the selected specimen area contains an undesired cell, this cell can be eliminated by a direct laser shot (Fig. 4B).

(c) Increased laser energy catapults the target specimen into the cap of a micro-fuge tube. Even large cell-membrane stacks (e.g., 290 × 369 μm) can be catapulted (Figs. 2D, 4E, and 5H). This results in empty patches within the cell culture (Figs. 2C, 3C, 4D, and 5G). The laser-catapulted cell-membrane stacks are well preserved and allow direct correlation with their templates in terms of shape, size, and original position (Figs. 2D, 4E, and 5H). Microdissection and catapulting of cell clusters in these large sizes takes less than 2 min. The manipulation of single living cells is done within seconds. Laser-assisted isolation is performed with cell clusters of about 10 cells (Fig. 3) and several tens (Figs. 2 and 4), as well as single cells (Fig. 5).

III. Step II: Collection of Catapulted Cells

A. Procedure

1. Cover the catapulted cells in the cap with 25 μl of Hanks' solution.
2. Close the cap with the remaining tube and store for up to 30 min at room temperature.
3. Centrifuge the tube for 1 min at 8000 × *g* and discard the supernatant.
4. Resuspend the cells in 20 μl trypsin–EDTA solution and incubate for 10 min at room temperature to detach the cells from the membrane.
5. Centrifuge for 1 min at 8000 × *g*.
6. After centrifugation the trypsinized cells (pellet) are resuspended in 15 μl (<10 cells) or 20 μl of supplemented medium (4 parts medium:1 part conditioned medium). *Note*: If there are fewer than 10 catapulted cells in the tube follow protocol step 7; if there are 10 or more cells in the suspension proceed with step III.
7. With fewer than 10 cells in the suspension form a so-called "hanging drop-let": place the 15 μl drops inside the lid of a culture dish.
8. Turn the lid and mount it on the remaining culture dish. (This results in "hanging droplets" where single cells come in close contact and start aggre-gation; Figs. 3D and 5I.) Incubate the cells within the "hanging droplet" overnight at 37 °C in a gassed incubator.

9. The next day flip the lid around again and add 5 µl of supplemented medium to the droplet (total: 20 µl) and proceed with step III.

IV. Step III: Recultivation of Catapulted Cells

A. Procedure

1. Place the 20-µl droplets of the cell suspension with a sterile micropipette onto the bottom of a culture dish, either alone or with sufficient space from other droplets.
2. Incubate overnight at 37 °C.
3. 24 h later mark the droplets on the bottom of the culture dish with a permanent pen. This facilitates tracing of the cells.
4. Remove the medium and fill the culture plate with 4–10 ml of fresh supplemented medium, depending on the diameter of the culture dish.
5. The growing of the cells can be observed during next days (Fig. 2F) and weeks (Figs. 3F and 4F). A regular change of cell culture medium is recommended.

B. Notes to Step III

(a) Because of the round shape of a droplet the aggregated (Figs. 3D and 5I) or the nonaggregated cells (Fig. 2) are forced to move in the tension-free center of the droplet (Figs. 3E, 2F, and 4F, respectively). After about 3 h these cells begin to spread out on the culture dish now showing a more flattened cell shape with extending filopodia (Fig. 2E). After 1 day in culture the cells show the typical flat shape of epithelial cells (Figs. 2F, 3E, and 4F).

(b) Depending on the density of the plated cells, after 3 days in culture most cells enter the cell cycle (more than 30 cells) and start proliferation (see Fig. 2G–I). Fewer than 10 cells need more time before they proliferate (Fig. 3F). In our case this took more than 3 weeks. Proliferating EJ28 cells show typically round-shaped cells sitting on top of the flat cell layer (Fig. 2H and I). We were unable to get a "clone" from one single isolated cell, even with supplemented medium, but succeeded with single pooled cells. The cells were grown further and observed for up to 4 weeks.

V. Future Applications

The new approach of the described microdissection protocol for live cells depends on the special procedure of preparation and recultivation of cells. Now the laser microdissection includes the processing steps "select-microdissect-eject-and-cultivate." This opens the way for "going *in vivo*" with a broad spectrum of applications, for example, the establishment of homogeneous cell populations out of heterogeneous cells (i.e., after transfection experiments with green fluorescent protein (GFP) chimeras). In addition, specific cells within a mixed cell population can be identified by their morphology, isolated, and cultivated separately. This protocol fulfills an old dream of cell biologists to isolate and recultivate living cells. The just-described method has been applied to cell cultures but can be extended in the future to tissue cultures and to unfixed unfrozen sections of living tissues, for example, slice cultures from biopsies. Such isolated live cells can be examined, manipulated, and potentially cloned in an effort to provide more biological material for analysis and studies of normal and aberrant biological processes.

Acknowledgments

The authors thank H. Pösl, R. Schütze, and A. Starzinski-Powitz for technical assistance and instrumental and conceptional support. A. Mayer was supported by a grant from the Boehringer Ingelheim Stiftung given to A. Starzinski-Powitz, Goethe Universitaet Frankfurt am Main.

References

Bonner, R. F., Emmert-Buck, M., Cole, K., Pohida, T., Chuaqui, R., Goldstein, S., and Liotta, L. A. (1997). *Science* **278,** 1481.

Emmert-Buck, M. R., Bonner, R. F., Smith, P. D., Chuaqui, R. F., Zhuang, Z., Goldstein, S. R., Weiss, R. A., and Liotta, L. A. (1996). *Science* **274,** 998.

Kubo, Y., Klimek, F., Kikuchi, Y., Bannasch, P., and Hino, O. (1995). *Cancer Res.* **55,** 989.

Lahr, G. (2000). *Lab. Invest.* **80,** 1.

Lahr, G., Stich, M., Schütze, K., Blümel, P., Pösl, H., and Nathrath, W. B. J. (2000). *Pathobiology* **68,** 218.

Meier-Ruge, W., Bielser, W., Remy, E., Hillenkamp, F., Nitsche, R., and Unsöld, R. (1976). *Histochem. J.* **8,** 387.

Schindler, M. (1998). *Nat. Biotechnol.* **16,** 719.

Schindler, M., Allen, M. L., Olinger, M. R., and Holland, J. F. (1985). *Cytometry* **6,** 368.

Schütze, K., and Lahr, G. (1998). *Nat. Biotechnol.* **16,** 737.

Schütze, K., Pösl, H., and Lahr, G. (1998). *Mol. Cell. Biol.* **44,** 735.

CHAPTER 31

Fluorescence *In Situ* Hybridization of LCM–Isolated Nuclei from Paraffin Sections

Douglas J. Demetrick,[*,†,‡,§,||] **Lisa M. DiFrancesco,**[*,||] **and Sabita K. Murthy**[¶]

[*]The Departments of Pathology and Laboratory Medicine

[†]Oncology

[‡]Biochemistry and Molecular Biology

[§]Medical Genetics

[||]The University of Calgary, and Calgary Laboratory Services

[¶]Consultant Geneticist and Head of Molecular Cytogenetics Unit
Genetics Center, Al Wasl Hospital, Dubai, UAE

I. Introduction

The widespread growth of genomic and transcriptomic analysis techniques has increased the need to clearly separate specific target cells from contaminating cells, in order to identify unique markers associated with the target cells. The need for this is even more important for diagnostic purposes, where artifacts from contaminating cells can lead to erroneous results. Consequently, laser capture microdissection is still important as a method for separation of specific cells from other cells in complex or tissue mixtures.

TECHNIQUES IN CONFOCAL MICROSCOPY
489
DOI: 10.1016/B978-0-12-384658-7.00031-X

Currently, many protocols for performing fluorescent *in situ* hybridization (FISH) on cells from tissue make use of nucleus-only preparations. Where specific cells must be evaluated from frozen or formalin-fixed, paraffin-embedded tissue, LCM–FISH to isolate whole nuclei is still useful (DiFrancesco *et al.*, 2000). During the past several years, improvements to the nucleus isolation from fixed tissue such as optimizing it for B5 tissue (Schurter *et al.*, 2002), or small improvements to the general technique (Murthy and Demetrick, 2006) may increase the applicability of the LCM–FISH technique. As well, the coexistent use of LCM and FISH has also been applied to situations where the FISH is performed prior to the LCM (FISH–LCM), particularly in the case of microdissecting bacterial pathogens (Klitgaard *et al.*, 2005), which are more easily visible when fluorescently stained, or to specific cells identified by fluorescence (Gjerdrum *et al.*, 2001). Nonetheless, the techniques to isolate whole nuclei by LCM are still quite specialized and still require optimization to the specimen to achieve best effectiveness.

FISH was first utilized to determine the cytogenetic map position of genes (Pinkel *et al.*, 1986). This technique has also proved to be very useful in assessing gene copy number from clinical specimens, and such information may be used, for example, to predict the response of a cancer to treatment (Bitran *et al.*, 1996; Houston *et al.*, 1999). The method usually works best with fresh or frozen tissue and offers localization of a genomic abnormality to a specific cell, very valuable in cancer specimens that are commonly contaminated with normal tissue elements such as inflammatory cells or fibroblasts. Heterogeneity of gene copy number may be a useful additional feature to assess in cancer specimens (Sauter *et al.*, 1993), as evidence suggests that metastases likely occur via specific dominant clones in the primary tumor (Simpson *et al.*, 1996; Theodorescu *et al.*, 1991). FISH of paraffin sections has been described, and commercial kits are currently available to perform such analyses. If the FISH target is highly amplified, one can often directly identify amplification within the cancer cells of paraffin sections. Tissue or fixative autofluorescence, however, can seriously compromise interpretation of low-level amplification, or of deletion (McKay *et al.*, 1997; Szollosi *et al.*, 1995). High-throughput methods of paraffin FISH, (Richter *et al.*, 2000; Schraml *et al.*, 1999) using tissue microarrays, (Kononen *et al.*, 1998) have also been described but are subject to the same caveats as above. Purification of nuclei from paraffin sections for FISH could decrease the effects of tissue autofluorescence; however, the advantage of *in situ* verification of the source of the cells is lost. The use of laser capture microdissection to prepare nuclei from breast carcinoma cells for both FISH and flow cytometric analysis has been described (DiFrancesco *et al.*, 2000). This technique, LCM/FISH, allows detection of normal interphase copy numbers, and thus detection of low-level amplification or even single copy deletions, which would be very difficult in paraffin section FISH. Some minor refinements to the method have also improved the yield and reproducibility of the technique.

II. Materials and Methods

A. Laser Capture Microdissection Procedure

Specimens of breast carcinoma were fixed in 10% neutral buffered formalin and embedded in Surgiplast matrix. Thirty-micron sections were cut from the paraffin blocks, stained with Harris' hematoxylin, dehydrated through a graded ethanol series to xylene (three xylene washes), and then allowed to air dry. Staining with eosin yielded wide-spectrum background fluorescence that interfered with subsequent fluorescent analysis and is strongly discouraged. Dehydrated sections were subjected to laser capture microdissection using a PixCell II instrument (Arcturus Engineering, Inc.) using the typical conditions:

30-μm beam: power 70–80 mW; pulse duration 6–7 ms

15-μm beam: power 40–50 mW; pulse duration 3–5 ms

The 7-μm beam is very difficult to use with thick sections due to contamination of the captured cells with adjacent uncaptured tissue or poor adhesion of the captured cells. The spotting laser is usually very difficult to see because of the thickness of the tissue section, so the position of the beam is marked on the video monitor with a dry-erase marker. Cells and cell nests are fixed to the polymer film on CapSure LCM caps and placed tightly into 0.5-ml microfuge tubes. Some experimentation is usually needed for the tissue at hand and multiple bursts to allow optimal adhesion to the polymer may also be needed to firmly fix the cells. The LCM caps in their tubes can be stored up to 1 year at room temperature prior to nucleus extraction.

B. Nucleus Isolation Procedure

The following procedure details the dissolution of the CapSure LCM cap membrane matrix with sequential organic solvents, followed by rehydration of the tissue and proteinase K-mediated release of the nuclei. *All organic extractions should be performed in an appropriate fume hood to minimize occupational solvent exposure and care must be taken to avoid ignition of flammable materials.*

1. 100 μl of fresh, high-quality chloroform is pipetted into an Eppendorf 0.5-ml microfuge tube and capped with a CapSure LCM cap containing a microdissected specimen. The tube is inverted for 10 s, then microcentrifuged at 3000 $\times$ g for 30 s to release the tissue specimen from the "capture" polymer of the LCM cap.

2. The LCM cap is removed and 200 μl of anhydrous ethyl ether is added and mixed with the chloroform by inversion. This step is necessary to lower the density of the solvent to allow successful centrifugation of the tissue fragments while ensuring that the polymer "capture" medium is still soluble.

3. Following microfuge centrifugation, the supernatant is removed by pipette suction with a narrow pipette tip and the pellet is washed two times as above with 300 μl of xylene to remove dissolved LCM cap polymers, taking care to not disturb the fragile tissue pellet.

4. The sample is rehydrated by washing once with 400 μl of absolute ethanol to remove xylene, followed by one wash with 95% ethanol, one wash of 70% ethanol, and two washes with TE (10 mM Tris, 1 mM EDTA, pH 8.0) for rehydration, always leaving approximately 50 μl of residual solution to minimize accidental removal of tissue.

5. The sample is finally washed once with 400 μl of proteinase buffer (50 mM Tris, 10 mM NaCl, 10 mM EDTA, pH 8.0) and approximately 100 μl is left in the tube.

6. Proteinase K is added (50 μl at 0.015% to a final concentration of 0.005% Techniques in Microscopy). The digest is then incubated for 30–60 min at 37 °C in a water bath, with gentle finger vortexing approximately every 10 min. For FISH analysis, the sample is diluted with 350 μl of TE, then centrifuged at 10,000 × g for 2–3 min followed by careful removal of as much supernatant as possible without disturbing the pellet.

7. The pellet is then gently resuspended in approximately 20 μl of 10 mM Tris pH 8.0 by pipetting with a wide-bore pipette tip, and the suspension is pipetted onto clean microscope slides (1–20 μl depending on the concentration of nuclei). Circling the site of sample application with a diamond pencil on the underside of the slide is helpful to later visualization. The slides are allowed to thoroughly air dry, then fixed in fresh methanol: glacial acetic acid (3:1) for 5 min, air dried, and baked at 37 °C for at least 4 h prior to hybridization or storage. Slides are stored at −20 °C in slide boxes sealed within hybridization bags containing Drierite dessicant. *It is important to bring the slides to room temperature prior to opening the storage bag.*

FISH is performed on the LCM and touch preparations according to our previously published methodology (Demetrick, 1996; DiFrancesco *et al.*, 2000). Fluorescent images are captured with a PXL 1400 cooled CCD camera (Photometrics, Phoenix, AZ) using Electronic Photography software (BioDx, Pittsburgh, PA). Bright-field and phase contrast images are captured with a Spot 1 digital camera (Diagnostic Instruments, Inc.).

III. Discussion

Evaluation of the preparation following isolation of the nuclei using phase contrast microscopy is highly recommended (Fig. 1). A loss of up to half the nuclei during the FISH procedure is not uncommon and is likely due to detachment from the glass slide during DNA denaturation. The size of the nuclei significantly affects the optimal thickness of the specimen. If the specimens are too thin, nuclear slicing occurs and the yield of whole nuclei following LCM is very low. FISH evaluation

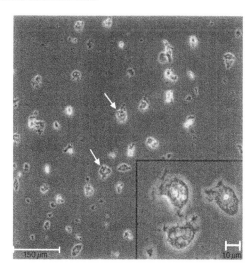

Fig. 1 LCM preparation of "naked nuclei." Cells were harvested from 30-μm sections of hematoxylin-stained breast carcinoma by LCM. The nuclei were purified using organic solvents, rehydration, and proteinase K digestion as described in Section II.B, dropped onto clean glass slides, and fixed. The specimens were photographed under phase contrast with the "naked nuclei" indicated by arrows. Inset shows a higher power view of the nuclei. Scale bars are as indicated.

allows good visualization of normal gene copy numbers in interphase nuclei with very low background or autofluorescence (Fig. 2). Gene amplification is obvious (Fig. 3). We have used this technique on either formalin-fixed paraffin-embedded normal breast or carcinoma sections, and it has proved to be reproducible among different investigators in our laboratory. We have also utilized paraffin tissue microarrays for FISH. While the paraffin microarrays save considerable time and effort with an optimal gene target, in our experience detection of low-level gene amplification or assessment of gene deletions was very difficult (data not shown). As well, the use of a weak probe will also make interpretation of tissue microarray FISH difficult. Although a more laborious technique than tissue microarray FISH, LCM/FISH interpretation is much more obvious for low copy number assessment.

The choice of tissue will dictate the thickness of the paraffin sections required for LCM/FISH, as large nuclei will necessitate thicker sections to preserve yield. We have been able to use 15- to 20-μm sections with *some* specimens, which offers superior morphology for dissections, and still had an acceptable yield, but *30-μm sections give a good yield with almost all carcinoma specimens that we have tried.* When establishing this technique in one's laboratory one should validate the technique first on thicker sections and then, if desired, on thinner sections (15–25 μm). Bear in mind that some microtomes may be significantly inaccurate when specifying specimen thickness during cutting of the paraffin blocks, and section thickness errors will lead to decreased nuclear yields.

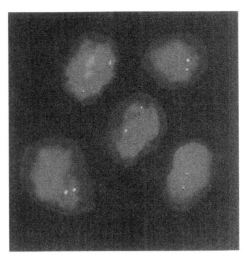

Fig. 2 LCM/FISH of breast carcinoma cells with normal *p53* copy number. Thirty-micron paraffin sections of hematoxylin-stained breast carcinoma were subjected to laser capture of carcinoma cells. The nuclei were prepared and subjected to FISH using a PAC genomic DNA probe containing the *p53* gene (red signals). DAPI-stained (blue) nuclei show two *p53* signals per interphase nucleus. (See Plate no. 16 in the Color Plate Section.)

IV. Troubleshooting Tips

1. Do not use adhesive or other coated slides to pick up the paraffin sections from the microtome, or you will not be able to remove the microdissected cells from the glass slide! Follow the Arcturus protocols for deparaffinizing and hematoxylin staining the sections, then dehydrate back through xylene. Dehydration and delipidation seem to be important in ensuring good adhesion to the LCM polymer film, so use fresh xylene in at least the last deparaffinization bath for every batch (5–10) of slides.

2. It may also be helpful for histology novices to have a picture of the histologic appearance with selected areas for LCM demarcated by a morphology expert, to help guide the microdissection. Visualization of the morphology of thick sections can be difficult. Although the use of the PixCell diffuser improves visualization during dissection, one risks scratching the tissue section with the cap holder, or wasting a cap to evaluate appropriate areas to capture before starting the dissection. A thin sheet of good quality photocopier paper or a thin business card laid on top of the slide will allow good visualization of the section without requiring use of the diffuser (D. Billings, Arcturus, 2001, personal communication).

3. As well, check that the section thickness is thick enough to avoid nuclear slicing (incomplete nuclei), that the capture polymer is dissolving in the ether and that the released tissue sections are not floating and thus being removed with solvent supernatants. One can place a sample (2–5 μl) of the purified nuclei on a

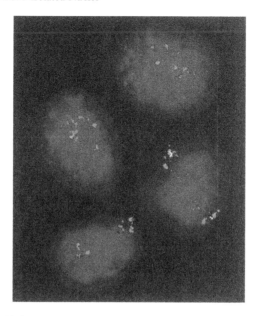

Fig. 3   LCM/FISH of infiltrating duct carcinoma cells. Thirty-micron paraffin sections from a case of infiltrating duct carcinoma were subjected to laser capture of cells or cell groups from formalin-fixed breast cancer tissue. The nuclei were prepared and subjected to FISH using *c-erbB-2* (red signal) and an α-satellite centromeric probe for chromosome 17 (green signals). Normal copy numbers would be 2–4 individual small signals per interphase nucleus. Although both probes show an amplified signal, the α-satellite sequences are normally present in many copies, while the *c-erbB-2* gene is abnormally amplified. (See Plate no. 17 in the Color Plate Section.)

slide and examine it after air-drying under phase-contrast optics or stain it with hematoxylin to evaluate the quality of the preparation, prior to using it for FISH. The proteinase K digestion allows a great deal of latitude. If the yield of nuclei is poor, try digesting the tissue longer or with more enzyme. It is not uncommon for components of the capture polymer to remain undissolved. Leave them in the tube for the washes, but when resuspending the cells for harvest (and leaving the capture polymer in the tube), add more TE up to 700 μL, or perform two pipette extractions (e.g., 350 μL each) to maximize the yield of nuclei, and then centrifuge the nuclei slightly longer (6 min instead of 5 min) to pellet them again from the diluted supernatant.

4. One must try to avoid capturing large amounts of collagen during the LCM. The thickness of the section ensures that extraneous material will be removed with the LCM cap if a strong band of collagen is captured. Although adhesion to adhesive tape may help remove this extraneous material from the LCM cap, avoiding the capture of nonessential collagen is the best strategy.

5. Ensure that centrifugation speed is enough to pellet the nuclei but not so much as to distort or destroy the nuclei.

6. Store slides for several days prior to FISH as this ensures both better retention of nuclei to the glass and better quality (higher signal-to-noise) FISH signals. As well, photobleaching under fluorescence may be lessened and the fluor signal to noise better, if the slides are left at 4 °C overnight prior to visualization.

7. Use very dilute DAPI (e.g., 40 pg/ml in McIlwaine's buffer) to counterstain the FISH slides, as overly bright nuclei will interfere with FISH interpretation. Interphase nuclei may stain differently from these published conditions because of both the preparation and the particular source of DAPI, so be prepared to optimize the staining experimentally by varying DAPI concentrations.

Acknowledgments

This work was funded by the Canadian Breast Cancer Research Initiative and the Alberta Cancer Board. D.J. Demetrick thanks the CIHR (Clinician–Scientist Program) for salary support and also gratefully acknowledges the Canadian Foundation for Innovation, the Alberta Science and Research Authority, and the Terry Fox Foundation for equipment funding. The authors are also grateful to Calgary Laboratory Services for salary support to D.J. Demetrick and L.M. DiFrancesco, as well as for access to the clinical specimens necessary to develop this technology.

References

Bitran, J. D., Samuels, B., Klein, L., Hanauer, S., Johnson, L., Martinec, J., *et al.* (1996). Tandem high-dose chemotherapy supported by hematopoietic progenitor cells yields prolonged survival in stage IV breast cancer. *Bone Marrow Transplant.* **17**(2), 157–162.

Demetrick, D. (1996). The use of archival frozen tumour tissue specimens for fluorescence in situ hybridization. *Mod. Pathol.* **9**, 133–137.

DiFrancesco, L. M., Murthy, S. K., Luider, J., and Demetrick, D. J. (2000). Laser capture microdissection-guided fluorescence in situ hybridization and flow cytometric cell cycle analysis of purified nuclei from paraffin sections. *Mod. Pathol.* **13**(6), 705–711.

Gjerdrum, L. M., Lielpetere, I., Rasmussen, L. M., *et al.* (2001). Laser-assisted microdissection of membrane-mounted paraffin sections for polymerase chain reaction analysis: identification of cell populations using immunohistochemistry and in situ hybridization. *J. Mol. Diagn.* **3**(3), 105–110.

Houston, S. J., Plunkett, T. A., Barnes, D. M., Smith, P., Rubens, R. D., and Miles, D. W. (1999). Overexpression of c-erbB2 is an independent marker of resistance to endocrine therapy in advanced breast cancer. *Br. J. Cancer* **79**(7–8), 1220–1226.

Klitgaard, K., Molbak, L., Jensen, T. K., *et al.* (2005). Laser capture microdissection of bacterial cells targeted by fluorescence in situ hybridization. *Biotechniques* **39**(6), 864–868.

Kononen, J., Bubendorf, L., Kallioniemi, A., Barlund, M., Schraml, P., Leighton, S., *et al.* (1998). Tissue microarrays for high-throughput molecular profiling of tumor specimens. *Nat. Med.* **4**, 844–847.

McKay, J. A., Murray, G. I., Keith, W. N., and McLeod, H. L. (1997). Amplification of fluorescent in situ hybridisation signals in formalin fixed paraffin wax embedded sections of colon tumour using biotinylated tyramide. *Mol. Pathol.* **50**, 322–325.

Murthy, S. K., and Demetrick, D. J. (2006). New approaches to fluorescence in situ hybridization. *Methods Mol. Biol.* **319**, 237–259.

Pinkel, D., Straume, T., and Gray, J. W. (1986). Cytogenetic analysis using quantitative, high-sensitivity, fluorescence hybridization. *Proc. Natl. Acad. Sci. USA* **83**(9), 2934–2938.

Richter, J., Wagner, U., Kononen, J., Fijan, A., Bruderer, J., Schmid, U., *et al.* (2000). High-throughput tissue microarray analysis of cyclin E gene amplification and overexpression in urinary bladder cancer [In Process Citation]. *Am. J. Pathol.* **157,** 787–794.

Sauter, G., Moch, H., Moore, D., Carroll, P., Kerschmann, R., Chew, K., *et al.* (1993). Heterogeneity of erbB-2 gene amplification in bladder cancer. *Cancer Res.* **53**(10 Suppl), 2199–2203.

Schraml, P., Kononen, J., Bubendorf, L., Moch, H., Bissig, H., Nocito, A., *et al.* (1999). Tissue microarrays for gene amplification surveys in many different tumor types. *Clin. Cancer Res.* **5,** 1966–1975.

Schurter, M. J., LeBrun, D. P., and Harrison, K. J. (2002). Improved technique for fluorescence in situ hybridisation analysis of isolated nuclei from archival, B5 or formalin fixed, paraffin wax embedded tissue. *J. Clin. Pathol. Mol. Pathol.* **55**(2), 121–124.

Simpson, J. F., Quan, D. E., Ho, J. P., and Slovak, M. L. (1996). Genetic heterogeneity of primary and metastatic breast carcinoma defined by fluorescence in situ hybridization. *Am. J. Pathol.* **149,** 751–758.

Szollosi, J., Lockett, S. J., Balazs, M., and Waldman, F. M. (1995). Autofluorescence correction for fluorescence in situ hybridization. *Cytometry* **20,** 356–361.

Theodorescu, D., Cornil, I., Sheehan, C., Man, S., and Kerbel, R. S. (1991). Dominance of metastatically competent cells in primary murine breast neoplasms is necessary for distant metastatic spread. *Int. J. Cancer* **47,** 118–123.

INDEX

Printed and bound by CPI Group (UK) Ltd, Croydon, CR0 4YY

03/10/2024

01040310-0012